*Low-Temperature Microscopy
and Analysis*

Low-Temperature Microscopy and Analysis

Patrick Echlin

University of Cambridge
Cambridge, England

Plenum Press • New York and London

Library of Congress Cataloging-in-Publication Data

Echlin, Patrick.
    Low-temperature microscopy and analysis / Patrick Echlin.
        p.   cm.
    Includes bibliographical references and index.
    ISBN 0-306-43984-0
    1. Cryomicroscopy. 2. Cryopreservation of organs, tissues, etc.
I. Title.
QH225.E34   1992
578'.4--dc20   91-39738
                                                                                    CIP

Figures 5.11, 9.4, 9.6, and 9.7 from *Journal of Electron Microscopy Technique*; reprinted by permission of John Wiley and Sons, Inc.

ISBN 0-306-43984-0

© 1992 Plenum Press, New York
A Division of Plenum Publishing Corporation
233 Spring Street, New York, N.Y. 10013

Printed in the United States of America

To my wife, Shirley

Very high and very low temperatures extinguish all human sympathy and relations. It is impossible to feel affection beyond 78° or below 20° of Fahrenheit: human nature is too solid or too liquid beyond these limits

—SYDNEY SMITH, 1836

I have gathered a posie of other men's flowers and nothing but the thread that binds them is my own.

—MICHEL MONTAIGNE, 1592

Foreword

The frozen-hydrated specimen is the principal element that unifies the subject of low-temperature microscopy, and frozen-hydrated specimens are what this book is all about. Freezing the sample as quickly as possible and then further preparing the specimen for microscopy or microanalysis, whether still embedded in ice or not: there seem to be as many variations on this theme as there are creative scientists with problems of structure and composition to investigate. Yet all share a body of common fact and theory upon which their work must be based. *Low-Temperature Microscopy and Analysis* provides, for the first time, a comprehensive treatment of all the elements to which one needs access.

What is the appeal behind the use of frozen-hydrated specimens for biological electron microscopy, and why is it so important that such a book should now have been written? If one cannot observe dynamic events as they are in progress, rapid specimen freezing at least offers the possibility to trap structures, organelles, macromolecules, or ions and other solutes in a form that is identical to what the native structure was like at the moment of trapping. The pursuit of this ideal becomes all the more necessary in electron microscopy because of the enormous increase in resolution that is available with electron-optical instruments, compared to light-optical microscopes. On the size scale below one micrometer, frozen-hydrated specimens offer the hope of escaping from the dilemma that the "unlimited" resolution of electron optics can, on the one hand, often be wasted by inadequate specimen preparation, while light microscopy can give perfect specimen preparation, but only inadequate resolution. In this context, the time has certainly come in which a comprehensive and unified coverage of low-temperature techniques can strongly influence the continued development of biological electron microscopy and microanalysis.

The pursuit of improved, if not ideal specimen preparation by low-temperature techniques has developed steadily over more than 25 years. Methods have been developed at the level of cellular fine structure (notably freeze fracture and freeze substitution), microanalysis of diffusible substances, the structure of macromolecular assemblies (notably freeze-drying and shadowing), and most recently even the internal structure of macromolecules. At the highest resolution, low temperature is needed not so much to preserve the native, hydrated state, which it does admirably well (but so do other techniques, such as glucose embedment), as for the extra margin of protection which it provides against radiation damage.

The development of all of these techniques has been aided not just a little in the past by the author of this book, Patrick Echlin, through his effort in organizing a series of four International Meetings on Low-Temperature Microscopy, beginning in 1977, and through the emphasis that he has given to the field in his role as editor of the Journal of Microscopy. The next logical step, given the maturity of development of the subject, would have to be nothing else than to write this book. It is a volume that will speed access to existing techniques and greatly expand awareness of related work for all who seek a unified presentation of low-temperature methods and their underlying theoretical foundation. Publication of this book therefore makes a truly important contribution toward advancing the process of learning what goes on in biology at a level below what can be seen with the light microscope.

Robert M. Glaeser

Department of Molecular and Cell Biology
University of California at Berkeley

Preface

Water is the most abundant and most important molecule in the biosphere and outer lithosphere. As a vapor it forms a vital envelope around our planet; as a liquid it covers about 75% of the Earth's surface and dissolves almost everything. As a solid it is permanently present at the Poles and on many mountain peaks and is a seasonal reminder of the changing climate of our environment. Liquid water is vital for living organisms. It is both a reactant and the medium in which reactions occur and their products are transported. Water is the most abundant and least expensive building block of living matter and when converted to the solid state can provide the perfect matrix in which to study the structure and *in situ* chemistry of hydrated material. This book considers the nature of this solid matrix, its constituent components, and how it may be formed, manipulated, examined, and analyzed.

This volume has grown from the firm belief that low-temperature microscopy and analysis is the only way we may hope to obtain a true picture of the fine structure and composition of ourselves and our water-filled environment. I will discuss the physical basis and the practical aspects of the different procedures we need to use, the problems that occur, and the advantages that accrue. The conversion of liquids (primarily water) to their solid phases (primarily ice) forms a central feature of this book. The text falls into four unequal parts. The first three chapters consider water in the liquid and solid states. The next four chapters discuss the various manipulations we may make to the solidified matrix. There then follows three chapters that show what we may hope to see in the frozen samples by means of photons and electrons, and another chapter considers the processes we need to use to analyze their constituent elements and molecules. The final chapter contains updated information on the whole subject.

The book provides a number of well-tested procedures that will enable the novice to cryomicroscopy and analysis to get started. It also gives a detailed background from which future developments can take place. The reader is provided with sufficient general information on how to implement a particular low-temperature process and the reasons why it should be used. A comprehensive bibliography at the end of the book provides the provenance and specific details of existing practices. Low-temperature microscopy and analysis is not the sole preserve of biologists and those interested in hydrated organic samples—although these types of samples present both the greatest challenge to the existing technologies and the only hope of solving the unresolved questions posed by such samples. The processes that will be discussed can be used to study and analyze the solid state of all the liquid and gaseous materials which exist on our planet (with the possible exception of helium). Low temperatures provide an important way to study radiation-sensitive and labile samples, hydrated organic systems, and phenomena that only exist at temperatures at which living processes stop.

Low-temperature microscopy and analysis is not without potential dangers, and it is important that experimentalists are fully aware of safety issues in the laboratory. Cryogenic liquids can cause severe burns, and exposed parts of the body must be protected with the appropriate clothing and face masks when using these materials. Liquid nitrogen should only be used in a well-ventilated laboratory, as 1 liter of the liquid expands to nearly 700 liters of an inert tasteless gas which can cause asphyxia. Some secondary organic liquid cryogens have very low flash points and form dangerously explosive mixtures with oxygen, which may condense from the atmosphere. Some of the resins used in low-temperature embedding may cause contact dermatitis and, like all laboratory chemicals, should be handled with gloves. An electron microscope laboratory is replete with potential hazards associated with vacuum and high-pressure equipment and containers, high voltages and ionizing radiation, and toxic and inflammable chemicals. Reputable industrial companies provide safety information about their products and supplies; responsible governments issue safety legislation about laboratory practices. These warnings should be heeded, because by understanding the potential difficulties and adopting sensible laboratory procedures, low-temperature microscopy and analysis can take place in a safe, productive (and happy) environment.

The arguments for adopting cryotechniques and low-temperature microscopy and analysis are secure and proven. It should come as no surprise that so many people are now using one or more of these methods to examine and analyze hydrated, liquid, and gaseous specimens. It is, however, astonishing that in 1991, the one-hundredth anniversary of the electron, anyone should continue to use an electron beam instrument at ambient temperatures.

Patrick Echlin

Cambridge

Acknowledgments

This book would not have been possible without the help of many people. I am priviledged to have been working in this field for the past 25 years and am indebted to many people for listening to my ideas, correcting my errors, and providing practical advice. Much of this interaction has come through my editorship of the *Journal of Microscopy* and from the four International Conferences on Low-Temperature Microscopy and Analysis I have organized in Cambridge, in 1977, 1981, 1985, and 1990. I am most grateful to the many people who have provided illustrations for this book and who are acknowledged in the text. I appreciate their willingness to help, in many instances providing better and more updated illustrations than those I requested.

I am particularly grateful to Felix Franks in Cambridge, Albert Saubermann at Stony Brook, and Tom Hayes in Berkeley for their continuous constructive criticism and for providing facilities in their laboratories to jointly test out new ideas, improve existing procedures, and use low-temperature microscopy and analysis to solve problems in biology and medicine. My thanks also to Jerry Whidby and the Directorate of the Philip Morris Research Center in Richmond, Virginia for inviting me to their laboratory as a Visiting Scientist for several summers in the mid-1980s and for providing unparalleled facilities for low-temperature scanning microscopy and x-ray microanalysis. Finally, this book would not have been possible without help from Ruth Hockaday with some of the typing and from Peronel Burge, Peter Boardman, and Stephan Morris for the design and execution of many of the illustrations.

Contents

The Properties and Structure of Water

1.1. INTRODUCTION

Although low-temperature microscopy and analysis are not solely associated with hydrated systems, most studies are concerned with systems where water, either as ice or amorphous water, is an important structural component. For this reason, water merits special attention and it is necessary to appreciate some basic facts about the compound, first at normal temperatures and pressures, and then at low temperatures.

It is not intended to discuss all the properties of water as these are well documented in the literature, but instead to provide an overview of the more general features of the compound and show how they may relate to the processes and practices of low-temperature microscopy and analysis. For those readers who wish to know much more about the subject, reference should be made either to the comprehensive multivolume series edited by Franks (1972–1982), or to the somewhat less compendious publications by Eisenberg and Kauzmann (1969) and Franks (1983, 1985).

1.2. THE PROPERTIES OF LIQUID WATER

Water, one of the simplest molecules, is in many respects a curious substance, with some unexpected properties. At 273.16 K (the triple point of water) the solid, liquid, and vapor phases coexist, and whereas the hydrides of the other elements close to oxygen in the Periodic Table (e.g., HF, H_2S, NH_3, and CH_4) only exist as gases at normal temperatures and pressures, water is a liquid. The apparently anomalous liquid state of water is reflected in the central role it had in terrestrial biogenesis and continues to have for the presence of life on Earth.

Many of the characteristics of water are well known, and Table 1.1 summarizes the properties that relate to the processes involved in the conversion of water to ice. These properties are due either to the presence of hydrogen bonds, which affect the internal chemical energy of the molecule, or to the solvent properties, which relate more to the processes by which water is associated with other molecules.

The relatively low density of water is due to the fact that the hydrogen bonds are continually breaking and reforming. Because energy can be stored in hydrogen bonds, water has a high specific heat and can lose or store large amounts of thermal

Table 1.1. The Physical Properties of Water

	Liquid	Solid
Density		
$(kg\,m^{-3})$	1000 (277 K)	978 (239 K)
Latent heat of fusion		
$(J\,g^{-1})$	334 (273 K)	235 (253 K)
Self-diffusion coefficient		
$(m^{-2}\,s^{-1})$	2.2×10^{-9} (273 K)	1.0×10^{-14} (273 K)
Thermal conductivity		
$(J\,s^{-1}\,m^{-1}\,K)$	0.58 (273 K)	2.1 (273 K)
Specific heat		
$(J\,g^{-1}\,K^{-1})$	4.2 (273 K)	2.1 (273 K)
Viscosity		
$(N\,s^{-1}\,m^{-2})$	0.23 (273 K)	$10^{7}/10^{15}$ (273 K)
Isothermal compressibility		
$(N\,m^{-2})$	4.9	2.0

energy with only a small change in temperature. Such a feature has a moderating influence on our climate and is an important factor in the temperature control of homeothermic organisms. It can also act as a restraint when we attempt to cool hydrated samples rapidly. The large heat of vaporization, due to the fact that extra energy has to go into breaking hydrogen bonds, provides the basis for an effective cooling mechanism for many terrestrial plants and animals, and in combination with the high cohesivity and tensile strength of water, provides terrestrial plants with both the pathway and the mechanism to transport dissolved minerals from the roots to the aerial structures. The viscosity and self-diffusion coefficient of water are properties that determine the rate at which water molecules are displaced from their temporary position at equilibrium. Both properties are strongly temperature dependent and have a large influence on the kinetics of crystallization. The shear viscosity of water at 1 bar* increases by a factor of 2 between 298 and 273 K, but when water is undercooled to 193 K at a pressure of 2 kbar the viscosity increases by a factor of 1500 (Lang and Ludermann, 1980). The self-diffusion coefficient of water in ice at 273 K is five orders of magnitude lower than in the liquid at the same temperature.

The solvent properties of water in aqueous solutions are independent of the chemical properties of the molecule, but are related to the number of solute molecules. The relatively high melting and boiling points of water have provided two of the historical cardinal points on our thermometers, and it was discovered early that the presence of solutes in water can depress the melting point and vapor pressure and elevate the boiling point. The presence of solutes lowers the free chemical energy of

---

*Many different units are used for the citation of pressure. In this book, the bar, which is an acceptable SI unit, will be used to describe pressures above normal atmospheric pressure, and the pascal (Pa), which is the appropriate SI unit, will be used to describe pressures below normal atmospheric pressure. In addition, the term torr will also be used to describe pressures below normal. The use of three units, which is contrary to the SI system, is in line with the common usage of these three units in low-temperature microscopy and analysis. The relationship between the different units is as follows: $1.0\,Pa = 1.0\,Nm^{-2} = 1.0 \times 10^{-5}\,bar = 7.5 \times 10^{-3}\,torr = 9.87 \times 10^{-6}\,atm$.

water and provides the basis for osmotic pressure whereby water molecules in contact with a semipermeable membrane move from a more dilute solution to a more concentrated solution.

Finally, water is the life support system of our planet. For aquatic organisms it is their total environment, and terrestrial organisms have evolved extraordinary mechanisms for maintaining a hydrated interior. Water is the stuff of life. It is the sole medium in which metabolism occurs and in which the products are transported. In some notable instances, it is the precursor of metabolic reactions. Water is intimately involved in structure at all levels, from the three-dimensional conformation of a protein within a chloroplast membrane to the maintenance of turgor in the leaves containing the photosynthetic tissues.

1.3. THE STRUCTURE OF LIQUID WATER

The structure and dynamic properties of water have been investigated by means of a multiplicity of techniques. Scattering methods using radiation, such as Raman and infrared spectroscopy and x rays, and particles such as neutrons, provide information about the intramolecular structure, and the combined use of hydrogen and oxygen isotopes with nuclear magnetic resonance gives information about the transport processes of the molecule. Because water molecules so readily link with each other and with dissolved materials, it has until recently been difficult to study the structure of isolated liquid water molecules or even small groups of water molecules. This has presented many problems, particularly as the properties and behavior of the bulk liquid depend on details of its molecular stucture. More definitive information about individual water molecules has been derived from studying water vapor by means of molecular beam scattering and from water molecules adsorbed onto selected molecular sieves.

The structure of pure liquid water is not fully understood and our knowledge of it in relation to biological, organic, and inorganic systems is even more problematic. Much of the unusual behavior of water arises from the presence of hydrogen bonds due to electrostatic interactions between charges within the water molecule. It is now generally accepted that the single water molecule is a tetrahedron with the oxygen atom at its center and the hydrogen atoms at two of the corners. Such a molecule would fit within a sphere of a radius of 0.28 nm. The covalent OH^- bonds connecting the two hydrogen atoms with the oxygen atom are not in a straight line, but at an angle of about 105°. This angle is at 109° in ice but is not constant in liquid water due to an average sharing of electrons and distribution of charge. The postulated structure of a single water molecule is shown in Fig. 1.1.

The water molecule has two positive charges, which can be related to the position of the two hydrogen atoms, and two negative charges, which can be related to the two lone pairs of electrons within the structure. This separation of charge within the molecule is the cause of the high molecular dipole moment (1.87×10^{-18} electrostatic units) of the water molecule. Although the net charge on the molecule is zero, an immediate consequence of such a charge separation is that the oxygen in the water

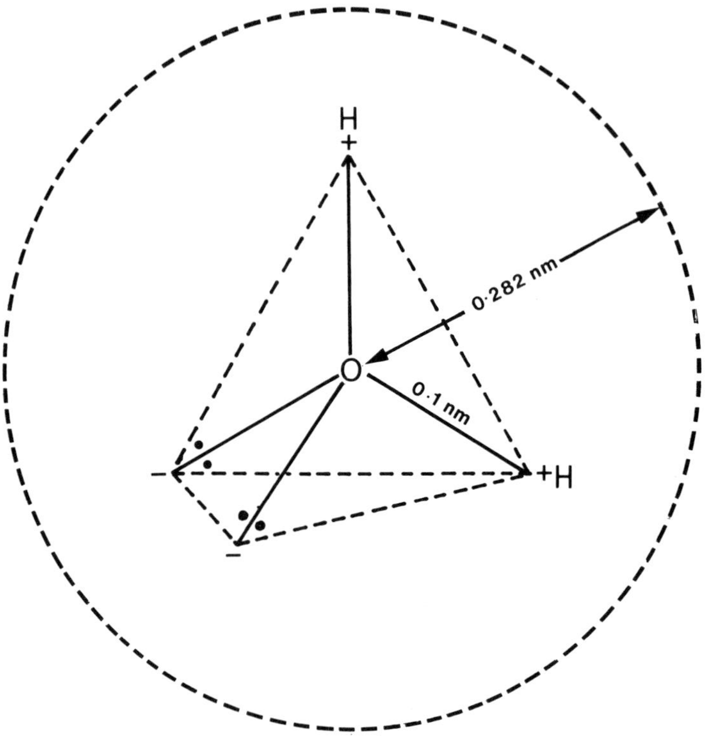

Figure 1.1. Four-point charge model of a single water molecule. The oxygen atom is at the center of a regular tetrahedron, the vertices of which are occupied by two positive charges (hydrogen atoms) and two negative charges. The O–H distance is 0.1 nm. The distance of closest approach of two molecules (van der Waals radius) is 0.28 nm. Redrawn from Franks (1985).

molecule is very electronegative and the hydrogen acts almost like a bare proton which is strongly attracted to electronegative atoms such as the oxygen in a neighboring water molecule. This attraction takes the form of hydrogen bonds between the two lone pairs of electrons and the protons on the two hydrogen atoms so that each individual water molecule can form hydrogen bonds with its four nearest neighbors as shown in Fig. 1.2. Further interactions take place to give an extensive and highly dynamic, three-dimensional network of water molecules in which the H-bonded tetrahedral order remains, although it is distorted by strained and broken hydrogen bonds (Fig. 1.3).

It is important to remember that *hydrogen bonds* are relatively weak and that they occur in many substances other than water. The strength of the bond between the hydrogen atom of one molecule and the negative charge on some part of another molecule varies but has a dissociation energy in the range 10–$40 \, \text{kJ mol}^{-1}$. In liquid water, the energies of the hydrogen bonds between water molecules vary between 20 and $35 \, \text{kJ mol}^{-1}$. The hydrogen bonds between the water molecules are much stronger

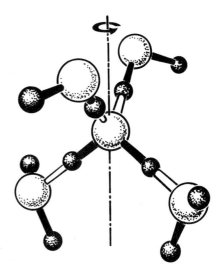

Figure 1.2. Diagram of a small group of water molecules. Hydrogen-bonded structure of water containing five water molecules. The small spheres represent hydrogen atoms; large spheres, oxygen atoms; the white rods, hydrogen bonds. Redrawn from Walfern (1964).

than the *van der Waals attractive forces*, which are the forces more usually associated with the structure of liquids such as the neutral, nonpolar molecules in liquid hydrocarbons, membrane lipids, and the internal parts of proteins. The van der Waals forces are between 1 and 5 kJ mol^{-1}. In comparison, *ionic bonds* in which electrons move from one atom and become attracted to another, are much stronger, and a

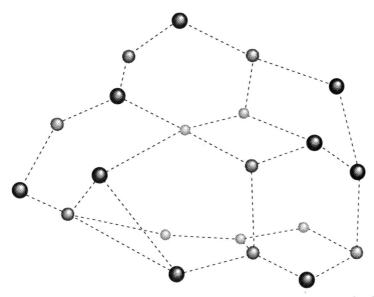

Figure 1.3. A "snapshot" view of the structural elements of liquid water showing polygons of various sizes and nonlinear hydrogen bonds. Only the positions of the oxygen atoms are shown. Redrawn from Angell (1982).

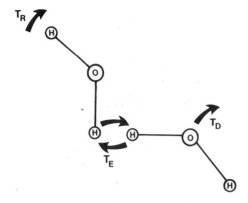

Figure 1.4. Dynamic processes in water. T_D is a correlation time for translational diffusion, i.e., a measure of how long a water molecule stays in one place $(10^{-12}\,s)$. T_R is a correlation time for reorientation of a water molecule $(3 \times 10^{-12}\,s)$. T_E is the lifetime for an exchange of protons between water molecules $(1 \times 10^{-3}\,s)$. Data from Packer (1977).

typical value would be $770\,kJ\,mol^{-1}$, found in NaCl. The *covalent bonds*, in which electrons are shared by two atoms, are of intermediate energy, and the bond energy between oxygen and hydrogen in water is $460\,kJ\,mol^{-1}$.

The extensive hydrogen bonding in water accounts for many of the special properties of the liquid, and because these bonds are relatively weak they are continually breaking and reforming. The relative time scale for the molecular movements within liquid water are shown in Fig. 1.4. Pure water, by its very nature, is a highly dynamic structure.

1.4. STRUCTURAL MODELS OF LIQUID WATER

Several models have been proposed to explain the dynamic structure of bulk water. Toward the end of the last century, Wilhelm Rontgen (1892) suggested that liquid water at ambient temperatures contained ice molecules, but it was the major conceptual advance of Bernal and Fowler (1933) that showed that water molecules do not exist in isolation but bind strongly to each other. They proposed that the hydrogen bonding found in ice was also present to a certain extent in liquid water, although the extensive rigid three-dimensional network of the solid had disappeared. In light of more recent work, the Bernal and Fowler model is now considered to be too highly ordered in spite of the findings of Dore (1985), which provided evidence that the highly disordered tetrahedron structure can persist above the melting point. Pauling (1959) suggested that groups of hydrogen-bonded water molecules could form pentagonal dodecahedra bonded together as a series of cages or "clathrates" forming an open structure. Guest molecules, including water, could reside within these cages. This model, which is shown in Fig. 1.5, is also now considered to be too rigid and has been modified to one in which the clathrates are continually breaking up and reforming with a mean lifetime of 100 ps $(10^{-10}\,s)$.

Water is now no longer considered to be a mixed solution of distinguishable species but a continuum of molecular configurations. This change in emphasis is seen in the model originally proposed by Frank and Wen (1957), which, together with its modifications, more closely incorporates both the dynamic and structural aspects of

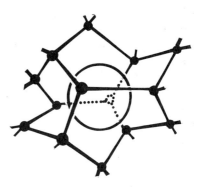

Figure 1.5. A diagrammatic representation of a water cage or clathrate formed around a nonsolute particle. The configuration of the water molecule is governed by the tetrahedral bonding patterns that must be maintained. Redrawn from Franks (1985).

water. In this model, the initiation of hydrogen bonds between water is highly cooperative and favors the formation of further hydrogen bonds and thus maintains the structure of water. The reverse is also true and the breakage of bonds leads to further disruption and the disintegration of the water structure. This "flickering structure" is in fairly good agreement with many of the properties of water, and the model proposed by Nemethy and Scheraga (1962), shown in Fig. 1.6, conveys a good impression of the dynamic structure. The liquid is composed of five-, six-, and seven-membered polygons, which are in turn hydrogen bonded with varying degrees of ease and stability into a random three-dimensional network. There will be a large proportion of strained and broken bonds and there is little evidence to suggest regions of more ordered structure.

1.5. PERTURBATIONS OF THE LIQUID WATER SYSTEM

The dynamic state of water, with relaxation times measured in picoseconds, has several important consequences. The ability of water to readily form hydrogen bonds

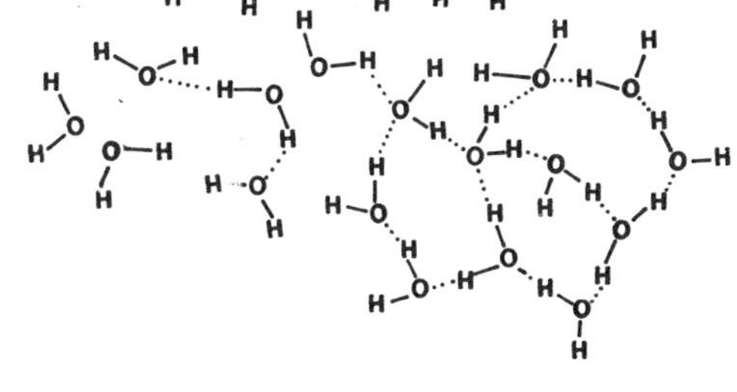

Figure 1.6. Diagrammatic representation of the "flickering cluster" model for the structure of liquid water. The tetrahedrally coordinated structural clusters and the nonstructured regions of the unbonded molecules form and break very rapidly. Redrawn from Nemethy and Scheraga (1962).

means that its structure is easily disrupted by changes in temperature and by the presence of soluble materials. The fact that water so readily interacts with such a wide range of other molecules means that we must also consider the structure of water in relation to other substances. Low-temperature microscopy and analysis is rarely applied solely to the solid phases of pure water; nearly all investigations are directed toward the more complex associations of water in cells and tissue, soils, foodstuffs, and organic compounds. Although the homogeneous nucleation temperature has been accurately and reproducibly measured for pure water, similar measurements are more difficult to make and are less accurate in the more complex aqueous systems containing dissolved solutes and hydrophilic surfaces.

There are three main classes of association between water and solute molecules:

1. *Ionic interactions* between water and the hydration shell which surround cations and anions. A simple example would be a solution of NaCl, but more complex interactions are found between the inorganic cations and anions in soils.

2. *Polar interactions* between water and nonionized solutes which are capable of extensive hydrogen bonding. A sucrose solution would be a good example and similar polar associations play an important role in the revolving phases of metabolism in living systems.

3. *Apolar associations* where the solutes have a low aqueous solubility and do not form hydrogen bonds with water. In these circumstances, the water molecules are reorganized to form cavities around the apolar molecules with the O–H groups directed away from the center of the cavity. This phenomenon is referred to as hydrophobic hydration by Franks (1985), and examples would include interactions between water molecules and hydrocarbons, and between alkyl derivatives such as alcohols, ketones, ethers, and amines, all of which make a substantial contribution to the maintenance of the conformational integrity, structural stability, and biochemical specificity of biological materials such as proteins. These apolar associations are not just limited to biopolymers but influence the way water interacts with any materials with which it cannot form hydrogen bonds. The extent and nature of these apolar associations and the restrictions they impose on the configuration and dynamic properties of water have important consequences in cryomicroscopy. Any biological (organic) material that relies on extensive hydrophobic interactions for the maintenance of structure is likely to become more unstable at low temperatures (Franks, 1985).

Although it is convenient for purposes of explanation to discuss each of these types of interactions separately, in the complex systems that will concern us here—e.g., biological samples, soils and clays, and polymers—all three types of interaction will be found, with each type of interaction influencing the effectiveness of the other two. The situation that exists in a muscle cell may help to demonstrate the complexity of these interactions. There will be ionic interactions between the water and the dissolved electrolytes; polar interactions between water and the low molecular weight compounds participating in the different metabolic pathways; and the apolar associations between water and the constituent biopolymers of the genome and the cytoskeleton.

1.6. DIFFERENT BINDING STATES OF WATER

So far in our discussion we have been concerned primarily with the properties and structure of pure water; a substance that probably does not exist outside the confines of a research laboratory. It has been shown that water is both reactive and interactive and we need to briefly consider the nature of some of these interactive forms of what is best referred to as "prefix" water. References to these different "forms" of water abound in the literature as "vicinal" water, "interfacial" water, "surface-modified" water, "bound" water, "free" water, "bulk" water, and "unfreezable" water. We need to understand something of their nature and ascertain whether some of the named forms are synonyms derived more from semantics than scientific reasoning. The degree to which water is bound to surfaces or ordered structures is influenced by the presence of solutes that affect the kinetics of the phase transition from liquid to solid and from liquid to vapor—important factors in the preparation, examination, and analysis of materials by low-temperature microscopy.

1.6.1. Surface-Modified Water

The terms "vicinal," "surface-modified," and "interfacial" water appear to be synonyms and describes water in contact with a solid surface. The water close to a protein in a cell, or a piece of humus in the soil, is considered to have properties that are significantly different from the bulk water surrounding the material. The innermost layer of water at the interface is pure water and is considered to be more ordered than the bulk liquid. The further one moves away from this interface and into the bulk liquid, the more the concentration of any dissolved solutes increases and the orderliness of the water decreases. Franks (1979) considers that in these situations three types of water may be distinguished (Fig. 1.7):

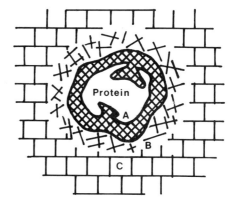

Figure 1.7. Surface-modified water. Diagrammatic representation of the environment surrounding a protein in solution. The protein can be considered to have a large hydration shell A in which molecular motions are limited. The region C is the unperturbed water and the region B is intermediate between A and C, in which some water molecules will be perturbed by the protein hydration shell. Redrawn from Franks (1979).

1. The relatively immobile internal water, which is an integral part of the material and intimately involved in its structural integrity and function. A good example would be the role water plays in maintaining the tertiary structure of proteins.
2. Water molecules distributed at hydrophilic regions on the outside of the material. This water is considered to be exchangeable with the "bulk" or "free" water, which constitutes the larger amount of water surrounding the material.
3. A poorly defined outer hydration region where the effects of the introduced material compete with the unperturbed structure of the bulk water.

It is generally agreed (Israelachvili, 1985) that the effect of surface modification does not extend much further than one or two layers of water molecules. The number would be greatly increased in solutions containing proteins and carbohydrates, and in particular in the instances where water molecules play an important role in the internal structure and functionality of enzymes. At a higher level of organization, Resing (1972) showed that although the median mobility of water on plant cell walls is about 20 times lower than in bulk water, it is still much closer to that seen in liquid water than to that in ice.

1.6.2. Perturbed Water

The term "perturbed water" is used to describe the water next to hydrated ions, molecules, and macromolecules and the adsorbed layers around colloidal particles. The water molecules are considered to assume a more ordered configuration, which, as we will see, may approach the structure seen in the crystalline state. The molecules are more densely packed, the molecular concentration is increased, and the random movements restricted, all of which affect the bond strengths and hence their reactivity. This type of water is frequently (and it is suggested, erroneously) described as "bound" water.

Although in some well-characterized systems water may have significantly different properties (Cooke and Kuntz, 1974), the use of the term "bound water" can be misleading. In some instances, the term "bound water" has been interpreted to imply that the water molecules involved in these types of interactions are irreversibly attached to the surface in question. This is most unlikely to be the case, because (1), as discussed earlier, the properties of such bound water will even in the very short term change with increasing distance from the interface, and (2) in the long term even the water molecules closest to the interface will exchange very rapidly with molecules in the bulk material. Thus the exchange rates of water molecules between a fish swimming in a pond will be very high. If the fish is now removed from the pond and dried, the exchange rate will be very much lower. Water can only be said to be bound to other molecules when there are no other water molecules with which it may exchange. The use of the term "bound water" must imply a certain sense of instability and be accompanied with defining parameters such as the nature of the surface, bond lengths, and for how long the binding has occurred. Packer (1977) has reviewed data relating to the time scale of dynamic changes in water molecules. In bulk water at 290 K, molecular motions are of the order of 1–10 ps, while in undercooled water modified at an interface such as a protein, these motions are of 1 ns at 266 K. Molecular exchange between water molecules bound at a surface and the bulk liquid are of the order of 100 μs. It is these relatively fast movements of water that we must consider when we come to discuss the procedures of rapid cooling. There is, however, a suggestion that some of the water in very small pores or capillaries will not freeze at all (Cooke and Kuntz, 1974) and that the amount of this so-called unfreezable water may be a good measure of the amount of "bound water" in the system. Franks (1986) does not agree, because he considers that this implies that the water is in an equilibrium state with a long residence time. This is not to suggest that unfrozen water does not exist, because Packer (1977) and Derbyshire (1982) have shown that after freeze concentration, about half the water in colloidal solutions of proteins and carbohydrates and in biological solutions remained unfrozen and mobile even at temperatures as low as 266 K.

The term "unfrozen water" has implications in the preparation of specimens for low-temperature microscopy, for within the molecular dimensions occupied by perturbed water, instead of the usual separation of the solute and solvent phases below the eutectic point (see Chapter 2), the solution may undercool and be transformed directly into an amorphous glass. There is considerable variation in the amount of unfrozen water depending on the nature and dimensions of the system being studied and the temperature at which the investigations are carried out. In biological systems studied between 263 and 253 K the amount of unfrozen water is between 10 and 35% of the water in the cells and tissues. It is, however, important to remember that pure unfrozen water at low temperature is a special case and usually limited to quite small volumes.

1.6.3. Unperturbed Water

The term "unperturbed water," more commonly referred to as bulk or free water, is used to describe the principal part of water in systems where there are no obvious interactions other than those with the water molecules themselves. The implication is that such water is free of any structural restraints imposed by the system and that it is free to interact with any components within the system. As will be shown in Section 1.8, it is now generally accepted that nearly all the water associated with biological systems is in a more or less perturbed state. It is probably more helpful to consider perturbed and unperturbed water as two opposite extremities of a continuum. As one moves away from the inextricably "bound" state, the remaining water may be more readily converted to the frozen (crystalline) state; movement away from the unfettered unperturbed state makes this process less likely. It is, however, important to appreciate that noncrystalline water can exist at low temperatures and that we can take advantage of this naturally occurring or artificially induced state in the preparation of specimens for low-temperature microscopy and analysis.

1.7. UNDERCOOLED WATER

The term "undercooled water" refers to water that remains liquid well below its equilibrium freezing point without ice nucleation and crystallization. Undercooled water, which is frequently referred to as supercooled water and sometimes as sub-cooled water, is included here because it is the metastable liquid state of water intermediate between normal liquid water found above 273 K and the various solid forms of water that exist below this temperature.

Provided adequate precautions are taken to prevent ice nucleation, small quantities of pure water can be undercooled to 233.5 K and still remain in the liquid state. At increased pressures, the water remains liquid to even lower temperatures (Fig. 1.8). Although samples as large as 1.0 ml have been observed to remain liquid at 243 K, undercooled water becomes progressively more difficult to observe as the sample size increases and the water purity and temperature decrease, because of the increased probability of nucleation. In an extensive series of studies by Angell and his co-workers (see Angell, 1982, 1983), it was discovered that many of the dynamic and thermodynamic properties of water appeared to diverge at about 228 K. The density and diffusion coefficient diverges and the compressibility, vapor pressure, and viscosity increases (Fig. 1.9). There seems to be no satisfactory explanation for these phenomena, although it is known that water can be made to behave more normally

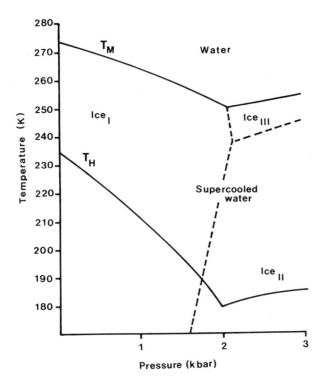

Figure 1.8. Metastable state of undercooled water. The figure shows the range for water in the 1–3-kbar range defined by emulsion nucleation temperatures. T_M is the equilibrium freezing temperature and T_H is the homogeneous nucleation temperature. Broken lines represent approximate boundaries. Redrawn from Angell (1983).

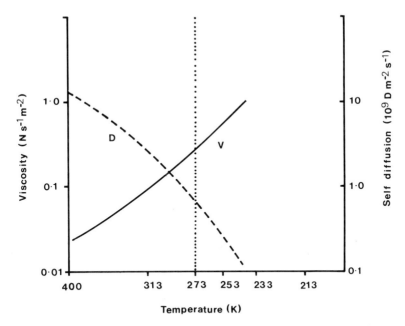

Figure 1.9. The viscosity (*V*) and self-diffusion (*D*) of water as a function of temperature. Redrawn from Angell (1982).

by the addition of some solutes such as urea to the undercooled liquid. The anomalies associated with undercooled water are preserved in dilute solutions of physiological strength, i.e., 200–300 mosmol. The anomalies in undercooled solutions of high solute concentration only disappear at high pressures; they cannot be avoided simply by rapid cooling.

Undercooled water is becoming increasingly more important in the preparation of samples for low-temperature microscopy. Under special circumstances, the water in small droplets and thin films and layers, when rapidly cooled, will first undercool, and in the absence of any nucleating agents will be transformed to a noncrystalline, glasslike state. The existence of this phenomenon has received some support from the studies of Sceats and Rice (1982), who postulate that there is a continuity of state between the glasslike form of ice and undercooled water.

The exact conditions for successful undercooling of biological tissues in nature are now beginning to be understood, and Sakai and Larcher (1987) consider that the following conditions should exist for undercooling to occur: (1) small sample size; (2) small intercellular spaces to diminish nucleation; (3) low water content; (4) an absence of intracellular nucleators; (5) barriers against extracellular nucleation; (6) compartmentation into small units, which can freeze independently; (7) the presence of compounds (as yet, not precisely identified) that depress nucleation. Such conditions are usually only found in organisms that have become acclimatized to living at low temperatures.

The amount of undercooled water in a given sample will show a large variation and is related closely to the probability of nucleation. This depends on the size of the space occupied by the water, its purity, and the temperature regimen to which it has been subjected. With the exception of the special cases mentioned above and in samples that are permeated with antifreeze agents, it would be quite wrong to suggest that we can substantially undercool the large volumes of water normally associated with most cells and tissues. Nucleation events and crystallization quickly intervene.

Extracellular solutions exhibit various degrees of undercooling depending on the effectiveness of available nucleators and the nature of the surfaces surrounding the water. Although some water in plants may undercool to as low as 226 K (George *et al.*, 1982), the general extent of this effect depends on the size of the spaces in which the undercooling occurs. Water in the large cavities in the vessels and tracheids of the vascular system of plants shows little undercooling, whereas water in the small capillaries within the walls of living cells of the buds of some woody species shows substantial undercooling (Ashworth and Abeles, 1984). These aspects of undercooling are discussed in detail by Grout (1987) in reference to the survival of plants at low temperatures.

Undercooled water and solutions are also of interest to biochemists in studies on the kinetics of enzyme reactions. As Douzou (1977) and Douzou *et al.* (1978) have shown, the rates of reaction are considerably slowed down in undercooled solutions and cryosolvents and enable metabolic pathways to be studied in greater detail. These aspects of undercooled water are discussed by Franks (1985).

1.8. WATER IN CELLS AND TISSUES

Water is the most abundant and mobile component of biological systems, and before leaving this chapter on the properties and structure of liquid water, we should briefly consider the special case of the state of water in living cells and tissues. The principal reason for including this topic is that a large proportion of the effort in low-temperature microscopy and analysis is directed toward biological material, most of which is strongly hydrophilic. We must enquire whether any of the water in the cells behaves exactly like a pure liquid, and establish the degree to which dissolved electrolytes and molecules, and macromolecular surfaces, influence the structure and properties of water.

Early work on cell and tissue water envisaged it as a homogenous, dilute aqueous solution in which the water was for all intents and purposes in a free and unperturbed state. A closer examination of this premise showed that this view is not correct, because water in biological systems exhibits different physicochemical properties from the bulk material. There are, for example, anomalies in the expected values from osmotic studies and from freezing point and vapor pressure depression, which seemed to suggest that some of the water in the cell is in a perturbed state. More recent work using modern techniques has confirmed this view.

A consequence of this renewed interest in cell water has, not unexpectedly, resulted in the formulation of a number of different models to explain the state of the water in the cell. While this is not the place to discuss these models in great detail, it is appropriate that we briefly consider the main features of the different models.

The most generally accepted view is that the majority of the water in the cell—and here we take the term cell to mean the cytoplasm and its attendant organelles—exhibits properties similar to the unperturbed water in the bulk aqueous phase. The remaining cell water, which might account for 10% of the total, is present as a hydrated phase immediately next to all the surfaces that exist in the cell. These would range from free ions and metabolites to macromolecules and the structurally organized components of the cytoskeleton. In such a model, the water could function as a solvent, as a substrate for and a product of metabolism, and as a component of the cellular structural macromolecules.

An alternate view is the association–induction hypothesis proposed by Ling and co-workers (see Ling, 1970, 1984 as source references). This model proposes a more ordered structure in which virtually all the water is considered to be in a series of polarized multilayers, which extend outwards for a considerable distance from the monolayer of water surrounding extended protein surfaces. In this state, water has a reduced osmotic activity and a reduced solubility for hydrated ions and large molecules. It is envisaged that ions and solutes are variably excluded from the polarized water layers and that they are instead associated with fixed charges on biopolymers such as protein. Although this is an imaginative model, there is only fragmentary evidence for the *in situ* localization of the extended protein layers that would be necessary to bind ions and be associated with the multilayer polarization of the water.

There are a number of other models, details of which have been concisely summarized and clearly illustrated by Clegg (1985). None of the models are entirely satisfactory, although all of them indicate that some of the water in cells is in a perturbed state. The disagreements center on whether a structural and physicochemical basis exists in cells that would modify the form of cellular water. There is considerable structural evidence obtained by a wide range of microscopical techniques and cytochemical methods that shows that the cytoplasm of all eukaryotes (higher plants, animals, and fungi, but not bacteria) is filled with a vast network of fibers and filaments. This cytoskeleton, as it is referred to, is composed principally of proteinaceous microtubules, microfilaments, and intermediate filaments, together with a vast array of more specialized proteins, which in turn assemble and interlink, disassociate, and depolymerize the three main components. This filamentous array is a dynamic structure and its constituent components appear and disappear at different times during the cell cycle.

The whole array of the cytoskeleton and attendant specialized proteins is sometimes referred to as the microtrabecular network. This network is less well defined at the structural level, because of the difficulty of preserving the smaller more specialized molecules, which maintain the structural integrity of the three well-characterized components. The cytoskeleton–microtrabecular network is considered by some workers (Porter, 1984; Ling, 1980) to form the structural basis for the organization of water in cells. The evidence for this assertion is ambiguous and is based on the assumption that a large proportion of the structural components are hydrophilic. A somewhat iconoclastic account of this hydration view of cell water may be found in the recent book by Negendank and Edelmann (1989).

The cells of mature plants present a special case, for unlike most animal cells, a large proportion of the cell volume is taken up by the vacuole. Vacuoles are the largest of plant organelles and one of the principal structural components of higher plants. Up to 90% of the volume of mature cells may be made up of a relatively inert and metabolically inexpensive acidic (pH 5–6.5) solution osmotically equilibrated with inorganic and organic solutes (Boller and Wiemken, 1986). Although vacuoles do not contain the structural proteins of the cytoskeleton, they do contain between 1% and 10% of the total cellular protein. About half is associated either as structural or enzymatic components of the vacuolar membrane, the tonoplast, and the remainder as enzymes associated with the proteolytic activity of the vacuolar sap. In addition, vacuoles contain many of the secondary products of cell metabolism, including alkaloids, saponins, and glycosides. Some compounds, including proline and glycine betaine (Leigh et al., 1981), are at a very low concentration in the vacuole relative to the cytoplasm. Such compounds are of interest because they appear to act as natural cryoprotectants in plants (see Chapter 3, Section 3.3.3).

The presence of such a large array of dissolved materials would suggest that the water in the plant vacuole is also perturbed although probably to a lesser degree. It remains to be seen whether the water that occupies the cell free space or apoplast, which exists exterior to the limiting cell membrane and between the cells of higher plant tissue, is similarly modified. This is in no sense an inactive zone as it plays an important role in the transport of water and dissolved inorganic and organic materials. The presence of these dissolved substances would also confer a degree of order on the water. Similar extracellular spaces exist in animal cells, as collecting ducts for both cell excretions and the products of cell metabolism. The composition of these biofluids are well known and they all exist as dilute solutions of electrolytes, molecules, and macromolecules, although they lack the structural components of the cytoplasm discussed earlier.

In accepting that cell and tissue water is, on a time-averaged basis, perturbed, it is also important to ask whether different degrees of modification may exist in different compartments of cells and tissues. Water is presumably more perturbed in the cytoplasm and less so in plant vacuoles, the apoplast, and the extracellular spaces in animal tissues. Such differences in the amount of surface modification of water will have an important bearing on the way we prepare cells and tissues for low-temperature microscopy and analysis.

1.9. SUMMARY

Water, the most common liquid on our planet, has a number of remarkable properties, which are a consequence of its molecular structure. The molecule is strongly polarized owing to the separation of the charges on the oxygen (−) and hydrogen (+). The constantly shifting weak hydrogen bonds which can form as a consequence of this charge separation are responsible for the dynamic structure of liquid water and for its interactions with both dissolved materials and intact macro-scopic surfaces. The hydrogen bonding of water influences the properties of the liquid, many of which are intimately associated with the processes of life. The ionic, polar, and apolar associations of water with a wide range of substances indicate that pure liquid water rarely exists, and the different ways in which the water associates with materials appear to modify the bulk liquid. This affinity of water molecules for surfaces has given rise to some misunderstanding as to what is meant by the terms "bound" and "free" water. The complex interactions between liquid water and the varied components of cellular systems reveal that some of the water associated with biological material may be in a modified form. These variations in the strength and extent of the bonding between water and substrates will influence the amount and form of the solid phases to which liquid water may be converted when cooled during many of the preparative methods of cryomicroscopy.

2

The Structure and Properties of Frozen Water and Aqueous Solutions

2.1. INTRODUCTION

Following discussion of the structure and properties of liquid water, we need now to consider the ways by which water molecules may link together into the apparently more stable and solid phase of ice. We are all familiar with the appearance and properties of ice, and for our everyday use of such things as ice cubes we have little concern about the structure of the material, the size of the crystals, or even the processes by which it is formed. But when we turn from the domestic use of ice to its use as a matrix in low-temperature microscopy and analysis, it will become quickly apparent that the structure and properties of the material assume a much greater significance.

It will be useful first to consider briefly the structure and properties of ice before considering the far more complex situation where water is in combination with other compounds and/or hydrophilic surfaces. One of the more unusual properties of water is that it can exist in a large number of different solid forms depending on the pressure and temperature prevailing at the time of its transition. We will, however, confine our attention principally to the polymorphs of ice which form at normal pressures and a few tens of degrees below the melting point and the ice that is produced when liquid water and water vapor are cooled rapidly. We will discuss the physicochemical processes involved in these transitions and show how they may be used to advantage in cryomicroscopy and analysis.

Much is already known about the structure and properties of ice, and a more detailed account of the properties and processes which now follow may be found in the standard texts by Fletcher (1970) and Hobbs (1974).

2.2. THE DIFFERENT FORMS OF ICE

As liquid water is slowly cooled to below its freezing point, the random and dynamic structure in which a tetrahedron configuration predominates assumes a more stable form. The liquid is rapidly changed by a first-order phase transition into

19

a solid material in which the basic tetrahedral structure is now extensively cross-linked into hexagonal crystals, which at normal pressures are characteristic of this stable polymorph of ice. The extensive hydrogen bonding, which maintains the structure of the ice, is very labile, and any changes in pressure during the cooling process, or changes in temperature of the solidified material, can readily affect the crystal structure.

Depending on the temperature and pressure, solid water can exist in at least nine different stable crystalline forms, one metastable crystalline form, and one (or two?) metastable amorphous forms. A phase diagram of many of these different forms in relation to the liquid state is shown in Fig. 2.1, and some of their properties and interconversions are given in Table 2.1. In low-temperature microscopy and analysis, we are, with one exception, primarily concerned with those forms of ice that form at normal pressures.

Much of our discussion will center on the two forms of ice I, hexagonal ice (I_H) and cubic ice (I_C), and on amorphous ice (I_A), all of which have a lower density than water. At pressures in excess of 2 MPa (2 kbar), eight other forms of ice can exist, each within its own temperature and pressure range. The high pressures push the water molecules closer together and the hydrogen bonds are consequently bent

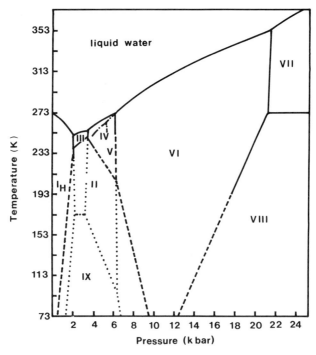

Figure 2.1. Solid–liquid phase diagram of water. Broken lines represent approximate and dotted lines represent estimated phase boundaries. I_H to I_X are the nine stable crystalline forms of ice. Redrawn from Hobbs (1974).

Table 2.1. Physical Properties and Interconversions of the Different Polymorphs of Ice[a]

Ice phase	Pressure (kbar)	Lower temperature (K)	Upper temperature (K)	Transition at upper T
I_H	0	0	273	Melts
I_C	0	0	183	$I_C > I_H$
I_V	0	0	153	$I_V > I_C$
I_{II}	3	0	243	$I_{II} > I_{III}$
I_{III}	3	183?	253	Melts
I_{IV}	5	133?	248	Melts
I_V	5	123?	263	Melts
I_{VI}	14	123	323	Melts
I_{VII}	25	275	383	Melts
I_{VIII}	25	0	275	$I_{VIII} > I_{VII}$
I_{IX}	3	0	168	$I_{IX} > I_{III}$

[a]Data from Hobbs (1974) and Fletcher (1970).

and distorted. These high-pressure forms of ice are denser than liquid water and have rhomboidal, tetragonal, and monoclinic crystal systems. Less is known about these high-pressure ice polymorphs, which although of interest to planetary geologists (Hemley et al., 1989), do not exist in nature, even at the depths of the polar ice caps. With one exception, they need not concern us further as they appear to have no application in low-temperature microscopy and analysis.

The exception is the form of ice that is considered to be produced as a result of using the high-pressure freezing methods that have been developed during the past 20 years by Moor and his colleagues (see Moor, 1987, for a summary). An examination of the phase diagram in Fig. 2.1 will show that at approximately 2 MPa (2 kbar) and between 253 and 183 K, it should be possible to undercool water without the formation of either of the two polymorphs of ice I. If heat is removed from water kept under these conditions, the liquid will slowly crystallize to form either ice II or ice III, two of the high-density, high-pressure polymorphs of ice. However, it is important to appreciate that the advantages of the high-pressure freezing method (see Chapter 3, Section 3.6.6), are derived not from the formation of ice II and ice III, but from the substantial amount of undercooling that can occur in liquid water at these elevated pressures.

We will now briefly consider some of the properties and structure of the three forms of ice that exist at normal pressures. These forms of ice, which appear identical when examined in the microscope, may be readily distinguished from each other by the differences in their electron or x-ray diffraction patterns. Figure 2.2 shows electron micrographs and diffraction patterns of the three forms of ice most commonly encountered in low-temperature microscopy and analysis.

2.2.1. Hexagonal Ice (I_H)

Hexagonal ice is the polymorph of ice that forms when water is cooled, even quite rapidly, and is energetically the most favorable structural arrangement of water

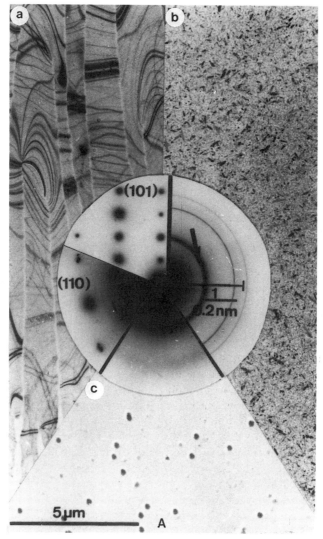

Figure 2.2. Electron micrographs and electron diffractograms of three forms of ice in pure water (A) and in frozen-hydrated sections (B). (a) I_H obtained by rapidly cooling a 50–80-nm layer of water spread on a carbon film. The diffractogram shows the (110) and (101) planes. (b) I_C obtained by warming a 70-nm layer of I_V formed by deposition of water vapor onto a surface below 150 K. The diffractogram contains a small contribution of the (100) form of I_H. (c) I_V obtained by deposition from water vapor on a thin film held at below 150 K in the electron microscope. The layer is approximately 70 nm thick and there are no strong reflections in the accompanying diffractogram. Micrograph courtesy of Dubochet *et al.* (1982). (B) Frozen-hydrated sections with different states of solidified cellular water and their corresponding electron diffraction patterns. (a) Cryosection of a high-pressure frozen apple leaf. The cellular water is vitrified as demonstrated by the broad ringed diffraction pattern. It transforms upon irradiation to cubic ice (I_C) as shown in (b) with the accompanying single-ringed electron diffraction pattern. Severe structural alterations (segregation patterns) appear when the sample is cooled slowly, and the characteristic electron diffraction patterns of hexagonal ice are visible in (c). Micrograph courtesy of Michel *et al.* (1991).

a

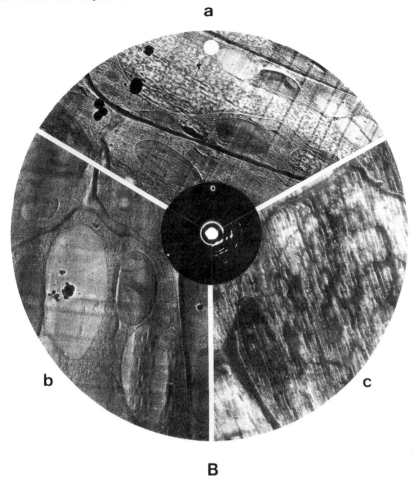

b

c

B

Figure 2.2. Continued.

molecules. It is the ordinary ice we encounter in ice cubes, snow flakes, and frozen peas, and to a large extent is generally the ice we find in frozen, hydrated specimens prepared by low-temperature methods. The most striking example of hexagonal ice is seen in the sixfold symmetry of the constituent ice crystals in snow flakes, most of which are thin hexagonal plates together with hexagonal prisms and needles. The detailed symmetry of I_H has been established (Hobbs, 1974; Fletcher, 1970; Eisenberg and Kauzmann, 1969) and need not concern us here, although we need to briefly consider the position of the atoms within the unit cell of the crystal and their relationship with one another. The unit cell for the hexagonal lattice is a prism containing four oxygen atoms set on a rhombic base and is what one might have expected from the tetrahedral structure of the liquid water molecule. Each oxygen atom is at the center of a tetrahedron formed by four oxygen atoms, and each water molecule is hydrogen bonded to its four nearest neighbors to form an open lattice.

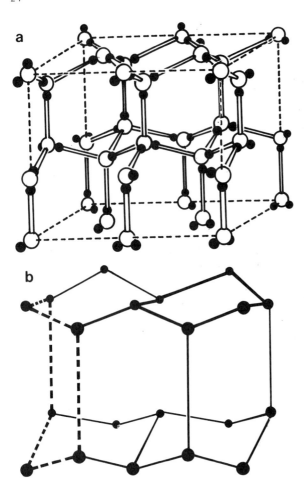

Figure 2.3. The crystal structure of hexagonal ice (I_H). (a) Diagram showing the regular lattice of oxygen atoms (open circles) with one hydrogen atom (solid circles) lying on each 0–0 axis. Each oxygen atom is hydrogen bonded to four other oxygen atoms placed at the vertices of a regular tetrahedron. Note the crinkled sheet arrangement normal to the c-axis. (b) Lattice structure of I_H showing the six-membered rings of water molecules in the "chair" configuration (thicker solid line, upper right) and the "boat" configuration (dotted line to the left). The hydrogen atoms are not shown. Redrawn from Hobbs (1974).

It can be seen from Fig. 2.3, that when the lattice is viewed in one direction, the hexagonal rings of water molecules are arranged in wavy or crinkled layers and are said to have the conformation of a "chair." When viewed at right angles to the first direction, the hexagonal rings are formed from three molecules in one crinkled layer and three molecules in an adjacent layer. These hexagonal rings are said to be in the "boat" configuration. Eisenberg and Kauzmann (1969) point out that an important characteristic of this structure is the open spaces or "shafts" running through the crystal and which account for the fact that the I_H is less dense than the liquid from which it is formed. These spaces are clearly demonstrated in the model of the arrangement of oxygen atoms in hexagonal ice shown in Fig. 2.4. Although the position of the oxygen atoms in the lattice is firmly established, there are a vast number, ca. 10^{23}, of possible arrangements for the hydrogen atoms in the crystal. The closer I_H is to its melting point, T_M (273 K), the greater the chance there is for changes in the position of the hydrogen atoms in the crystal. As the temperature of the ice decreases, these changes in hydrogen positions become less frequent until a temperature is reached when the lattice structure is virtually fixed. The temperatures

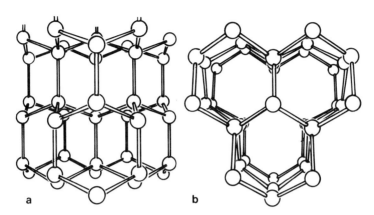

Figure 2.4. The arrangement of oxygen atoms in hexagonal ice (I_H). The hydrogen atoms are not shown. (a) View perpendicular to the c-axis. (b) View along the c-axis showing the open structure of ice. Redrawn from Hobbs (1974).

at which these fluctuations in molecular orientation occur are right in the middle of the temperature range we work with in many of the techniques used in low-temperature microscopy. This can have several important consequences, for the changes in molecular configuration affect the properties of I_H and, in turn, the ways we might wish to manipulate our samples.

I_H is formed primarily as liquid water is cooled and is the major matrix component of most specimens prepared for low-temperature microscopy and analysis. As we will show, the actual temperature at which crystallization occurs depends on a number of factors, but once the hexagonal ice is formed and although the crystals may grow, they remain in the hexagonal form, even when cooled down to the temperature of liquid helium. Indeed, as the discussion in the previous section revealed, the crystal structure of I_H is stabilized at low temperatures. I_H can also be formed under the special circumstances when cubic ice I_C is warmed to approximately 193 K. Details of this and other transformations will be considered later in this chapter.

Much of what we know about the properties of ice has been obtained from studies on I_H and the values obtained vary according to the ice thickness, temperature, and method of measurement. The values that are given here are for comparative purposes only and reference should be made to the appropriate chapters in Hobbs (1974) and Fletcher (1970) for further details of the properties of ice and how they are measured. The article by Kuhs and Lehmann (1986) provides a useful up-date on the structure and properties of hexagonal ice.

A considerable amount is known about the transport properties of ice. I_H has a maximum bulk dc electrical conductivity of 5 $\mu\Omega$ m^{-1} at 238 K and which decreases by at least an order of magnitude as the temperature is lowered to 198 K. The comparable figures for copper and wood at 293 K are 0.017 $\mu\Omega$ m^{-1} and 10^2 to 10^5 $\mu\Omega$ m^{-1}, respectively.* The surface conductivity of ice is two orders of magnitude

*Copper and wood have been chosen as examples so that the value for ice may be put into a better conjectural framework. The values for wood are useful as they are close to the values one might expect to find in biological material and polymers.

higher than the bulk conductivity, but is of little importance at temperatures below 253 K. The electrical conductivity of ice is critically dependent on purity and there is a dramatic increase in the conductivity when impurities at concentrations as low as 1 ppm are introduced into the crystal lattice. The relative electrical conductance of ice is an important factor we need to consider when examining and analyzing frozen specimens at low temperatures in electron beam instruments.

At temperatures close to the melting point (T_m) of I_H, the thermal conductivity coefficient is 2.2 W m^{-1} K^{-1}, which is approximately four times greater than the value for water at the same temperature. The thermal conductivity of ice increases as the temperature is lowered, and at the temperature of liquid nitrogen (77 K) is approximately five times greater than at 273 K (Fig. 2.5). The comparable figures for copper and wood at 273 K are 400 W m^{-1} K^{-1} and 2.4 W m^{-1} K^{-1}, respectively. The values for most hydrated biological materials are between 0.2 and 0.5 W m^{-1} K^{-1}. The specific heat of water at 273 K is approximately 4.2 J g^{-1} K^{-1}, and for undercooled water at 238 K the value increases to 5.8 J g^{-1} K^{-1}. The specific heat of ice at 238 K is 1.8 J g^{-1} K^{-1} (Fig. 2.6). The values for copper and wood at 293 K are 0.38 J g^{-1} K^{-1} and 1.8 J g^{-1} K^{-1}, respectively. The specific heat for hydrated animal tissue is approximately twice as high. These thermal properties are important when we come to consider the processes involved in the initial cooling of specimens, sectioning frozen materials, and in situations where frozen hydrated samples are examined at low temperatures in electron beam instruments.

It is frequently necessary to fracture and section frozen specimens to reveal subsurface details, and for this reason it would be instructive to understand

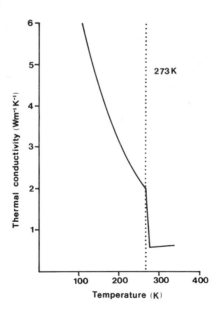

Figure 2.5. Thermal conductivity of water and ice at atmospheric pressure as a function of temperature. Redrawn from Bald (1987).

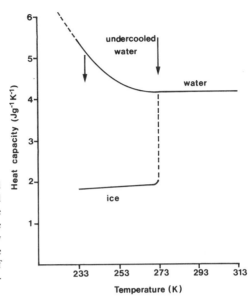

Figure 2.6. Heat capacity of water, undercooled water and ice as a function of temperature. Solid lines are from experimental data, broken lines are extrapolations. The arrow at 273 K indicates the phase transition at equilibrium. The second arrow at 237 K indicated crystallization from the undercooled state. At 273 K the heat capacity of undercooled water increases with decrease in temperature. Data from Angell (1982).

something about the mechanical properties of ice. In many respects, ice can be regarded as an engineering material in that it has recognizable properties that relate to elasticity, creep, and plastic flow. Point defects and dislocations may be recognized in samples of ice examined by cryomicroscopy. In common with any material, the properties of ice depend on its purity and substructure. While most of the impurities in liquid water are excluded from the solid phase, minute amounts of material are in true solid solution in the ice crystals. It is these impurities that give rise to the point defects in ice, which in turn, give rise to stacking faults and dislocations. Ice invariably exists in a polycrystalline form and grain boundaries are formed where adjacent crystals touch each other.

The high number of hydrogen bonds in ice means that the material can absorb a large amount of energy by reorganization of the protons. The structural model of I_H shows that the material is a stack of crinkled sheets and one of the principal deformations occurs as these sheets slide over each other. Gold (1958) has shown that within the temperature range 243–233 K ice will show elastic behavior provided the stress is brief, ca. 10 s, and kept below a certain value. The Young's modulus of polycrystalline ice between 223 and 93 K is 9.5×10^5 N. This should be compared to the value for epoxy resin, which is 2.1×10^2 N at 293 K. Imperfections, dislocations, and foreign inclusions within the ice will contribute to the extent of deformation, which is also temperature dependent. At higher temperatures, if the stress exceeds some critical value, the ice will either fracture or exhibit plastic deformation and creep (a time-dependent plastic deformation under conditions of constant stress) owing to the movement of the dislocations in the material. As the temperature decreases, there is less plastic deformation and creep, and at 173 K ice becomes a very brittle material. These elastic properties of ice are important in situations where the material is to be sectioned or fractured.

The purpose of this section has been to show that ice is not just a cold block of relatively hard material. Depending on the temperature and environmental stress, the solid may exhibit several dynamic properties, which can have an important influence on the role the material may play as the principal matrix component of frozen specimens. The values have all been given as median figures within the temperature range usually experienced in cryomicroscopy and compared with the same parameter for better well known materials. This has been done quite deliberately, so that the properties of ice and its reaction to environmental factors may be more readily appreciated in relation to the procedures and practice of microscopy and analysis at ambient temperatures.

2.2.2. Cubic Ice (I$_C$)

Hexagonal ice (I_H), is considered generally to be the only form of ice that occurs under natural conditions on our planet. It has been suggested by Whalley (1983) and Whalley and McLaurin (1984) that water in the upper atmosphere may sometimes freeze first to cubic ice (I_C) and form Scheiner's halo, a phenomenon that occurs at 245 K and is occasionally seen at a low viewing angle around the sun or the moon. This is one of the few examples of I_C occurring in nature.

Cubic ice is a metastable form of ice I which exists between 193 and 123 K, and may be formed in the laboratory under the following conditions:

1. When water vapor is allowed to condense on a clean substrate maintained at below 143 K.
2. When amorphous ice (I_A) is warmed to above 123 K.
3. When high-pressure forms of ice are cooled to liquid-nitrogen temperatures, the pressure reduced to atmospheric, and the samples then warmed to 157 K for a few minutes.

Cubic ice cannot be formed by simply cooling hexagonal ice to below 193 K, although two special exceptions to this may be seen in some recent work by Lepault et al. (1983) and Mayer and Hallbrucker (1987). Lepault and co-workers were able to transform I_H and I_C into I_A by cooling samples to liquid-nitrogen temperatures and irradiating the thin layers of ice in a cryo-electron-microscope with electron doses of $100-500e \cdot nm^2$. Mayer and Hallbrucker have produced I_C directly by spraying 3-μm water droplets at high speed onto a copper plate maintained at 190 K. The material gave the characteristic x-ray diffraction patterns for I_C and the material transformed to I_H at ca. 198 K, although at a rate lower than I_C prepared by vapor deposition.

Within the context of low-temperature microscopy and analysis, cubic ice is usually only encountered when amorphous ice is heated. The formation of I_C is most certainly the exception rather than the rule, and its presence in specimens should be taken as a warning that something is wrong, either with the process of specimen preparation or that there is excessive water vapor contamination inside the column

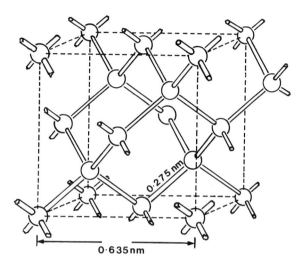

Figure 2.7. The structure of cubic ice (I_C) as shown by the position of the oxygen atoms. The hydrogen atoms are not included. Note the difference in the dimensions of the unit cell. Redrawn from Fletcher (1970).

of the cryo-microscope. An example of the former situation would be where an examination of amorphous ice is being carried out at temperatures higher than 113 K. The presence of I_C could be taken as a clear indication that specimen warming had taken place. In the latter case, contaminating water vapor, which may have initially condensed as I_A, may also have warmed and been transformed into the cubic crystal state.

From the structural point of view, the positions of the oxygen atoms in cubic ice are very close to that encountered in hexagonal ice, with each oxygen atom being surrounded by four other oxygens (Fig. 2.7). The unit cell dimensions are different and cubic ice contains eight water molecules rather than the four molecules found in hexagonal ice. There are small differences in some of the other properties, although much less is known about I_C compared to I_H. There are only very small (ca. 0.1 nm) dimensional changes as cubic ice is transformed to hexagonal ice, unlike the much larger structural changes that occur when liquid water is converted to I_H and, as we will now show, then I_A is transformed to I_C.

2.2.3. Amorphous Ice (I_A)

Amorphous ice is a noncrystalline form of ice, which, from the cryo-microscopist's viewpoint, is the ideal state of water. The term "vitreous ice" (I_V) is also used to describe the noncrystalline form of ice, and the debate continues as to whether the two materials are one and the same thing. In cryo-microscopy, although the two terms are assumed to be synonyms, they are also used in conjunction with the two different procedures used to form noncrystalline ice.

Noncrystalline ice can be formed by condensing water vapor onto a surface maintained at below 113 K, and this form of ice, in which no long-range order can be detected by diffraction techniques, is usually referred to as I_A. Alternatively, a noncrystalline form of ice can be formed by rapidly cooling minute (μm^3) quantities of liquid water. This material, which is referred to as vitreous ice, I_V, was first produced by Bruggeller and Mayer (1980) and Mayer and Bruggeller (1982). Pure water and dilute aqueous solutions were propelled as a fine stream of liquid, at high pressure and supersonic speeds, into a melting cryogen. Under these conditions, the liquid passed through the undercooled temperature region and to below the homogeneous nucleation temperature (T_{hom}) and glass transition point (T_g), before solidification occurred.

The glass transition temperature T_g marks a change in the physical properties of water which still retains the random arrangement of molecules. At temperatures below T_g the water is a highly viscous undercooled liquid which will slowly begin to crystallize (devitrify) as the temperature is raised. Samples must be maintained below their T_g to ensure that they remain in a vitreous state. T_g for pure water is ca. 135 K, but the presence of dissolved solutes will raise this temperature. It is difficult to give a precise figure for the T_g of water, as the studies of Johari *et al.* (1987) have shown that glassy and liquid water follow an almost reversible path through a temperature range of 113–148 K in which structural states remain thermodynamically continuous.

The processes that lead to the formation of I_A from the slow deposition of water vapor at liquid-nitrogen temperatures are understood. Water molecules landing on the cold surface quickly lose their energy and remain more or less at their point of initial contact. There is insufficient energy to migrate over the surface to positions of minimal potential energy which favor crystallization. The processes that lead to the formation of I_V are less clear, but the stochastic approach would suggest that the tetragonal arrangement of water molecules that is considered to predominate in the liquid state becomes locked in position as the liquid is rapidly cooled. As the temperature rises, the water molecules in both I_A and I_V have sufficient energy to migrate to thermodynamically more favorable positions on a crystal lattice, and both forms of ice are slowly transformed, initially to I_C and then to I_H. In this respect I_A and I_V are identical. Amorphous ice can also be formed by irradiating thin films of cubic ice maintained at below 70 K, either with high-energy electrons (Lepault *et al.*, 1983) or by ultraviolet photons (Kouchi and Kuroda, 1990).

The situation is complicated by the fact that it appears that two forms of I_A may exist (Narten *et al.*, 1976; Sceats and Rice, 1982). Experiments by Mishima *et al.* (1984) produced a glasslike ice when I_H, cooled to liquid-nitrogen temperatures, was subjected to pressures in excess of 10 MPa (10 kbar). The material remained in the amorphous state as the pressure was released at the low temperatures. This high-pressure form of amorphous ice $I_A(H)$ has a high density of 1.31 g cm^{-3}, while the low-pressure form of amorphous ice $I_A(L)$, which has already been described, has a density of 0.91 g cm^{-1}. X-ray diffraction studies showed small differences, and the two forms would transform one into the other with a measurable production of latent heat. Studies by Hemley *et al.* (1989) show that this high-pressure, high-density amorphous ice can be converted back to a crystalline solid by applying a pressure of 1 GPa (1 Mbar) at 77 K. Heide and Zeitler (1985) have shown that at very low temperatures, i.e., 8–20 K, I_H, I_C, and $I_A(L)$ all change into $I_A(H)$ when irradiated with an electron beam, and remain in this state at these low temperatures without further irradiation. At between 30 and 70 K, I_H, I_C, and $I_A(L)$ remain stable and

$I_A(H)$ spontaneously transforms to $I_A(L)$. At higher temperatures, the now familiar transformations take place. It would appear that $I_A(H)$ is a special case and we will exclude it from further consideration here as it appears only to be formed either under high-pressure conditions or when ice at very low temperatures is subjected to ionizing radiation.

The two remaining questions that should concern us here are whether I_V and $I_A(L)$ are one and the same material, and more importantly, whether I_V, *sensu stricta*, can be formed by rapidly cooling liquid water. The first question is more easily answered, for although Angell (1982a) has showed that I_V and I_A have different physical properties, such differences are of minimal significance in any discussions of low-temperature microscopy and analysis. However, for the sake of consistency, the term I_A will be used to describe the noncrystalline ice that is formed when water vapor condenses onto a very cold surface, and I_V will be used to describe the glasslike ice that is formed when water is cooled rapidly.

Before attempting to answer the second question, we should perhaps try to more closely define what is meant by vitrified water. We have already established that the term "vitrification" may be used to describe the process whereby liquid water of low viscosity is transformed into a liquid of high viscosity, which is referred to as a glass. A glass is a material with a viscosity of ca. 10^{14} N s m^{-2}, which for water would mean that it would take about a day for a water molecule to diffuse a distance equivalent to its own diameter. Vitrified water is a very viscous liquid in which there are negligible molecular movements. Franks (1985) provides an alternative description of a glass as a solid in which molecules can oscillate about a set position but cannot move from that position, i.e., cannot form a periodic pattern. From a practical point of view, a glass could also be considered to be a microcrystalline solid in which the crystallites are less than five molecular dimensions. In the case of water this would be between 1 and 2 nm.

Rasmussen (1982a, b), Angell (1983), and Franks (1985), on the basis of theoretical calculations and experiments, originally cast some doubt on the possibility of being able to truly vitrify pure water. It was also suggested that impurities in the water, and inclusion of the cooling agent in the water to form a clathratelike structure, might prevent vitrification. This latter objection is no longer tenable, as Bachmann and Mayer (1987) have produced I_V by spraying droplets of purified water onto a highly polished metal plate maintained at temperatures below 113 K.

There is an increasing amount of evidence that strongly supports the claims that a substantial amount of the water in small droplets and ultrathin film can be vitrified [see Mayer (1985, 1988) and Dubochet *et al.* (1987) for a summary of the evidence]. It is even claimed that the surface of very small blocks of tissue may be vitrified (Hutchinson *et al.*, 1978; McDowall *et al.*, 1983). In all cases the claims are supported by x-ray and/or electron diffraction data which reveal the characteristic diffuse bands associated with I_V. When such materials are warmed above the glass transition temperature, the characteristic diffraction patterns of I_C begin to appear. Mayer (1985) considers that both I_C and I_H generally appear as contaminants in the vitreous ice, and it remains uncertain whether this is a consequence of incomplete vitrification or contamination of the sample after vitrification has occurred. The I_H contaminant can be readily identified; the I_C is much more difficult to identify, and Mayer is careful to point out that the only way to exclude I_C as a contaminant is to devise rigorous experimental procedures to ensure that it cannot form.

The important question as to whether vitreous ice or microcrystalline ice is formed in rapidly cooled liquid water appears to center on the lowest limits of resolution that can be used to obtain diffraction patterns. The actual size of the crystallites depends on the system being examined and the techniques being used. It is generally possible to obtain an electron diffraction pattern with electron doses about three orders of magnitude lower than is used to obtain an electron micrograph. Nevertheless, it is known that damage can occur even at these low irradiances. The usual procedure to check whether I_V is present is to examine the electron diffraction pattern as the vitrified material is slowly warmed. In the vitreous state, although the diffraction rings are diffuse and indistinct, it is possible to make a rough estimate of the size of any microcrystallites that may be present. It is generally accepted that the limit is about 2 nm, i.e., there may be microcrystallites that are smaller than this figure but they cannot be detected. As the temperature continues to rise, the I_V begins to transform to I_C and the diffraction rings appear much sharper and extend further from the center. This is an indication of increased resolution, and although diffraction patterns equivalent to a spatial resolution of between 0.3 and 0.4 nm have been found in some circumstances, the more general figure remains at about 2 nm. (Stewart and Vigers, 1986). Riehle (1968) considered the cutoff point to be 5 nm, and that ice with crystallites smaller than this dimension would be assumed to be in a vitrified state. In light of more recent work, the figure of 5 nm is too high and a figure of about 2 nm is now generally accepted.

In the context of our discussions here, we will assume that a sample is vitrified if it satisfies the following two criteria:

1. There is an absence of the diffraction rings associated with any of the crystalline forms of ice.
2. The characteristic diffraction rings of I_C appear as the temperature of the vitrified ice sample is raised above the devitrification temperature, T_d.

2.3. CONVERSION OF LIQUID WATER TO ICE

Most of what will be discussed in this section will center on I_H as this is the form of ice most frequently encountered in cryo-microscopy and indeed in most other processes that rely on converting liquid water to ice. The freezing of liquid water to hexagonal ice involves three distinct but closely related processes. In the techniques that are most commonly used in low-temperature microscopy and analysis, the three processes are applied very rapidly and in the following order: heat removal, ice nucleation, and crystal growth. We will discuss each of these processes in turn and conclude by comparing crystallization with vitrification.

2.3.1. Removal of Heat from the System

During cooling, the low temperature in the immediate environment of the specimen acts as a heat sink and removes thermal energy until the specimen is in equilibrium with its surroundings. The rate of heat removal is dependent on both the temperature differential between the surroundings and the sample, on the properties of the sample being cooled, and on the cryogen. Such properties would include the thermal conductivity, specific heat, latent heat of crystallization, and the shape and form of the sample. As we have already seen (Section 2.2.1), the values for thermal conductivity and specific heat show some variation with temperature. The latent heat of crystallization also shows variation with temperature. It has a value of $334\,J\,g^{-1}$ at 273 K and falls to 170 J g^{-1} at 233 K if the specimen can be undercooled sufficiently so that freezing occurs. Because these properties vary between different samples, they are usually combined together into a single function, the surface heat transfer coefficient (h), in the following manner:

$$h = 0.023(vp)^{0.8} \times u^{-0.5} \times D^{-0.2} \times Cp^{0.3} \times K^{0.7} \tag{2.1}$$

where v is the relative velocity of the specimen and fluid cryogen, p is the density of the fluid cryogen, u is the viscosity of the fluid cryogen, D is the characteristic dimension of specimen, Cp is the specific heat capacity at constant pressure, and K is the thermal conductivity.

The prime objective of any cooling technique used in low-temperature microscopy and analysis must be to ensure that the conduction of heat within the cooling medium and across the specimen/coolant interface is faster than heat conduction within the specimen. The relationship between the internal and external factors can be calculated and is expressed as the Biot number (B), which to a first approximation is a measure of the relative rates of convective and conductive heat transfer. The Biot number is dimensionless and is related to the surface heat transfer coefficient (h), the sample thickness (d), and the heat conduction of the sample (k) in the following manner:

$$B = \frac{hd}{k} \tag{2.2}$$

Other factors such as the sample shape and its surface characteristics are also going to affect the final value. At high Biot numbers, e.g., 30, the cooling rate of the sample is specimen limited; at low values, e.g., 0.1, the cooling rate is coolant limited. In the former case there is little that can be done to increase the rate of specimen cooling, but in the latter case improvements to the heat extraction process will improve the sample cooling rate.

Diller (1990) has analyzed further the conditions under which the theoretically fastest cooling rates can be achieved and provides evidence to suggest that the maximum cooling rate occurs *inside* the sample rather than at the surface. His calculations show that the maximum depth in the sample at which a threshold cooling rate of

10^4 K s^{-1} can be sustained, is about 45 μm for vitrification and 145 μm for crystallization. Diller proposes that for each position in the specimen there will be a unique time at which a unique maximum cooling rate occurs.

The heat dissipation from the sample will vary depending on whether the heat is removed by conduction or by convection. Conduction relies on direct contact between the object and its surroundings, and the energy flow (be it heating or cooling) involves minimal molecular displacement of either the object or its surroundings. This is the process that operates when small droplets of liquid or solid samples are rapidly brought into contact with a cold metal plate.

Convective cooling relies on the mass transfer of heat by the circulation of a cryogenic liquid past the surface of the specimen, and unlike conductive cooling involves mass displacement of the coolant molecules. This circulation can be natural convection, which occurs either when there is a variation in the density of the liquid due to changes in temperature or as a result of the action of gravity. Alternatively, the circulation can be forced convection, which occurs when there is an additional force applied to the liquid cryogen. This occurs when a small sample is rapidly plunged into a stationary liquid cryogen or when liquid cryogen is forced to flow rapidly past a stationary specimen. The extent to which heat transfer takes place with convective cooling is critically dependent on the flow characteristics of the cryogen, which may be laminar, turbulent, or a transitional. Although much is known about the hydrodynamics of natural convection, this is now rarely used as a cooling procedure for cryomicroscopy as the heat transfer processes are too slow. Turbulent and laminar flow convection systems are very complex, and it is only in the last few years that any serious attempt has been made to understand these flow characteristics in relation to convective cooling.

Bald (1984) has shown that in most experimental cooling techniques the Biot number is quite low and that convective heat transfer from the sample to the coolant is slow compared to conductive heat transfer. At high Biot numbers, the opposite would be true and the surface temperature of the sample is always the same as the coolant. Some work by Zasadzinski (1988) confirms that convective heat transfer from the coolant to the specimen is the limiting step in the rapid cooling of small samples. Much of what is known of these heat transfer processes and how this information can be applied to improving the cooling procedures used in conjunction with low-temperature microscopy and analysis may be found in the recent book by Bald (1987). Diller (1990) has made similar analytical predictions on the distribution of temperature and cooling rate during quench cooling of biological samples.

Many of the hydrated systems prepared for cryomicroscopy are frozen using liquid cryogens, which involves a combination of conductive and convective cooling. When material at ambient temperatures is placed in a liquid coolant such as propane cooled by liquid nitrogen, there is an immediate conductive transfer of heat from the high temperature of the sample to the low temperature of the coolant. This energy transfer raises the temperature of the coolant so that it moves away from the object being cooled and is replaced by further coolant molecules. There is a mass motion by the coolant molecules, which effectively removes thermal energy from the object being cooled. This is a very simplistic account of what happens and ignores how the cooling can be seriously affected by the viscosity, specific heat, and thermal conductivity of the cryogen, and by variations in the size and shape of the specimen. We will return to these matters in Chapter 3 when we consider the properties of cryogens and the different systems that have been devised to cool samples.

There are many consequences arising from the removal of heat from aqueous systems. The physical and chemical processes of all liquid systems, from simple solutions to the complex mixtures found in cell sap, are affected. In biological systems the reaction rates in metabolic pathways are slowed down, all by varying degrees. The molecular motions of the water molecule are slowed sufficiently to give rise to centers of nucleation and the water moves from a disordered state to a more ordered state.

2.3.2. Nucleation Phenomena

We have already seen that it is not possible to undercool large volumes of pure water much below 273 K without freezing, although small volumes of water can show substantial amounts of undercooling and still remain in the liquid state (Fig. 2.8). This undercooling cannot, however, proceed indefinitely, because as more heat is removed, the random thermal movement of the water molecules produces clusters of the right proportions and correct orientation, which favor the condensation of further water molecules.

Water molecules in the interior of the cluster have a negative volume free energy because they have a large number of near neighbors with which to react. In contrast, water molecules at the surface of the cluster at the interface with the bulk liquid, have a positive surface free energy because there are fewer near neighbors with which they may interact. If the water cluster is small, the positive surface free energy predominates and the cluster will break up. As the cluster size increases, the number of water molecules in the interior of the cluster increases faster than at the surface until the volume free energy equals the surface free energy. At this point the total

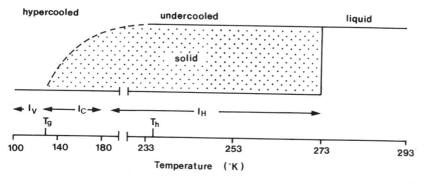

Figure 2.8. The relationship between the forms of stable and metastable water and ice which can exist at subzero temperatures. Modified and redrawn from Franks (1985).

free energy of the system is zero and the cluster is of a critical size to act as a nucleus of an ice crystal.

In brief, the following phenomena occur as the temperature of the water decreases:

1. The magnitude of the surface free energy *decreases*.
2. The magnitude of the internal free energy *increases*.
3. The critical cluster size for nucleation *decreases*.
4. The average cluster size *increases*.
5. The probability that a cluster of critical size will be formed *increases*.
6. The lifetime of clusters *increases* (via slow diffusion).

Nucleation is a statistical phenomenon and the lifetime of a given cluster is inversely proportional to temperature and directly proportional to its size (Fig. 2.9). At high temperatures much larger clusters are required for nucleation. Thus at 263 K, 16,000 molecules are needed to form an effective nucleus; at 233 K only 200 are required. At higher temperatures, the probability of such a small cluster existing is very low and its lifetime would be very short (Dufour and Defay, 1963; Franks, 1985; Fletcher, 1970). Although the rate of nucleation increases rapidly with decreasing temperature, there are only a few good estimates of how fast the event occurs. Hobbs (1974) considers that in pure water, with droplets measuring a few micrometers in diameter, the probability of nucleation becoming appreciable only arises when the liquid is undercooled to between 253 and 223 K. At 223 K, Mason (1957) estimates that the nucleation rate increases by a factor of 6 for each further degree fall in temperature; Franks (1985) considers the figure to be nearer 20 for each degree drop in temperature. Measurements have also been obtained from aqueous solutions (Michelmore Franks, 1982) and from cells undercooled in oil–water emulsions (Mathias *et al.*, 1984). In both cases, the nucleation temperature was lower than expected from pure water alone. The whole process whereby water, in the absence

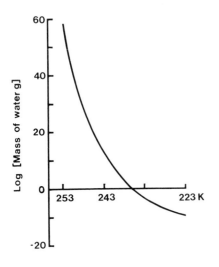

Figure 2.9. Mass of water (g) likely to contain at least one nucleus of the critical size for homogeneous nucleation as a function of temperature. Redrawn from Franks (1985).

of foreign bodies and solid surfaces, provides its own nucleation sites, is known as *homogeneous nucleation* and is the only nucleating mechanism that occurs in pure water and aqueous solutions. It must be emphasized that nucleation is a probability event and it is difficult to give a precise temperature at which it will occur. Thus at 273 K it is most unlikely that nucleation will occur; at 233 K, nucleation is a virtual certainty.

There is, however, a specified temperature range at which this probability event will occur—the homogeneous nucleation temperature T_{hom}—and as we have already discussed, it is possible to undercool water to 233.5 K, at which point the water clusters are of the correct form and dimension to initiate crystallization of the whole volume of liquid being cooled. It is generally assumed that the process of ice nucleation involves a stepwise growth of nuclei until a critical dimension is reached and spontaneous crystallization occurs. The instability of undercooled water and the divergence in its properties as the temperature is decreased has led Franks (1985) to suggest that a more realistic model for ice nucleation in undercooled water may need to involve cooperative growth of ice nuclei.

Water containing dissolved solutes will undercool to much lower temperatures, because T_{hom} is lowered according to the following equation:

$$T_{hom} = 38.1 - 1.8T_{dep} \qquad (2.3)$$

where T_{dep} is the melting point depression for the solution in degrees kelvin. This forms the basis of procedures used to artificially lower the T_{hom} of solutions by adding antifreeze agents, a matter we will return to in Chapter 3.

If we are concerned only with the low-temperature microscopy and analysis of aqueous systems, then the experimental protocols should be designed to initiate the freezing event as close as possible to the homogeneous nucleation temperature. At T_{hom}, the number of critical nuclei will be at a maximum. If at this point, heat is now removed from the system more rapidly than it is being produced by the latent heat of crystallization, there will be minimum ice crystal size and growth. Such crystals would not be visible in the electron microscope.

We have already noted that pure water is a rarity, and it is for this reason alone that water, particularly where it contains even small traces of foreign solid material, will undergo heterogeneous nucleation before homogeneous nucleation can occur. *Heterogeneous nucleation* occurs when a particulate substance within the aqueous matrix acts as a catalyst and increases the probability that a water cluster of critical dimensions will form. The probability for a heterogeneous nucleating event occurring increases with the increased volume of water and a decrease of temperature (Fig. 2.10). There is considerable debate as to what constitutes a good heterogeneous nucleating material. Franks (1985) and Taylor (1987) have suggested a number of factors that might determine the nucleating efficiency of a substrate:

1. It should have a small crystal lattice mismatch with I_H.
2. It should have a low surface charge.

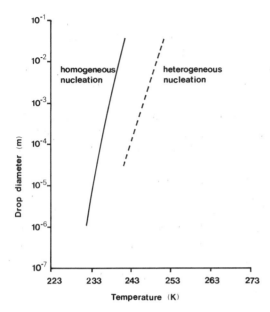

Figure 2.10. Homogeneous and heterogeneous nucleation temperatures for water as a function of sample volume. Data from Angell (1983).

3. It should have some degree of hydrophobility, i.e., water will not spread on the surface.
4. It should have a diameter slightly larger than 10 nm.

It has been claimed by Meryman and Williams (1981) and Rall *et al.* (1983) that biological material does not contain any structures that could act as catalysts for ice nucleation. However, the fact that no cells are capable of undercooling naturally to 233 K has led Franks (1985) to suggest that biological material must contain heterogeneous nucleating sites. The question now arises as to the exact chemical nature of these materials and where they may be located in the cell. We have already suggested that a large proportion of the cell water is perturbed, primarily by macromolecules. Although this perturbed water is likely to diminish the number of potential heterogeneous nucleation sites within living systems, it is generally believed that sufficient molecular species remain to act as nuclei for ice crystal formation. In this respect, it is important to remember that in biological systems, ice nucleation first takes place outside the cell and that even under conditions of fast cooling the cells can only tolerate a small amount of undercooling before nucleation events are initiated inside the cell. This is supported by the suggestion of Mathias *et al.* (1984) that the catalytic site for heterogeneous nucleation in red blood cells is located on the internal surface of the plasma membrane. Although the chemical nature of the catalyst remains unknown, the results suggest that proteins must be a prime candidate.

Organic molecules would appear to be the best candidates for heterogeneous nucleating agents, as Fletcher (1970) considers that molecularly dispersed solutes are too small to act as effective heterogeneous nucleating agents. Some additional information comes from studies on the mechanisms that enable plants, animals, and microorganisms to avoid freezing damage when exposed to seasonal low temperatures. Bacterial and higher plant polysaccharides, insect and higher plant glycerol, and fish glycopeptides and polypeptides all emerge as possible chemicals that will lower the nucleation temperature, T_{het}. Some of the best studies of these are the antifreeze polypeptides and glycoproteins from Arctic and Antarctic fish. Yang *et al.* (1988) have determined the structure of an alanine-rich polypeptide from the winter flounder and consider that it depresses the freezing point of water by binding to the surface of the ice nuclei and inhibiting the formation of large ice crystals. There appears to be no appreciable effect on either the equilibrium freezing point or melting temperature. The articles by Franks (1985), Clarke (1987), and Grout (1987) provide additional information on the avoidance of freezing damage in living organisms. However, in nearly all instances the natural antifreeze compounds are only produced in specific organisms and in advance of the environmental stress, and it is by no means certain that these or similar compounds are of common occurrence in cells and tissues that never experience such environmental stresses. We will return to the use of artificial antifreeze agents in the following chapter.

Heterogeneous nucleation is the principal nucleation event in nearly all dilute solutions and biological samples because at any given subzero temperature the number of nuclei formed per second per unit volume by heterogeneous nucleation is greater than the number formed by homogeneous nucleation. Figure 2.10 shows the profiles of the two events as a function of sample volume. A considerable amount of work has been carried out on developing the theory behind the nucleation of water and we have only touched on those aspects of the process that are of prime concern for the procedures by which water and solutions are frozen for low-temperature microscopy and analysis. A comprehensive treatment of the subject may be found in the books by Hobbs (1974), Fletcher (1970), Franks (1985), and Franks and Mathias (1983).

2.3.3. Ice-Crystal Growth

Following the nucleation event, the liquid to crystal phase transition is completed by the growth of the nuclei into hexagonal ice crystals of various shapes and sizes. The force driving the crystallization processes is the difference between the higher free energy of the molecules in the liquid aqueous environment, relative to their energy in the bulk crystal. The mechanisms that determine the rate and formation of ice-crystal growth are

1. The transport of water molecules to the point of crystal growth.
2. The accommodation of water molecules at the growing interface.
3. The transport of heat away from the interface.

The transport of water molecules to the point of growth is a diffusion process and is governed by the activation energy and viscosity of the liquid, both of which increase as the temperature is lowered. The net effect is that the transport of water to the crystal face is slowed down as the temperature drops. This is clearly one of the advantages of undercooling the water prior to crystallization.

The accommodation of water molecules at the growing crystal interface is complex, and the interested reader is referred to the appropriate chapters of Fletcher (1970) and Hobbs (1974) for a detailed explanation of the processes involved. Single water molecules are loosely associated at the crystal surface because of the small number of neighboring molecules. A more stable arrangement is found when two or more molecules are associated at the liquid/crystal interface and at the molecular level. The radius and smoothness of the surface are also important factors which determine whether a water molecule should remain at the crystal interface.

If water is cooled slowly, the ice crystals will first form around the water clusters initiated by the nucleation events. With further cooling, it is these initial ice crystals that tend to grow, rather than new crystals forming at other nucleation sites. This is due simply to the fact that it requires less energy for a water molecule to attach to an existing ice crystal than to become involved in a new nucleation process. Very small ice crystals have high surface energies and are thermodynamically unstable and tend to fuse with larger, more stable ice crystals.

The rate of ice-crystal growth and their form are critically influenced by the degree of undercooling of the liquid and the rate at which heat is removed from the system. The maximum rate of ice-crystal growth is at 260 K; at 143 K there is no growth. We can distinguish four main types of crystallization processes:

1. With very slow cooling, i.e., less than 1 K/min, and in conditions where there is little undercooling to favor numerous nucleation events, a small number of large ice crystals are formed. This is the process that is believed to occur as fresh water lakes freeze and give rise to ice crystals measuring approximately 30×80 mm.

2. With faster cooling, i.e., more than 1–100 K/sec, and in conditions where some degree of undercooling has occurred, there will be more nucleation events and more ice crystals. These crystals will be a mixture of small crystals and larger crystals formed as a consequence of recrystallization events involving the small metastable ice crystals.

3. With fast cooling, i.e., 10^3 K/s, more undercooling can occur and many small ice crystals will be formed, as there is less time for ice-crystal growth, and recrystallization to take place.

4. Under conditions of ultrarapid cooling, i.e., 10^4–10^5 K/s, where there is substantial undercooling, there will be a very large number of nucleation events but virtually no time for crystallization, and the result will be at best I_V, at worst microcrystalline I_H. Bald (1986) has proposed a theoretical model to derive a quantitative relationship between the critical cooling rate and the average crystal size (Fig. 2.11). While this model affirms one of the central tenets of the cryomicroscopist's faith, namely, that fast cooling results in small ice crystals, it should only be used as a guide to the actual crystal sizes one might expect to find in a given sample cooled at a given rate.

For the microscopist, any success in diminishing the effects of crystallization are, of course, highly dependent on the rapid removal of the heat from the water. As we have discussed, this is a complicated phenomenon, which could at first glance be exacerbated by the heat production due to the latent heat of crystallization. If the rate of heat release during the phase change *equals* the rate of heat removal by

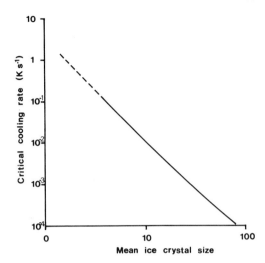

Figure 2.11. The relationship between the critical cooling rate and mean crystal size. The mean crystal size is given by dividing the mean crystal diameter by twice the radius of the critical nucleus. The solid part of the line is derived from experimental data, the dotted part by extrapolation. Data from Bald (1986).

conduction within the sample before the phase transformation is complete, the sample will warm sufficiently to allow ice crystals to grow. Figure 2.12 shows some theoretical cooling curves for water at different rates of heat removal, calculated by Stephenson (1956). These curves show that a cooling rate of 33.6 GJ m^{-3} s^{-1} (3×10^3 K s^{-1}) would be necessary to avoid any rewarming and recrystallization at the growing ice-crystal front. In most hydrated specimens that are prepared for low-temperature microscopy and analysis, this does not present a problem. The rate of crystallization compared to that of pure water is much lower than the rate at which heat can be removed from the sample.

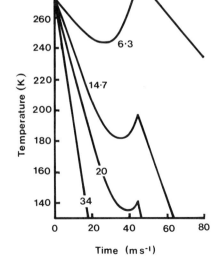

Figure 2.12. Theoretical cooling curves for water at different rates of heat removal measured as kW cm^{-3}. The lowest curve shows partial vitrification and no rewarming due to crystallization. The upper three curves show that the heat is not removed faster than it is being produced by crystallization and that some warming is evident. Redrawn from Stephenson (1956).

The size and surface area/volume ratio of the specimen are major factors influencing the removal of heat from the freezing front, and the freezing process produces a continuum of crystal sizes and forms. At one extreme there are the large dendritic crystals which are formed when water is cooled very slowly. At the other extreme, vitreous ice is formed under conditions of rapid heat removal. In spite of vigorous attempts to push the freezing process closer to the production of vitreous ice, most of the cooling procedures we will encounter in low-temperature microscopy and analysis produce ice crystallites of nanometer to micrometer dimensions which fall within the middle of these two extremes.

Because even the smallest volume of water does not undergo instant crystallization, the process of ice-crystal formation in aqueous solutions must inevitably lead to some degree of phase separation, i.e., as the cooling proceeds there will be a mixture of the solid and liquid phases of water before the entire volume of liquid is transformed to the crystalline state. This is of little consequence in the case of water alone, but as we will see, can cause major problems during the cooling of aqueous solutions and cell and tissue fluids. The process of vitrification does not involve a phase separation.

2.4. AQUEOUS SOLUTIONS AT SUBZERO TEMPERATURES

The conversion of water and aqueous solutions to a solidified state involves removing heat from the system. It is important to distinguish between the effects of cooling *per se* and the effects due to freezing. We will concern ourselves primarily with the latter effects, but it should not be forgotten that lowering the temperature of an aqueous solution in the absence of freezing (i.e., undercooling) will affect transport processes and the position of chemical equilibria. Because ice is formed exclusively from water molecules, the process of crystallization must at some stage lead to a separation of the solute and solvent components. This separation can have a number of secondary effects, some of which can seriously jeopardize the usefulness of cryotechniques for microscopy and analysis.

2.4.1. Phase Equilibria and Phase Diagrams

The process of ice formation in aqueous solutions is no different from the process that occurs in pure water, although the presence of dissolved materials will perturb the kinetics. Intracellular ice formation can be followed in real time by observing the process by light cryomicroscopy, and a direct correlation can be made between the visually detected ice and cell survival (McGrath *et al.*, 1975; Schiewe and Korber, 1982). The effect of removing water from the aqueous phase is to produce a more concentrated solution of the dissolved material that surrounds the growing ice crystals. This buildup of solute concentration can also be seen by light cryomicroscopy as a darkening within the liquid at the ice/solution interface. Korber *et al.* (1983) have developed a system capable of quantitative measurement of solute concentration in a well-defined freezing situation.

   The phenomenon of phase separation may be more easily understood by considering what happens as a dilute solution of table salt in water is cooled slowly enough to allow equilibrium conditions to prevail.* The salt solution is made up of two components, sodium chloride and water, which, depending on the temperature, can exist in a number of distinct phases. A phase is considered to be a homogeneous, but not necessarily continuous, region of a component. For example, a glass of iced water contains the solid, liquid, and just above the surface, vapor *phases* of the *component* water.

   As the salt solution is slowly cooled, the water will begin to form ice, a process that excludes sodium chloride from the growing crystal lattice. This gradual cooling concentrates the remaining unfrozen salt solution and depresses its freezing point until it reaches a critical concentration, at which point the salt crystallizes as well. This process is usually referred to as freeze concentration and is the principal cause of the freezing artifacts in quench-cooled water-rich biological material. Because there is virtually no solid–solid solubility between the two components, a microscopic examination of the now frozen, concentrated salt–water mixture would show that it is made up of small ice crystals surrounded by crystals of salt. The mean size of the crystallites is governed by the nucleation characteristics of the two components. There are still two components, although their phases have changed.

   This frozen mixture of salt and ice crystals is known as a eutectic, which is a term used to describe a mixture of two or more components in which the constituents are in such proportions that the solidification (or melting) point is at the lowest possible temperature for such a mixture. This critical concentration is referred to as the eutectic concentration and the temperature at which the solidification occurs is known as the eutectic temperature. During cooling, the eutectic mixture freezes last, which means that it will melt before the ice as the system is warmed up. This type of phase separation, which occurs under equilibrium cooling conditions, is the simplest case and serves only to illustrate the general features of this phenomenon.

   The cooling rates used in specimen preparation for low-temperature microscopy and analysis are substantially higher. Under these conditions, nonequilibrium cooling takes place, and although phase separation still occurs, the situation is complicated by substantial undercooling and solute supersaturation. In solutions of high viscosity and under very fast cooling conditions, eutectic phase separation may not occur at all.

   The behavior of the cooled salt–water solution discussed earlier may be quantitatively described by means of phase diagrams such as the one shown in Fig. 2.13. This is a simple, binary phase diagram and does not, for example, consider the effect of pressure on the system or the addition of another component, which would immediately add another dimension to the phase diagram. A good introduction to phase

*A distinction should be made between *equilibrium* and *nonequilibrium* cooling. Equilibrium cooling is only achieved with very slow cooling rates and is more commonly associated with cryopreservation procedures than with low-temperature microscopy. At, or just below, 273 K ice and water are in equilibrium, and as the temperature drops, ice crystallization occurs without any significant undercooling. There can be no vitrification. Nonequilibrium cooling occurs when the sample is cooled so rapidly that substantial undercooling occurs before nucleation. In extreme conditions, the solutions may be vitrified without phase separation.

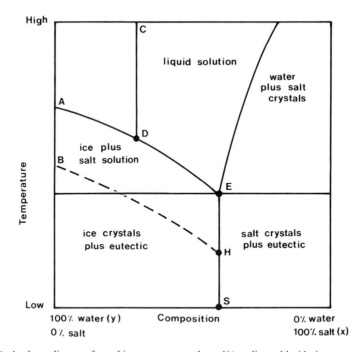

Figure 2.13. A phase diagram for a binary system such as 1% sodium chloride in water cooled under equilibrium conditions. The sodium chloride (x) and the water (y) form a homogeneous solution and there is negligible solid/solid solubility. As the salt solution is cooled slowly it follows the line C–D. At D, crystals of water (y) will be formed and as the temperature decreases, the cooling curve follows the line D–E and the solution becomes more and more concentrated with respect to the sodium chloride (x). Eventually the cooling curve reaches the eutectic point E, where the solution of composition S is in equilibrium with pure ice and pure salt. Further cooling is followed by simultaneous freezing of both the salt and the water. In biological systems, the eutectic separation of solid phases is not often observed because the multiphase cellular solutions contain many different components which influence the freezing process. For the cryomicroscopist, the important part of the phase diagram is the line D–E. If the water (y) is allowed to freeze slowly it will push a solution in front of it which becomes increasingly more concentrated with respect to the salt (x). This solution only solidifies at the eutectic point. At the fast, nonequilibrium cooling rates we seek to use in low-temperature microscopy and analysis, the saturation line A–D–E is passed without either the water or the salt solidifying. The solution will pass through a metastable, supersaturated state and depending on when the nucleation events intervene, the cooling curve follows the supersaturation line B–H. Redrawn from Robards and Sleytr (1985) and Menold *et al.* (1976).

diagrams in aqueous systems is given by Taylor (1987), and additional information on more complex systems is given by MacKenzie (1977) and Shepard *et al.* (1976).

The eutectic temperatures and eutectic concentrations of simple solutions are well documented, and some of these properties are given in Table 2.2. Many solutes do not form true eutectics, and one of the best examples is glycerol. As a dilute aqueous solution of this compound is cooled there is a progressive ice freezing until a glycerol concentration of about 65% is reached. At this point the water is in a (thermodynamically) metastable state (Franks, 1986), and its free energy is higher

Table 2.2. Eutectic Temperature and Concentrations of Simple Solutions of Some
Inorganic and Organic Compounds

Compound	Eutectic temperature (K)	Eutectic concentration (%)
Potassium nitrate	270.1	10.9
Sodium chloride	251.2	23.6
Calcium chloride	218.0	29.8
Glucose	268.0	32.0
Sucrose	259.0	62.5
Glycerol	226.5	67.0
Potassium citrate	257.4	12.8

than the value at equilibrium under the same conditions of pressure, temperature, and composition. Beyond this point, further cooling produces a homogeneous glass, i.e., it vitrifies rather than forming a mixture of two crystalline components.

An alternative halfway state between complete vitrification and complete crystallization can also exist. Depending on the cooling rate and the ease of crystallization of hydrated compounds, the cooled solution may attain a partially crystalline state in which an array of very small ice particles are embedded in a glasslike matrix. Macfarlane (1987) refers to this as a partially crystallized glass and most biological materials are probably transformed to this state after either slow or rapid cooling.

If we briefly return to the binary salt–water system we were discussing earlier, the addition of glycerol would make this a three-component system in which there is increased undercooling and supersaturation. For example, the addition of 40% glycerol to a dilute salt–water solution, lowers T_{hom} to 203 K and T_m to 257 K. The eutectic temperature for the salt–water mixture would be lowered from 252 K to 193 K. It is possible to exploit this phenomenon as one of the routes to vitrification.

In low-temperature microscopy and analysis, we are exclusively concerned with specimens that have been cooled rapidly. In these circumstances, the conventional phase diagram is less useful than the so-called supplemented phase diagram (McKenzie, 1977), which attempts to interrelate thermodynamic properties such as T_m with kinetic properties such as T_g, T_{hom}, T_{het}, T_d, and T_r. The supplemented phase diagrams combine information obtained by differential scanning calorimetry or differential thermal analysis with data from samples cooled under both equilibrium and nonequilibrium conditions. These diagrams can provide data on both vitreous and crystalline material, and an examination of Fig. 2.14 reveals the interrelationship of these properties. Below the glass transition temperature, T_g, the water molecules are capable of vibrational movement only. As the temperature rises, the molecules become sufficiently mobile to move from the random vitreous arrangement to the ordered structure of the crystal. This occurs at the devitrification temperature, T_d, which is an exothermic event that may be measured. As the temperature increases further, large crystals begin to grow at the expense of smaller, less stable crystals. The recrystallization temperature, T_r, marks the point at which this occurs. Finally, at a much higher temperature, the crystals will melt at the melting temperature, T_m.

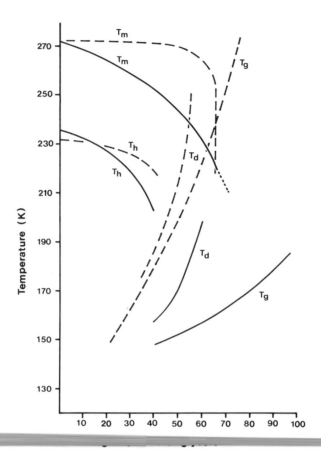

Figure 2.14. Supplemented phase diagram for water–glycerol (solid line ——) and water–polyvinylpyrrolidone (broken line – – – –) mixtures as a function of concentration and temperature. The diagram interrelates thermodynamic quantities such as melting (T_m) and kinetic quantities such as glass transition temperature (T_g), devitrification temperature (T_d) and homogeneous nucleation temperature (T_{hom}). Based on data from MacKenzie and Rasmussen (1972), Franks (1977), and Luyet and Rasmussen (1968).

Although these supplemented diagrams have only been worked out for simple binary solutions, they provide useful information on the changes in the dynamic properties of the solid phases of water as it is cooled and reheated. Thus, glass transitions and melting are seen in both rapidly cooled and slowly cooled samples; devitrification is only seen in rapidly cooled samples. The diagrams also give a measure of the relative amounts of ice and unfrozen water in solutions of varying concentrations cooled under different conditions. Such information is useful in designing cooling strategies for more complex solutions such as biological fluids.

Although these models are useful in understanding the principles of phase systems, they really do little more than provide a hint of what occurs when complex, multicomponent systems are frozen. Virtually nothing is known about the details of the phase equilibria of cell and tissue fluids, which do not behave as ideal solutions because they are composed of a mixture of dissociated, nondissociated, agglutinating, and colloidal substances. In addition, the presence of local concentrations of sugars, sugar alcohols, and proteins tends to prevent the crystallization of other solutes.

2.4.2. Consequences of Phase Transitions During Freezing

The appearance of a burst water pipe following a cold spell of weather is tangible evidence that water undergoes phase changes as it cools. The specific volume of water is at a minimum value just above the freezing point ($18.016 \times 10^3 \, mm^3 \, mol^{-1}$ at 277 K) and increases as it cools (or warms). This results in a 9–10% increase in the volume of water as it changes from the liquid to the solid state. The extent of the damage this expansion may cause will depend on the mechanical properties of the surrounding matrix. Thus elastomers are less affected than more rigid components, although it is frequently difficult to unambiguously assign specimen damage solely to the expansion of water. The effects of solute concentration are more serious. In extreme cases, one or more of the dissolved salts may be precipitated ("salting out"), which can lead to changes in pH and in turn to changes to the conformation, state of aggregation, and functionality of macromolecules. The increased solute concentration will lower the freezing point of the water, which in turn will influence the rate of ice crystal growth and the mobility of ions. As the ice crystals grow, they can draw water from other parts of the system and results in local regions of dehydration and in turn, further increased concentration of solutes.

2.5. CELLS AND TISSUES AT SUBZERO TEMPERATURES

All the effects described in the previous section on aqueous solutions are present in biological systems. Indeed, it can be argued that the situation is potentially even more serious because the sheer complexity of cells and tissues makes valid experiments difficult and the construction of predictive models virtually impossible.

Whenever ice is formed in biological systems it will increase the concentration of dissolved and dispersed material in the cells and tissues. Franks (1985) considers that the ratio of concentrations before and after freezing can be used to quantify the damage that may have occurred. However, the freeze concentration factor does not take into account the sizes of the different segregation compartments, but is only a measure of the total volume of ice in the system. In low-temperature microscopy and analysis, the size and distribution of the segregation compartments is the critical factor for good preservation. The freeze concentration effect is far more pronounced for solutions that are initially dilute than for more concentrated solutions. Figure

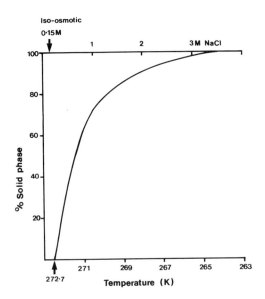

Figure 2.15. The effect of freeze concentration on a 0.154 *M* NaCl solution used as a suspending medium for red blood cells. As the temperature is lowered the solution freezes at 272.7 K and with further cooling ice separates out and the solution becomes more concentrated. At 270 K there is already a tenfold increase in the concentration and at 267 K, 90% of the liquid water has been converted to ice. Redrawn from Franks (1985).

2.15 shows what happens as red blood cells are cooled. At the equilibrium freezing point, the initial isotonic concentration is 0.15 *M*. At 270 K the solute concentration is 1 *M*; at 263 K, it is 3 *M*. At the eutectic temperature of 250 K, the solute concentration has risen to 4.5 *M*.

The two main approaches that have been made to try and understand what happens to cell structure at subzero temperatures are derived either from the more reductive analytical approach of biochemistry or from the long-established and more generalized discipline of cryopreservation (Franks, 1985). Both approaches, together with microscopic techniques, are beginning to shed some light on the changes that may occur in cell structure during cooling, although the results are frequently ambiguous.

It is a great pity biologists have come to cryomicroscopy relatively late in investigations on cell structure, for our judgement of what is a realistic representation of structure using low-temperature methods is strongly influenced by the dogma derived from images obtained under the most horrifying conditions of specimen preparation. Cell structure must be a manifestation of the slowly revolving phases of metabolism, and whereas biochemical methods and cryopreservation techniques are cognizant of this dynamism, such processes may only be imaged at low resolution. In this respect, we should remember the finding of Farrant *et al.* (1977), who demonstrated that the general methods that result in the preservation and subsequent recovery of viability, i.e., allow cells and tissues to be frozen and thawed and still remain alive, result in a poor preservation of ultrastructure as considered by the consensus criteria we use to judge these images. Conversely, many of the techniques we use in low-temperature microscopy and analysis kill the cells and tissues. In spite of an increasing amount of structural evidence from correlative optical and analytical methods, there is an uncomfortable feeling that our perception of cell structure might be quite wrong.

But to return to more soluble problems, low temperatures and freezing can influence both cell structure and metabolism. Phase separation in the complex mixture of solutions that exist in cells and tissues can trigger a host of interrelated reactions. It is possible to separate the deleterious effects of cellular ice formation from the solution effects that occur ahead of the freezing front as it moves through the cells and tissues. This is the most critical time, for all the tissue contents are in a liquid state and even a small change in the composition of this environment can have a profound effect on the functionality of the tissue. It is at this stage that much of the damage occurs, for apart from the mechanical effects of crystallization and the possibility of intracellular freezing being seeded by extracellular ice formation, once the solid state has been achieved the only further short-term damage that can occur is during melting and recrystallization.

As a result of extensive work on this subject, it is possible to separate the root causes of the deleterious effects into those that are a direct consequence of dehydration and those that have their origins in low temperatures *per se*.

2.5.1. Dehydration Effects

Ice crystals grow at the expense of the water in their immediate surroundings, and at low cooling rates an unfrozen cell surrounded by ice will shrink as the water is removed. This occurs because the vapor pressure of water in the cell is higher than ice at the same temperature and the cell water will diffuse through the plasma membrane to the extracellular ice. As the freezing front progresses through the sample, it is preceded by a concentration gradient. The consequent decrease in cell volume can be readily observed in the light microscope. The shrinkage will reduce the effective surface area of the plasma membrane and can alter the distribution of the membrane proteins (Skaer, 1987).

2.5.2. Biochemical Effects

In addition to these structural changes, the removal of water from the cells can result in a whole series of biochemical changes, which can have the following effects:
1. There will be an increased concentration of solutes which may in turn:

   a. Lower the freezing point of the solution.
   b. Depress the nucleation temperature.
   c. Change the viscosity and thermal diffusivity of the solution.
   d. Change the mobility of water molecules.
   e. Affect the dissociation constants and ion binding equilibria of acidic and basic groups.
   f. Affect the pH of the cell fluids.

2. There will be changes in the conformation of macromolecules which depend on water for their higher-order structure. This will have the following effects:

   a. Their structural integrity will be compromised.
   b. The efficacy of active sites on enzymes will be diminished.
   c. The permeability of membranes will change.

These effects do not work in isolation. For examples, the precipitation or "salting out" of a single electrolyte may affect the pH of the cell fluid and the dissociation constants of other electrolytes. This in turn may affect the binding capacity of small molecules and the configuration of macromolecules. The knock-on effect of these changes is to affect the rates of biochemical reactions and, in turn, the optimal functioning of the cells.

2.5.3. Low-Temperature Effects

The intrinsic effects of low temperature, as distinct from the effects of freezing, have already been briefly mentioned in the previous chapter. Lowered temperatures result in decreased rates of reaction and changes in pH. For example, the neutral point of water is pH 7.0 at 298 K, but at 273 K it is 7.5 (Robinson and Stokes, 1968), and in undercooled water at 238 K it is close to 8.5 (Taylor, 1987).

The extent of damage brought about in cells and tissues by the processes that occur ahead of the freezing front are directly influenced by the rate of cooling. Slow cooling generally results in greater structural damage than do higher cooling rates. The ultrahigh cooling rates that we have now come to associate with vitrification cause much less damage, presumably because all the cell components from ionized solutes to structural macromolecules remain more or less in their relative natural positions that exist in the living cell. The profound changes that can occur in biological tissues as their cell fluids are cooled prior to freezing should act as a cautionary note against using prolonged and extensive undercooling as a preparatory step to ice formation. It is a question of maintaining a balance between the desirable effects of undercooling, i.e., lowered nucleation temperature and smaller ice crystallites, and the undesirable effects, which center on a perturbed metabolism. We know that the rates of metabolic reactions are lowered; we have no idea how such changes may affect the ultrastructure of the constituent cells. Depending on the spatial resolution of the microscope or analyzer, these low-temperature induced changes will profoundly influence our ability to measure accurately local concentrations of chemicals as well as obtaining a correct image of the cell structure. In addition, we have little hope of understanding the complex ramifications of phase separation in biological systems. We should seek to avoid them entirely.

2.6. VITRIFICATION

Whether or not a liquid vitrifies depends on the nucleation frequency, ice-crystal growth rate, and the rate of heat removal from the system (Angell and Choi, 1986).

The process may be more easily understood in terms of the time it takes for the relevant events to occur. In water at 273 K, the molecules undergo between 10^{11} and 10^{12} rotational and translational movements per second. As the temperature is decreased, the movements slow down so that at 123 K it takes about 15 min for a molecule to complete a rotational or translation movement. Concurrent with this time scale for molecular movement, there is a second time scale which relates to the rate of crystallization. The actual rate is not known, but various estimates (Sargeant and Roy, 1968; Turnbull, 1969; Fletcher, 1971, and Uhlmann, 1972) give a time of 10^{-6}–10^{-10} s, to transform 50% of the water.

Starting with water at 273 K, the rate of crystallization at first increases to 260 K and then decreases, as the temperature is lowered, because the force needed to drive water molecules to the growing crystal interface increases. The minimum time for a volume fraction to crystallize is thus a compromise between the driving force for crystallization and the speed of molecular movements, both of which are slowed by decreasing temperature. In order to achieve vitrification it is necessary to adopt one or more of the following strategies:

1. Avoid or suppress heterogeneous nucleation (virtually impossible in biological systems).
2. Slow down the crystallization process by the addition of antifreeze agents (these compounds may have a deleterious effect on both the structure and biochemistry of the cell).
3. Use a cooling regimen faster than the crystallization time (very difficult to achieve).
4. Use very small samples.

Heterogeneous nucleation may only be avoided either by using very pure solutions or by subdividing the samples into small droplets, some of which will be free of nucleating agents. MacKenzie (1977) calculated that the probability of a nucleation event varies with the volume of the drop. Droplets of the required size may be prepared either in water–oil emulsions (Rasmussen and MacKenzie, 1973) or in the form of aerosols. These procedures are impracticable for large samples and most biological tissue but give satisfactory results with single cells.

Crystallization is a kinetic process derived from a complicated combination of nucleation frequency and crystal growth, and may be slowed down by undercooling the solutions and/or adding antifreeze agents such as glycerol to the solution. The actual cooling rates to achieve vitrification have yet to be measured, but they are assumed to be at least as high as the molecular relaxation rates. Plattner and Bachmann (1982) found crystallites in 200-μm droplets cooled at a rate of 10^4 K s^{-1}. Fletcher (1971) and Uhlmann (1972) estimate that 1-μm droplets or layers should vitrify when cooled at a rate of between 10^7 and 10^{10} s^{-1}; 10-μm layers would require cooling rates of between 10^{10} and 10^{13} K s^{-1}.

Whether true vitrification, as qualified earlier in Section 2.2.3, can every be achieved, or whether a partially crystallized glass (see Section 2.4.1) is the best we can every hope to achieve in practice still remains unanswered. Leaving aside the continuing theoretical discussions, the crucial question from the microscopist's point of view must be the quality of the micrographs. Ice-crystal-free specimens must be the ultimate goal of any low-temperature specimen preparation schedule, for under such conditions all the ionic and molecular species and all the structural components should, theoretically, remain unaltered from the liquid state.

We have already shown that most vitrified samples will undergo a crystallization phase change involving both nucleation and crystal growth when warmed. The same high solute concentration solutions which can be completely vitrified during cooling can be returned to the liquid state without crystallization, provided the heating rate is high enough. Under these conditions, the devitrification temperature, T_d, is so close to the melting temperature, T_m, that there is insufficient time for crystallization to occur before the melting temperature is reached. There are, alas, no practical consequences of this phenomenon for low-temperature microscopy and analysis, because the high concentration of solutes necessary to achieve this noncrystallizable state are invariably damaging to the other components in the mixed aqueous systems. While such high solute concentrations are a particularly acute problem with biological samples, which are normally composed of dilute aqueous solutions, it may provide a means of examining the structure and distribution of other materials that may exist in an aqueous matrix. In passing, it is of interest to note that in some Northern arboreal woody plants, the content of free sugars such as raffinose and stachyose may be as high as 27% dry weight during the winter months. Under these conditions, the water in the living cells slowly cools and vitrifies (e.g., glass transitions have been measured) and the plants are claimed to survive the low winter temperatures (Hirsh et al., 1985; Hirsh, 1987).

2.7. RECRYSTALLIZATION AND MELTING

Recrystallization is generally associated with solutions that have been cooled rapidly under nonequilibrium conditions, for in addition to presenting problems of heat removal from the sample, it also results in the formation of many small metastable ice crystals. The rate of recrystallization is temperature dependent and becomes progressively faster as one approaches the melting temperature of a given solution (Fig. 2.16). The process of recrystallization, which involves the growth of crystals, is quite distinct from devitrification, which describes the process of passing from the amorphous to the crystalline state. In contrast to devitrification, recrystallization is more difficult to measure by calorimetry because the changes in free energy are much smaller. The amount of recrystallization is more readily measured by comparing the size and numbers of ice crystals in systems cooled and warmed under controlled conditions; see, for example, Schweie and Korber (1982). Three types of recrystallization have been identified by Luyet (1965), and each is important in low-temperature microscopy and analysis.

2.7.1. Spontaneous Recrystallization

This type of recrystallization invariably occurs during the routine slow cooling procedures associated with cryopreservation and is directly related to the energy

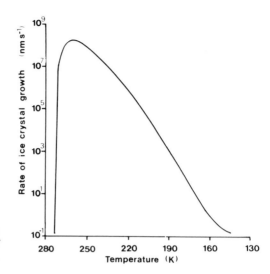

Figure 2.16. The rate of ice-crystal growth in relation to temperature. The maximum rate of growth is at 260 K. Data from Riehle (1968).

released as the latent heat of crystallization. If this heat is not removed fast enough from the sample, the highly localized rise in temperature causes transitory melting followed by immediate recrystallization. As already discussed in Section 2.3.1, this is not a serious problem provided the heat is removed rapidly from the system. This form of recrystallization can cause further redistribution of dissolved substances and can only further exacerbate the problems we have already encountered during the formation and growth of ice in aqueous solutions (Section 2.4).

2.7.2. Migratory Recrystallization

This occurs at a much lower rate and describes the process whereby water molecules migrate from the smaller less stable ice crystals to the larger more stable crystals, even at temperatures as low as 143 K (Dowell and Rinfret, 1960). The rate of recrystallization increases with rising temperatures, although it does not occur at any appreciable rate until the temperature is above 183 K. In general, the recrystallization temperature is higher in the presence of high molecular weight solutes, such as those found in biological tissues. In a study on yeast cells, Bank and Mazur (1973) found no ice crystal damage in samples stored at 223 K for 24 h, although some damage was seen after 30 min at 228 K and extensive damage was seen after 5 min at 253 K. It is, however, very difficult to calculate T_r in biological systems because of the complex nature of the solutions in cells and tissues. For this reason, caution dictates that frozen samples should be stored at liquid-nitrogen temperatures. Migratory recrystallization can present a problem in situations where frozen samples are sectioned or fractured at high temperatures. We will return to these matters in Chapters 4 and 5.

2.7.3. Irruptive Recrystallization

This is more commonly seen in light-optical microscopy and in association with the slow rewarming used in cryopreservation procedures. Ice crystals, which have been formed during cooling, resume crystal growth when the temperature reaches a certain narrow range during slow rewarming. The crystal growth can be followed in the light cryomicroscope and will cause a sudden decrease in the opacity of the sample.

Recrystallization must be avoided wherever possible. Spontaneous recrystallization will be negligible under conditions of fast cooling and migratory recrystallization can be prevented by keeping samples at the lowest possible temperature during all the preparative procedures following the initial act of freezing.

2.7.4. Melting

The process of melting occurs as crystalline material is transformed from the ordered solid to the disordered liquid. It is an endothermic reaction and the effect may be measured by differential scanning calorimetry (DSC) and differential thermal analysis (DTA) but not by electron diffraction methods, which only operate at high vacuum. The melting temperature of pure water is 273 K, but the presence of dissolved solutes can lower this temperature by 1.86 K/mol. One of the most obvious consequences of melting is that the phase separations seen in frozen aqueous solutions immediately disappear and the prefrozen state is restored. In simple solutions, such as the table salt dissolved in water, it would be impossible to detect any difference between the before and after freezing states. It would be quite wrong to think that the same thing happens in more complex solutions and in biological tissues. As Fig. 2.17 shows, melting is not the exact opposite of freezing and cannot be expected to undo or restore the damage that may have been caused as a result of phase separations during the initial freezing process. In this respect, cells and tissue are like water pipes: The damage is only apparent on thawing.

It may appear that we need not be concerned unduly with melting and recrystallization as all the preparative procedures designed for low-temperature microscopy and analysis are aimed at cooling samples as rapidly as possible and then manipulating, storing, examining, and analyzing them at the lowest possible temperatures. However, some of the subsidiary procedures, such as freeze substitution and freeze drying, involve the removal of water from specimens either by solvation or sublimation, and sectioning exposes samples to high mechanical stresses. These processes are more easily carried out at the higher temperatures where recrystallization and even melting can become a problem.

2.8. STRATEGIES FOR LOW-TEMPERATURE SPECIMEN PREPARATION

It is appropriate to conclude this chapter by considering the broad strategies we should adopt in order to obtain optimum specimen preparation using low-temperature techniques, before going on to discuss, in some detail, the tactics we should use

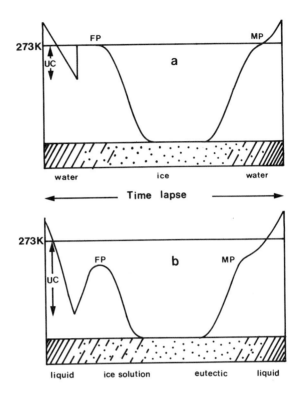

Figure 2.17. Temperature and phase changes during freezing and thawing of (a) pure water and (b) a dilute aqueous solution. UC, undercooling; FP, freezing point; MP, melting point. Redrawn from Sakai and Larcher (1987).

to achieve this end. It is assumed that our ultimate goal is complete vitrification of the specimen and that partially crystallized glasses and microcrystallized samples are a second best option. In simple physicochemical terms our procedures must aim at reaching as close as possible T_{hom} before nucleation and keeping below T_{hom} during solidification. Because this is only possible in very small samples, we should be more realistic and aim at reducing the critical temperature interval during which crystallization can occur in the sample. In pure water this temperature interval is from 273 to 130 K; in most hydrated samples and aqueous solutions the interval is from 273 to 173 K, and in most biological material the figure is from 270–265 to 183 K. There are several strategies we can adopt to reduce the size of this critical temperature interval:

1. Freezing point depression. Some aqueous solutions of simple molecular solutes tend to supersaturate and undercool before ice crystallization occurs. This tendency is more evident at high solute concentrations and in biological systems that have a greater molecular complexity. Most biological systems do not contain natural antifreeze agents and it is usually necessary to rely on artificially introduced foreign chemicals, which must not damage or even perturb the specimen.

2. Undercooling. The degree of undercooling that can be achieved is critically dependent on the sample. Pure water will, exceptionally, undercool to 233 K; in most unperturbed biological material 270–250 K is the limit.

3. Rapid cooling. If the heat can be removed by nonequilibrium cooling procedures faster than the ice crystals form, then the sample will vitrify. Equilibrium cooling should be avoided.

4. Small samples. Large samples lose heat more slowly and will approach equilibrium cooling rates that favor crystallization. A small sample size is not always practicable, and is only an option for thin films and microdroplets of liquid.

5. Low-temperature storage. Once the cooling has been achieved, every attempt must be made to prevent either recrystallization or melting during subsequent procedures of manipulation, examination, analysis, and storage.

Our ultimate goal of vitrification will only occur if the cooling is fast enough to extract the thermal energy from the sample without allowing the necessary time for the formation of water clusters of a critical size. With a few exceptions the cooling rates are too low and the specimens are too large and too dilute to allow this to occur. In the absence of artificial cryoprotection, we may have to be content with the second best. But for most low-temperature microscopy and analysis this may be good enough.

2.9. SUMMARY

This chapter is concerned, primarily, with an examination of the processes of sample cooling and the strategies we can adopt to successfully convert unstable hydrated specimens to a stable solidified state. When water and aqueous solutions are cooled to a sufficiently low temperature, depending on the environmental conditions, they will form one or another of the many forms of ice. In low-temperature microscopy and analysis, we are only concerned with I_H, I_C, and I_A (I_V) as the other ice polymorphs only exist under conditions of high pressure. Much is known about I_H, the principal form of ice found in most frozen specimens. An appreciation of the thermal, electrical, and mechanical properties of this material are important in understanding how the ice is formed initially and how the frozen material may be subsequently treated during specimen manipulation examination, and analysis.

The conversion of aqueous solutions to the solid state involves three distinct, but closely integrated processes. The initial removal of heat is largely specimen dependent. As cooling proceeds, homogeneous and heterogeneous nucleation events intervene to produce the water clusters which grow into ice crystals. The nucleation and the subsequent formation and growth of the ice crystals are kinetic phenomena which are strongly influenced by temperature. Under exceptional circumstances, the water may be converted to a glasslike state which preserves the random molecular arrangements characteristic of the liquid state.

The formation of ice and the consequent phase separations within the previously fluid state can severely affect the structural integrity and chemical activity of the systems being examined. This problem is particularly acute in biological systems. Although it is possible to quantify the phase changes in tertiary systems, virtually nothing is known about the complex phase separations that occur in cell and tissue fluids. Such phase separations are best avoided entirely by ensuring that the sample is converted to a vitrified state. In many specimens this is not possible and it will be necessary to accept that the microcrystalline or partial glass state is an acceptable alternative for most specimens being examined and analyzed at low temperatures.

3

Sample Cooling Procedures

3.1. INTRODUCTION

At the end of the previous chapter we discussed the different strategies we might adopt to successfully convert water and aqueous solutions to a solid state suitable for microscopy and analysis. There are two main advantages to be gained from this phase change: the provision of a solid matrix suitable for manipulation and microscopy, and an effective immobilization of dynamic processes. It should now be clear that if our purpose is to convert the water in a sample to a vitreous or even a microcrystalline state, solely by lowering the temperature, then we must limit ourselves to small samples, which should be cooled as rapidly as possible. The low thermal conductivity of water prevents high cooling rates in the center of a specimen much thicker than 25 μm, and adequate preservation will be achieved throughout larger specimens only if the properties of water are changed either by physical or chemical means.

There are five principal ways this rapid cooling may be achieved: spray cooling, jet cooling, plunge cooling, impact cooling, and high-pressure cooling. Each of these methods and their derivatives will be considered in turn, both in terms of efficiency and utility. In spite of these manifestly efficient cooling procedures, many of the samples we wish to study by low-temperature microscopy and analysis are much too large to be effectively frozen without the intervention of chemical additives to reduce the size of the ice crystals. For this reason, it seems appropriate to consider first what we can do to the sample before the cooling takes place with a view to optimizing the effectiveness of the cooling process.

3.2. SPECIMEN PRETREATMENT

We will consider here some procedures and interventions that may be applied to the samples before they are cooled rapidly. These processes, which may be either benign or invasive, may be used to advantage to improve the quality of the cryopreservation.

3.2.1. Chemical Fixation

On the assumption that chemical fixatives act by cross-linking and stabilizing macromolecules, some mild fixatives have been used in association with low-temperature preparation techniques for morphological studies. Such studies fall into three main categories: freeze etching, some aspects of cryosectioning, and low-temperature embedding procedures. Much of the earlier work on freeze etching made use of mild fixatives, such as some of the organic aldehydes, to stabilize and strengthen cell structures and cause sufficient changes in membrane permeability to permit easier access of cryoprotecting agents. These fixatives play no part in minimizing ice crystal formation and growth, but because they strengthen organic material, they may help to minimize ice crystal damage.

One of the main problems associated with the use of chemical fixatives, which are multifunctional agents designed to cross link some proportion of the molecules in the sample, is that they will also affect the hydration shells surrounding macromolecules and the sol-gel state of the cytoplasm. Figure 3.1, from Kellenberger (1987), provides a compelling argument against the use of chemical fixation in low-temperature microscopy and analysis. It is now generally accepted that these chemicals are responsible for creating their own artifacts (Sleytr and Robards, 1982; Plattner and Bachmann, 1982) which are seen mainly in the ultrastructure of biomembranes. This is rather alarming, because much of our knowledge about membrane ultrastructure has been derived from freeze-fractured samples which have been chemically treated prior to rapid cooling. These matters are discussed in some detail by Menco (1986) and Skaer (1987). More recent work on membrane ultrastructure prudently avoids the use of chemical fixatives.

Brief and mild chemical fixation is still employed in connection with the low-temperature embedding procedures that are discussed in Chapter 7 of this book. As

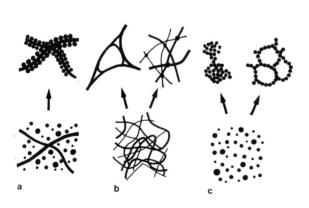

Figure 3.1. Diagrammatic representation of cytoplasm before and after chemical fixation. (a) Cytomatrix composed of globular and vesicular elements. (b) Cytomatrix composed of filamentous elements and fibrous macromolecules. (c) Cytomatrix composed of globular, vesicular, and fibrous macromolecules. In each case the chemical fixatives and organic solvents precipitate, coagulate, and cross-link the various components and move them from their original location in the cell. [Modified and redrawn from Kellenberger (1987).]

a   b   c

will be shown, these methods appear to preserve ultrastructural integrity with high fidelity, although there is a nagging uncertainty that the images that have been obtained may not truly represent what is present in the living tissue. For morphological studies, chemical fixatives are best avoided until such times as we know exactly what these materials are doing to macromolecules in an aqueous environment.

In some circumstances, such as when it is necessary to infiltrate specimens with low-molecular-weight antifreeze agents, it would be appropriate to consider using chemical fixation prior to sample cooling. Brief exposure for 1–2 h at 277 K to a 2.5% solution of one of the organic aldehydes in an isotonic buffer at an appropriate pH would appear to be the best answer. Oxidative and metallic fixatives should be avoided.

Chemical fixation is a relatively slow process in comparison to the molecular movements that occur in cells and tissues. Echlin (1991) has made some simple calculations on the effectiveness of rapid cooling procedures. If it is assumed that we need to preserve a piece of biological tissue to a depth of 10 μm, at a cooling rate of 40 kK s^{-1} solidification is achieved in 0.5 ms. By comparison, chemical fixation would take 5–10 s (Mersey and McCulley, 1978). This rate-limiting step is a function of both the chemical reactions of the fixative and the relatively slow diffusion of the chemicals through membranes and internal barriers. The situation is further complicated by the fact that the buffers in which the fixatives are maintained may enter the cells ahead of the fixative and cause protein extraction and intracellular translocations. The average time for physiological events is 1–5 ms and the movement of electrolytes in the aqueous phase of cells is ca. 2.0 μm s^{-1}; in cytoplasm it may be tenfold slower (Echlin, 1991). Jones (1984) has calculated that the resolution time for tissue rapid cooling studies of physiological changes is of the order of 250 μsec. Events faster than this will not be arrested by rapid cooling procedures. The processes of nucleation and ice crystal growth that occur during rapid cooling take place at an even faster rate.

The relative slowness of chemical fixation is best illustrated by following the effect of a fixative on living cells observed in the light microscope. It may take as long as half a minute before trans-cytoplasmic movements are arrested, which makes chemical fixation of questionable value in any studies on the kinetic processes in cells and tissues. The detrimental effects of fixatives are not limited to cells and tissues, as many of the same effects are seen in aqueous solutions of biopolymers. The perceived need for chemical fixation appears to be confined to morphological studies of biological samples and their derived polymers. Nonbiological samples are usually sufficiently robust not to require these processes.

In analytical studies where one is concerned with measuring and localizing diffusible substances, the situation is quite unambiguous. No chemical fixatives should be used. There is an overwhelming body of evidence that shows that even the mildest and briefest chemical fixation can cause dramatic changes in the permeability of cells. This would mean that many of the soluble constituents are either relocated in, or irretrievably lost from, the tissues. These changes are not universal for *all* elements in *all* tissues. Insoluble crystalline deposits and covalently bonded elements are much less mobile than ions and small molecules.

3.2.2. Artificial Nucleation Promoting Agents

One important factor in determining the size of ice crystals is the rate at which water molecules form clusters of the appropriate form and size to initiate ice nucleation. The higher the nucleation rate, the smaller the ice crystals. In the previous chapter, it was made clear that it is frequently not possible to vitrify untreated specimens of a size convenient and practical for most cryomicroscopy and that we should accept specimens in which the water has been converted to microcrystalline ice.

From the cryomicroscopists point of view there would appear to be two ways to approach the problems associated with nucleation. We could either try to delay nucleation by rapid cooling or attempt to promote multiple nucleation at temperatures higher than the usual T_{hom} or T_{het} associated with the system being studied. The former approach leads to a reduction in the size of nuclei that are stable and results in the formation of many small ice crystals. The latter approach may provide a means of quickly producing microcrystalline ice at somewhat higher temperatures. With this in mind, it has been suggested that artificial nucleators should be introduced into cells to promote nucleation in the same way that silver iodide crystals are used to seed rain clouds to produce rain. Unfortunately, the most likely chemicals for biological artificial nucleators are either toxic, e.g., chloroform, or difficult to use, e.g., liquid CO_2. In any case, it is doubtful whether artificial nucleators would be of much use in the cytoplasm of cells, because there is already a large number of macromolecules, inclusions, and surfaces of the appropriate size and dimension to act as nucleation sites.

3.2.3. Natural Nucleation Promoting Agents

It is known that a number of organisms are capable of producing compounds whose natural function is to supress undercooling and facilitate ice nucleation *in vivo*. These compounds are generally proteins or carbohydrates, and their effectiveness as nucleating agents is related to their dimension, wettability, and molecular symmetry. The best known of these natural ice nucleating agents is the proteinaceous material isolated from the common phylloplane microorganism *Pseudomonas syringae* (Schnell and Vail, 1972), which catalyzes the nucleation of ice on leaves. Pooley and Brown (1990) have isolated active cell free ice nuclei from the outer membrane of *Ps. syringea* and show that they are active in promoting nucleation at 268 K and above. Welch and Spiedel (1989) provide structural evidence for the location of ice nucleation sites on the surface of the bacterium. The Afro-alpine plant, *Lobelia teleki*, which grows in conditions where there is a daily fluctuation of 263–283 K, produces a carbohydratelike ice-nucleating material, which effectively suppresses undercooling (Krog *et al.*, 1979). The same material also works *in vitro* and has been shown to promote ice nucleation in saline microdroplets. Similar results have been reported by Zachariassen and Hammel (1988) using an ice-nucleating agent found in the haemolymph of the beetle *Eleodes blanchardii*.

Although the detailed chemical nature of these compounds is not known and the mechanisms by which they act are not clearly understood, the phenomenon of biological nucleation warrants further study. It would, for example, be useful to accelerate microcrystalline ice formation in the large aqueous vacuoles and intercellular spaces of mature plant tissue and the cell free space and collecting ducts that exist in many animal tissues.

At present, the use of artificial nucleators in cryomicroscopy is probably a lost cause. This is primarily due to a lack of knowledge about the materials, the wide variability of results, and the difficulty of introducing the agents into cells without altering their physiological function. In contrast, the problem of intracellular ice formation in large specimens has been approached from the opposite direction and ways have been devised to prevent such ice crystals from forming. Such methods will be discussed in the following section.

3.3. CRYOPROTECTANTS

Although the ultrarapid cooling of chemically untreated specimens is the method of choice for low-temperature microscopy, this approach fails to prevent the formation of large ice crystals inside pieces of tissue much larger than 50 μm in diameter and 25 μm thick, unless they are subjected to high-pressure cooling (see Section 3.6.6). If the internal fine structure of large specimens is to be studied by cryomicroscopical means, they generally must be infiltrated with an antifreeze agent to prevent the formation of ice. Cryoprotectants are chemical agents that were originally employed to protect cells from damage during the very slow equilibrium cooling (less than 100 K min^{-1}) procedures used for the cryopreservation of cells and tissues. These processes, although not within the primary context of this book, are now sufficiently well advanced to permit storage of viable mammalian tissues in a vitreous rather than a frozen state (Armitage and Rich, 1990). It is important to realize that although these same chemicals are effective in preventing ice crystal damage in ultrastructural studies when used in conjunction with high cooling rates, they become much less effective in preserving cell vitality. The converse is also true.

At high cooling rates, chemical cryoprotectants are thought to act in the following general manner:

1. Many are polyhydroxy compounds (Franks, 1983) whose most important property is their ability to interact with water, presumably by means of their hydrogen bonding potential.
2. They are all very soluble in water and their presence will perturb solutions to the extent that they can no longer be considered to be in an ideal state.
3. They suppress the anomalous structuring of water and promote undercooling before ice nucleation is initiated. This activity results either in the formation of many small ice crystals rather than a few large crystals, or complete vitrification.
4. They increase the viscosity of the medium, which decreases molecular mobility and slows down the rate of ice crystal growth.

Different cryoprotectants have different combinations of these properties; nearly all are alien to cells and tissues. For this reason alone, care must be taken in interpreting the ultrastructural data obtained using these materials. However, not all cryoprotectants are artificial. We have already discussed in Chapter 2 (Section 2.3.2) how a number of organisms produce natural cryoprotective agents, including glycerol. Cutler *et al.* (1989) infiltrated the internal air spaces of a number of species of leaves with dilute solutions of the white flounder antifreeze protein. This resulted in a significant depression in the spontaneous freezing temperature relative to water-infiltrated controls. Although this suggests that antifreeze proteins from one species may act to improve the cold hardiness of another, much further work is needed before we could hope to use these compounds as natural cryoprotectants in low-temperature microscopy and analysis.

Cryoprotectants can be divided into those that work from inside cells, i.e., penetrate the cell membrane, and those that have their effect without penetrating the cell. This is not to suggest that all the penetrating cryoprotectants enter cells freely. For example, glycerol freely enters some cells but in other cells will only cross the cell membrane after it has been altered by chemical fixation. The different modes of action of the two classes of cryoprotectants have been considered in some detail by McGann (1978).

3.3.1. Penetrating Cryoprotectants

The most commonly used penetrating cryoprotectant is glycerol, although dimethyl sulphoxide (DMSO), methanol, ethanol, ethylene glycol, and some sugars have all been used. DMSO, ethanol, and methanol all have toxic side effects and are known to produce artifacts (McIntyre, 1974) and are less widely used in ultrastructural studies. Glycerol is largely nontoxic, and although a long list of glycerol-induced artifacts is accumulating, it still remains the cryoprotectant of first choice for structural studies. The specific mode of action of the penetrating cryoprotectants is considered to be related to the following effects:

1. They lower the equilibrium freezing point.
2. They lower the homogeneous nucleation temperature, T_{hom}.
3. They enhance undercooling.
4. They raise the recrystallization temperature.
5. They act as a solvent, thereby keeping potentially harmful salts in solution during freezing.

In common with all cryoprotectants, glycerol must be used at fairly high concentrations, ca. 20%–25%, and care must be taken of any osmotic effects when processing samples in this material. Because of its variable permeability into cells, and in particular, its limited permeability into the intracellular compartments of plant cells, glycerol is frequently used in combination with mild chemical fixation. Small pieces of tissue are immersed or infiltrated with a suitably buffered 2%–3% glutaraldehyde solution for 1–2 h, followed by a slow stepwise infiltration with a 10%–30% buffered glycerol solution over a period of 1–5 h. The time of infiltration is critical and a balance must be made between the time necessary for complete infiltration and the damaging effects of the cryoprotectant. The infiltration procedures are best carried out at 277 K and with continuous gentle agitation. Plant material has been successfully infiltrated with glycerol by growing specimens in a 20% solution of the chemical in a well-aerated environment. Robards and Sleytr (1985) provide a series of recipes for using penetrating cryoprotectants in the preparation of biological material.

Glycerol is simple to use, and in spite of the large catalogue of artifacts, its use is associated with a wide range of verifiable ultrastructural information. However, it must be remembered that all penetrating cryoprotectants alter membrane permeability and are thus unsuitable for studies designed to measure the distribution of soluble ions and molecules using high-energy beam microanalytical techniques and cytochemical methods. It is also difficult to completely freeze dry or freeze etch (i.e., remove a surface layer of ice by sublimation) tissues that have been infiltrated with glycerol. Initially the water component of a glycerol–water mixture can be removed by sublimation, but a limit is set by the low vapor pressure of the glycerol, which is unetchable to any extent except at very high vacuum.

3.3.2. Nonpenetrating Cryoprotectants

This class of cryoprotectants consists of such chemicals as serum albumin, dextran, polyethylene glycol (PEG), sucrose, polyvinylpyrrolidine (PVP), and hydroxyethyl starch (HES). The use of PVP and HES in low-temperature microscopy and analysis has been investigated in Cambridge by Echlin, Franks, and Skaer (Franks and Skaer, 1976; Echlin et al., 1977; Franks et al., 1977; Skaer et al., 1977, 1978, 1982; Franks, 1980; Skaer 1982). These studies, which covered a wide range of plant and animal tissues, showed that buffered solutions of PVP and HES gave acceptable preservation as demonstrated in frozen sections, freeze-fracture images, and microanalytical studies. Because these polymers do not penetrate the tissues, they allow ice to be readily sublimed from the interior of cells. This is particularly useful for examining the fracture faces of bulk frozen material, for although it is possible to see the cell outlines in unetched specimens, removal of a surface layer of ice facilitates visualization of cellular detail.

Although these polymers have less effect than glycerol on cellular functions such as movement, locomotion, and excitability, they may not be suitable for all specimens as there are indications that they may adversely affect the physiological processes of some tissues. The polymers are only effective at high concentration, ca. 20%–25%, and they may exert a noncolligative osmotic pressure, resulting in shrinkage and a redistribution of water across the cell membrane (Allan and Weatherbee, 1978, 1979, 1980; McGann, 1978). There is a wide variation in the tendency of tissues to shrink in the presence of these two polymers. It is thus important to strike a balance between the minimum concentration of the polymer necessary for adequate cryoprotection and the time the tissues are exposed to it, and the need to ensure that the surface of the sample is thoroughly wetted with the material. In the case of plant cells, Wilson and Robards (1982) have calculated that PVP is likely only to coat the outermost cell layer of a given tissue as the polymer molecule (mol. wt. ca. 70,000) is unable to penetrate the cellulose cell walls. Skaer (1982) has shown that there is a substantial increase in ice crystal damage within intra- and intercellular compartments such as plant vacuoles and the luminal spaces in gut, vascular, and excretory tissue where the highly aqueous phases are inaccessible to the incubating polymer. Barnard et al. (1984) found that the rate of fluid secretion by *Calliphora* salivary glands is reduced by adding high-molecular-weight cryoprotectants such as PVP and dextran to the Ringers bathing solution. This reduced activity does not, however, appear to be related either to interference with the stimulating hormone or to the increased osmolarity of the medium as the ionic activity remains essentially unchanged. Because the salivary glands show recovery to normal secretion rates after washing in Ringers solution, it is suggested that the cryoprotectants are only having a physical effect on the system. Toxic effects have been reported in cells after prolonged exposure to these polymers. Skaer (1982) found PVP to be toxic to some cells, and Lucy (1970) found that PEG can cause cell fusion. Unlike the use of glycerol in association with morphological studies, there does not appear to be an ideal nonpenetrating cryoprotectant and it is necessary to establish the optimum conditions for each polymer with each tissue that is examined. In a typical experiment with plant tissue, a 20%–25% solution of PVP or HES is made up in a suitably buffered salt solution. The pieces of tissue are placed in this mixture for a few minutes and gently stirred to ensure that the tissue surface is thoroughly wetted with the polymer solution.

The method by which nonpenetrating cryoprotectants exert their action is not clearly understood, although they probably act in the same way as their penetrating counterparts. The nonpenetrating cryoprotectants have high solution viscosities, which effectively lowers the diffusion rate of water and retards crystallization, permitting the cell cytoplasm to undercool. Our studies in Cambridge have shown that the polymer solutions in which the tissues are embedded must be vitrified during the cooling process to ensure effective cryoprotection of the cells. This suggests that prevention of freezing *outside* the cells delays nucleation *inside*, so that ice crystal formation is only initiated at the nucleation temperature of the cell cytoplasm (Skaer *et al.*, 1977, 1978). This is in agreement with the much earlier suggestion of Mazur (1970) that the formation of extracellular ice can somehow seed the growth of intracellular ice. The fact that the polymeric cryoprotectants do not penetrate cell membranes and that it is not necessary to chemically prefix cells before cryoprotection makes these materials particularly useful in preparing tissues for the analysis of ions and small-molecular-weight species (Echlin *et al.*, 1980a, 1980b, 1982).

Provided the samples have been fixed prior to cryoprotection, sucrose is considered to be an effective cryoprotectant (Barnard, 1987). Prior fixation is necessary because osmotically active cells and tissues will be severely plasmolyzed by the high, ca. 2.0 *M*, concentrations of sucrose necessary to obtain adequate cryoprotection. It remains uncertain whether the sucrose enters the cells even though the permeability barriers have been disrupted by chemical fixation. It is accepted that.there is a net flow of water from the cell into the surrounding medium, which would in turn lower the nucleation temperature of the cytoplasm. The high viscosity of the sucrose in the external medium would affect the transport processes of water and in turn the initiation and growth of ice crystals. In this respect sucrose would act like a nonpenetrating cryoprotectant.

By using cryoprotectants it is possible to avoid entirely the formation of ice inside large aqueous specimens, even if the suitably marinated samples are cooled at low cooling rates. But a price has to be paid for achieving the vitrified state by these means. Penetrating cryoprotectants are usually only effective when accompanied by some form of chemical fixation, and the cryoprotectants themselves produce their own catalogue of artifacts. Nonpenetrating cryoprotectants can cause cell shrinkage and it is less easy to judge the time required for effective exposure to the chemical. At present there seems to be no way around these problems, and as long as there is a need to study large, highly aqueous samples by low-temperature microscopy and analysis we are either going to have to compromise the specimen by chemical intervention before cooling or accept some measure of ice crystal damage in the partly frozen samples.

3.3.3. Vitrification Solutions

A possible way forward might be to consider the use of cryoprotectant cocktails. This is the approach taken by cryobiologists whose principle concern is vitrification with the view of long-term preservation and revival. The rationale behind the approach is summarized in the recent papers by MacFarlane and Forsyth (1990) and Armitage and Rich (1990). Fahy et al. (1987) discuss some of the properties of vitrification solutions that do not freeze when cooled, at moderate rates and elevated pressures, to very low temperatures. Valdez et al. (1990) have used a vitrification solution composed of glycerol and propylene glycol in a continuous cryoscanning electron microscope (SEM) study of mouse blastocysts.

3.4. EMBEDDING AGENTS

It has long been known that it is useful to embed pieces of tissue in 10%–20% gelatin prior to freezing. The gelatin provides additional support to frozen samples and allows them to be more easily sectioned and fractured. Following these earlier studies, serum albumin, methyl cellulose, PEG, and dextran have also been used for the same purpose. Effective encapsulation has also been achieved using low-temperature agarose (Watanabe et al., 1988) and the polymeric cryoprotectants PVP and HES (Echlin et al., 1977; Pihakaski and Seveus, 1980). The use of PVP and HES have the added advantage that solutions of these materials readily vitrify when cooled. Such vitrified solids are more easily sectioned and fractured than crystalline materials. Experience has shown that plant tissues invariably require encapsulation in order that they may more easily be sectioned and fractured in the frozen state. The presence of large watery vacuoles and a cellulosic cell wall tends to make frozen plant cells very brittle and more difficult to cryosection and fracture smoothly.

The materials used as embedding agents also make excellent low-temperature glues. It is often necessary to attach a piece of frozen tissue to a specimen holder prior to transfer to another piece of preparative equipment. A small droplet of PVP or HES applied at the junction of a frozen specimen sitting on a precooled specimen holder cools rapidly and creates an effective bond. A number of low-melting-point organic materials have also been used as low-temperature glues and we will return to this matter later in this chapter.

3.5. NONCHEMICAL PRETREATMENT

The precooling preparative procedures outlined in the preceding section all suffer from the disadvantage that they involve the addition of foreign substances to the samples. It is, however, possible to carry out some precooling manipulations that are compatible with the life processes in biological systems and which are unlikely to affect the chemical integrity of nonbiological samples.

One of the biggest problems in cooling samples is, of course, extracting heat from specimens with low thermal conductivity. From this it follows that the size and shape of the specimen is an important parameter when it comes to cooling, and we should, wherever possible, seeks ways of maximizing the surface to volume ratio of the sample. It is no accident that microcrystalline ice and vitrification have only been obtained routinely in microdroplets, in thin films of material, or at the very surface of bulk specimens. In many instances, we have no opportunity of dictating the size and shape of the specimen, but we should seek to use samples as small as is compatible with the physiological and morphological aims of the experiment. The simple expedient of cooling the sample to a few degrees above the freezing point of water will lessen the temperature range through which the sample must pass to arrive below the recrystallization point of ice. Such precooling must not, of course, affect the normal activity of any biological specimens, and it is important to ensure that such precooling does not cause a phase change in materials such as the lipids in biological specimens.

Isolated cells, cell components, viruses, macromolecules, and aqueous particulate suspensions are usually most conveniently prepared either as a thin film suspension or as a pellet of concentrated material following centrifugation. Cryoprotectants may be mixed with the pellet if this is considered necessary. Some samples such as airborne particulate material, spores, and pollen grains are quite dry and it may be necessary to first wet the sample surface with a dilute detergent solution. Alternatively, such dry material may be suspended in a small amount of a nonaqueous liquid such as mineral oil.

Most biological material, and plant material in particular, contains large amounts of water, and it is important to ensure that sample dehydration is kept to an absolute minimum during specimen preparation. This is a particular problem in the preparation of thin (ca. 100 nm) film suspensions used in cryo-TEM, and elaborate procedures that are taken to prevent dehydration are discussed in Chapter 9 (Section 9.2.3). In a series of low-temperature SEM and microanalytical studies on tobacco leaves (Echlin and Taylor, 1986), it was found necessary to sample leaves in a moist chamber to ensure that the tissue remained in a fully hydrated state. The small strips of material were quench cooled within 10 s of being excised from the whole leaf. These problems are discussed in more detail in Chapters 10 and 11 (Sections 10.9 and 11.6). There are no quantitative data regarding the extent of accidental dehydration of biological material prior to quench cooling, but all the evidence suggests that excised organs and tissues lose water at an alarming rate.

In many multicellular and heterogeneous hydrated specimens there may be problems associated with exposing areas of interest and excising sufficiently small samples for optimal cooling. Surface contamination by cell and tissue fluids will increase the path length through which the cryogen has to absorb heat. Wherever possible such fluid layers should be mopped up or aspirated before quench cooling. There is a critical balance to be achieved between having a sample too wet or too dry. The general rule is to excise and/or expose the tissue as rapidly and as gently as possible and to proceed to cool it as quickly as possible. Robards and Sleytr (1985) provide a series of practical methods that may be used in sampling and specimen pretreatment.

Rebhun (1972) described a series of experiments in which different specimens were coated with a variety of organic insulating substances or finely powdered nucleating materials prior to quench cooling. He found that these surface modifications increased sample cooling rates by promoting nucleate boiling at the cryogen–sample-surface interface. In light of the increased interest in low-temperature microscopy it would seem that these materials and processes are worthy of extended reinvestigation.

The general rule that emerges from investigations on the pretreatment of biological specimens is to aim for the minimal intervention that would be consistent with an optimal experimental result. Compromises will have to be made with large samples, for there is little we can do to biological material without perturbing its metabolism and physiology. One thing is, however, certain: Ice crystal damage is reduced by rapidly cooling the sample, and we will now proceed to discuss this, the most critical step in low-temperature microscopy and analysis.

3.6. RAPID COOLING PROCEDURES

A number of different terms are used to describe this process, including terms borrowed from metallurgy (quench cooling), terms derived from conventional microscopy (cryofixation, cryoimmobilization), and simple hyperbole (ultrarapid cooling and hyperquenching). There is little to choose between the various terms, all of which imply the rapid removal of heat from the sample. In some instances the suffix "freezing" is used in place of cooling—e.g., "fast freezing." This should not be encouraged, as the term "freezing" is used to describe the process by which a liquid is converted to a crystalline solid, a process we are studiously trying to avoid.

In any discussions on rapid cooling, we need to consider three closely related matters:

1. The cryogens used to extract heat from the sample.

2. The mechanisms used to effect this heat removal.
3. A measure of the efficacy of the cooling processes.

A large number of ingeneous devices have been constructed to facilitate rapid cooling of small samples, i.e., samples no greater than 0.25 mm^3. It is not intended to provide a comprehensive review of all the methods that have been developed, but to concentrate instead on the tried and proven techniques that are currently being used and giving consistently good results. Those readers interested in a more general account and in the development of rapid cooling methods are referred to the publications of Rash (1983), Robards and Sleytr (1985), Gilkey and Staehelin (1986), Sitte *et al.* (1987), Mayer (1988), and Elder (1989). The review article by Menco (1986) is particularly useful, for in addition to providing an overview of the main methods of rapid cooling, it also contains a list of the manufacturers who make equipment for low-temperature microscopy and analysis, together with a comprehensive list of applications of the assorted techniques. The book by Bald (1987) is also useful, for it provides for the first time a detailed and quantitative account of what is thought to occur during the cooling process. It lists the physical properties of many of the cryogens and some representative biological materials, and provides formulas, equations, and even computer software for calculating the rate of cooling under a number of conditions.

3.6.1. Liquid and Solid Cryogens

A suitable cryogen for the purposes of low-temperature microscopy and analysis should have the following properties:

1. A low melting point and a high boiling point to minimize film boiling at the surface of the specimen.
2. High thermal conductivity and thermal capacity.
3. High density and low viscosity at its melting point.
4. It should be safe, inexpensive and readily available.

None of the materials we will consider have all these properties, but it should become apparent that some materials are better than others. Table 3.1 gives some of the thermophysical properties of a number of different cryogens. The average cooling rate achieved by some of the liquid cryogens is shown in Table 3.2.*

Liquid Helium (LHe$_2$)

The low melting point of this element (2.2 K), together with its high heat conduction and superfluidity, gives it the potential to be an excellent primary cryogen. Unfortunately, film boiling occurs too readily and immediately surrounds any

---

*The cooling rates described in this chapter are an average of the values that have been obtained for the cryogen concerned, measured between 273 and 173 K. These figures should be used for comparative purposes only, because, as will be shown later, they are subject to a great deal of variation.

Table 3.1. Thermal and Physical Properties of a Number of Liquid Cryogens[a]

Cryogen	M Pt	B Pt	Sp Ht	Th Co	R.C.E.	V at MP	Th In
Ethane	90	184	2.27	0.24	1.3	9×10^{-3}	0.060
Freon 12	115	243	0.85	0.13	0.5	1.4×10^{-3}	0.045
Freon 22	113	232	1.09	0.14	0.7	1.5×10^{-3}	0.053
Isopentane	113	301	2.29	0.18	0.8	—	—
Liquid helium	2.2	4.3	4.4	0.02	[b]0.01	0.2×10^{-3}	0.010
Liquid nitrogen	63	77	2.1	0.13	0.1	1.5×10^{-3}	0.052
Slush nitrogen	63	77	2.1	0.13	0.2	—	—
Propane	84	231	1.92	0.22	1.0	87×10^{-3}	0.54

[a]Data for the relative cooling efficiencies from Sitte (1987). The cooling efficiencies are relative to propane at 1.0. M Pt, melting point (K); B Pt, boiling point (K); Sp Ht, specific heat ($J g^{-1} K^{-1}$); Th Co, thermal conductivity ($J M^{-1} s^{-1} K^{-1}$); R.C.E., relative cooling efficiency; V at MP, viscosity at melting point (poise); Th In, thermal inertia ($J cm^{-2} K^{-1} s^{-0.5}$).
[b]Estimated by extrapolation.

Table 3.2. The Mean Cooling Rate and the Range of Measured Rates of Number of Liquid Cryogens

Cryogen	Temperature (K)	Mean cooling rate ($10^3 K s^{-1}$)	Range of cooling rates ($10^3 K s^{-1}$)
Ethane	90	13–15	0.8–500.0
Freons	90	6–8	0.2–98.0
Liquid nitrogen	77	0.5	0.03–16.0
Slush nitrogen	63	1–2	0.9–21.0
Propane	83	10–12	0.2–100.0

warmer object with an insulating layer of helium gas (Leidenfrost phenomenon), which effectively lowers the rate of heat removal from the sample to about $0.1 \times 10^3 K s^{-1}$. However, as we shall see, LHe₂ is a most effective coolant for the high-purity metal blocks used in impact cooling. With the notable exception of liquid nitrogen, the other liquid inorganic gases are not used for cooling because they are either too dangerous, e.g., LH₂ and LO₂, or even less effective and more expensive than LN₂, e.g., LNe₂, and LAr₂.

Liquid Nitrogen (LN₂)

In spite of its low cost, nonflammability, and ready availability, liquid nitrogen at its equilibrium boiling point (77 K) is not a good primary cryogen. Like LHe₂, the material suffers from the disadvantages of the Leidenfrost phenomenon, which reduces the cooling rate to about $0.5 \times 10^3 K s^{-1}$. LN₂ is more frequently used to cool secondary cryogens such as some of the organic gases.

Before dismissing LN₂ as a primary cryogen, there are some circumstances where it may be used. It can be used to cool chemically cryoprotected samples where fast cooling rates are not critical. It is also the primary cryogen in the high-pressure cooling process that we describe later in this chapter.

Undercooled Nitrogen or Nitrogen Slush (SN₂)

The effective cooling rate of LN_2 may be approximately doubled to 1.2×10^3 K s^{-1} if the liquid is converted to a mixture of the liquid and solid. This material can be prepared easily in the following manner: A polystyrene drinking cup containing three or four cocktail sticks is 2/3 filled with LN_2 and placed in a vacuum desiccator. The desiccator is evacuated via a wide bore tube to a pressure of 10–15 kPa (ca. 1×10^{-2} Torr), and as the pressure is reduced, the liquid begins to boil and lose heat. A layer of solid nitrogen will form on the surface of the LN_2 and this may be mixed into the underlying liquid to form a slush. The cocktail sticks prevent violent boiling, which can splash the cooling LN_2 into the desiccator. It may be necessary to break the vacuum and refill the plastic cup a couple of times, but within a few minutes there should be about half a cup of SN_2 at 63 K, the triple point of nitrogen. Air may then be leaked into the desiccator and the sample quickly plunged into the slush. The slush should be used as quickly as possible because its temperature begins to rise as soon as the vacuum is released. The temperature rise may be delayed by putting a piece of metal in the bottom of the cup (three predecimalization British pennies are ideal), which provides a suitable thermal mass. The slush is not stable and there are considerable variations in the cooling rates that have been recorded using this method. It is also only suitable for small specimens as its heat capacity is low. The main advantage of the technique is that the heat from the sample is first taken up by melting the solid nitrogen before the cryogen becomes liquid. An interesting modification of the SN_2 method is suggested by Sybers *et al.* (1983), who found that the addition of powdered graphite to the slush improved its cooling capacity.

A more sophisticated piece of equipment has been described by Umrath (1975), where the SN_2, which is formed in an outer chamber, is used to undercool LN_2 kept at ambient pressure in an inner chamber. The only advantage is that the undercooled LN_2 stays cooler for a longer period of time. Undercooled nitrogen is not the primary cryogen of first choice for isolated pieces of tissue. It may be more useful for cooling specimens in small holders, which, in turn, have to be closely fitted into another piece of equipment kept under liquid nitrogen. If one of the liquefied organic gases has been used as the primary cryogen, solid cryogen will form on the surface of the specimen holder when it is assembled under LN_2. The solid material will have to be chipped away under LN_2 to make sure the specimen holder will make good thermal contact with the other part(s) of the equipment.

Supercritical Liquid Nitrogen

On the basis of theoretical studies, Bald (1983) calculated that LN_2 maintained near its melting point of 63 K at a pressure of 34 bar should produce cooling rates equal to or better than those produced by the liquid organic gases. The high cooling capacity of supercritical LN_2 is related to its high liquid density, low viscosity, and the finding that the Leidenfrost phenomenon is diminished because boiling at the surface of the specimen is reduced. The sample is placed in a pressure vessel above a second pressure vessel containing LN_2, the whole assembly pressurized to just above the critical point (35 bar), and the specimen fired rapidly into the supercritical LN_2. Cooling rates of 30×10^3 K s^{-1} have been demonstrated on the equipment, which is commercially available from Oxford Instruments as the CS 5000 Cryospeed (Fig. 3.2). There is no information on the practical application of this equipment.

Liquefied organic gases are the most commonly used liquid cryogens for low-temperature microscopy and analysis, and include the halogenated alkanes (halocarbons) such as dichlorodifluoromethane (Freon) 22, and the alkanes, propane, propane–isopropane mixtures, and ethane.

Liquefied Halocarbons

At the time of writing, these compounds are still available, although their involvement in the depletion of the ozone layer in the upper atmosphere over the poles may restrict their future use. The halocarbons may be condensed by passing the gas through a tube, to the bottom of a metal container half immersed in LN_2. The condensed liquid may be poured into a metal container surrounded by LN_2, which will cause it to solidify. Just before the sample is to be cooled, a copper or aluminum rod is pushed into the solid cryogen to partially melt the solid and form a pool of liquid cryogen surrounded by the solid. As this pool begins to refreeze, the cryogen is ready to use. It is important that the pool of cryogen is large enough to accommodate the specimen and that it remains as close as possible to its freezing point. The practical problems associated with this method are that the sample may become trapped in the cryogen if it solidifies, and that samples will become coated with solid cryogen if they are stored in LN_2. If elemental analysis is to be subsequently performed on the cooled tissue, the halocarbons have been found to leave chlorine contamination (Echlin and Saubermann, 1977; Gupta, 1979). The advantages of the halocarbons as cryogens are their low viscosity close to their low freezing point and their nonflammability. The cooling rates are in the range $6–8 \times 10^3$ K s^{-1}, which is lower than may be obtained with other organic gases. The refrigerant halocarbons, which are chemically related to dry cleaning fluids and some anesthetics, are thought to be toxic.

Propane and Ethane

The liquefied form of these two compounds gives the highest cooling rates. They may be readily condensed from the gaseous state by the same procedures used for the halocarbons. Because of their lower melting points, it is helpful to precool the incoming gases before they are condensed in a high-conductivity metal container half immersed in LN_2. Melting pools of either propane or ethane may be used to cool samples, although both liquids become more viscous than the halocarbons as they approach their melting points. The addition of 30% isopentane to propane lowers the melting point to below 77 K (LN_2). If bottled gas (ca. 96% propane) is used

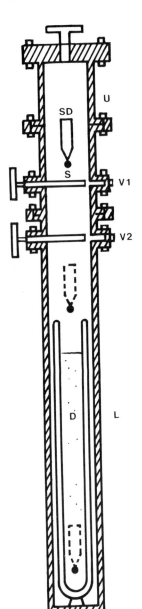

Figure 3.2. Simplified diagram of a supercritical liquid-nitrogen rapid cooling device. The sample (S) is placed on the specimen dart (SD) in the upper chamber (U) and liquid nitrogen in the Dewar (D) in the lower chamber (L). The upper and lower chambers are pressurized to above the critical pressure of 35 bar, valves V1 and V2 are opened, and the specimen dart is fired rapidly into the supercritical nitrogen. [Modified and redrawn from Bald (1987).]

instead of pure propane, it is only necessary to add up to 4% isopentane to depress the freezing point to below 77 K (Ward and Murray, 1987). Silvester *et al.* (1982) and more recently Ryan *et al.* (1990) have shown that ethane can give a higher cooling rate than propane, and as Sitte *et al.* (1987) have pointed out, ethane has a comparatively high vapor pressure at low temperatures, which makes it easy to eliminate residues of the material from vitrified specimens which must remain below 133 K. Propane and ethane have good cooling capacities and are inexpensive and easy to use. The cooling rate for ethane is 13×10^3 K s^{-1} and 10–12×10^3 K s^{-1} for propane. Both gases are very inflammable, and care must be taken when using them, even in the low-temperature liquid state. The flash point for propane is 169 K, and 143 K for ethane. These two compounds are best used in an atmosphere of nitrogen because oxygen from the air can condense onto the surface of the liquid hydrocarbon and form a potentially explosive mixture. Only small, ca. 20-ml, quantities of the liquid should be used at a time, and care must be taken to dispose of the material following use. Sitte *et al.* (1987) and Ryan *et al.* (1987) provide a series of safety measures that should be followed when using these, and indeed any, cryogens.

Solid Cryogens

Solid coolants are normally blocks of very pure metals such as silver or copper with a highly polished, contamination-free surface, which are cooled either by liquid or gaseous nitrogen at or near 77 K or helium at 4 K. These solid cryogens form the basis of the so-called impact cooling devices that we will discuss in Section 3.6.5. The very low temperature and high thermal capacity of these materials makes them excellent cryogens, although their effectiveness is only really apparent when they are incorporated into a more complex piece of equipment. Some of the properties of these materials are given in Table 3.3.

Table 3.3. Thermal and Physical Properties of Some High-Purity Materials that May Be Used as Solid Cryogens[a]

Material	Specific heat (J g^{-1} K^{-1})	Thermal conductivity (J m^{-1} s^{-1} K^{-1})	Thermal inertia (J m^2 K$^{-0.5}$)
Aluminum	3.7×10^{-1} (77 K)	410 (77 K)	1.9
	8.4×10^{-3} (20 K)	—	1.6
	2.6×10^{-4} (4 K)	15,000 (4 K)	0.3
Copper	2.1×10^{-1} (77 K)	570 (77 K)	3.2
	8.0×10^{-3} (20 K)	10,500 (20 K)	2.6
	9.1×10^{-5} (4 K)	11,300 (4 K)	0.3
Gold	9.6×10^{-2} (77 K)	252 (77 K)	2.6
	1.6×10^{-2} (20 K)	1,500 (20 K)	2.4
	1.6×10^{-4} (4 K)	1,710 (4 K)	2.3
Sapphire	6.3×10^{-2} (77 K3	960 (77 K)	1.6
	2.0×10^{-2} (20 K)	15,700 (20 K)	0.6
	8.0×10^{-6} (4 K)	410 (4 K)	0.1
Silver	1.5×10^{-1} (77 K)	471 (77 K)	2.9
	1.3×10^{-2} (20 K)	5,100 (20 K)	2.7
	1.3×10^{-4} (4 K)	14,700 (4 K)	0.4

[a]The thermal inertia term is a measure of the ability of material to absorb heat without an increase in temperature. Data from Bald (1987) and Robards and Sleytr (1985).

Bald (1987) has calculated that the high cooling rates that may be obtained, ca. 100×10^3 K s^{-1}, are achieved by choosing a block material with the largest value for thermal inertia.* Figure 3.3 shows that the thermal inertia of a copper block maintained at 25 K has one of the highest values for any material and should therefore provide the highest cooling rates for use with the impact cooling technique.† Bald (1983) has proposed that high cooling rates would be achieved by using a composite block composed of a copper–gold matrix coated with a thin layer of silver. However, the cooling rates within the sample might be slower than either copper or silver alone. On the basis of finite element calculations, Bald (1983, 1987) comes to the conclusion that a metal block precooled to liquid helium temperatures, ca. 4 K, will reduce, rather than enhance, the effective cooling capacity of these materials owing to changes in the thermophysical properties of the metals at very low temperatures. Bald (1987) has also calculated that the cooling efficiency of a silver block maintained at 4.2 K with LHe_2 is only marginally better than that of a copper block at 77 K using LN_2. These calculations are contrary to many experimental results, which show on average that the ice-crystal-free zone in a specimen prepared on a copper block cooled with LHe_2 is twice the thickness of the zone found in material prepared on a block cooled with LN_2. It is important that the results of these theoretical predictions of cooling rates are confirmed by experimental data from cooled samples.

Somlyo *et al.* (1985a, b) have taken advantage of the fact that Freon 22 ($T_m = 113$ K) is solid at LN_2 temperatures, and pressed solid blocks of the material against the specimens they wished to cool rapidly. Heat flow from the sample was absorbed,

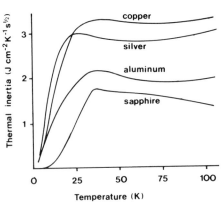

Figure 3.3. Variations in the thermal inertia of different high-conductivity materials used as solid cryogens [redrawn from Bald (1987)].

*The term "thermal inertia," which is a combined function of density, specific heat, and thermal conductivity, is a measure of the heat-absorbing capacity of a material. A high thermal inertia implies that more heat can be taken from a warm specimen before the temperature of the absorbing material begins to rise.

†Solid type IIa and IIb diamond has the highest known thermal conductivity of any material (approximately 4–6 times better than copper). It is now possible to artificially produce polycrystalline diamond in commercial quantities and it would be interesting to know how effective such material would be as a solid cryogen.

Table 3.4. Comparison of Mean Cooling Rates and the Range of Measurements Recorded in the Literature of the Five Main Methods of Sample Cooling

Method	Mean rate of cooling $(10^3\,\mathrm{s}^{-1}\,\mathrm{K}^{-1})$	Range of measurements $(10^3\,\mathrm{s}^{-1}\,\mathrm{K}^{-1})$	Depth of ice-free zone $(\mu\mathrm{m})$	Best cryogen (K)
Plunge	10–12	2–500	5–10	Ethane (93)
Spray	50	8–100	20–30	Propane (83)
Jet	30–35	7–90	10–15	Ethane (93)
Impact	25	6–55	15–25	Copper (20)
Pressure	0.5	—	500	Nitrogen (77)

initially, by the latent heat of the solid up to its melting point, followed by the latent heat of the melting cryogen.

There are several procedures by which the cryogens may be used to cool samples. They are all designed to transfer a sample at ambient temperature, ca. 293 K, to well below the recrystallization point of water, ca. 133 K, as rapidly as possible. In many cases the equipment is linked to devices that enable samples to be quench cooled while in known physiological or chemical states. The various types of equipment are designed to cool different types of specimen, and there is a wide range in the cooling rates that can be achieved. Table 3.4 gives an average value for the cooling rate achieved by each of the five main procedures, together with a measure of the depth of the ice-free zone that may reasonably be expected in a hydrated specimen. Each of the five procedures will be discussed in turn.

3.6.2. Immersion or Plunge Cooling

This is one of the most popular, effective, and easy to use methods of sample cooling and is best suited for the occasions where the entire volume of very small or thin specimens and droplet suspensions is to be adequately cooled. The technique is simply to plunge the sample into a suitable liquid cryogen such as propane or ethane, as rapidly as possible. Samples may be thin $(0.1–1.0\ \mu\mathrm{m})$ suspension films on copper electron microscope grids (Dubochet et al., 1982b); 1–10-μm-thick tissue sections or cell suspensions mounted between streamlined metal foils (Ryan and Purse, 1985); droplets suspended at the end of a wire or a pair of fine forceps; or pieces of tissue mounted at the leading edge of a low weight holder for cryomicroscopy (Fig. 3.4). It is important that the sample and its holder are as small as possible and that the specimen support is of low mass and high thermal diffusivity as Ryan and Purse (1985) have shown that cooling is relatively slow in specimens mounted on solid metal supports. The geometry of the specimen and its holder should be configured to ensure that the sample is the first object to enter the cryogen.

Although small is beautiful as far as rapid cooling is concerned, the very size of these minute specimens brings another set of problems. The high surface area to volume of small specimens means they can rapidly dry out. This is a particular problem with thin films and suspensions, and every attempt should be made only to

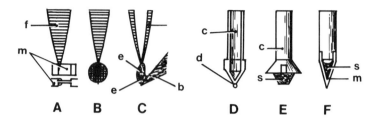

Figure 3.4. Different types of specimen holders for use with immersion or plunge cooling procedures. (A) Thin metal foil (m) sandwich containing flat specimens or liquid samples held by fine forceps (f). (B) Electron microscope grid held by fine forceps. (C) Droplet suspension at the end of a pair of fine forceps (f). Excess liquid (e) is blotted (b) away. (D) Low-weight carrier (c) for a small droplet (d), which may be protected by a thin metal foil. (E) Solid specimen (s) at the end of a low-weight carrier (c). (F) Sharp pointed metal holder of thin metal foil (m) covering a droplet suspension (s) for use with high-speed injection into the liquid cryogen. [Modified and redrawn from Sitte *et al.* (1987).]

carry out quench cooling in a chamber that provides an environment of independently controlled temperature and humidity. Details of the construction and use of these chambers may be found in Chapter 9 (Section 9.2.3).

Costello *et al.* (1984) and Ryan and Purse (1985) provide details of the optimal conditions for plunge cooling samples sandwiched between two metal foils. The cryogen, for example propane, should be in a metal container of high thermal conductivity sitting in a bath of LN_2. The liquid phase of the cryogen should be about 30 mm deep to provide the specimen with a reasonable path length before it comes to rest. Thermal gradients in the cryogen may be avoided by the use of a magnetic stirrer, which also helps maintain the cryogen at or near its melting point. The distance between the sample and the cryogen is of critical importance. If the sample is too close, it may slowly precool and form large ice crystals. If it is too far away, it takes longer for the sample to enter the coolant. Ryan and Purse (1985) showed that when a bare thermocouple was plunged through a 110-mm region above LN_2, a temperature drop to 173 K was recorded entirely within the cold gas layer. This problem may be alleviated by maintaining the liquid level of the cryogen as close as possible to the top of the container and by keeping the container covered with an insulating lid, which is only removed momentarily before rapid cooling is initiated. Murray *et al.* (1989) describe a countercurrent cooling device in which samples are plunged through a smooth upwelling organic cryogen stream which facilitates rapid heat transfer from the sample. The device is marketed by Oxford Instruments as the CQ 6000 Cryoquench.

Optimal cooling rates are determined not only by the sample size and cryogen thermal characteristics. The best results also depend to a large extent on the speed at which the specimen enters into and moves within the cryogen. It is not sufficient just to immerse the sample rapidly below the surface; it must also continue at the same velocity well into the depths of the cryogen in order to maximize forced convective cooling, which is the most efficient process of heat transfer. The path length within the cryogen has to be calculated to ensure that the specimen is at the same temperature as the cryogen by the time it comes to rest. This will ensure that the smallest ice crystals are formed within the specimen and that recrystallization will not occur. The forced convection device recently described by Ryan *et al.* (1990) ensures that the sample is plunged into a cryogen pool nearly 2 feet deep. Plunging by hand is not fast enough and it is necessary to use some sort of spring-loaded mechanical device to give injection velocities of ca. 2–4 m s^{-1}. Injection velocities higher than 5–6 m s^{-1} produce only a marginal improvement in the cooling rate of the sample (Costello *et al.*, 1984; Ryan *et al.*, 1990) but increases the chance of specimen damage on impact with the cryogen. Damping devices should be fitted to prevent the injector bouncing back at the end of its traverse into the cryogen. The average cooling rate for plunge cooling is 10–12 × 10^3 K s^{-1}.

A large number of immersion cooling devices have been described and their general features may be seen in Fig. 3.5. A number of these devices are commercially available (Reichert–Jung KF 80, and Oxford Instruments, CQ 6000). Most immersion devices can be easily constructed in a laboratory workshop, and for more specific details on their construction, mode of operation, and relative efficiency, reference should be made to the following papers: Costello and Corless (1978), Costello *et al.* (1984), Elder *et al.* (1982), Robards and Crosby (1983), Robards and Sleytr (1985), Ryan and Purse (1985), Ryan *et al.* (1987), Sitte *et al.* (1987), and Cole *et al.* (1990). The device described by Akahori *et al.* (1987) and the Leica KF 80 may be used both for immersion and impact cooling. A careful reading of these and other articles will show that although the different devices all give reasonably good tissue preservation, they quote a very wide range of cooling rates. This is a matter we will return to later.

3.6.3. Jet Cooling

In the immersion cooling method, the specimen is moved through the cryogen. In the jet cooling method the opposite occurs and high-velocity (10 m s^{-1}) jets of cryogen are moved past the surface of the sample. The method that was originally suggested by Burstein and Maurice (1979) has been modified in different ways by Muller *et al.* (1980), Van Venetie *et al.* (1981), Espevik and Elgsaeter (1982), Plattner and Knoll (1984), and Haggis (1986). To use the technique, propane is first cooled down by LN$_2$ to 93 K in a pressurized container. The cold liquid propane reservoir is then pressurized with cold nitrogen gas, which causes the cryogen to emerge from one (or two opposing) jets at high velocity onto the specimen, which, because of the high pressures involved, must be sandwiched between thin metal foils. A simple diagram of the apparatus is shown in Fig. 3.6. The sample and the specimen holder have a very low thermal mass, and it is most important that they are dropped into

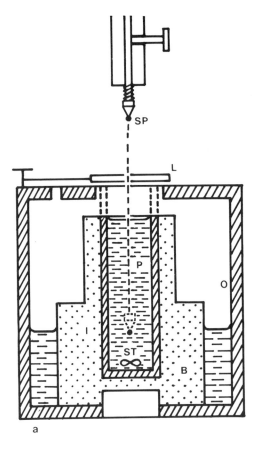

a

Figure 3.5. (a) Simple plunge cooling device. The inner chamber (I) is ca. 100 mm deep and made of thin-walled copper and sits snugly in a large aluminum block (B) cooled by liquid nitrogen in the outer insulated chamber (O). The aluminum block contains heater elements, thermocouples, and a thermostat to control and measure the temperature. The secondary cryogen is liquid propane (P), which should fill the inner chamber to within 2–3 mm of the top. If the inner chamber is no more than 20 mm in diameter the secondary cryogen will not need stirring. If needed, a magnetic stirrer (ST) can be fitted at the base of the inner container. When the propane has reached the desired temperature, the inner chamber is raised to a point level with the top of the outer insulated chamber, the lid (L) swung aside, and the specimen (SP) mechanically injected into the cryogen. (b) Comprehensive cooling device. After Sitte (1987). The secondary cryogen (S) is in a container (C) made of a high-conductivity metal. This container is deep and narrow so that the temperature of the cryogen remains constant without the need for stirring. A metal sleeve (SL) prevents continuous contact between the liquid nitrogen (LN$_2$) in the insulated outer container (I) and the secondary cryogen. The secondary cryogen chamber is surrounded by a metal block of aluminum (A) in which a heating element (H) and a thermocouple (T) are embedded. Just prior to cooling, the top of the secondary cryogen chamber is raised from position 1 to position 2, which is 5 mm from the top of the insulated outer chamber at position 3. The dry gaseous nitrogen (G) prevents contamination of the secondary cryogen. When immersion cooling is to take place, the shutter (SH) is moved to one side and the specimen (SP), placed at the end of a spring-loaded injector rod (R) guided by block (B), is plunged into the cryogen. [Modified and redrawn from Sitte (1987).]

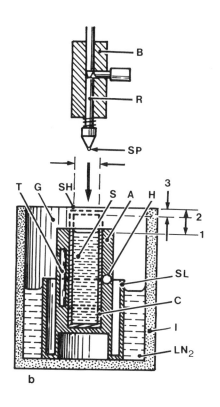

Figure 3.5. Continued.

liquid nitrogen as soon as the jet cooling is completed. The technique, which requires precise sample alignment and synchronization of the cooling jets, gives cooling rates in the range $30\text{--}35 \times 10^3$ K s^{-1}, and is limited to small tissue and cell suspensions and thin liquid films. Propane jet equipment is available from Balzer Union, as the Jet Freezing Device JD 030. Zierold and Schafer (1987) have modified the propane jet method to a propane shower device, which sprays a stream of pressurized liquid propane directly onto samples being examined by light microscopy. Zierold (1991) has used this device to quench cool single amoeba in a defined state of locomotion with a time resolution of ca. 0.5 s. In a series of studies carried out by Echlin and Chapman (unpublished results), cooling rates of 20×10^3 K s^{-1} were measured when 50-μm thermocouples were rapidly thrust into a jet of LN$_2$ pressurized to 1 bar. While we were unable to successfully cool small pieces of tissue (they flew all over the laboratory!), this simple technique might prove useful on samples suspended between thin metal foils.

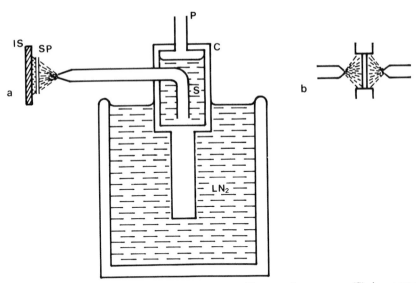

Figure 3.6. Single (a) and double (b) jet-cooling apparatus. The secondary cryogen (S), i.e., propane or ethane is cooled in a closed chamber (C) cooled by liquid nitrogen (LN_2) in an outer insulated container (I). The liquid secondary cryogen is pressurized (P) and a jet of liquid cryogen is applied to the specimen (SP). In the simple jet method (a) the sample is covered with a thin metal foil and rests on an insulated support (IS). In the double jet method (b) the thin-suspension sample is sandwiched between two metal foils and the cryogen jets applied simultaneously from both sides. [Modified and redrawn from Sitte (1987).]

3.6.4. Spray and Droplet Cooling

In the spray-cooling technique small, ca. 20-μm, droplets of low-viscosity aqueous cell or particulate suspensions are sprayed, either on the surface of propane cooled to its melting point, or directly onto the surface of a cooled metal block. The droplets are easily produced using a nebulizer or an artist's air brush. The small size of the droplets means that they cool very rapidly assuming that the heat flux across all surfaces is sufficiently well balanced. It is usually convenient to place any liquid cryogens in a depression in a metal block cooled by LN_2, and following cooling, the cryogen may be removed by vacuum evaporation leaving the frozen droplets in the cold metal block for further processing (Fig. 3.7). The method, which is limited to single cells, tissue suspensions, and suspended particulate matter, gives a very high cooling rate of about 50×10^3 K s^{-1}, and is usually used in conjunction with freeze-drying and/or resin-embedding techniques. The disadvantages of spray freezing are that there may be changes in the surrounding medium due to evaporation; there may be deleterious effects of shear forces on labile macromolecules and cells, and premature sample cooling on the cold gas layer above the surface of the cryogen.

The technique, which was originally introduced by Williams (1954), has been extensively modified by Bachmann and Schmitt-Fumian (1973), and Plattner and

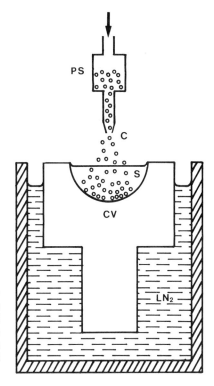

Figure 3.7. Spray-cooling apparatus. A particle suspension (PS) is pressurized and passed through the opening of a small capillary (C). Microdroplets are formed and are rapidly cooled in the pool of secondary cryogen (S) in the copper block (CU). The copper block is cooled by liquid nitrogen (LN₂) held in an insulated container (I).

Bachmann (1982, 1984). Samples of water have been vitrified by spraying microdroplets under reduced pressure onto a polished metal plate maintained at between 15 and 77 K (Mayer, 1985a, b). A further modification of the spray-cooling method has been devised by Mayer and Brugeller (1982) and Mayer (1985b), who projected a 10-μm jet of water at supersonic speed and high pressure into a liquid cryogen and produced vitrified water. It is difficult to give a cooling rate for this modification of the spray-cooling method, but an extrapolated rate would be in the region of 100×10^3 K s^{-1}. A spray-cooling device is available from Balzers Union.

A derivative of the spray-cooling method is to be seen in the emulsion droplet cooling procedure originally suggested by Buchhein and Welsch (1978). Small, ca. 10-μm, aqueous droplets containing the sample are formed as an emulsion in liquid paraffin. Small droplets of the emulsion are cooled rapidly and it is thought that differential thermal contraction of the oil and water phases pressurizes the contents of the aqueous droplets, lowering the ice nucleation temperature. In this respect it is a form of high-pressure cooling. Brugeller and Mayer (1980) have used the same approach to rapidly cool water dispersed as 5-μm droplets in n-heptane. The microdroplets of water vitrified when a fine jet of the emulsion, at a pressure of 100 bar, was directed into liquid ethane cooled to 90 K.

3.6.5. Impact or Slam Cooling

This method, which is sometimes referred to as metal mirror cooling or metal block cooling, is proving to be one of the most useful ways of cooling the surface regions of quite large samples. The material to be cooled is pressed rapidly against the highly polished surface of a metal block maintained either at 15–25 K by means of LHe$_2$, or at 77 K using LN$_2$. Figure 3.8 shows the general features of an impact-cooling device. The main advantage of this method is that although metals such as copper and silver have a thermal capacity about the same as organic liquids, they

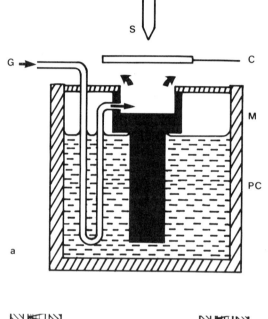

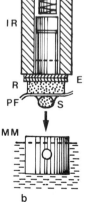

Figure 3.8. Impact-cooling apparatus. (a) The metal block (B) is cooled in a Dewar by the primary cryogen (PC), which is either liquid nitrogen or liquid helium. The highly polished surface of the metal block is continually flushed with precooled dry gas (G), (N$_2$ or He$_2$) to prevent surface frosting. The cover (C) is removed immediately prior to sample cooling and the sample (S) pressed rapidly against the cold metal surface. Below the main diagram are details of the air-cushioned injection system developed by Sitte and colleagues before (b) and after (c) injection has taken place. The sample (S) is mounted on a thin plastic film (PF), which sits on a soft rubber support (R) lightly stuck to the leading edge of the injector rod (E). The sample is moved rapidly toward the metal mirror surface (MM) with the spring-loaded injector rod (IR). After contact (c) the rubber foam deforms around the specimen and then hardens with cooling. The sample remains pressed against the metal mirror and is not deformed because in the initial phases of contact and rapid cooling, pressure is built up slowly as the damping air bag (DB) slowly empties. [Modified and redrawn from Sitte (1987).]

have a much higher thermal conductivity, and consequently the same amount of heat is transferred 10,000 times faster through copper than through propane. This would produce cooling rates at the surface of the sample of between 80 and 100×10^3 K s^{-1}, although the cooling rate a few micrometers below the surface would be closer to 25×10^3 K s^{-1}. The method, which was originally devised by Van Harreveld and Crowell (1964), has been modified and improved primarily by Heuser et al. (1979), Escaig (1984), Heath (1984), and Phillips and Boyne (1984).

The metal surface must be scrupulously clean as contaminants such as condensed water vapor will substantially lower the surface thermal conductivity. This cleanliness is achieved by continually flushing the metal surface with dry helium or nitrogen gas. The purity of the metal is also important, as even the smallest trace (ca. 200 ppm) of some materials will reduce the conductivity of 99.999% copper by 10%–15%. The effects impurities have on the cooling rates that may be obtained with high-purity metals are discussed by Heath (1984). Any impact cooling device should ensure that there is immediate and continued contact between the specimen and the cold metal surface as any specimen bounce will reduce significantly the effective cooling rate. Heuser (1977) and Heuser et al. (1979) suppressed bouncing by the use of strong retaining magnets, while Boyne (1979) and Phillips and Boyne (1984) used a hydraulic damping system on the specimen injector. Coulter (1986) claims to have solved the problem of bounce suppression and tissue damage by decelerating the rate of approach of the tissue to zero velocity at the surface of a block of sapphire maintained at 17 K with LHe$_2$. It is, however, unclear why Coulter chose to use sapphire as the block material, for although the thermal diffusivity (heat flow) of the material is marginally higher than copper at 17 K, the thermal inertia is much lower.

The results that have been obtained with the impact-cooling technique are impressive, and it has been possible to demonstrate ice artifact free zones as deep as 25 μm in chemically substituted tissues. Although the debate will doubtless continue, the potential advantages of cooling the metal block with LHe$_2$ rather than LN$_2$ do not appear to have been confirmed either by the few comparative studies that have been made (Boyne, 1979; Philips and Boyne, 1984) or supported by theoretical and thermodynamic calculations (Kopstad and Elgsaeter, 1982; Bald, 1987). Sitte et al. (1987) calculated that the temperature gradient in the first 1.0 μm of a sample pressed against a metal mirror cooled to 15 K would reach a maximum at 288 K μm^{-1}. A comparable figure for a mirror cooled to 77 K would be 226 K μm^{-1}. At a distance 10 μm from the surface, the figures are 29 K μm^{-1} for a LHe$_2$-cooled mirror and 23 K μm^{-1} for a LN$_2$-cooled mirror. These calculations assume perfect and instantaneous contact between the sample and the metal mirror.

The effectiveness of the cooling depends, to some extent, on the speed at which the sample makes contact with the mirror surface and the thrust developed by the system following the initial contact. Best results have been obtained at initial speeds of 5 m s^{-1}. At faster speeds there may be deformation or total destruction of the sample. The sample is cushioned against some of the mechanical forces by being placed on a soft foam rubber support or a thin layer of gelatin, with a thin intermediate plastic layer for easy separation following cooling. The sample is mechanically pushed against the cold metal mirror, and after the initial contact, the soft support deforms and surrounds the specimen. This provides gentle but continuous contact between the specimen and the mirror resulting in optimal cooling with minimum compression artifacts in the specimen. The initial elasticity of the soft support will also help compensate for any differences in parallelism between the specimen surface and the metal block.

Impact-cooling equipment is relatively complicated, and although a number of laboratory models exist, machines are available from LifeCell Corp, CF-100; Reichert–Jung KF-80; Med-Vac Inc, Cryopress; and Ted Pella, "Gentleman Jim." A very much simpler model can be made as follows: A high-purity, copper block is cooled in a Dewar of LN_2, and immediately prior to use, the polished metal surface is raised just above the surface of the LN_2. This is to ensure that the surface is not wetted with LN_2; the nitrogen gas from the boiling LN_2 will keep the metal surface free of frost. The specimen is then quickly pressed onto the metal surface and the metal block and specimen immediately returned below the surface of the LN_2. This procedure gives variable but acceptable results provided the block surface and the specimen surface make and retain near perfect planar and thermal contact. Simple impact-cooling devices are described by Heath (1984) for LHe_2, and by Bearer and Orci (1986), Allison et al. (1987), and Padron et al. (1988) for LN_2. A derivative of this procedure may be seen in the method by which the sample is cooled between the opposing jaws of precooled pliers (Eranko, 1954; Wollenberger et al., 1960; Hagler and Buja, 1984). Squeeze freeze devices are available from Ted Pella Inc. (Cryoplier) and Gatan Inc. (Cryosnapper).

3.6.6. High-Pressure Cooling

An examination of the phase diagram of water will show that vitreous water should form when liquid water is rapidly cooled under high pressure. As water freezes it increases in volume, and high pressures will hinder such expansion and, in turn, hinder crystallization. This effect is manifest in a decrease in the freezing point and a reduction in the nucleation rates and the rate of ice crystal growth. A decrease in crystallization means that less heat has to be extracted per unit time of cooling. Because the procedure takes advantage of a reduced critical cooling rate, high-pressure cooling should not be considered a rapid cooling procedure per se. Moor (1987) estimates that the cooling rate at the center of a 500-μm specimen would be only 0.5×10^3 K s^{-1}. These relatively low cooling rates may not be fast enough to stabilize dynamic events deep in the sample (Echlin, 1991) and may still allow phase changes to occur in membrane lipids (Knoll et al., 1987). The freezing point of water at approximately 2.1 kbar is reduced to 251 K, although undercooling will continue on down to 183 K before nucleation events intervene. At these elevated pressures, heat may be extracted from the specimen while it is in a highly undercooled state. Diminishing the temperature range over which crystallization can occur increases the chance of vitrification or at the very worst, the production of microcrystalline ice. Dahl and Staehlin (1989) have estimated that in practical terms, the application of 2.1 kbar pressure has the cryoprotective effect equivalent to about 20% glycerol.

For the better part of 20 years, Moor and his colleagues in Zurich have been seeking ways to exploit this phenomenon and apply it to the rapid cooling of biological specimens. The work has resulted in a high-pressure cooling device that is available from Balzers Union as the HPM 010. The simple diagram in Fig. 3.9 shows the principal features of this equipment. The samples are subjected to pressures of about 2 kbar by the application of a high-pressure jet of warm liquid propanol a few milliseconds before the sample is cooled by a jet of LN_2. A description of the apparatus and its use is given by Moor (1987) and Dahl and Staehlin (1989). The results that are now beginning to appear from the application of this technique are impressive. Pure water can be vitrified to a depth of about 10 μm, and animal tissue can be prepared without apparent ice crystal damage down to a depth of 600 μm if cooled from both sides. Some quite spectacular results have been obtained from plant tissues, which contain a higher amount of water. Root tips show excellent preservation to a depth of ca. 100 μm (Craig and Staehlin, 1988; Kiss et al., 1990),

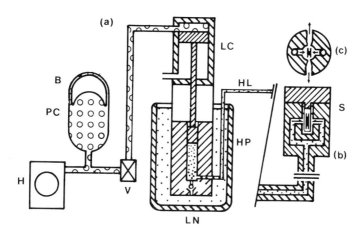

A

Figure 3.9. High-pressure cooling apparatus and an example of the results. A: Apparatus. (a) A hydraulic pump (H) forces oil into a pressure container (PC) containing a rubber balloon (B). The pressurized oil is released by opening the main valve (V), which forces the piston down in the low-pressure cylinder (LP). The lower part of this piston is in the high-pressure cylinder (HP) immersed in a liquid-nitrogen Dewar (LN). This movement compresses liquid nitrogen, which is driven through the insulated high-pressure line (HL) into the sample chamber shown in more detail in (b). The metal foil sandwiched sample (S) is momentarily bathed in pressurized iso-propyl alcohol before the two jets of pressurized liquid nitrogen are directed onto the opposite sides of the specimen shown in more detail in (c). This delays cooling by about 15 ms and allows the sample to be exposed simultaneously to high pressure and low temperature. [Redrawn from Moor (1987).] B: Results. (a) Transmission electron micrograph of an apple leaf section after high-pressure cooling, freeze substitution, and embedding. Cell wall (CW), mitochondria (M), thylakoids in a chloroplast (T), a nucleus with nuclear pores (NP), and golgi complex (G) all show strong contrast. There is good congruence of the dimensions of the organelles compared to the frozen hydrated section shown in (b), although the contrast is less good. The vacuole (V) shows considerable internal structure. [Picture courtesy of Michel et al. (1991).]

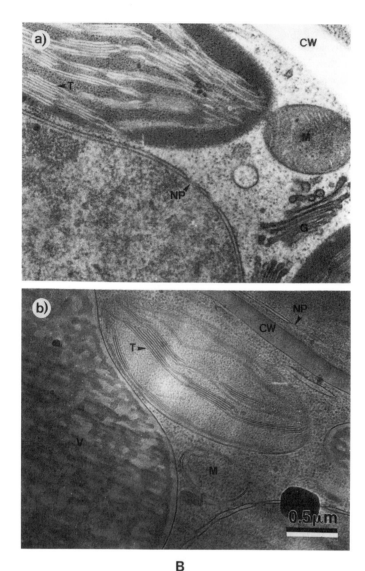

B

Figure 3.9. Continued.

and fungal-infected leaf segments vacuum infiltrated with tap water prior to freezing showed good preservation to a depth of 600 μm (Knauf and Mendgen, 1988). Walther *et al.* (1988) showed that the yield of well-preserved specimens increased to 80% if 8% methanol was added to the infiltration solution. Studer *et al.* (1989) and Michel *et al.* (1991) have obtained high yields by replacing extracellular fluids with 1-hexadecene. Hexadecene is not a cryoprotectant, but provides an incompressible liquid in air spaces that would otherwise collapse. The osmotic effects are negligible because the hexadecene is strongly hydrophobic (Studer *et al.*, 1989). Additional applications of high-pressure cooling may be found in the paper by Dahl and Staehelin (1988). The results open up a new perspective for the low-temperature microscopy and analysis of large samples.

Apart from the technical difficulties of applying high pressures to a sample moments before it is blasted with a jet of LN_2, the method still has a number of problems. These center primarily on the effects of rapidly subjecting biological and hydrated organic material to high pressures. In addition to the possible destruction of the specimen, it is accepted that phase transitions may occur in lipids and membranes, although it appears that proteins are unaffected. There are scattered reports in the literature that there are subtle changes in the appearance of biological tissues prepared by high-pressure cooling, i.e., membrane tears, folds, and blebbing, and reverse depolymerization of cytoskeletal components. There is insufficient evidence to say whether these changes are caused by the high pressure or by the additional processing, which must necessarily follow the initial cooling event. The pressure also raises the solubility of gases in water and decompression effects may occur during postcooling procedures such as freeze substitution. Martin Müller (ETH, Zurich, personal communication) has found it is more difficult to obtain large replicas from high-pressure cooled samples

Dahl and Staehelin (1989) find that high-pressure cooled samples are very brittle and contain random cracks. The source of these cracks is unknown, but they may arise from two processes.

1. From the high shear forces acting on the sample during pressure cooling.
2. From the reduction in volume of ice_{II} and ice_{III} which are formed when water is cooled under high pressure.

As samples are freeze-substituted for long periods at 213 K and atmospheric pressure, the ice_{II} and ice_{III} crystals slowly convert to the more voluminous ice_{I} (I_c). Some of the more destructive effects of high pressure have been overcome by protecting the specimen in a sandwich of thin-walled, high conductivity metal shields. A small space between the shields allows pressure to be applied to the sample. Bald (1987) has recently suggested that it should be technically feasible to combine high-pressure cooling and impact cooling. It is postulated that such a device could rapidly cool thin, flat specimens, and produce ice crystals no bigger than 10 nm down to a depth of 200 μm. Such a device would not alleviate any of the problems associated with either of the two constituent techniques.

3.6.7. Cryoballistic Cooling

This technique was one of the earliest approaches to rapid cooling of biological tissue and involved firing a cold projectile into the sample (Monroe, 1968). A simpler device has been proposed by Von Zglinicki *et al.* (1986), in which a hollow needle, chilled in liquid propane, is plunged into the living tissue to provide both simultaneous excision and rapid cooling. The technique combines the advantages of impact cooling and liquid cryogen cooling as the cold metal needle is coated with a thin layer of cold liquid propane.

3.6.8. Directional Solidification

Although not a fast cooling method, directional solidification is included here because it is a precise method of providing constant uniform cooling rates from 263 to 233 K for samples that are subsequently examined by light microscopy and SEM for the effects of ice crystal damage. Samples 1–3 mm thick are placed between two constant temperature sinks, one above and one below the phase transition temperature being studied. The sample is moved at a constant rate across the gap between the two heat sinks and enables a linear temperature gradient to be imposed upon the sample (Bischof *et al.*, 1990).

3.7. COMPARISON OF SAMPLE COOLING TECHNIQUES

It is very difficult to compare the different methods and come up with a single technique that would be best suited for all specimens. A somewhat pessimistic view is held by Bald (1986), who concludes that for most samples cooled rapidly at ambient pressure, none of the critical cooling rates are high enough to produce the microcrystalline ice that would be necessary if the specimen is to be examined by electron microscopy. At 2.5 kbar, ice crystals, <10 nm, may be produced at cooling rates in the range $3–20 \times 10^3 \, \text{K s}^{-1}$. This somewhat draconian view needs modification as good preservation can be routinely achieved in very small droplets, ultrathin films, and in the surface regions of larger samples. Nearly 20 years ago Moor (1971) calculated that cooling rates of $10 \times 10^3 \, \text{K s}^{-1}$, would be necessary to preserve small hydrated specimens in which the ice would be in a microcrystalline form (10 nm?). Most of the cooling procedures we have been discussing give these sorts of cooling rates, and ice crystals of these dimensions would be acceptable for most electron microscopy and analysis of diffusible elements.

Because of the fundamental limitation of poor heat transfer through hydrated specimens, there is general agreement that only small specimens, up to 50 μm in diameter, can be adequately cooled by the methods we have been discussing. Larger specimens would not be adequately cooled much below 25 μm from the surface. It is, however, not sufficient just to use the criterion of cooling rate as an indicator of the best cooling procedure to adopt. We need to consider the shape, form, and properties of the sample, how it is to be processed after cooling, and the reasons for wanting to cool the sample in the first place, e.g., low-resolution microscopy or high-resolution analysis.

For most low-temperature microscopy and analysis of tissue blocks and droplets no bigger than 0.25 mm^3, plunge cooling into melting propane using a mechanical injector will probably give the most acceptable results. Isodiametric specimens much larger than 0.25 mm^3 cannot be uniformly frozen, and some ice-crystal damage will be unavoidable in larger uncryoprotected specimens. The immersion method is simple to use and the equipment inexpensive to construct or purchase. Impact cooling and jet cooling are better suited for the surface regions of samples and for thin, flat specimens and suspensions. It is unlikely that there will be any significant ice-crystal damage in the first 20–25 μm of such samples. Spray and droplet cooling appear ideal for very small, ca. 20 μm, samples. Although high-pressure cooling is beginning to produce some impressive results, in many instances they are no better than can be obtained by more conventional means and at considerably less expense. As mentioned earlier, high-pressure cooling is not a rapid cooling procedure and should not be used to arrest fast events in cells and tissues. When deciding on a particular cooling protocol it is important to bear in mind the objectives of the experiment. The cooling procedures required for the optimal preservation of the interior of a sample to be examined at ambient temperature in a transmission electron microscope are quite different from the procedures we would need to analyze the outside of a hydrated sample examined at low temperatures in a scanning electron microscope.

3.8. THE SIGNIFICANCE OF COOLING RATES

We have shown that minimum ice-crystal growth and good quality cryopreservation can only be achieved in noncryoprotected samples if the cooling rate, in the region of interest in the sample, is at least 10×10^3 K s^{-1}. A great deal has been written about the different ways it is possible to measure the rates of cooling achieved by the various procedures we have been discussing. In many instances, the figures quoted are either the mean rates at the surface of a sample, or the result of model and/or *in vitro* experiments that bear little relation to what is occurring inside the sample.

Mean cooling rates of *specimens* are difficult to interpret because they are a function of the size of the measuring probe and its position within the specimen. Plattner and Bachmann (1982) have assembled an impressive table summarizing the range of cooling rates recorded in the literature. Depending on the size of the thermocouple and/or sample, and the method of measurement, the cooling rates recorded for a given cryogen range over four orders of magnitude. The problems of direct temperature measurement in the type of small sample we ideally like to use in low-temperature microscopy and analysis begin to appear even more daunting.

In an attempt to understand what is going on inside a specimen, Costello (1980) and Costello *et al.* (1984) have suggested that a thermocouple coated with a 10-μm layer of epoxy resin might provide a more useful model system to compare the different cooling procedures used for the cryopreservation of hydrated samples. Although epoxy resins do not exhibit the phase changes seen in water when it is cooled, i.e., latent heat of crystallization, they have a number of thermal characteristics that are similar to biological material. The good mechanical properties of epoxy resin mean that the same thermocouple may be used repeatedly. Ryan and Purse (1985) and Ryan *et al.* (1990) have performed the same type of experiments using gelatin-coated thermocouples. Both groups of investigators have derived cooling rates that appear to be closer to what may be occurring inside pieces of biological material. However, it would be naive and misleading only to rely on these absolute measurements as a clear indicator that a particular rapid cooling protocol will give an ice-free sample. A simple example will illustrate this point. An epoxy-coated, 50-μm copper-constantan thermocouple mechanically plunged into melting propane might easily record a cooling rate of $10–20 \times 10^3$ K s^{-1}. It would be unwise to believe that the same cooling rate would be achieved inside a 0.5-mm^3 piece of muscle cooled by exactly the same procedure. There are obvious differences in size, shape, material properties, and consistency that would militate against an equivalence of cooling rates. In addition, there will be a variation in the time different regions of the sample are frozen, and the cooling rate at which the freezing occurs.

It is now generally accepted that, at best, the measurements that have been obtained using thermocouples should only be used on a comparative basis, and we should briefly consider other approaches that could be used to establish an effective cooling protocol. Instead of trying to find out what sort of cryopreservation might be expected with a given cooling rate, it might be more sensible to first set the limits of cryopreservation that are required, i.e., crystallite size and depth of ice-free layer, and then determine what cooling rates would be necessary to achieve this objective.

Jones (1984), using a simple physical model, has estimated freezing times during rapid tissue cooling by describing heat transfer at the moving boundary between frozen and nonfrozen tissue. The calculations indicate that cooling rates of 40×10^3 K s^{-1} are associated with freezing times of less than 0.5 msec at a depth of 10 μm in the sample. Bald (1987) has approached the problem in a similar fashion, and his largely theoretical calculations can be used to provide a positive way forward in our understanding of what happens to a sample during rapid cooling. Bald (1987) suggests that we should obtain as much information as possible about the specimen and the equipment before attempting any cooling procedures. Data should be collected on the thermophysical properties of the sample and the cryogen(s), the method(s) available for rapid cooling, and the expected upper limit for the size of ice crystals for the type of microscopy or analysis to be carried out. These data can provide the basis for direct calculations, or more conveniently, can be fed into a purpose written computer program CRYOFIX. The equations or the program calculates the cooling rate at the propagating freezing front throughout the specimen cross section and compares these values with the critical cooling rate needed for a given ice crystal size. The computer program allows this to be carried out as an iterative process and the experimental parameters may be continually modified in order to obtain the procedure(s) that should be adopted for good cryopreservation of the given sample.

Although this largely theoretical approach puts much less emphasis on measuring cooling rates, it is still useful to have some idea of the comparative rates that may be obtained with different cryogens and different pieces of equipment. Cooling rates may easily be measured directly by using thermocouples, or indirectly and with more difficulty by measuring changes in the electrical properties or the fluorescence of the sample during cooling.

3.8.1. Thermocouples

Temperature changes may be measured on the basis of the voltage that is generated at the junction of two dissimilar metals such as copper–constantan (a copper–nickel alloy) and chromel–alumel (nickel–chromium and nickel–aluminum alloys). Robards and Sleytr (1987) provide full details of the construction and properties of these standard thermocouples and how they may be calibrated. The sensitivity of the standard thermocouples is only satisfactory down to about 60 K. Escaig (1984) gives details of gold–iron and chromel thermocouples and silicon diode temperature sensors that are sensitive in the range 4–80 K. Thermocouples for measuring the cooling rates of liquid cryogens are best formed as spark-welded beads at the end of two wires. Such thermocouples may be less useful for measuring temperatures inside the pieces of equipment used in low-temperature microscopy and analysis. A few years ago, our group in Cambridge (Clarke et al., 1976) made thin, ca. 10-μm, film thermocouples. These were extremely fragile and were not designed to measure cryogen cooling rates, but were used to measure the temperature of specimens sitting on the cold stage of a scanning electron microscope. Escaig et al. (1977) have, however, used similar thin film thermocouples to record very high cooling rates of ca. 100×10^3 K s^{-1} in liquid cryogens.

The size of the thermocouple has a direct bearing on the cooling rate it measures, and if they are going to be used to provide comparative cooling rates, the following parameters should be observed:

1. The same small thermocouple should be used for all the measurements.
2. The cooling rate should be measured over the same temperature range, usually 273–173 K.
3. Wherever possible, cooling rates should be measured either for different cryogens in the same piece of equipment, e.g., propane, ethane, halocarbons, in a plunge-cooling device, or for the same cryogen in different cooling devices, e.g., propane by jet and immersion cooling.
4. Details should be given of the size of the thermocouple, its location in the equipment, and if appropriate its entry velocity and depth of immersion into the cryogen.

It is much easier to compare the cooling rates of different cryogens than to compare different methods of cooling. Strictly comparative cooling rates have been obtained by Costello and Corless (1978), Elder et al. (1982), and Ryan et al. (1987, 1990). Their data are summarized in Table 3.5, and although comparisons within a group are instructive, comparisons between groups have little meaning other than the indication that certain trends begin to emerge. A less accurate comparison of the cooling rates which may be obtained by different techniques, was given earlier in Table 3.4. The data presented in these two tables show that although there is a very wide range in the cooling rates of different cryogens and different cooling methods, certain trends do emerge from the experiments. It is for this reason that ethane and propane are the most favored cryogens for immersion, jet, and spray cooling (Table 3.6).

Table 3.5. Comparative Immersion Cooling Rates Measured Using Different Procedures, Thermocouples, and Cryogens[a]

	Mean cooling rate ($K^{-1} s^{-1} \times 10^5$)					
Cryogen	Rate A	Rate B	Rate C	Rate D	Rate E	Rate F
Liquid nitrogen	16	0.80	0.70	0.70	0.13	—
Slush nitrogen	21	1.7	—	1.0	—	—
Freon 22	66	9.0	6.3	0.8	2.6	1.5
Propane	98	19.1	8.3	1.8	4.1	1.7
Ethane	—	—	12.0	—	—	2.1

[a]All rates are cited between 273 and 173 K. Rate A recorded with a 70-μm thermocouple with a spring-assisted velocity of 0.5 m s^{-1} (Costello and Corless, 1982). Rate B recorded with a 300-μm thermocouple with a spring-assisted velocity of 1.4 m s^{-1} (Elder *et al.*, 1982). Rate C recorded with a 300-μm thermocouple with a spring-assisted velocity of 2.3 m s^{-1} (Ryan *et al.*, 1987). Rate D recorded with a 50-μm thermocouple rapidly hand dipped (Zierold, 1980). Rate E recorded with a 360-μm thermocouple allowed to free fall by gravity (Schwabe and Terracio, 1980). Rate F recorded with a 25-μm thermocouple covered with 415 μm of 20% gelatin at a spring-assisted velocity of 3 m s^{-1} (Ryan *et al.*, 1990). (The rate F data are probably the most significant as the recorded rates most closely represent the cooling rates one might expect *inside* a 0.5-mm-thick hydrated specimen.)

Table 3.6. Cooling Rates from Model Samples Plunged into Different Coolants at a Constant Rate[a]

Cryogen	Thermocouple A	Thermocouple B	Thermocouple C
Liquid nitrogen	80	34	43
Slush nitrogen	317	79	79
Freon 12	714	63	37

[a]Thermocouple A, bare 500-μm thermocouple. Thermocouple B, 500-μm thermocouple inside a 2.4-mm phosphor-bronze bead, low Biot number. Thermocouple C, 500-μm thermocouple inside a 2.4-mm PTFE plastic bead, high Biot number. The thermocouple inside the phosphor-bronze bead has a low Biot number ca. 0.002; the thermocouple inside the PTFE has a high Biot number. The figures show that the cooling rates for large samples are very low. The results obtained with the PTFE coated thermocouple are close to those one might expect from a large frozen biological sample. Ice is a poor conductor of heat, and the cooling rate in the sample will be determined by the thermal conductivity of the ice rather than the heat transfer properties of the specimen. (Data from Robards and Sleytr, 1985.)

3.8.2. Electrical and Fluorescence Methods

Heuser *et al.* (1979) and Van Harreveld and Trubatch (1979) describe methods based on changes in the electrical capacitance and resistive current flow of water as a freezing front progresses through a hydrated sample. Such measurements can be used to calculate the *freezing* rate inside the sample. This method has not been sufficiently widely used to be able to evaluate its usefulness. The fluorescence method is based on the finding that certain dyes fluoresce at different wavelengths depending on their temperature. Aurich and Foster (1984) have used this technique to measure cooling rates within specimens.

3.9. EVALUATING THE RESULTS OF SAMPLE COOLING

The only sure and certain way to judge the effectiveness of a given cooling procedure is to examine the quality of the final product. In thin sections and thin liquid suspensions, evidence for vitrification can only be obtained by examining the x-ray or electron diffraction patterns in fully frozen-hydrated samples. This type of evidence is not available from frozen hydrated bulk sample and thick sections, and here one must rely on the size of ice-crystal "ghosts," which only become apparent after the ice is removed from the sample. An alternative criterion for sample preservation is met when the grain size of the background ice is so fine that individual ice crystals cannot be detected in the imaging system being used.*

Bilayer lipids exhibit phase changes in the temperature range 313–273 K which are sensitive to the rate of cooling. The higher the cooling rate, the better the preservation of the high-temperature phases. Costello *et al.* (1984) have summarized how these changes, which may be observed in freeze-fracture replicas, can be used to evaluate the effectiveness of rapid cooling procedures. Favard *et al.* (1989) provide an ingeneous method of using magnesium oxide crystals as a means of surface labeling to recognize freezing artifacts at varying depths in rapidly cooled gel samples. They were able to show that the average size of ice crystallites at the initial point of the impact cooling at the surface of a 2% gelatin gel was ca. 35 nm, which increased in size to ca. 1000 nm, 25 μm below the surface.

Many of the samples we use for low-temperature microscopy and analysis are subjected to further processing, such as freeze-drying, freeze substitution, and freeze etching. In these circumstances, the extent and size of ice-crystal ghosts, the evidence of incipient prefreezing dehydration, might be useful guidelines in judging the quality of the preservation. In addition, more general features such as the preservation of structural relationships between organelles, visualization of membranes, maintenance of immunochemical properties, and the retention of small water soluble solutes, might also be useful criteria. The presence of "halos" around mitochondria is now generally accepted as being a sure sign of poor preservation.

In many instances, the best cryopreservation for morphological purposes is considered only to be found in ultrathin sections of resin embedded material which has been freeze substituted following rapid cooling. It is judged as being the best because it most closely resembles the images we obtain following conventional, ambient temperature, chemical fixation. However, such fidelity of ultrastructure is not seen in the images obtained from chemically unsullied, fully frozen-hydrated, thin sections, examined at low temperature in a transmission electron microscope. In this respect, the quality assessment is closer to one of the more important criteria we use to judge the standard of cryopreservation for analytical purposes, where the best

---

*It is generally assumed that the ice-crystal ghosts seen in the microscope are derived from a multitude of small individual ice crystals. There is evidence to suggest that these ice-crystal ghosts are not derived from individual crystals but are the dendritic branches of a single ramifying crystal (Ryan *et al.*, 1987, 1990). This would account for the observation that there are reduced ice-crystal profile sizes at the center of quench-cooled specimens (Van Venrooji *et al.*, 1975). Dubochet *et al.* (1988) provide evidence from electron diffraction that there is but a single dendritic ice crystal in a cell frozen by rapid cooling.

results are considered to have been achieved when the sample is virtually featureless! However, even these criteria are subject to uncertainty as the morphological images and analytical results that are presented must also be an expression of the different metabolic states of cells and the local variations in the water content of the sample. We will return to these matters in more detail when we consider the artifacts and damage that are caused by particular low-temperature preparation procedures.

Philips and Boyne (1984) have suggested that chromatin adjacent to the nuclear envelope might provide a sensitive indicator of freezing damage. It was found that optimal cooling procedures usually produced compact, electron dense, and sharply delineated chromatin. A honeycomb appearance was associated with poor rapid cooling methods. Work by Sitte *et al.* (1987) has confirmed this suggestion. Knoll *et al.* (1987) consider that a measure of the quality of cryopreservation may also be provided by observing cellular endo- and exocytotic processes. Membrane fusions are apparently only seen in well-preserved material.

Irrespective of the way the sample has been quench cooled, it is necessary to have some objective assessment of the effectiveness of the cooling procedure. Calculated cooling rates are of much less value than the quality of the images observed in the microscope. Dalen (1991, unpublished data) has shown that the fine structure of the intracristal spaces of mitochondria is a sensitive and simple way of judging the quality of cryofixation. In well-preserved samples this space is open; in poorly preserved samples the space is occluded with electron-dense material (Fig. 3.10).

3.10. LOW-TEMPERATURE STORAGE AND SAMPLE TRANSFER

The rapid cooling processes we have been discussing are the first part of a chain of procedures leading to the examination and analysis of samples in some form of microscope or analytical system. The first and most important concern about the specimen once it has been cooled is that it must remain below 133 K, the recrystallization temperature of water. At this temperature any vitrified phases are preserved, and the activation energy for the rearrangement of water and solute molecules is no longer available. The sample is, for all intents and purposes, unperturbable, and storage in LN_2 is the simplest way this may be achieved. Specimens for low-temperature microscopy and analysis should not be stored in low-temperature mechanical refrigerators. The temperatures may not be low enough for long-term storage and thermal gradients may develop within the cold container. In addition, electrical and mechanical faults in refrigerators are more likely to occur than someone forgetting to top up a storage Dewar with LN_2. It is also frequently necessary to transfer specimens, at low temperatures, to other pieces of precooled equipment without the sample either warming up or becoming contaminated.

3.10.1. Sample Storage

A large storage Dewar of LN_2 is a convenient and inexpensive way of storing specimens, and Rigler and Patton (1984) and Williamson (1989) give details of the

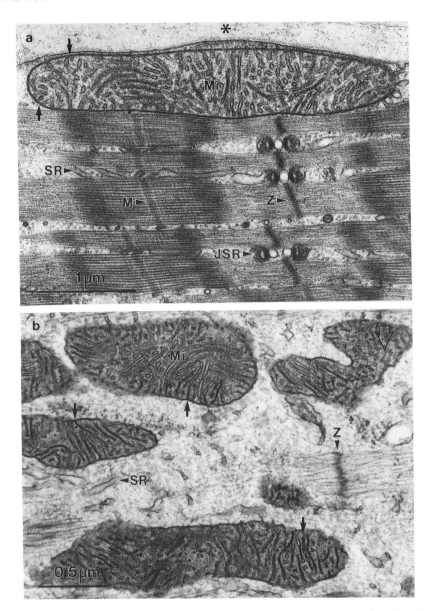

Figure 3.10. Transmission electron micrographs of liquid helium impact-cooled samples of frog skeletal muscle, freeze substituted in an acetone + osmium mixture, embedded in resin, and sectioned at ambient temperature. (a) Section cut from the region at the point of impact (*) with the cold copper block. Note that the intracristal spaces of the mitochondria are fully open, the Z and M bands are clearly delimited, and the SR and JSR are well preserved. (b) Section cut from a region within the specimen approximately 20 μm from the initial point of impact. The intracristal spaces are occluded with electron-dense material at several points (arrows), the Z bands are indistinct and the M bands are missing, and the SR and JSR are poorly preserved. The direction of cooling is from the top to the bottom of the pictures. Z and M, muscle myofibrils; Mi, mitochondria; SR, sarcoplasmic reticulum; JSR, junctional sarcoplasmic reticulum. (Pictures courtesy Helge Dalen, University of Bergen.)

types of containers that are best suited for the small samples that have to be stored. Small low-density polythene bottles or 35-mm film containers with a piece of string knotted into the screw cap, are the simplest and least expensive containers. It is important that samples are stored in such a way that mechanical damage is avoided as the samples are moved around. Most frozen material is notoriously brittle and it is all too easy for samples to be broken or dislodged from the specimen holders used later in the chain of sample preparation. The most satisfactory arrangement is to screw or slot the specimen holders into prefabricated containers, which may be easily loaded and unloaded under LN_2. The specimen holder and their containers must be clearly labeled, preferably by direct metal engraving. Unmounted samples may be stored in aluminum foil packets or the small plastic tubes that form part of the readily available dip-stick storage containers for use with LN_2 Dewars.

Although unmounted samples may be attached to cryomicrotome specimen holders or scanning electron microscope specimen holders by mechanical means, the brittle nature of the frozen material makes this a difficult task. It is usually easier to use a low-temperature glue such as toluene ($T_m = 178$ K) or Tissue-Tek. The polymeric cryoprotectants PVP and HES also act as good cryo-glues (Echlin et al., 1977). Although it would be far more convenient to load the samples into or on specimen holders before rapid cooling, it must be recognized that the additional thermal mass of the holder may result in inefficient cooling. This may be partially ameliorated by designing streamlined specimen holders and making sure the sample is the first object to enter the cryogen.

It has always been assumed that specimens stored in LN_2 would be safe for an indefinite period. There are, after all, thriving concerns in California who are storing human heads in LN_2 in the hope that medical science will eventually produce remedies for the ills from which the heads (and bodies?) died. Leaving aside these intimations of immortality, it has been suggested (Rebiai, 1983) that the unexpectedly high solubility of water in cryogenic liquids could lead to desiccation of hydrated samples stored for a long time in direct contact with LN_2. These predictions are probably of little consequence here, for most storage Dewars contain LN_2 saturated with water owing to repeated exposure to the atmosphere. Nevertheless, it would probably be unwise to expose thin sections or thin films of hydrated material to dry LN_2 for long periods of time.

3.10.2. Low-Temperature Sample Transfer

If the sample is to be examined in a fully frozen-hydrated state on the cold stage of an electron or photon beam instrument, care must be taken in transferring the sample either from the rapid cooling equipment or from the storage containers. A number of these devices are commercially available and they either form part of a more complicated piece of equipment for low-temperature sample manipulation, or are specifically designed for a particular microscope or analyzing system. Specific details are available from the various enterprises that make these devices.

The underlying principle behind the construction of all the devices is quite simple. The sample is maintained at or near LN_2 temperatures by having a close mechanical and hence, thermal contact between the sample holder and a large piece of precooled metal such as copper. The sample holder is loaded into the transfer device under LN_2 and the thoroughly cold specimen is quickly drawn into a well-cooled shroud or sleeve, which prevents atmospheric water vapor and volatile materials condensing on the surface. The more sophisticated devices allow the shroud to be evacuated, thus further diminishing the chance for surface contamination. The sample may be quickly transferred to the precooled cold stage of the next piece of equipment in the process line, via an air lock if this is necessary. Any dead volumes of air within vacuum valves or within the shroud can be removed by flushing with cold dry nitrogen.

The sample, which has been rapidly cooled in such a way as to produce the optimum cryopreservation, now either rests safely in the frozen depths of a storage Dewar or sits, somewhat more precariously, on the cold stage of a microscope, freeze-dryer, or freeze-etching apparatus, at the business end of a cryomicrotome or in some other part of the process of low-temperature preparation. We now need to consider the ways we can further manipulate the frozen specimen prior to its examination and analysis. The following four chapters will consider the physical processes of sectioning, fracturing, etching, and drying before finally considering the processes of substitution and embedding, which involve chemical intervention.

3.11. SUMMARY

The initial act of rapidly cooling the sample is probably the most critical part of all the preparative procedures used in low-temperature sample preparation. Provided the cooling rate is at least $1 \times 10^4 \, K \, s^{-1}$, a sample no larger than 25–50 μm will be preserved with ice crystals no bigger than 10 nm. There are, however, severe physical limitations to the rate at which heat may be removed from most of the specimens we commonly use in low-temperature microscopy and analysis and we must accept that only the smallest samples will be preserved without any ice-crystal damage. Chemical fixation and cryoprotection are best avoided in morphological studies of all but the larger specimens and should not be used in conjunction with studies involving the analysis of low-molecular-weight, water-soluble materials.

Specimens may be cooled rapidly by a number of different liquid and solid cryogens in a variety of different ways. Although most of the methods give cooling rates of about $1 \times 10^4 \, K \, s^{-1}$, there is wide variation and no single method gives good results with all samples. Immersion fixation into liquid propane is a good general method; impact and jet cooling give satisfactory results for surface layers and thin suspensions; and spray cooling seems best suited for the cryopreparation of microdroplets.

The best approach to adopt when setting up procedures for the rapid cooling of a sample is to first establish the upper limits of ice crystal damage that may be tolerated in the experimental system. Provided sufficient data are available about the thermophysical properties of the specimen, cryogen, and cooling technique, it should be possible to simply calculate the parameters for the rapid cooling procedure that will give the best preservation.

Cryosectioning

4.1. INTRODUCTION

The internal features of any object may be exposed, either by more or less randomly fracturing or breaking open the specimen, or by sectioning it at more precisely determined places. It is possible to distinguish between *fracturing* and *cutting*. The former is a brittle breakage of the crystalline components with plastic deformation only occurring at a molecular level, while the latter is a flow process involving extensive plastic deformation of the crystallites and possibly melting due to friction and/or local pressure. In fracturing, the initial crack propagates some distance ahead of where the force is being exerted. In cutting processes, the separation of the sample occurs at the knife edge itself. The exposed internal surfaces may either be replicated or studied directly by a wide variety of techniques, many of which are discussed in an earlier book by Echlin (1984).

Sectioning and fracturing processes in low-temperature microscopy and analysis will be discussed in this and following chapters. The procedures have been divided, somewhat arbitrarily, into those that involve minimal chemical intervention and those that require somewhat more chemical intervention. The former group, which includes sectioning and fracturing techniques, relies on the solidified phases of water providing the mechanical strength to otherwise soft objects to allow the dissecting processes to take place. In the latter group, which will center primarily on freeze substituion, low-temperature embedding, and freeze-drying techniques, the water is removed either by physical or chemical means and replaced by waxes or resins, which in turn are hardened or polymerized to materials sufficiently strong to be sectioned.

4.2. CRYOSECTIONING

Cryosectioning, or cryomicrotomy, describes the process whereby samples are sectioned at temperatures below 273 K to produce thin slices of material, which may then be examined and analyzed by some form of transmitted radiation (illumination). The procedure is a derivative of the sectioning processes that microscopists have been using for centuries on wax- and, latterly, plastic-embedded material. A summary of these procedures is given in the book by Reid (1974) and in a more recent paper by Sitte (1984).

The actual molecular and physical processes of cutting are still not completely understood. Kellenberger *et al.* (1986) and Acetarin (1987), on the basis of extensive work with resin-embedded material, consider that both fracturing and cutting at ambient temperatures are cleavage phenomena in which material is pulled apart. As material is cleaved, it first passes through an elastic phase in which the material elongates in a reversible manner before undergoing irreversible plastic flow. The plastic flow continues until the material ruptures, after which the elastic elongation may be reversed but the plastic deformation remains unchanged. The relative amounts of elastic elongation and plastic flow vary according to temperature and the chemical composition of the material. The important difference in cryomicrotomy is that the mechanical strength of the object has been increased by taking advantage of low temperature, which, in all substances, decreases their viscosity and plastic flow and if the temperature is low enough will convert essentially liquid materials to solids. In the course of this chapter we will examine the sectioning process and see how closely it resembles what occurs at ambient temperatures.

Cryomicrotomy is not confined to frozen specimens where the nonaqueous components can be thought of as being embedded in a water matrix. The technique has important applications in the preparation of plastics, polymers, and elastomers many of which are either liquid or highly flexible at ambient temperatures. The advantages for biological materials are that it provides a simple process to cut objects, hardened by cooling, without the use of deleterious fixation, dehydration, and embedding. Frozen sections provide a favorable basis for x-ray microanalysis of electrolytes and soluble cell constituents, and for the histochemical and immunocytochemical localization of molecules. The absence of elaborate and time-consuming preparative procedures means that cryosections can (usually) be produced quite rapidly.

4.2.1. The Different Types of Cryosections

Frozen sections can be divided into those that are thin or ultrathin (50–100 nm) and examined and analyzed in the transmission electron microscope, thick sections (0.1–1.0 μm), which are studied by scanning transmission electron microscopy, and thicker sections (8–50 μm), which are studied by light microscopy. It is also possible to classify sections on the basis of their water content. In biological and hydrated organic material, nearly all cryosections are cut from frozen material where the water is present, most usually, as ice crystals of varying dimensions. If such sections remain in a fully hydrated state during subsequent preparation, examination, and analysis without either drying or melting, they are referred to as being *frozen hydrated*. Such sections are quite distinct from *freeze-dried* sections, where the water is carefully removed by sublimation after the cutting process.

There are advantages and disadvantages to each of these types of sections. Until recently, frozen-hydrated sections were used almost exclusively for analytical purposes, even though there are a number of problems associated with the use of such material. One major problem is that there is very little contrast in the transmitted electron image from frozen-hydrated sections because the electron scattering coefficient of ice is very close to that of the organic material held within the ice matrix.

Such sections are of little use for morphological studies alone, although this appears to be less of a problem in the few examples of thin frozen-hydrated sections that have been cut from vitrified samples. The absence of morphological information makes it difficult to recognize cell compartments with any consistent degree of certainty, although with practice this does become a little easier. There is a remarkable improvement in sample morphology if the frozen-hydrated section is freeze-dried (Figs. 4.1 and 4.8). There is no electron scattering from ice crystals and a marked increase in contrast may be seen between the remaining organic matrix and the voids or "ice-crystal ghosts." Sections that contain large ice crystals show a larger contrast change than those with small crystallites.

A second problem associated with frozen-hydrated sections is that the local concentration of soluble constituents may be relatively small when compared to the dry weight concentrations. This is understandable in cells composed of 90% water. In x-ray microanalysis, elements in low concentrations present problems of signal-to-noise discrimination and it becomes difficult to determine accurately the characteristic x-ray counts. Soft x-rays from such elements as Na and Mg are absorbed by the water matrix, which reduces their chance of detection. For any quantitative studies it is important to have a fairly accurate measure of the hydration states of the sample. These are matters we will return to in Chapter 11 (Section 11.8.1). The third problem associated with frozen-hydrated sections is that they are considerably more prone to radiation damage in the electron beam than their freeze-dried counterparts (section 9.9.5).

4.2.2. The Cryosectioning Process

There has been much discussion on the nature of the cryosectioning process, particularly on whether the sections are produced by local and transient melting, akin to the movement of an ice skater over the surface of ice, or by fracturing and cleaving without any significant increase in temperature. This is a matter of some importance to our basic understanding of cryomicrotomy. If the sectioning process involves a transient melting, then there would be a serious risk of redistribution of diffusible substances during the cutting process and the subsequent ice recrystallization would have a deleterious affect on ultrastructure. If, however, the sectioning is more akin to a continuous fracturing process, then this might help explain many of the curious artifacts that are frequently associated with cryomicrotomy.

There can be no doubt that a finite amount of energy is generated by the interaction of the extremely fine edge of the knife, which initially may be of molecular dimensions, and the solid material that is being cut. It is assumed that this mechanical energy is converted to heat, and there is no way of telling whether this heat is absorbed by the frozen block face, the section, or the knife. It is still uncertain whether the small amount of heat generated during cryosectioning is a necessary prerequisite for the process to work, i.e., the transient melting of ice acts as a lubricant for the cutting edge as it passes through the frozen sample. It is unclear whether this phenomenon would appear at all temperatures, for an examination of the phase

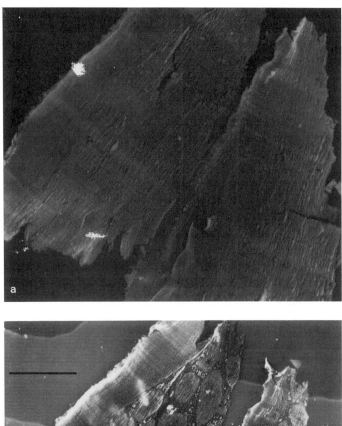

Figure 4.1.(a) Secondary electron image of a frozen-hydrated section (0.5 μm thick) of rat renal cortex cryosectioned at 220 K. Note the albumin layer (upper left), which sectioned with less discontinuous chip formation than the tissue, demonstrating differences in material properties between tissue and albumin. It is difficult to clearly discern any recognizable structures in this fully hydrated sample. (b) Secondary electron image of the same section after the ice has been removed by freeze-drying, in which it is possible to recognize details of the kidney tubules. Bar marker = 100 μm. [Picture courtesy Saubermann *et al.* (1986b).]

diagram of water (Fig. 2.1) shows that the lowest melting point of water is 251 K, which is only achieved at a pressure of 2100 bar.

In some of the early work on this subject, Thornberg and Mengers (1957) calculated that there was indeed a zone of melting at the point of sample–knife interaction. Their calculations showed that cryosectioning at 222 K resulted in a 30–50 nm melt zone at the knife edge, which rapidly refroze as the cold knife advanced further into the specimen. Calculations by Hodson and Marshall (1972) confirmed this view, but emphasized that the melting zone was likely to be much smaller than that postulated by Thornburg and Mengers. If all the mechanical energy involved in cutting at 180 K was transferred as heat to the frozen section, Hodson and Marshall calculated that the melt zone would only be a few nanometers thick. Studies by Saubermann et al. (1977a) suggest that this zone would be no more than 0.5 nm when cutting 0.5–2.0-μm cryosections. The possible significance of a melt zone at the knife edge needs to be considered in relation to the total thickness of the frozen section. A thin transient melt zone on each face of the typical 0.5–1.0-μm section used for x-ray microanalysis would be likely to have much less effect than a similar melt zone on each face of a 50-nm section used for high-resolution studies. The only situation where transient melting might be a matter of concern would be where frozen sections are being used for immununocytochemical techniques. Unlike x-ray microanalysis, which collected information throughout the depth of the section, immunocytochemical techniques only detect antigens at or very close to the section surface. Transient melting might cause the relocation of antigenic sites on the surface of the frozen sections.

Saubermann et al. (1977a) have carefully measured the forces that are developed during cryosectioning by mounting miniature strain gauges attached to a load cell fitted to the arm of the microtome. They found that at 240 K, the sectioning process consumes very little energy and that relatively smooth sections are produced. At 190 K, a much higher energy input is necessary for cryosectioning, and the sections that are produced have a much rougher appearance. They could find no evidence of transient melting. Frederick and Busing (1981) and Frederik (1982) have studied the problem of melting during cryosectioning by examining the shape and form of ice crystals both in intact cryosections and in replicas of their surfaces. They too could find no evidence that the cutting process caused either local melting or recrystallization. Karp et al. (1982) have sectioned solid toluene at a temperature just one degree below its melting point of 178 K and found that any temperature increases due to cutting were too small to melt the 100-nm sections. This is a particularly encouraging piece of evidence, because far less energy is required to melt toluene compared to ice because of the significantly lower heat of fusion of toluene (72 J g^{-1}) compared to ice (334 J g^{-1}) and the fact that the thermal conductivity of ice is approximately four times greater than toluene at the temperatures used to cut the sections. McDowall et al. (1983) have cut vitrified sections at very low temperatures without any suggestion of either melting or phase transformations of the ice, i.e., they could find no evidence that the I_v had been irreversibly transformed to either I_c or I_h. The bulk of the evidence suggests quite strongly that section melting and ice-crystal growth is not a major problem during cryosectioning at temperatures below 240 K.

Although melting does not occur, there is some evidence that phase changes may result as a consequence of cutting vitrified material. Chang *et al.* (1983) observed a series of bands in cryosections cut at below 143 K from material that they considered to have been vitrified during the initial cooling. Unlike McDowell *et al.* (1983), Chang and his colleagues considered that the banding represented cutting-induced devitrification and melting, an observation that was subsequently confirmed by Frederik (1986). Small changes in the cutting procedure can have a considerable effect on the final quality of the section.

A number of workers (Saubermann *et al.* 1977); Saubermann, 1980; Frederik *et al.*, 1984; and Zierold, 1987 have sought explanations for the process of cutting frozen sections by comparing it to the process of chip formation during metal machining. Saubermann (1980), who has worked primarily with thicker sections, considers that there is both compression and stress at the knife edge as it is forced into the frozen block. These forces are relieved either by the sample rupturing along the planes of least resistance or, depending on the ductility of the material, by plastic deformation of the sample. If the frozen sample is brittle, a feature of lower temperatures, there will be little plastic deformation and the compressive forces will cause the sample to rupture along the shear line of least resistance and produce a series of discontinuous chips. At higher temperatures, the frozen material would be much less brittle and there would be increased plastic flow at the knife edge. This would result in the formation of a continuous chip as plastic deformation occurs in the shear zone. The chip, or frozen section, would tend to curl upward as it is cut because of the bending forces imposed by the position of the knife. Saubermann (1980) and Saubermann and Heyman (1985) consider the shear zone temperature to be the most critical factor in cryosectioning. They found that at a block temperature between 243 and 213 K, liver tissue was compressible, but that at 183 K it fractured under compression. Saubermann and Heyman consider that liver tissue may be cryosectioned only at temperatures below 183 K if the cutting process caused a sufficient temperature rise in the shear zone so that the frozen material became ductile. At this temperature, the energy needed for sectioning would be decreased and the frozen material would undergo plastic deformation rather than fracturing.

Frederik *et al.* (1984) examined the lines of deformation that frequently occur on cryosections as well as on metal chips, and resin and wax sections, and concluded that they were probably caused by pressure due to shearing forces that developed perpendicular to the surface of the knife. Such forces would result in a periodic build up and loss of sectioned material. Sitte (1984) and Zierold (1984) believe that a similar process occurs during the ultramicrotomy of plastic sections and that the high-frequency deformation lines or "chatter" that can be observed on sections are a result of a periodic build up of pressure followed by a material sliding away at the knife edge. This high-frequency deformation can be removed by changing the parameters of the cutting process (Fig. 4.2). These are matters we will return to later in this chapter.

On the basis of the work of Saubermann, Frederik, and Zierold and colleagues, it is possible to construct theoretical models to help explain cryosectioning, at least of pure water, if not the material it embeds (Fig. 4.2). It is assumed that the knife

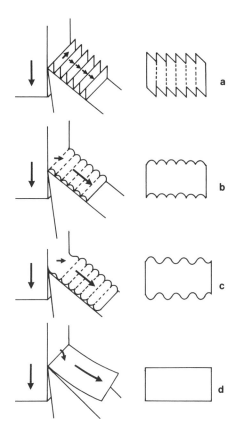

Figure 4.2. Diagrams to illustrate some of the different theoretical models proposed to explain the process of cryosectioning. (a) *Discontinuous fracturing or chip formation.* Pressure at the knife edge causes the brittle, noncompressible sample to repeatedly fracture and form a series of overlapping chips. (b) *Folding model.* Pressure at the knife edge causes material that is less brittle to first fold and then slip away. There is little variation in section thickness. (c) *Slip and stick model.* Variations in friction between the partially ductile block material and the knife edge are sufficient to cause variations in the thickness of the cut section, which is partially relaxed as the section slides away. There is variation in section thickness. (d) *Continuous chip formation.* The material is sufficiently ductile in the shear zone at the knife edge to allow plastic flow over the knife surface and gives rise to a relatively smooth, even thickness section. [Modified and redrawn from Saubermann (1980) and Zierold (1987).]

does not have an infinitely sharp edge, which will result in a degree of pressure perpendicular to the cutting edge and cause successive rupture and dislocation of chips of material from the block face. Depending on the brittleness and/or the ductility of the sample, this could result in a series of inclined chips and would probably represent what would occur as a series of ice crystals were fractured. The shear stress would go along the crystal planes of the ice (I_H) and the successive layers would fall away like a series of inclined books or playing cards. This would represent the *discontinuous fracturing model* of Saubermann (1980) and is what one would expect to occur when cutting thick sections at temperatures below 193 K where ice shows an increased hardness and much less plasticity. The work of Zierold (1987) provides the basis of another model for cutting at somewhat higher temperatures than 193 K.

In the *folding model* of Zierold, the material being removed from the block face undergoes plastic deformation to form a series of undulations which produce a continuous section with only a small variation in thickness. The *fracture–friction* model recently proposed by Richter *et al.* (1991) is similar in many respects to the folding model proposed by Zierold. Richter and colleagues have found that ultrathin sections of vitrified material have a smooth and a rough face. They proposed that the smooth face results from friction with the knife, whereas the rough surface, which they consider to be the former block face, is a result of the release of stresses caused by compression and bending during the sectioning process.

Frederik *et al.* (1984) consider sectioning as a flow process with the inherent characteristics of fracture, compression, and deformation. They propose the *slip and stick model*, in which a fracture occurs between the section and the knife, which causes the shear plane to oscillate. With increased friction, the shear plane moves toward the block face, resulting in thick sections, and as the friction decreases, the sections become thinner. This results in sections with a periodic thickness variation.

One final model is proposed by Saubermann (1980) and occurs when there is plastic deformation in the shear zone due to heat generated during cutting work. This caused the material to become ductile and it flows over the knife edge to produce a *continuous chip*. It should be emphasized that if this cutting process, which is characteristic of what occurs at higher (240 K) temperatures, takes place, there is no suggestion that it involves any melting at the knife edge. Saubermann *et al.* (1977) calculated that the energy involved in cutting a 500-nm cryosection would be 1.3×10^{-4} J at 243 K rising to 5×10^{-4} J at 193 K. As stated earlier, this would cause a miniscule rise in temperature at the knife–section interface. However, Frederik *et al.* (1984), in spite of experimental evidence to the contrary, have revived the earlier notion that a very thin film of water may be responsible for diminishing friction at the knife edge. They consider this to be an alternative explanation for why relatively smooth sections can be obtained at higher cutting temperatures.

The most convincing argument against transient melting during cryosectioning comes from the work of Saubermann and colleagues (1981). Artificial electrolyte gradients established in gelatin were quench frozen and sections were subjected to x-ray microanalysis to determine whether there was any elemental displacement. No significant differences were found in serial sections cut at 243 K and 193 K, even though their earlier calculations (Saubermann *et al.*, 1977) showed that under the "worst case" conditions there might be enough energy generated during the sectioning process to cause transient melting. Saubermann *et al.* (1981) found no difference in the mean size of the ice crystallites (0.4-μm) in sections cut at 243 K and 193 K, which was interpreted to signify that there was no recrystallization at the higher cutting temperature. This may not necessarily always be the case. Seveus (1979) has suggested that it would be difficult to measure any changing in ice crystallite size, if the crystallites are large to begin with. It would thus seem necessary to repeat this experiment on vitrified or at least microcrystalline material.

Although the presence of dissolved solutes, particularly in biological material, would raise the recrystallization temperature of water, it would be unwise to suggest that no recrystallization can occur at 243 K. Saubermann *et al.* (1986a, b) have cut flat, 0.5-μm sections, with a minimum of discontinuous chip formation, from kidney tissue maintained at an average temperature of 220 K. There appeared to be no difference in the distribution and concentration of elements and the local concentration of water in the proximal tubules from material cut at 220 K and 193 K.

These models for cryosectioning are all based on pure water with no regard to whether it is in a crystalline of amorphous form. Biological and hydrated materials are far more complex and consist of varying amounts of organic molecules and inorganic ions embedded in a matrix of either vitreous or microcrystalline ice. The size of ice crystals, which vary depending on the effectiveness of the sample cooling, will also affect the sectioning process. Wendt-Gallitelli and Woeburg (1984) were able to cut thin sections from the surface regions of muscle tissue that had been impact cooled at 4 K. This was not posssible from regions in which 50-nm ice crystals were visible. Vitrified solutions can be cut more easily and at lower temperatures than crystalline materials. A vitrified solution of 25% PVP can be sectioned as easily at 150 K as resin sections can be cut at ambient temperatures.

A complete understanding of the cryosectioning process in hydrated material is still some way off and it will require integrating what we know about the physicochemical and material properties of ice with what we know about the properties of organic materials at low temperatures. We know, for example, that the hardness of ice, its elastic and tensile properties, and impact strength all increase with decreasing temperature. This is consistent with the general finding that cryosectioning of biological specimens embedded in a polycrystalline ice matrix becomes more difficult at lower temperatures. In addition, it is known that well-frozen samples, characterized by a clear appearance, are much easier to section satisfactorily than samples that are poorly frozen and have a cloudy or opalescent appearance. This further supports the suggestion that ice-crystal size is important in the sectioning process. So too is temperature, although here it will be necessary to balance the ease of sectioning at higher temperatures with the increased chance of recrystallization. The difficulty in reaching a consensus agreement as to what might be happening during the sectioning process is due to the large number of variables in the process and the paucity of experimental data in which one or two of these variables, e.g., section thickness and temperature, have been analyzed in detail. For example, it would be instructive to relate the forces measured during cutting at different temperatures, to the image quality of thick and thin sections cut from a standard sample (gelatine–salts solution?) cooled at different rates.

Although the theory of cryosectioning is not completely understood, a number of general features begin to emerge:

1. Hydrated specimens which are prepared with a minimum of phase separation of the aqueous and nonaqueous phases, either by infiltration with cryoprotectants and/or by rapid cooling, can be sectioned more thinly, more easily, and at lower temperatures.

2. In samples with large ice crystals, smooth sections can only be obtained at an increased thickness and higher sectioning temperature.

3. The lower the amount of unperturbed water in a sample, the better the cooling and the easier it is to cut frozen sections. Thus an elastomer such as rubber will cut easily at 143 K and produce smooth sections, while it is virtually impossible to consistently produce smooth sections at nearly any temperature from an untreated vacuolate plant cell.

4. Provided the cutting speed is low and the specimen temperature is no higher than 233 K, cryosections may be produced without any damage from transient melting and recrystallization.

5. For any given sample, the specimen temperature and cutting temperature, both of which have a major influence on the quality of the frozen section, should be set at a value as low as possible to give consistently smooth and reproducible sections.

6. Thick sections (0.5–1.0 μm) are usually more easily cut at higher temperatures and thin sections (50–100 nm) are more easily produced at temperatures below 193 K. The best ultrathin sections have been cut at below 143 K from vitrified samples.

Cryosectioning is very much a practical procedure in which the success of the operation depends equally on the skill of the operator and the sophistication and ergonomic design of the equipment. We will now proceed to discuss how these processes may be carried out.

4.2.3. Equipment for Cryosectioning

The basic equipment for cutting frozen sections consists of some form of microtome in which the region surrounding the knife, specimen, and section may be kept at controlled low temperature. The basic design of the microtomes, and here we include the whole range of microtomes and ultramicrotomes, is little different from the instruments that are used to cut sections and plastic- and wax-embedded material. It is not intended to discuss the design and operations of microtomes and ultramicrotomes. Those readers who want further information on these topics should refer to the book by Reid and Beesley (1991) on ultramicrotomy and ultramicrotomes and the book by Pearse (1980) on microtomes for light microscopy. In addition, reference should be made to the technical literature available from the instrument manufacturers.

The instruments range from the slightly modified conventional microtomes to instruments that have been especially designed for cryomicrotomy. The latter designs recognize the fact that the cryomicrotome is but one step in the chain of specimen preparation for low-temperature microscopy and analysis and due provision is made for getting frozen samples and sections in and out of the microtomes. These ancillary devices are an important part of successful cryomicrotomes. Most of what we will discuss in this section will be concerned with the equipment needed for cutting frozen sections in the 50-nm to 1–2-μm range. Before discussing these so-called cryo*ultra*microtomes, we should briefly consider the cryomicrotomes that are used for cutting thicker sections for light microscopy.

4.2.3.1. Cryomicrotomes

Such instruments produce frozen sections in the range 1–50 μm at temperatures between 273 and 238 K, which are then usually melted and examined in the light microscope for pathology, histochemistry, and fluorescent studies. The high sectioning temperature, ca. 250–260 K, introduces too many recrystallization artifacts to make sections useful for examination and analysis in electron beam instruments. The cryomicrotomes for light microscopical studies either operate with the whole instrument in a cold box (the so-called cryostat instruments) or where only the knife and specimen holder are cooled either by a Peltier cooling device, by dry ice (solid CO^2), or by adiabatic expansion of liquid CO_2 to temperatures between 200 and 250 K. There are a large range of cryomicrotomes that can be used for the production of frozen sections for light microscopy. They are based on rocker, sledge, or rotary mechanisms, and include instruments that will reduce animals as large as small monkeys to a series of 50-μm frozen sections!. The Shandon OT Cryostat (Fig. 4.3) or the Reichert–Jung 2800N Frigocut would be good examples of microtomes that

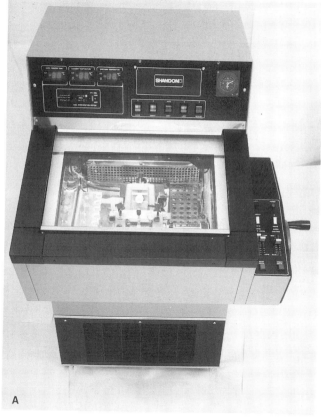

A

Figure 4.3. Cryostat microtome in which the whole microtome is placed in a controlled low-temperature box. The Shandon OT cryostat shown in (A) is used primarily for cutting thicker (0.5–30 μm) frozen sections at 240–270 K for pathology, surgery, histochemistry, and fluorescent studies. Details of the cutting region are shown in (B), (k), knife; (s), specimen, and (a), antiroll plate. The same principle may be used for producing the thin (50–500-nm) sections used for electron microscopy. The microtome is replaced by an ultramicrotome and the cryostat would be required to operate at much lower temperatures. In both types of instrument, the actual cutting process is controlled from outside the cryostat box. (Pictures courtesy Shandon Southern Products Ltd.)

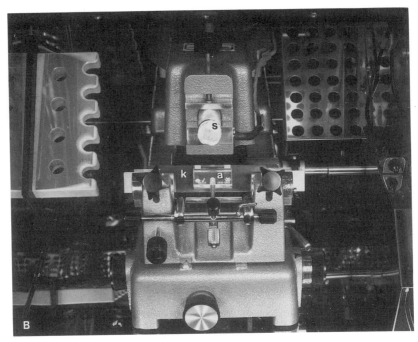

Figure 4.3. Continued.

operate from within a cold box. Further details of the cryomicrotomes and the procedures used in cutting thicker frozen sections for light microscopy are available in the book by Pearse (1980).

4.2.3.2. Cryoultramicrotomes

The basic requirements of these instruments are listed below:

1. There should be good temperature control of the knife and the specimen.
2. The temperature should go down to at least 100 K, i.e., well below T_c for water.
3. The process of refrigeration should be vibration free.
4. The specimen cutting area should be protected from contamination, principally from water vapor.
5. The controls for the ultramicrotome, e.g., cutting speed, knife angle, section thickness, should be readily accessible and operate effectively over the whole range of usable temperatures.
6. It should be possible to quickly and firmly attach specimens to the microtome arm without the sample experiencing undue rises in temperature. The specimen supports must be mechanically rigid.

7. Adequate facilities should be provided for the efficient collection and handling of frozen sections. This usually requires having sufficient space around the sectioning area and the provision of special tools such as insulated forceps, grid holders, and clamps, and even miniature freeze driers.

Cryoultramicrotomes are of two general types: Either the whole ultramicrotome is placed in a cold box (cryostat ultramicrotome) or only the area around the knife and sample are kept at low temperature (cryoultramicrotome).

Cryostat Ultramicrotomes. These were developed from the pioneering work of Bernhard (1965) and later by Appleton (1974, 1978). The whole ultramicrotome, which has to be specially adapted to work at subambient temperatures, is placed at the bottom of a top-loading refrigerator, which is cooled by mechanical compressors supplemented by liquid nitrogen cooling. The Slee Type TUL cryostat ultramicrotome is a good example of this type of instrument. The advantage of this approach to cryomicrotomy is that it provides a very large working space around the sample and the temperatures are very stable. Although the equipment will operate down to 170 K and produce sections between 25 and 500 nm thick, personal experience has shown that it is rather difficult and awkward to use.

The microtome is approximately 50 cm down inside the cold box, and in order to mount the specimen, adjust the ultramicrotome, and collect the frozen sections, it is necessary to stand, bent over at the machine with ones hands encased in rubber gloves inside the cold chamber. Such procedures quickly lead to fatigue, frozen fringers, and frustration! The all important knife/sample interface can only be adequately viewed with long working distance binoculars, and after a while the Perspex cover to the cold box begins to frost over. Because of the large mass of equipment, it takes a long time for everything to equilibrate to a given temperature. Although the knife temperature may be separately controlled, the specimen remains at the same temperature as the cold box. The cryostat approach to cryoultramicrotomy has now been largely superseded by the cryoultramicrotome, in which only the immediate region around the knife and sample are maintained at an appropriate low temperature.

Cryoultramicrotomes. After a rather faulty starting during which there were problems of temperature and mechanical stability, a number of reliable instruments are now commercially available. They all more or less follow the same design and the only real differences center on whether the cooling is continuous or intermittant. Leunissen *et al.* (1979), Saubermann (1980), and Saubermann *et al.* (1981a) have expressed doubts whether thermal stability can be provided by an intermittant flow of coolant. Biddlecombe *et al.* (1982) suggest that this is not the case, for although there are thermal gradients within the small cryochamber, the knife and specimen temperatures remain remarkably constant.

The mechanical parts of the ultramicrotome are kept at ambient temperatures, while the specimen/knife region is enclosed in a small chamber, which can be cooled by either a gentle flow of cold dry nitrogen gas or a continuous trickle of LN^2. Cold nitrogen gas may be obtained most conveniently by boiling off liquid nitrogen from a separate Dewar flask and passing the gas through a heat exchanger to provide some measure of temperature control. In addition to acting as a coolant, the dry nitrogen gas provides a suitable frost-free environment, which essentially eliminates the problem of contamination. A continuous cooling system is generally accepted as being more satisfactory than systems that periodically fill the bottom of the cryo-chamber with liquid nitrogen and introduce wide temperature fluctuations. Depending on the type of instrument, there may be individual temperature controls for the knife, specimen, and cutting environment (Fig. 4.4).

Modern cryoultramicrotomes are convenient to use, and great care has been taken with the ergonomic design of some of the instruments. The operator can carry out the entire sectioning process while sitting comfortably, no small consideration when carrying out an intricate and sometimes painstakingly slow process. The cooling equipment is usually in the form of a kit, which may be fitted to a conventional ultramicrotome. Systems are commercially available from Leica Ltd. (the Reichart–Jung FCS and FC4E Low Temperature Sectioning System) and RMC Inc. (the CR 21 Cryosectioning System).

The RMC Inc. CR 21 may be used with a wide range of ultramicrotomes. The knife and specimen temperature may be controlled independently and the sectioning temperature range is 100–330 K. The Reichert–Jung FCS and FC4E has been specifically designed for use with the Ultracuts and E ultramicrotomes. It is a highly

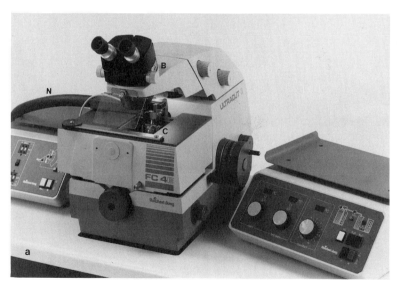

Figure 4.4. Cryoultramicrotome in which only the regions surrounding the knife and specimen are kept at a highly stable low temperature. In the Reichert-Jung FC 4E instrument shown in (a), a cryochamber (C), cooled by liquid nitrogen (N) surrounds the cutting area of an Ultracut ultramicrotome and allows thin sections to be cut at temperatures as low as 100 K. The cutting process is observed through a binocular microscope (B). The cutting process is highly automated and is controlled from outside the cryochamber. (b) Details of the cutting area, with a glass knife (K), sample (S) and a grid retaining device (G). (Pictures courtesy Leica UK plc.)

Figure 4.4. Continued

sophisticated piece of equipment, which will cut sections at temperatures as low as 100 K. There are separate controls for adjusting the temperature of the knife, specimen, and chamber, and the system has been designed to reduce water vapor condensation to very low levels. Both systems use liquid nitrogen/gaseous nitrogen for the coolant. Space will not permit a more detailed description of these and other models of cryoultramicrotomes, and in any event, it is far more satisfactory to obtain updated technical information direct from the manufacturer concerned. Robards and Sleytr (1985) provide a concise description of a number of different cryoultramicrotomes, including details of some of the earlier models, which although now superseded by later instruments, are still in service and providing eminently satisfactory results. Further information can also be found in the papers by Sitte (1984a) and Menco (1986).

Knives for Cryosectioning. The frozen sections may be cut using either glass, metal, sapphire, or diamond knives. Glass knives are more useful and certainly much less expensive than diamond knives. Griffith (1984) considers that the surface properties of glass are superior to diamond in that sections are more easily manipulated and have lower electrostatic charge. Seveus and Tarras–Wahlberg (1986) have shown that glass knives for cryoultramicrotomy should have a real angle of 45° rather than the 50° commonly used for ultramicrotomy. Tokuyasu and Okamura

(1959) have shown that such knives can be made by a symmetrical or balanced breaking process from perfectly flat-sided glass squares. Knives with the shortest score mark have the longest free break and potentially the sharpest edges, which are necessary to cut thin sections from hard frozen blocks. Such knives can now be produced using the commercially available knife makers (Stang 1988). The sharpness of glass knives is a very important factor in determining the quality of the cryosections. Tokuyasu (1980) showed that even with good glass knives, only the middle third of the knife edge would be suitable for cutting the best sections. The theory and practice of making good glass knives for cryosectioning are discussed in detail in the papers of Griffiths (1983, 1984) and Tokuyasu (1980, 1986). The quality of glass knives can be improved further by a thin coating of evaporated tungsten (Roberts, 1975). Special diamond knives with a triangular cross section are reported to give better ultrathin cryosections than glass or sapphire knvies, and although these types of knife are very expensive, more and more use is being made of them for cutting thin and ultrathin frozen sections. A number of manufacturers are now offering diamond knives designed especially for cryosectioning.

Metal knives, most usually in the form of razor blades, are frequently used to cut the thicker frozen hydrated sections that are used for x-ray microanalysis. Saubermann (1980) and Beeuwkes *et al.* (1982) give reasons why metal knives should be used for cutting thick frozen sections:

1. The metal blade has a thermal conductivity $40 \times$ higher than glass knives, which may be an important factor in helping to remove any heat generated at the cutting edge.
2. Razor blades have a small knife edge angle (25°) and are very flat in the region close to the cutting edge, which allows antiroll plates to be fitted more easily.
3. Razor blades have a large cutting edge and are thin enough to minimize the deflection of the frozen section after cutting. Saubermann and Scheid (1985) and Saubermann *et al.* (1986a, b) have used tungsten carbide knives to cut $0.5\text{-}\mu\text{m}$ frozen sections for x-ray microanalysis. Metal knives are not satisfactory for cutting ultrathin frozen sections. Mention has already been made of antiroll plates, which prevent sections curling up as they are cut. Figure 4.5 shows some of these advices.

Although most early work on cryomicrotomy stressed the importance of a temperature differential between the knife and the sample (the knife usually being at a lower temperature), this is now no longer considered so important. Beeuwkes *et al.* (1982) consider that the presumed temperature differential between the knife and the sample is probably illusory, since both the leading edge of the knife and the surface regions of the sample must both be very close to the temperature of the cryochamber, regardless of any control system. In addition, if the knife and specimen are at different temperatures, then contact between them during sectioning must result in the heating of one and the cooling of the other. This will give rise to differential expansion and contraction, which will cause dynamic dimensional changes, which would be inconsistent with rational sectioning.

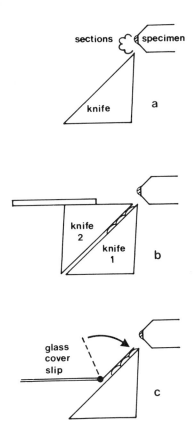

Figure 4.5. Antiroll plate systems for cryosectioning. (a) Dry cut cryosections tend to curl because of forces generated at the knife edge during cutting. The curled ribbons of sections are difficult to pick up and flatten on the specimen support. Various devices have been proposed to keep sections flat during the cutting process. (b) A carefully positioned second glass knife guides the emerging sections into a narrow slit between it and the knife used for cutting. (c) A glass cover-slip attached to a controlled support arm may be swung into position as the cutting process takes place. Any type of antiroll plate needs to be carefully positioned as the extent and direction of section curling show considerable variation. [Modified and redrawn from Robards and Sleytr (1985).]

Much of the success of cryomicrotomy comes from the way the sections are handled after cutting. As we will see, most frozen sections are cut dry, which makes them difficult to handle because they become electrostatically charged in the dry atmosphere in which they have been cut. Fine metal needles, antistatic devices, and conductive coatings on the knives at best only partially alleviate the problem. Other pieces of ancillary equipment include small presses to flatten frozen sections mounted on microscope grids, cutting tools to shape the block face prior to sectioning, transfer devices to convey the frozen hydrated sections to the cold stage of the microscope or freeze dryer, and adequate viewing and illumination system.

4.2.4. Sample Preparation for Cryosectioning

We have already discussed the importance of good cryopreservation as a prerequisite for good cryosectioning, and there are rarely any sectioning problems with small vitrified samples. The advantages and disadvantages of specimen pretreatment, e.g., fixation, cryoprotection, have been discussed in Chapter 3, and it should be clear that the requirements and expectations of the experiment will dictate the degree of chemical intervention that can be tolerated by the specimen. While the general rule of no chemical pretreatment for any samples that are to be used for analysis still applies, there is one possible exception, which involves cryosectioning procedures. Immunocytochemical labeling, using colloidal gold as an electron dense marker on ultrathin cryosections, is an important technique for localizing a wide variety of molecule in biological tissues, which must retain antigenicity and preservation of ultrastructural detail. Mild fixation and cryoprotection are necessary for the adequate preservation of ultrastructure. Such treatment does not appear to affect the antigenicity and allows thin frozen sections to be cut from the frozen blocks.

All samples for cryomicrotomy should be small and mounted in, or on, small specimen holders, which can be cooled as rapidly as possible and in turn easily inserted into the precooled specimen chuck of the cryomicrotome. Some of the best results have been obtained from samples suspended in 100-μm-diam droplets frozen at the end of the specimen holder. Most microtome manufacturers supply their own small specimen holders, but they may also be made from small 1 × 3 mm rivets of a suitable high-conductivity metal. The end of the holder that will come in contact with the specimen should be roughened to ensure good tissue adhesion. The writer has used small silver tubes 1 × 3 mm with an internal diameter of 600 μm as sample holders; they have the advantage that the specimens more easily remain in place. The different types of specimen holder are shown in Fig. 4.6.

Most samples will stick to the specimen holder by virtue of their own moisture content, but a firmer bond can be achieved by using one of the proprietary mounting media such a Tissue Tek. Alternatively, the sample may be held in position at the end of the sample holder using 6% methyl cellulose (Seveus, 1979); 25% PVP or HES (Echlin *et al.*, 1977); poly(vinyl) alcohol (Tokuyasu, 1989); 1% cross-linked bovine serum albumin, 20% gelatin, or concentrated solutions of dextran. Richter *et al.* (1991) recommend 40% ethanol in isopropanol. At 135 K it is sufficiently viscous to embed a precooled specimen; at 115 K it is sufficiently hard to be sectioned. Steinbrecht and Zierold (1984) have used *n*-heptane in the same way. All these materials should be used sparingly, and it is important to make sure they have no effect on the specimen. The embedding agents, and in particular PVP and HES, have the added advantage that as well as supporting the specimen, they aid with the sectioning process (Echlin *et al.* 1977; Tokuyasu, 1989). Prefrozen samples may be also be held in position in little wells at the end of the specimen chuck using a small drop of one of the "cryoglues" (Table 4.1) Robards and Sleytr (1985) and Zierold (1987) discuss the merits and disadvantages of the different "cryoglues."

Small pieces of rapidly cooled tissue may also be attached direct to the precooled microtome chuck. This is a rather cumbersome process and involves placing the

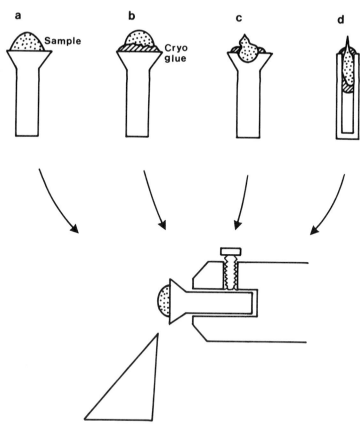

Figure 4.6. Different ways of mounting material prior to cryosectioning. (a) A suspension droplet is placed on the end of a roughened support and then quench cooled. (b) Pre-quench-cooled material is fixed to the support with a suitable "cryoglue." (c) Pre-quench-cooled samples are partially embedded in a depression in the end of the specimen support. (d) The specimen is inserted into a hollow specimen support, which is then quench cooled. The cold, small sample holders are placed into the microtome arm at low temperatures.

Table 4.1. Some of the Substances Used Either as "Cryoglues" or as Trough Fluids During Low-Temperature Sectioning

Liquid	Boiling point (K)	Melting point (K)
Toluene	384	178
N-Butyl benzene	456	185
Isopentane	301	113
N-Heptane	372	183
Dimethyl sulfoxide	465	292
40% Ethanol in isopropanol		115

frozen tissue into a hole at the end of the chuck and holding it in position with two or more opposing grub screws. The inherent brittleness and cooling-induced stress of frozens amples makes this a most delicate operation. Robards and Sleytr (1985) recommend using a thin film of the soft metal indium to fill the space around the frozen specimen and the microtome chuck. Such a procedure would provide additional mechanical support and ensure good thermal contact.

Sample orientation in relation to the plane of sectioning is important to ensure that the regions of the specimen that have the best cryopreservation are the regions that are going to be used for cryosectioning. Langanger and de May (1988) have cryosectioned cell monolayers parallel to their plane of growth by placing them on a thin gelatine substrate prior to rapid cooling. A similar process is described by Tvedt *et al.* (1988), in which cell monolayers grown on a Formvar film are coated with a thin layer of a PVP solution immediately before quench cooling.

4.2.5. Procedures for Cutting Cryosections

The description that follows is only a general account of the methods used to produce cryosections for electron beam instruments, and a number of practical details may well be omitted. There are substantial differences in the way sections are cut and handled in the different cryomicrotomes, and readers are urged to consult the appropriate instrument manual for a specific operational procedure. A good general account of the practical procedures of cryosectioning may be found in the book by Robards and Sleytr (1985), and in the papers of Christensen and Komorowski (1985) and Biddlecombe (1982).

4.2.5.1. Section Mounting and Trimming

The frozen sample should be mounted into the precooled microtome chuck without allowing it to warm up and must be handled only with precooled tools. It should be allowed to come to thermal equilibrium with the precooled microtome before any further manipulations are attempted. Depending on the type of microtome, this may take as long as an hour. It is now generally accepted that better sections, and in particular, serial sections, will only be obtained from sample blocks that have been trimmed, i.e., the opposite edges of the block are roughly parallel to each other. Most workers follow the now standard procedure for the ultramicrotomy of plastic embedded material and ensure that the leading and trailing edges of the sample block are parallel to the knife edge. The sides of the block are then trimmed to form either a square or a trapezoid with the leading edge slightly wider than the trailing edge. The size of the block is important and the cutting width should be not more than 250 μm in order to obtain thin, frozen hydrated, dry cut sections. The side of the truncated pyramid that forms the now shaped block face should slope away at about 30°–40°. If larger cutting forces are to be used, e.g., 0.5–1.0 mm, the sides of the block should have a much more gentle slope of about 10° (Ornberg, 1986). A variety of materials and methods have been used to trim the frozen blocks. They may be trimmed with metal blades or microcircular saws under liquid nitrogen before being transferred to the microtome, trimmed with precooled razor blades or files while fitted to the cold microtome chuck, or trimmed with special tools provided by the microtome manufacturer. The blocks may also be trimmed using the knife and knife holder used for cutting sections. A fresh knife must be used for cutting the actual frozen sections.

4.2.5.2. Sample Sectioning

For most work involving thin sections, temperatures in the region of the knife, and the specimen should be of the order of 140–190 K. The knife rake angle should be 40–45° with a 2–10° clearance angle (Fig. 4.6). The cutting speed should be as slow as possible, usually 1–2 mm s^{-1}, although the reported range is between 0.2 and 10 mm s^{-1}. High cutting speeds result in sample deformation and chatter lines. Leunissen *et al.* (1984) have shown that fractures appear in specimen blocks (and hence it is believed, in the derived sections) cut at 9 mm s^{-1} and below 190 K. At slow speeds (0.1 mm s^{-1}) fractures were only seen at below 150 K. Sitte (1987) considers that shorter cycling times also improve section quality. The cutting procedures are critical for each sample and it will most certainly be necessary to try a number of combinations before optimal cutting conditions are achieved. Homogeneous samples and well-frozen samples can usually be cut at temperatures below 140 K, while inhomogeneous, and less-well-frozen samples will need higher cutting temperatures. Thus Dubochet and McDowall (1984) were able to cut 100–150-nm sections from vitrified, cryoprotected insect flight muscle at 113 K, whereas the author has had to use temperatures as high as 230 K to cut 1.0-μm sections from frozen plant material. The sectioning temperature is equally important in analytical studies. Cameron *et al.* (1990) found that sectioning at 173 K prevented ion distribution in tissue, whereas sectioning at 233 K caused limited diffusion of Na^{+} and Cl^{-} but had no effect on K^{+}.

Frozen sections are nearly always cut "dry," i.e., they are not cut and floated off onto a trough liquid as is the normal practice with ambient temperature plastic-embedded microtomy. Trough liquids, if used, have to be chosen with care, and have included 50%–60% dimethyl sulfoxide (Bernhard, 1971); 40% glycerol containing 60% ethanol (Doty *et al.*, 1974); or cyclohexane at 153 K (Hodson and Marshall, 1970). Although the liquids appear to result in easier sectioning and help unfold crumpled sections, they are probably more trouble than they are worth. One should be concerned about what substances may be leaking out of the sections as they float on the trough liquid.

The block should be first faced by carefully taking a few thicker (200–500-nm) sections. These sections will probably have a frosty appearance and should be discarded. Further sections are cut at decreased section thickness settings until the block face appears shiny and transparent. An unpolished, cloudy block face is indicative of poorly frozen tissue, from which it will be impossible to obtain good frozen sections. The knife should be kept free of accumulated debris during these preliminary operations and the sample moved to a new region of the edge before cutting the thin sections that are required. The section thickness setting should be turned down to the point at which consistent sections are produced at each sectioning pass. Good sections have a clear, near-plasticlike appearance and show the usual range of interference colors in reflected light. It is difficult to cut frozen sections to predetermined thickness, but a setting of 100–150 nm will usually produce sections that have yellow green interference colors, and sections thicker than 250 nm tend to curl as they are sectioned. It is usually not necessary to use an antiroll plate with sections thinner than 200–300 nm; thicker sections will need the use of such devices if curling is to be prevented.

4.2.5.3. Section Collection.

Flat sections should be removed from the knife edge and transferred either to the side of the knife or directly to the microscope sample support, which has been previously coated with a suitable support film such as carbon-coated Formvar, Piloform, or nylon. Although TEM grids are the most popular supports for frozen sections, small metal collars with a metal-coated plastic film can also be used, provided they fit on the microscope stage. This type of specimen holder was developed by Saubermann and Echlin (1975) for use with thick (0.5–1.0 μm) frozen sections. The aluminized nylon films that were used remained flexible at low temperatures.

The best tool for moving frozen hydrated sections around is an eyelash glued to the end of a wooden applicator stick. It is for this reason that cryomicroscopists may be recognized by their balding eyelids. Once periocular alopecia has set in, hand-drawn 10-μm glass fibers (Ornberg, 1986), Dalmation dog hair (Barnard, 1982), or very thin metal needles may be used. It would appear that the coat hair of Polar Bears might also be useful as they are hollow and flexible (but first catch your Polar Bear). Thicker sections cut on metal knives may be moved in a similar fashion or with finely pointed badger or squirrel hair paint brushes. These are delicate procedures, which require a steady hand and the patience of Job, because the inevitable build up of static in the dry environment causes the sections to fly around the chamber the moment they are touched. Antistatic devices, such as the device made by Simco Corp[n], Hatfield, Pennsylvania, help prevent this aerial mobility.

The specimen support is patiently and progressively covered with frozen sections (although in practice two or three well-positioned good sections are considered enough!) and then gently moved to another part of the cryochamber to await further manipulations prior to sample transfer. Frozen sections that are to be used for immunocytochemistry can be picked up from the knife by means of a drop of a saturated sucrose solution and transmitted to a film-coated grid. Appleton (1973) devised an ingenious vacuum collection technique, which gently pulls the ribbon of frozen sections away from the knife edge and then gently deposits it on an underlying electron microscope grid. Depending on the microtome, it is sometimes possible to park, temporarily, the section-loaded grid in a container kept at a lower temperature than the sectioning temperature. Frozen sections are very sensitive to environmental conditions. Dry sections are damaged when exposed to moisture and frozen hydrated sections must be kept at below 173 K to avoid sublimation.

Post-sectioning manipulations will vary according to what is required from the frozen section and what equipment is available in the cryochamber. In some instances it has been found useful to flatten frozen sections either directly by means of a small weight with a highly polished surface in contact with the frozen section, or indirectly by using a folded grid or by placing a second coated grid over the first grid containing the frozen sections and pressing the two together with a polished metal rod (Biddle-combe, 1982). In practice, the frozen sections finish up by being distributed on both the upper and lower grids, so very little material is lost. Hagler and Buja (1984) have found that frozen sections may be secured by covering them with a Formvar film held in a metal ring. For those experimentalists who are particularly dexterous, cryoglues may be used to stick down the edges of frozen sections to the supporting grid. It must be remembered that all these procedures have to be carried out at low temperatures and that the samples must neither warm up or become contaminated with ice crystals. Figure 4.7 shows some of the methods that can be used to attach frozen sections to specimen supports. Frozen-hydrated cryosections should be trans-ferred to the cold stage of the microscope using a small portable cold chamber, which interfaces with both the cryomicrotome and the microscope.

Many of the problems associated with poor image contrast, radiation damage, and the transfer of frozen-hydrated sections can be avoided by using freeze-dried sections. They may be dried on a temperature-controlled platform in the microtome cold chamber and/or by exposing them to a gentle stream of dry nitrogen gas. Biddlecombe et al. (1982) dried 0.2–1.0-μm sections in the cryomicrotome at atmos-pheric pressure in the following manner. Sections on TEM grids were held for 3 h at 193 K, 30 m in at 213 K, and then transferred to a small desiccator containing Linde Co. 4 A moelcular sieve kept at 213 K. The temperature was allowed to rise to 293 K over a 1-h period and then finally heated to 313 K for 10 min. Frederik (1982) used a similar procedure, in which sections were kept in the cryochamber for 30 min at 183 K, after which the temperature was slowly raised to 293 K. Dried sections are very hygroscopic and must be maintained under anhydrous conditions. Sections may also be freeze-dried on the cold stage of the microscope, which, after the initial low-temperature transfer, is allowed to warm up in a controlled fashion. The later procedure has the advantage that the freeze-dried section is never exposed to the atmosphere. The electron optical contrast of freeze-dried sections is much higher than their frozen-hydrated counterparts, and they are also much more resistant to beam-induced mass loss. Frederik and Busing (1981) showed that the freeze-dried cryosection can be stabilized with osmium tetroxide vapor. The fixation procedure will, however, influence both the ultrastructural appearance and the distribution of elements. Accidental or premature drying out of frozen-hydrated sections inside the cold chamber of the microtome has been a matter of some concern, particularly where such sections are to be used for analytical studies.

If the frozen sections are only to be used for light microscopy or for morpholog-ical studies, then a certain amount of staining is permissible. Robards and Sleytr (1985), Griffiths (1984), Tokuyasu (1986), and Sjostrom et al. (1991) give details of how frozen sections on TEM grids, which should not be allowed to dry out, may be floated briefly on solutions of positive and negative stains. The staining times are

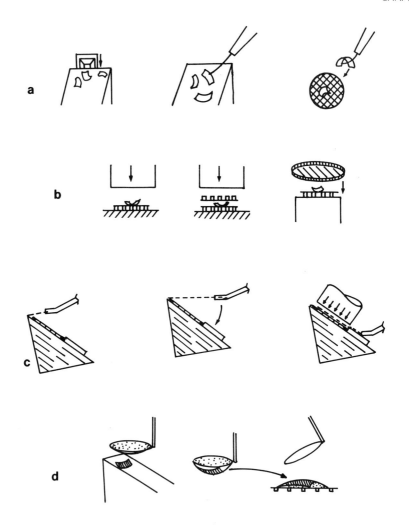

Figure 4.7. Different ways to pick up and mount cryosections. (a) The frozen sections are picked up with an eyelash probe and placed on a TEM grid. (b) The sections on the TEM grid may be flatted directly with a cold polished rod, flattened between a second TEM grid, or covered by a cold plastic film. (c) The sections may be guided by vacuum tweezers (Appleton, 1974) over a TEM grid attached to the shoulder of the knife and pressed flat with a cold polished metal rod. (d) Frozen sections may be picked up in a droplet of concentrated sugar solution and deposited onto a TEM grid. This method is suitable for frozen sections with are to be used for histochemistry or immunocytochemistry. [Modified and redrawn from Robards and Sleytr (1985).]

usually very short. In the procedures devised by Sjostrom and colleagues (1991), sections are cut dry from fixed and cryoprotected material, picked in a sucrose droplet, briefly rinsed, and then negatively stained with 2% ammonium molybdate.

4.2.6. Section Transfer

This is an important part of the procedure, and unless due care is taken, a well-prepared specimen can all too easily be ruined during the transfer to the microscope. The exact procedures that must be followed are dictated by the exigencies of the instrumentation, but the following general principles apply. Freeze-dried sections may be maintained in an anhydrous state by enclosing them between two carbon or plastic films and keeping them in a dry, although not necessarily cold, environment.

Frozen-hydrated samples should be maintained at below 130 K in a dry environment during transfer in order to prevent either water loss or gain by the sample. Such constraints require special transfer devices, examples of which may be seen in the papers by Saubermann (1981a, 1981b), Hax and Lichtenegger (1982), Sitte (1982), Biddlecombe (1982), Perlov *et al.* (1983), and Tatlock (1984). Speed is of the essence in the transfer of frozen-hydrated sections. Although Zierold (1987) claims that cryosections mounted on grids can be kept in a frozen-hydrated state for several weeks by storing them in capsules under liquid nitrogen, it would seem more prudent to use frozen sections as soon as possible after they have been cut. In addition to the various homemade transfer devices, many of the microscope manufacturers have their own purpose-built transfer devices. Oxford Instruments and Gatan Inc. make transfer devices that can be adapted for a number of different cryomicrotomes and microscopes.

4.2.7. Frozen Sections for Light Microscopy

The procedures for the preparation of the thicker (1–5 μm) frozen sections for light microscopy, i.e., for immunocytochemistry, autoradiography and cytochemistry, are essentially the same as those described for electron microscopy. The same type of equipment is used and it is probably best to first cut any thin (50–100 nm) sections before cutting the thicker sections. The thicker sections (1–5 μm) can be cut from larger block faces than are used for thin sections and are more easily produced using thin flat metal knives (razor blades) rather than the rather high-angle glass knives. Sections in the 0.5–1.0-μm range are usually cut dry at 223 K, while the 1–5-μm sections will require a somewhat higher temperature (243 K). Sections are picked up either by using a fine hair or a saturated sucrose droplet, and transferred to a precleaned glass microscope slide, which has been coated with a thin layer of 0.01% polylysine to help bind the frozen sections to the surface. Tvedt *et al.* (1989) describe a technique for collecting large cryosections needed for light microscopy survey studies of biopsy material. The frozen sections, cut at 140 K, are placed on a glass cover-slip covered with a thin layer of frozen 100% ethanol sitting on a cold stage maintained at the same temperature. When sufficient sections are collected, the glass cover-slip is gently heated until the alcohol just melts and the sections stretch in the viscous fluid. Further heating will evaporate the ethanol. The procedure has the advantage that it both flattens and fixes the sections prior to staining. Details of the cryosectioning techniques for light microscopy may be found in the papers of Christensen (1971), Christensen and Komorowski (1985), Reid (1974), and Tokuyasu (1986).

4.3. ARTIFACTS AND PROBLEMS WITH CRYOSECTIONING PROCEDURES

In addition to the general problems associated with the lack of image contrast and radiation damage in frozen-hydrated sections, there are a number of specific image perturbations that are recognized as artifacts associated with specimen preparation and, in particular, with the process of cryosectioning. Unlike plastic sections, the cutting distortions cannot be easily relaxed after sectioning and they frequently impair the electron microscope observation of thin cryosections.

For high-resolution ultrastructural studies, it is important to have smooth, thin cryosections. While smooth section are less important for analytical studies, the absence of any defects makes it much easier to identify regions of interest. Although the question of section melting during the cutting process appears to have been satisfactorily resolved, a number of other problems still remain. For work with freeze-dried sections, the major problem appears to be rehydration, whereas sectioning induced marks and defects appear to be the principal concern with frozen-hydrated sections.

4.3.1. Ice-Crystal Damage

Before discussing the artifacts that may be specifically related to the sectioning process, it may be useful to first consider the general problem of ice-crystal damage. This is very difficult to observe in fully frozen-hydrated sections but appears as a series of voids or "ice-crystal ghosts" in freeze-dried and/or frozen-substituted sections. The size of these voids generally increases with the distance from the initial freezing point at the sample surface. Frederik and Busing (1980, 1981) have shown that a large proportion of the ice-crystal damage is related to the initial cooling of the specimen and is not a consequence of ice-crystal growth in either the frozen tissue block or the frozen section. Many of the morphological effects of ice-crystal damage that could be observed in freeze-dried sections disappeared when the sections were rehydrated.

Zierold (1984) showed that the ratio of peak to continuum x-ray counts in freeze-dried sections was reduced by a factor of 2 as the mean ice-crystal size increased from 50 to 1000 nm. This could seriously affect the accuracy of x-ray microanalysis of freeze-dried sections. Hall (1986a) discusses this matter in some detail and concludes that accurate correction procedures may be difficult for ultrathin sections. See Chapter 11 (Section 11.7.2).

The artifacts associated with cryosectioning have received a lot of attention. Frederik et al. (1984) consider that the phenomena whereby material builds up at the boundary between the knife and the section can help explain some of the defects we find in cryosections. The characteristics of this buildup edge vary, i.e., periodically material is discharged towards the section and the block face, which leads to the production of deformations. This instability at the knife edge would, for example, explain the variation in section thickness that is sometimes found during cryosectioning. Many of the defects, other than compression and the artifacts introduced by imperfections at the knife edge, tend to disappear in sections cut below 133–143 K and are generally absent in vitrified material. The latter is not always the case, for Dubochet and McDowall (1984) found that not all cryoprotectants made samples less prone to cutting artifacts; e.g., polyethylene glycol seemed to increase the number of deep fractures going into the section. The sectioning defects and artifacts can be divided into a number of categories. These will be considered in turn and wherever possible an explanation given for their cause and a remedy offered for their alleviation.

4.3.2. Fold and Wrinkles

Folds and wrinkles are a well-known phenomenon of all sectioning procedures, indeed it is a common problem associated with trying to get a thin laminate to lie flat on a solid surface. (Who has ever been able to lay a beach towel on the sand without folds and wrinkles?) In sectioning these are serious defects and are probably the cause of most discarded sections. It is difficult to ascribe any single sectioning procedure that gives rise to this default, although a dirty knife edge and faulty section handling procedures often produce this artifact.

4.3.3. Crevasses and Furrows

A number of workers have reported a series of lines or furrows in both thick and thin sections running parallel to the knife edge, even when this is not perpendicular to the cutting direction (Fig. 4.8). It is thought that these furrows, or crevasses as our Swiss and French colleagues would like to call them, represent fractures which go deep into the section. They are found in nearly all cryosections, particularly if they are thick, although they are less frequent in vitrified material. Crevasses only appear on the block side of the cryosection and are absent from the block face. This defect is most probably caused by the frozen section curling up from the high-angle glass edge (Fig. 4.9). Saubermann and Echlin (1975) showed that thick sections invariably contain a series of continuous deep fractures. Later studies by Saubermann et al. (1981a) showed that they disappeared at higher sectioning temperatures. Crevasses can be minimized by using flatter metal knives, by cutting at higher temperatures, e.g., 230 K, or by cutting thin sections from adequately quench-cooled material.

4.3.4. Graininess

Frozen sections sometimes have a grainy appearance consisting of 30–100-nm particles uniformly distributed across the sample. This effect has been seen in both chemically fixed (Leunissen et al., 1984) and unfixed samples (Chang et al., 1983), and appears either to be associated with higher cutting speeds and higher cutting temperatures or to represent defects remaining on the block face from which the section was cut (Richter et al., 1991). The cause of this artifact is unclear. It may be a manifestation of the release of stress induced while cutting the previous section. It may represent transient surface melting during cutting followed by recrystallization. During the transient melting, salts dissolve and form small concentrates, which

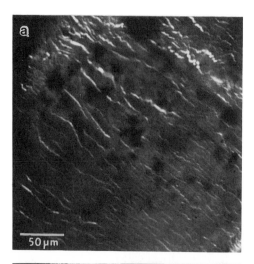

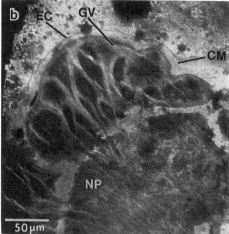

Figure 4.8.(a) Scanning transmission image of an unfixed unstained frozen-hydrated 0.5-μm section of leech ganglion. Note the virtual absence of any morphological information in the frozen-hydrated material and the series of furrows or crevasses running across the section from bottom left to top right. (b) The same section freeze-dried showing an increase of morphological information and the disappearance of the crevasses. EC, epithelial cells; GV, vacuolated glial zone; CM, capsular matrix; and NP, neuropil. [Picture courtesy Saubermann and Scheid (1985).]

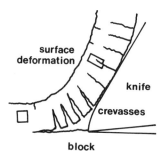

Figure 4.9. An explanation for the formation of crevasses in frozen sections. As sections are cut from partially ductile material they curl up from the high-angle glass knife edge. Although there must be some compression in the region of the section away from the knife edge, there is minimal visible deformation in this region. As the sections curl up over the knife edge, the stress that is induced in the cutting region causes small cracks to appear in the region of the section close to the knife. These cracks partially close as the section becomes flattened on the shoulder of the knife. [Modified and redrawn from Dubochet and McDowall (1987).]

appear as electron-dense deposits. Frederik (1982) has found that rehydrated freeze-dried sections have a distinctly granular appearance.

4.3.5. Chatter

The term "chatter," which was taken from machining tools, is used to describe a high-frequency vibration set up between the knife and the specimen during the cutting process. It is a feature of all ultramicrotomy and appears as a series of density variations in the section along the cutting direction (Fig. 4.10). The variations, which can have a periodicity between 100 nm and 5 μm, can be regular or irregular and are thought to be due to either mechanical vibration of the microtome or repeated stress and relaxation of the specimen. Frederik and Busing (1981) demonstrated a

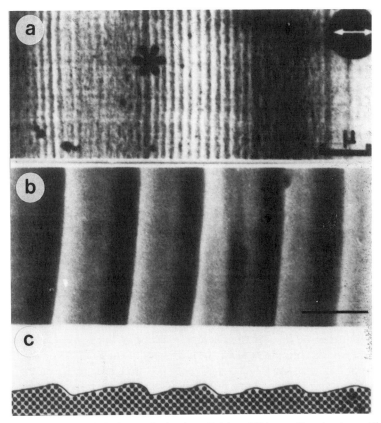

Figure 4.10. Chatter produced during sectioning in a Reichert FC4 cryoultramicrotome. The cutting direction is horizontal and the chatter is present as two vibration modes (a) and (b) with wavelengths of approximately 0.2 and 4.0 μm. Representation at the same scale of thickness of the section shown in (b), is shown in (c). The frozen-hydrated sections are 120 nm thick and cut at 125 K at a sectioning speed of 6 mm s^{-1}. Bar Marker = 1.0 μm. [Picture courtesy Chang et al. (1983).]

500-nm chatter artifact in both the section and the block face from which the frozen sections were cut. They considered that the chatter was due to the material property of the sample and that the relative amounts of vitrification versus crystallization in the specimen influenced the amount of chatter. At high cutting temperatures, the chatter has a saw tooth appearance; at low cutting temperatures, it is in the form of a sine wave. The amplitude of vibration in a given specimen can usually be reduced by decreasing the size of the area being cut, decreasing the section thickness, and further slowing the already diminished cutting speeds. The effect of chatter can be variable. In favorable cases it is virtually nonexistant, whereas in unfavorable cases the effect may be so marked that the frozen sections separate out as a series of parallel slices.

4.3.6. Compression

As is the case in resin sections, the solid matrix of the frozen sample can be compressed during sectioning (Fig. 4.11). Chang *et al.* (1983) have shown that most cryosections are compressed by a factor of 2 along the cutting direction. The primary cause is believed to be friction at the knife edge. It is generally more evident in thin sections, although Zierold (1986a) always finds compression lines in 70–100-nm sections cut at 173 K from well-frozen material. It may be related to sectioning speed and temperature. Ornberg (1986) found that compression could be virtually eliminated in thin sections cut at 110 K by reducing the cutting speed to 0.2 mm s^{-1} and the knife angle to 2°. Ornberg considers that static charge is one of the main causes of compression as it favors sections sticking to the knife edge during cutting. This is corroborated in recent studies by Richter *et al.* (1991), who found considerably less compression when an ion-spray antistatic device was mounted near to the knife edge. Because compression is a process of redistribution of matter, i.e., there is a shortening in the cutting direction without a significant widening, the process will result in a thickening of the sectioned material. The presence of compression can increase section thickness by between 20% and 50%. In less well preserved material with ice crystals larger than 200 nm, no compression is found, although other artifacts will intervene.

Compression is by no means uniform within a given section. Ornberg (1986), working with frozen chromaffin cells, found that the nuclei were unaffected, whereas the cytoplasm showed compression artifacts. Work by Jesior (1985) on plastic sections has shown that compression is a local event only and depends on the mechanical properties of the sectioned object and not on the surrounding matrix. In resin-embedded material, compression diminishes as the resin hardness increases. In a later paper, Jesior (1986) showed that compression is markedly reduced in resin sections that are cut somewhat thicker (50–90 nm) than the usual ultrathin sections, with a low-angle (30°) diamond knife set with the minimum clearance angle. Jesior concludes from these studies that lower sectioning temperatures for frozen samples will increase their hardness, which should in turn decrease section deformation and compression during cryomicrotomy. As we have already discussed, the ice matrix is

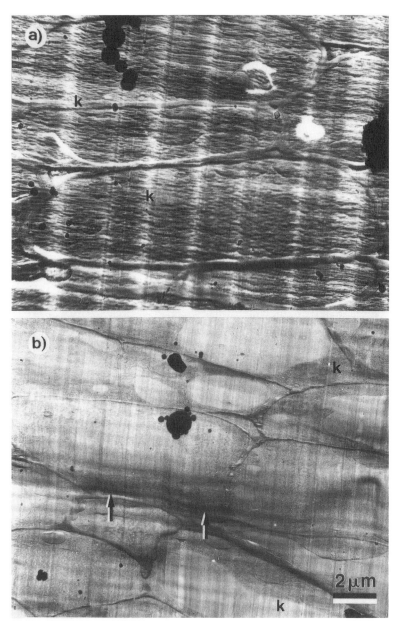

Figure 4.11. Effects of cryosection thickness. (a) Frozen-hydrated section with crevasses running across the picture together with knife marks (k) limit the structural analysis. (b) Only thin cryosections show a minimum of artifacts and permit structural analysis, but folds due to compression (arrows) and knife marks (k) may still be seen. [Picture courtesy Michel *et al.* (1991).]

more ductile at higher temperatures and would be more likely to exhibit plastic flow and deformation. This is contrary to the recent findings of Richter *et al.* (1991).

4.3.7. Block and Section Noncomplementarity

Chang *et al.* (1983) and Leunissen *et al.* (1984 have made carbon-metal replicas of both the frozen specimen block and the surface of the most recently cut section. At high cutting speeds and at temperatures below 200 K, the block face was made up of a series of crevasses, which were absent from the corresponding frozen sections. A gentle and continuous rippled appearance was seen on both sides of the sections independant of temperature. There is no complementarity between the morphology of the block face and the most recently cut frozen section, and it is obvious that some changes are being brought about in the section and/or the block by the cutting process. The changes are probably related to plastic deformation and multiple slippage (Saubermann *et al.*, 1981; Falls *et al.*, 1983). Frederik (1982) found that sections cut at 193 K showed more or less complementary surfaces, whereas sections cut at 168 K have a lack of complementarity, with one surface being smooth and the other with a series of saw tooth marks running parallel to the knife edge.

4.3.8. Knife Marks

This is another characteristic feature of all sectioning, and appears as a series of irregular marks running in the direction of the cutting (Fig. 4.12). The marks are usually only on the side of the section facing the block from which it was cut and are more usually associated with metal and glass knives. Knife marks are due to

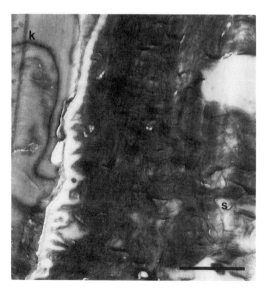

Figure 4.12. Knife marks (k) and surface undulations (s) in thin frozen sections cut at 133 K and then freeze-dried in the microscope column. Bar marker = 1.0 μm. [Picture courtesy Frederik *et al.* (1984).]

defects and dirt on the knife edge, and for this reason they should be changed frequently and their edges kept clean at all times.

4.3.9. Ripples and Bands

Series of bands and ripples have been observed in frozen sections by a number of different workers (Chang *et al.*, 1983); Karp *et al.*, 1982; Frederik *et al.*, 1982; Zierold, 1984, 1987). These bands, which run parallel to the knife edge in frozen sections, have a much finer periodicity than chatter marks and range from 10 to 100 nm (Fig. 4.13). Frederik *et al.* (1982, 1984) found that fine distortions occurred at sectioning temperatures between 123 and 193 K, but were generally absent from thinner sections. Chang *et al.* (1983) found that the bands were electron beam sensitive and consider them to be a cutting-induced devitrification artifact. Both Hodson and Williams (1976) and, later, Frederik (1984) consider that shear forces are the main cause of this deformation. Occasionally, shear forces at the specimen–knife interface would give rise to ripples whose frequency could be directly related to the cutting speed.

Leunissen *et al.* (1984) showed that ripples were present on both sides of sections cut from fixed and sucrose-cryoprotected material. These ripples were found at all sectioning temperatures (243–103 K) and at cutting speeds between 0.1 and 90 mm s^{-1}. They consider that the ripple structures seen at low magnifications most probably originate from plastic deformation at the frozen block–knife interface. They also raise the question whether transient melting might explain the high plasticity of the section at the cutting edge (Fig. 4.14).

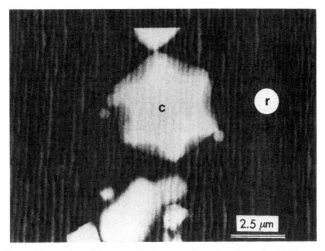

Figure 4.13. Ripples and bands in a 0.2-μm-thick frozen-hydrated section of rat kidney cut at 168 K. The ripples (r), which have a finer periodicity than chatter, run horizontally across the section; c, hexagonal ice crystal. [Picture courtesy Frederik *et al.* (1984).]

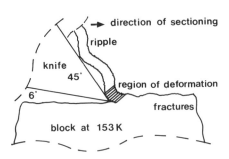

Figure 4.14. Explanations for the production of ripples in frozen sections. The rippled structure most probably originates from plastic deformation of the frozen material at or near the knife edge. An alternative explanation offered by Leunissen *et al.* (1984) is that at slow cutting speeds and at 153 K, for a fraction of the second, a small region of the section at the edge of the knife (indicated by the arrow) would melt. This slush would be deformed by friction and quickly refrozen to give rise to the undulating rippled section. The block face would remain frozen and unaltered during the whole process. [Modified and redrawn from Leunissen *et al.* (1984).]

Although the collected wisdom of cryosectioning procedures all points to using low cutting speeds as a means of diminishing sectioning artifacts, there are occasions where an increase in cutting speed increased the quality of the cryosections (Chang, 1983); Frederik *et al.*, 1984). This improvement is related to the factor that the build up of material at the knife edge is carried away at the higher cutting speeds, leaving an edge that favors the production of smooth sections. There is still some uncertainty whether or not a very thin melt zone is necessary for the production of smooth sections at these higher cutting speeds. There is no evidence to suggest that a superficial, transient melt zone does in fact form, and we do not know enough about the rheological properties of ice to predict whether it would undergo plastic deformation at the cutting edge.

4.3.10. Melting and Rehydration

Although the question of through-section melting during sectioning is more or less satisfactorily resolved, there are still questions about the melting of frozen-hydrated material and the rehydration of freeze-dried sections *after* sectioning has taken place. These matters have been investigated by Frederik and Busing (1981) and Frederik (1982), who catalogue a series of changes that could be observed in melted and rehydrated frozen sections. The structural changes brought about by melting involve a substantial rearrangement of cellular material, starting with the membranes and leading to the production of granules, globules, and eventually to a smearing of the subcellular structures (Fig. 4.15). Frederik and Busing (1981) find that rehydration also results in the rearrangement of membranes and the appearance of granules in mitochondria and other organelles (Fig. 4.16). Rehydration in a moist atmosphere obscures any ice-crystal damage by smoothing out the sharp outlines of ice-crystal ghosts. Rehydration can be a serious problem, for it will allow lipids to melt, and give rise to *in situ* redistribution of elements.

4.3.11. Accidental Section Drying after Cutting

There are contrasting views concerning this problem. Frederik (1982) considers that at temperatures between 173 and 193 K there is a real risk that frozen-hydrated

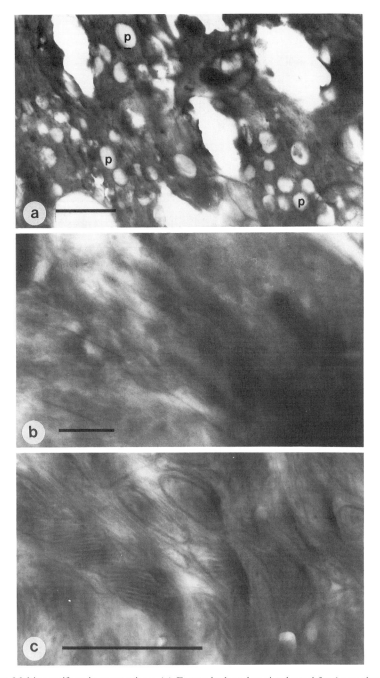

Figure 4.15. Melting artifacts in cryosections. (a) Frozen-hydrated section heated for 1 s to about 293 K. Circular profiles (p) are seen overlaying the section indicating melting artifacts. (b) Frozen-hydrated section heated for 1 min to about 293 K. The ultrastructure is severely disrupted by transient melting. (c) Control frozen-hydrated section not heated in which the ultrastructure is preserved. Bar marker: (a) and (b) = 1.0 μm; (c) = 0.5 μm. [Pictures courtesy Frederik *et al.* (1981).]

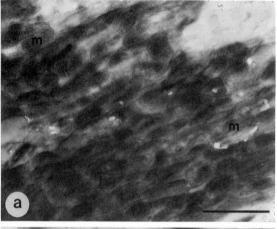

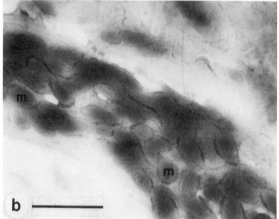

Figure 4.16. Rehydration artifacts in cryosections. (a) Control preparation. Transmission image of a freeze-dried cryosection cut at 193 K and fixed in osmium tetroxide vapor showing indistinct mitochondria (m). (b) Dry, unfixed sections exposed to a moist atmosphere for 5 min which resulted in a thickening of the outlines of the mitochondria. Bar marker = 1.0 μm. [Pictures courtesy Frederik (1982).]

sections may freeze-dry in the cold dry atmosphere in the cryochamber of the microtome. At 193 K, ice sublimes from sections at a rate of 22.5 nm min^{-1}; at 183 K, 3.3 nm min^{-1}; and at 173 K, only 0.6 nm min^{-1}. Frederik advises keeping freshly cut frozen-hydrated sections at 153 K where the drying rate is only 6×10^{-3} nm min^{-1}. This contrasts sharply with the finding of Chang *et al.* (1983), who were unable to measure any ice sublimation at 173 K from a layer of pure I_v but found that at 183 K the rate was 60 nm min^{-1}. The corresponding rate from frozen *sections* would be much lower. Chang and colleagues advise keeping frozen sections at 153 K to avoid the risk of dehydration.

The ice sublimation rate during and subsequent to cryosectioning depends primarily on the partial pressure of the phase boundary, gas diffusion, and convection. Zingsheim (1984) measured the sublimation rates of ice from sections kept in a cold nitrogen-cooled cryomicrotome. He found them to be 4–5 orders of magnitude lower than the rates measured during vacuum sublimation at the same temperature. This difference is a consequence of a much shorter mean free path length of water molecules in cold nitrogen gas saturated with water vapor and sections would lose water very slowly. Zingshein calculated that a 1.0-μm frozen section would lose water at the rate of 8 nm min^{-1} at 233 K, which is much slower than the rates measured by Frederik and Chang.

The available evidence suggests that it would be prudent to adopt the following course of action. Thin (50–100-nm) sections should only be kept temporarily in a cold nitrogen atmosphere at below 130 K and thicker (0.5–1.0-μm) sections below 160 K to avoid any significant water loss. If the sections are, additionally, to be stored under vacuum, the temperature should be below 120 K. Von Zylinicki and Zierold (1989) found that freeze-dried sections of isolated rat liver cells had poorer contrast and showed significant changes in elemental concentrations after exposure to room air for 2 min. Storage under vacuum for up to 12 h does not lead to such alterations, although storage for any longer periods leads to changes in elemental distribution and morphology.

4.3.12. Contamination

Contamination by ice crystals is the most common artifact in cryosectioning, and it is difficult to remove this effect entirely. It can be substantially reduced by careful cryohousekeeping and making sure that all low-temperature manipulations are carried out in dry environments. All other things being equal, it might be better not to attempt critical cryosectioning at times of high ambient humidity. Uncovered hands and fingers should not be placed in the microtome cryochambers; masks should be worn if the sectioning procedures involve the face coming into close contact with critical parts of the equipment. One of the insidious problems of ice-crystal contamination is that the pure ice crystals can grow on the surface of the frozen section and may cause incipient dehydration. The problem is less critical for frozen sections that are to be freeze-dried, although the presence of a thick crust of ice will only serve to delay the drying process. If the frozen sections are to be used in the fully-hydrated state, it is most important that they are kept dry and clean. Ice-crystal contaminants will completely obscure any structure detail in frozen sections and will nullify any analytical procedures that serve to express local concentrations on a wet weight basis.

Provided there is minimal energy input into the sectioning process, i.e., low cutting speeds, good heat conduction at the knife–specimen interface, and small ice crystals in the sample, then through-section melting should not be a problem in cryosectioning. There are now sufficient morphological and analytical data from frozen-hydrated sections to show that such sections can be produced from many different samples, some more easily than others, using a variety of sectioning equipment. Finally, many of the images we derive from frozen sections are unusual and sometimes difficult to interpret; correlative microscopy may be the only answer.

4.4. APPLICATIONS OF CRYOMICROTOMY

Much of what we have been discussing has an immediate and wide-ranging application to the biomedical sciences. Cryoultramicrotomy largely eliminates many of the problems associated with sample preparation, and when coupled to immunocytochemistry and histochemistry provides a powerful tool for linking the visualization of structure with molecular identity. This rapidly developing technique can be applied at both the light and electron microscope levels, and most of the cryosections that are cut are used for this purpose. The papers by Tokuyasu (1986), Boonstra *et al.* (1987), Leunissen and Verkleij (1989), van Bergen en Henegouwen (1989), and Boonstra (1991) provide an excellent introduction to these important analytical techniques. We have already discussed how cryoultramicrotomy can be used in connection with other analytical technique, particularly where soluble and labile substances are the target materials. Approximately 10% of frozen sections are used for this purpose. There are important light microscope applications in medicine and surgery where the speed of the technique allows thin sectioned diagnostic biopsy samples to be stained and ready for examination within 5 min of tissue excision.

Cryomicrotomy can be used to produce thin and ultrathin sections of plastics and polymers which are too soft to be cut at ambient temperatures. Polymers that have a T_g below 293 K are soft and can easily be hardened by cooling prior to sectioning, thus avoiding the use of chemical embedding and hardening techniques. The procedures for cutting polymers and plastics are no different from those used to cut frozen sections of biological and hydrated material. The considerably lower water content of most plastics means that cryosectioning these materials is easier than for many biological samples. Emulsions, paints, latex, and adhesives would probably provide interesting exceptions, particularly where the hydrated phase is in a microdroplet form which could be extensively subcooled prior to nucleation.

Polymers have been cryosectioned at temperatures between 253 and 77 K. Rubber has to be sectioned at 77 K, polystyrene at 160 K, polyurethane at 198 K, and polypropylene at 200 K. At higher temperatures, a variety of trough liquids have been used, but at lower temperatures sections must be cut dry. The problems of section collection and static charge remain as the principal difficulties associated with the technique. The book by Sawyer and Grubb (1987) on polymer microscopy should be consulted for further details of the cryosectioning techniques that can be used with these materials. Cryomicrotomy also has important applications in the food industry and in particular with the microscopy of dairy and meat products.

4.5. FUTURE PROSPECTS FOR CRYOSECTIONING

Robards and Sleytr (1985) quite correctly described cryosectioning for electron microscopy as a procedure in which "thin sections are difficult to obtain, difficult to maintain fully hydrated and uncontaminated and difficult to view without problems arising from radiation damage." The quite remarkable advances in both the instrumentation and our understanding of how to handle cryosections has now brought this technique to the status of a more or less routine procedure for preparing thick and thin, hydrated and dried cryosections for both analytical and morphological purposes. In spite of these advances, it is possible to identify a number of areas where improvements are needed, both to the instrumentation and the way frozen sections are handled.

1. The collection of thin, dry sections remains one of the weaknesses of the cryosectioning procedure, primarily because of problems of static buildup. Some cryoultramicrotomes are fitted with antistatic devices in the cryochamber, and they seem to help. Perhaps we should reexamine the technique devised by Hodson and Marshall (1970), who connected the tips of the forceps holding the metal support grid to an electrophorus via a power cable. When the copper grid is in close contact with the frozen sections on the knife edge, the metal plate of the electrophorus is first applied to an electrostatically charged insulator, then earthed, and finally lifted from the insulator. The excess charge develops and charges the sections, so when the plate is replaced on the insulator, the sections are attracted to the grid. The processes of cutting, maneuvering, and picking up frozen sections would be considerably facilitated if the frozen sections could be cut onto a trough liquid at 130 K. In ambient temperature ultramicrotomy, the aqueous trough liquid appears to lubricate the sections sliding off the knife. Could a suitable lubricant be found for frozen sections that will allow sectioning without causing damage to and extraction from the sections? Organic liquids are the most obvious choice, but it would perhaps be useful to explore the properties of rare gases such as krypton, which is liquid between 116 and 121 K and xenon, which is liquid between 161 and 166 K.

2. There could be a further improvement in the quality of knives used for cryomicrotomy. The purpose-designed, diamond knives are very good, but they are also very expensive. Metal-coated glass knives appear to have improved properties, which it is suspected are related more to an improved coefficient of friction than to any increase in intrinsic hardness. Ion implantation procedures are now being used to improve the hardness and cutting quality of machine tools; there seems to be no reason why the same techniques could not be applied to glass knives. We should seek ways of reducing the radius of curvature of frozen sections. It is not know whether this is due to plastic deformation which would be reversible or whether it is due to irreversible plastic flow. We also need to produce sturdy low-angle glass knives which could be used to produce flatter sections. The study by Lickfield (1985) has shown that a knife with a lower rake angle will produce thinner sections.

3. It would be useful to have more reproducible section thickness. Wolf and Schwinde (1983) produced a device that adjusted the advance of the ultracryomicrotome arm, and in later papers Wolf (1987) and Wolf *et al.* (1987) applied this to a cryoultramicrotome. The optical-electronic device using a reflection method is capable of reproducibly cutting 100-nm frozen sections with a precision of 85% at a knife temperature of 193 K. Such sections will be particularly useful for analytical techniques such as EPMA and EELS (Chapter 11, Sections 11.2.2 and 11.5.6), where precise section thickness is important.

4. As we will show in Chapter 9, greater use is being made of cryoelectron microscopy to obtain high-resolution morphological information about hydrated and biological materials. Ultrathin frozen sections below 50 nm cut from vitrified material are going to be necessary in order to achieve the spatial resolution required for these studies. Such sections will also have to be free of distortions and defects

and the long-range order sufficiently well preserved to enable signal/noise enhancement by Fourier-based image processing to be used effectively. Although there are a number of disadvantages to using very thin sections for x-ray microanalysis, such sections are becoming increasingly more important for electron-energy-loss spectroscopy, where multiple scattering can seriously diminish the effectiveness of the method.

5. Elimination of the many sectioning artifacts that are a common feature of cryomicrotomy is a matter of some urgency. The studies by Jesior (1985, 1986) on sectioning resin material show that there is reduced compression in harder objects which are cut at medium thickness (50–90 nm) using a low-angle diamond knife. We need to see if these same parameters apply to frozen material and to see if even lower sectioning temperatures will affect the hardness and cutting properties of ice. Lickfield (1985) has shown that section deformation is diminished if the rake angle of the knife is reduced to below 45°.

4.6. SUMMARY

This chapter discusses the first of several procedures that are used to expose the interior of frozen specimens for examination and analysis. Frozen sections, which may be ultrathin (50–100 nm), thin (100–1000 nm), and thick (1–2 μm) can be either frozen hydrated or freeze-dried. Frozen-hydrated sections retain their natural water content throughout preparation and examination; freeze-dried sections are hydrated after the sections are cut. The process of cryosectioning is described from theoretical, mechanical, and practical points of view. The actual process of cryosectioning is not completely understood, although there are several theories that claim to explain the way sections are formed. The cryoultramicrotomes and the cryostat microtomes needed for cryosectioning are now more or less standard. The practical procedures involved in sample preparation, mounting, and trimming, followed by sectioning, section collection, and transfer, are described in detail. Cryomicrotomy is not without its difficulties and problems, and a catalogue of recognizable and largely explicable artifacts is discussed. There are many applications of cryomicrotomy which provide a rapid, nonchemical procedure for the close examination and analysis of fluids and suspensions and the interior of polymers and hydrated organic and biological samples.

5

Low-Temperature Fracturing and Freeze-Fracture Replication

5.1. INTRODUCTION

Fracturing is best described as a separation of a specimen along a line of least resistance parallel to the applied force. It may be achieved either by applying tensile stress via impact with a blunt instrument, which will produce a single fracture plane, or by using a sharp tool, which will generate shear forces just aheaad of the knife edge and produce a series of chonchoidal fractures. Unlike sectioning, the fracturing tool should not make physical contact with the freshly exposed fracture face. This means that fractures have an undulating to rough surface in contrast to the smoother surfaces of sections. There are fewer surface defects and artifacts on a fracture face of a frozen sample, provided the fracturing process is carried out in conditions of low contamination, such as under liquid nitrogen and at a high vacuum and low temperature.

The process of fracturing is reasonably well understood. Completely brittle material will fracture without deformation, whereas completely ductile material does not fracture but will exhibit plastic deformation. Although there are few organic and hydrated materials that are completely brittle, even at 4 K, our aim in low temperaturing must be to obtain brittle fractures with a minimum of plastic deformation. In an ideal brittle fracture, deformation is limited only to the molecules that are actually pulled apart.

The applied force of the fracturing tool causes stress to build up in localized regions in the sample. This stress overcomes the cohesive properties of the specimen and then suddenly spreads rapidly through the sample, causing it to fracture. The way the sample fractures will depend on its brittleness, which is influenced not only by the propeties of the sample but by decrease in temperature, rate of stress application, and the presence of discontinuities such as minute cracks within the sample.

The fracture path follows the line of least resistance, but these pathways vary with different materials. It might be along the interface between two phases, i.e., particle–solvent, along regions where differential contraction has occurred during cooling, or as is now widely accepted, along the hydrophobic inner face of biological membranes. The situation is much less clear in single-phase material, and the fracture pathway may be related to differences in the local orientation of the constituent molecules (Zasadzinski and Bailey, 1989).

In an attempt to find an alternative to either fracturing or sectioning, Lechene *et al.* (1975) exposed the internal surfaces of samples to high-speed freeze sawing under liquid nitrogen. The exposed surfaces were flat, smooth, and featureless, but because no assurance can be given that local melting has not occurred during the sawing process, this method is generally considered not to be a suitable preparative technique for low-temperature microscopy and analysis.

In this chapter, we will consider two preparative techniques that involve fracturing. Both share many common features but there are important differences. In low-temperature fracturing, the actual fracture face is examined at low temperatures in an SEM. In freeze-fracture replication, frequently referred to as freeze-etching, a high-fidelity replica of the fracture face is examined at ambient temperatures in a TEM. Both procedures can be used for morphological and analytical studies, and because of the differences in procedure, can provide complimentary information about samples.

5.2. LOW-TEMPERATURE FRACTURING

Low-temperature fracturing is, basically, a very simple process and can easily be carried out with inexpensive equipment. However, in order to utilize the process effectively and to provide useful information about the sample, it is necessary to integrate the fracturing process into a system that allows the fracture faces to be processed further before being examined and analyzed in a deep frozen state in the SEM.

5.2.1. Simple Fracturing Devices

The simplest and most effective way is to fracture the sample under liquid nitrogen. The liquid nitrogen is best used in a double container surrounded by insulting material. The LN_2 in the outer container virtually eliminates nucleate boiling in the inner chamber, making it much easier to see the fracturing process. The frozen sample may be held firmly under LN_2 between two pairs of fine-nosed haemostat forceps with insulated handles, and then gently flexed until it snaps. Fractures usually appear in the most unexpected places and the process is accompanied by lots of debris. Alternatively, the sample may be held at a shallow angle under LN_2 and a scalpel blade pressed at right angles to the proposed fracture plane until it snaps. Care should be taken that the blade edge does not tough the freshly exposed fracture face and cause surface damage.

Once the fracture has been made, the fracture face should be "washed" in fresh liquid nitrogen to remove any debris. If necessary, the surface may be gently brushed with a fine camel or squirrel hair artist's brush. The fracture face is now ready for coating and examination at low temperatures, and it is from this point onwards that the simple fracturing procedures have their limitations. The fracture face needs protecting during transfer either to the microscope cold stage and/or to a cold stage in a coating unit without it melting, drying, or becoming contaminated. The simple shroud and capping devices discussed earlier (section 3.10.2) are reasonably effective; so too is transfer under LN_2, but the transfer is best done using a portable vacuum airlock device. The transfer of frozen samples at low temperatures is the weak link in the whole procedure and it is for the reason that most studies are now carried out using integrated fracturing, coating, and transfer devices.

5.2.2. Simple Integrated Cryofracturing Systems

A number of investigators including Echlin et al. (1970), Echlin (1971), Echlin and Moreton (1973), Gullasch and Kaufmann (1974), Saubermann and Echlin (1975), Fuhs and Lindemann (1975), Echlin and Burgess (1977), Marshall (1977a), Becaner et al. (1973), and Potts and Oates (1983) have built integrated cryopreparation devices as a part of existing pieces of equipment, while others, e.g., Robards and Crosby (1979), developed free-standing units. Although these devices played a significant role in the development of the procedure, they will not be discussed here because they are too varied and specialized and are not generally available outside the laboratories where they were constructed. Papers by Marshall (1987) and Beckett and Read (1986) contain valuable information of the development of cryofracturing equipment

The first dedicated cryofracturing units were developed by JEOL in the early 1970s, and most electron microscope manufacturers have, at one time or another, introduced cold stages, cryopreparation units, and transfer devices as part of their repetoire of instrumentation. Although some of these devices are still available, the emphasis is now for accessory supply companies to develop and manufacture cryopreparation units that can be bolted onto the side of a wide range of electron microscopes.

A number of commercial instruments are now available and they fall into two main classes: the *dedicated* units, which are only operational when attached to the side of the scanning microscope, and the *nondedicated* units, which allow sample preparation to be completed away from the microscope column. In the dedicated systems, the precooled sample is introduced into the cryochamber via an airlock and, after fracturing and coating, passed directly to the cold stage of the microscope via a second airlock. With nondedicated systems, it is usual for the sample to be cooled, fractured, and coated in a separate cryochamber and then transferred to the microscope cold stage using a portable vacuum transfer device.

All cryopreparative units are of the same basic design and have an LN_2-cooled cold stage fitted with a heater and thermostat, a fracturing knife or probe, metal- and carbon-coating devices, and viewing ports to watch the various manipulations. The dedicated units usually have their own rough pumping system but rely on the microscope to provide the high-vacuum conditions. The VG Microtech, Polaron SP200A Cryogenic System is a nondedicated unit, while the VG Microtech, Polaron E7400 and E7450 CryoTrans systems and the Oxford Instruments CT1500A and

CP2000 Cryotrans systems can operate both as dedicated and nondedicated cryopreparation chambers. In practice, most cyopreparation units are operated in a dedicated mode as this simplifies the operational procedures. Because the five commercial instruments all follow the same basic design, the description and operational procedures for the VG Microtech, Polaron CryoTrans device which now follows will serve as a general description for all the simple integrated cryopreparation units.

A photograph and diagram of the CryoTrans system is shown in Figs. 5.1 and 5.2. Samples can be loaded either directly onto the specimen holder at the end of a long insulated transfer rod and then cooled rapidly in melting nitrogen, or samples may be cooled separately and inserted, under LN_2, into holes in the specimen holder. The specimen holder, together with the frozen specimens, is withdrawn into the inner chamber of the specimen transfer device, which has previously been purged with dry nitrogen gas. The inner chamber is sealed by a trap door and the transfer device brought to the outer air lock of the cryopreparation unit, which is bolted to one of the side ports of the microscope. The outer air lock is evacuated by a separate rotary pump, and once a suitable pressure has been reached, the trap door to the inner chamber is partially opened to complete the pumpdown process. The whole cryochamber is pumped to a pressure of 1.33 Pa ca. 10^{-2} Torr and high-vacuum conditions

Figure 5.1. Side view photograph of the VG Microtech, Polaron E7400/ 7450 CryoTrans preparative unit. The unit interfaces to the SEM column by the airlock (a) to the right and samples are loaded via the airlock (1) to the left. The cold stage (c) and the fracturing knife (k) may be seen through the lower window and the sputtering head (s) is just visible through the upper window. The light and controls for fracturing the probing are on the top of the unit together with the needle valve to control the flow rate of argon during sputtering. The stage LN_2 Dewar may be seen at the rear of the picture. Picture courtesy BioRad Microscience Division.

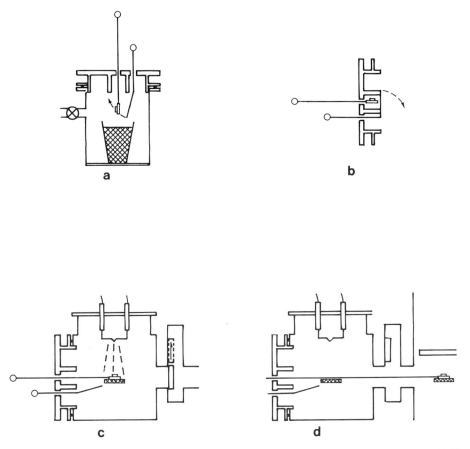

Figure 5.2. Sequence of events during the use of the VG Microtech, Polaron CryoTrans system. (a) The sample is quench cooled in nitrogen slush and then retracted into the vacuum transfer device. (b) The transfer device with the cold specimen under vacuum is moved to the front port of the CryoTrans unit, which is pumped out to admit the frozen specimen. (c) The specimen is fractured (etched) and sputter or evaporate coated in the CryoTrans unit before (d) being transferred to the SEM cold stage.

may be achieved by closing off the valve to the rotary pump and then opening up the second airlock between the cryochamber and the microscope column. The microscope vacuum system will easily maintain the cryochamber at a presssure of 1.33 mPa. Once high-vacuum conditions have been reached, the trap door to the inner chamber of the transfer device is opened fully and the specimen holder pushed forward on the transfer rod until it engages the slots on the precooled cold stage of the cryopreparation unit. It takes about 30 s to transfer a sample from liquid nitrogen outside the unit to the cold stage inside the unit. The cryopreparation unit cold stage, which is fitted with a small heater, is cooled by an external Dewar of LN_2 and the temperature controlled and held at any point between ambient and 100 K. The sample should be allowed to reach thermal equilibrium before any further manipulations are carried out.

The sample may be fractured either using a liquid-nitrogen-cooled exchangeable scalpel blade whose height may be adjusted, or using a cooled surface probe. An external illuminator and stereomicroscope allows a precise examination of the sample area. The height of the cold knife is adjusted and moved forward until it just makes contact with the frozen specimen. The fracture should be made so that the knife edge follows a slightly upward path through the sample thus avoiding the knife coming in contact wth the freshly exposed fracture face. The fracture face may be inspected with the stereomicroscope or examined uncoated in the scanning electron microscope. If the fracture has an exceptionally rough surface, the more protuding portions may be removed with a second or even third pass of the knife. Care must be taken not to contaminate the fracture surface with pieces of the fractured sample and it is important that pieces of frozen material that are removed during the fracturing process remain in good thermal contact either with the cold stage or the knife. If these fragments are allowed to warm up, the water will sublime and condense on the cold fracture face. Loose pieces of material can be removed from the fracture face by quickly withdrawing the specimen holder from the cold stage, and while turning the holder through 90°, gently tapping it so the pieces fall on the cold stage.

The fractured surface may be etched, i.e., a surface layer of water removed by sublimation, either by raising the temperature of the cold stage and the sample to between 210 and 200 K, or by using radiant energy from a heated tungsten wire located above the frozen sample. During the etching process the cold knife should be brought close to the sample surface, where it will act as an effective cold trap for the sublimed water. In the early days of low-temperature fracturing it was considered necessary to etch samples in order to obtain good topographical detail. Experience has shown that this is now not generally necessary because with a little practice sufficient detail can be seen in unetched samples. The rate of etching depends on a number of different factors, including the surface temperature of the specimen, the partial pressure of water vapor in the immediate environment around the specimen, and the temperature and distance of the radiation source from the frozen fracture face. Figure 5.3 shows the relationship between the sublimation of ice and temperature. It is difficult to achieve precise etching with the simple cryofracturing devices, and in practice the sample is either deeply etched or unaffected. Although it is difficult to etch to a precise depth, it is a useful way of removing contaminating ice crystals from the fracture face. Such material is in poor thermal contact and with a little care may be etched away with the radiant heater without affecting the underlying fracture face.

The final step in the preparative procedure is to coat the frozen fracture face with a thin film (10–12 nm) of a conducting material such as one of the noble metals or their alloys. The gate valve between the microscope and the cryochamber should be closed before the coating takes place, for the procedure is most easily carried out at a pressure of about 1.33 Pa using the diode sputter coating device located in the upper module of the CroyTrans unit. The cold stage should be allowed to cool to 110–120 K and the cryochamber flushed with argon. During sputtering the flow rate

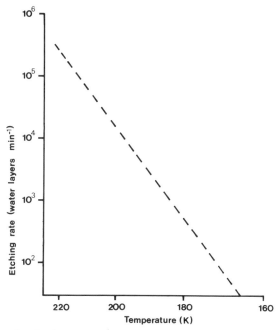

Figure 5.3. Graph showing the theoretical rate of ice sublimation in relation to temperature. [Redrawn from Robards and Sleytr (1987)].

of the gas should be adjusted to ensure a low and steady rate of metal sputtering to avoid heating the frozen surface. The multidirectional nature of sputter coating ensures that the rough fracture face receives an even coating layer. The CryoTrans unit also has a high-vacuum evaporation head which can be used to apply either a metal or carbon coating layer. Because of the unidirectional nature of evaporative coating, the frozen sample should be temporarily slipped off the cold stage and rocked from side to side during coating to ensure that the uneven fracture face is adequately covered. It will be necessary to open the gate valve to the microscope to ensure that the vacuum is good enogh to carry out the metal or carbon evaporation. The thermal mass of the sample will keep the sample sufficiently cold during this brief procedure. Carbon evaporates at a very high temperature, and Echlin and Taylor (1986) have shown that it is difficult to apply an evaporated carbon layer to a frozen surface without causing surface etching. Once the coating is complete, the sample may be moved onto the precooled cold stage of the microscope.

The microscope cold stage is a separate part of the CryoTrans system and is mounted via another port on the microscope specimen chamber. The cold stage fits onto the specimen stage of the SEM and is cooled by nitrogen gas passed through a Dewar of LN_2. The stage temperature can be controlled at any temperature between 273 and 110 K and cools down to working temperatures within 15–20 min. A small anticontamination plate is fitted above the cold stage and is connected directly to the LN$_2$ Dewar. This ensures that the anticontamination plate is at a lower temperature than the fracture face. Although the procedures for examining the frozen fracture faces will be considered in Chapter 10, Fig. 5.4 shows two examples of the type of image that may be obtained using the simple integrated cryofracturing units.

The other cryopreparation units operate in more or less the same way with small differences in the way the sample is transferred in and out of the cryopreparation unit. The Oxford Instruments system is virtually the same as the device, but the nondedicated VG Microtech, Polaron SP2000A is a little more elaborate. The frozen

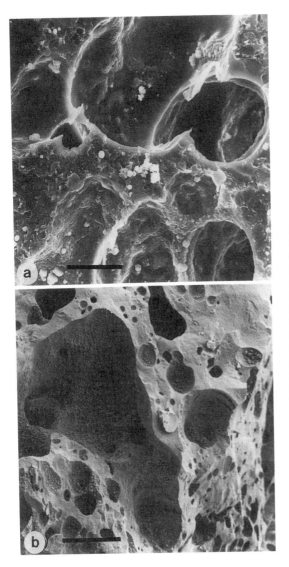

Figure 5.4. Two examples of low-temperature fractures made in a simple integrated cryofracture unit CryoTrans E7400). (a) Frozen-hydrated mouse lung which retains all the cellular and extracellular fluids in the lung tissue and shows how the bronchioles are covered with a thin layer of mucus. The air-filled alveoli show no sign of collapse. Bar marker = 10 μm. (b) Frozen-hydrated dessert chocolate mousse. The mousse is composed of a foamed emulsion of vegetable fat dispersed within an aqueous phase containing proteins, sugars, and a gelling agent. The form is initially stabilized by the presence of a layer of fat around the bubbles. The borders between the bubbles contain the alginate gelling agent, which rapidly cross-polymerizes to a three-dimensional structure and helps retain the structure of the mousse. The frozen-hydrated sample enables the gas, liquid, and solid phases to be retained and imaged. Bar marker = 200 μm. (Pictures courtesy Dr. A. Wilson, University of York.)

sample is mounted on a small copper rod, which is surrounded by a stainless steel shroud, which may be opened and closed by rotating an insulating sleeve at the end of the transfer rod. Both the sample holder and the metal shroud are cooled in liquid nitrogen, and this double protection system of a cold, shrouded sample within a low-vacuum environment, successfully protects the sample from contamination. The paper by Beckett and Read (1986) contains details of the operational and experimental procedures associated with this unit.

The simple integrated cryofracturing devices are modestly priced, are relatively easy to use, and are producing a lot of useful information about frozen specimens. Their main advantage is that they are easily fitted and demounted from the microscope column, thus allowing the microscope to be used for other purposes. The disadvantage of these systems is that they operate at a rather poor vacuum, and the design of the microscope cold stage only allows rather limited specimen movement.

5.2.3. Comprehensive Integrated Cryofracturing Systems

In addition to the simple, but nevertheless reasonably effective, cryofracturing systems that are available, there are a number of more complex instruments, which have additional features that have proved particularly useful in producing contamination-free frozen fractures.

Quite the most extraordinary of these devices is the nondedicated analytical vacuum cryogenic system, Vacuum Cryotom BK-02 being developed by Professor Boris Allakhverdov and Dr. Alexander Pogorelov and colleagues at the Institute of Biological Physics, Academy of Sciences of the USSR, Pushchino, Moscow Region. The main unit, constructed of stainless steel and titanium, is a large working chamber with an octagonal cross section fitted with a cold stage, anticontamination devices, high-vacuum pumping system, and control units. Various preparative modules may be attached to the working chamber via air locks, allowing a wide range of specimen manipulations. There are modules for specimen loading, cutting, manipultion, fracturing, electron beam, sputter, and evaporative coating, and a vacuum balance to control and measure the loss of water during sublimation. The working vacuum of the system is $13.3\,\mu$Pa (10^{-7} Torr), with a temperature range between 90 and 370 K. The cold specimens can be rotated and tilted during sample preparation. Few details of this equipment are available outside the Soviet Union, although a paper published in 1981 contains a general account of the instrument (Allakhverdov and Kuzminykh, 1981). The author has had a limited amount of experience with the prototype of this instrument and the results are most impressive.

Balzers Union AG have recently introduced a dedicated unit, the SCU 020, for use with low-temperature SEM. The unit, which is described by Muller *et al.* (1991), consists of two parts: (1) a high-vacuum preparation chamber equipped with a cold stage, motorized fracturing microtome, a sputter coater, quartz crystal thin film monitor, Meissner cold trap, and turbomolecular pumping; and (2) a cold stage residing in the SEM chamber separated from the preparation chamber by a high-vacuum gate valve. The SEM cold stage is fitted with anticontamination cold traps and a goniometer stage with motorized drives for x, y, and rotation. The rotation is

±45° and the tilt between 0° and 60°. The two stages are braid cooled from an LN_2 Dewar and fitted with devices to minimize vibration from the bubbling nitrogen. The unit allows samples to be fractured, etched, and examined at temperatures down to 123 K in a contamination-free environment of 200 μPa (ca. 2×10^{-6} Torr). A schematic diagram of this unit is shown in Fig. 5.5, and some results will be discussed further in Chapter 10.

One of the most versatile of these comprehensive integrated cryofracturing devices is the Biochamber, designed by Pawley and Norton (1978) and manufactured and marketed by AMRAY Inc. in the United States. This sophisticated piece of equipment has been used in various academic, medical, and industrial laboratories for more than 12 years and is the system used by the author since the late 1970s. The Biochamber, shown in Fig. 5.6, is a dedicated cryopreparation unit, which although designed primarily to work with AMRAY scanning electron microscope that is fitted with a cold stage. Details of the Biochamber and its mode of operation have been described by Pawley and Norton (1978), Echlin et al. (1979, 1980, 1981, 1982, 1983), Echlin and Taylor (1986), and Bastacky et al. (1985, 1987).

The Biochamber is a high-vacuum chamber with its own pumping system, attached to the side of an SEM and fitted with devices for fracturing, etching, and coating frozen hydrated specimens. With the specimen holder at 100 K in a vacuum of 50 μPa (4×10^{-7} Torr), samples may be fractured with a knife at 100 K. The fractured samples may be etched if necessary before being coated and transferred on a metal shuttle via an airlock to the cold stage of the microscope. The AMRAY microscope cold stage is quite unique and makes use of a Joule–Thompson refrigerator, which depends on adiabatic cooling using high-pressure nitrogen gas expanding through a fine nozzle. The thin, flexible gas supply lines allow the cold stage to be rotated by ±175°, tilted between −5° and +90°, as well as moving in the x, y, and z

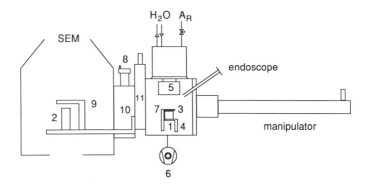

Figure 5.5. A diagram of the Balzers Union SCU 020 Cryopreparation unit. The SEM is fitted with a goniometer cold stage (2), anticontaminator plate (9), adaptor flange (10), and stage controls (8). The preparation unit has a cold stage (1), motorized microtome (3), cold trap (4), sputter coater (5), turbomolecular pump (6), thin-film monitor (7), and high-vacuum gate valve (11). The manipulator allows the sample to be loaded and moved within the unit and the endoscope allows the fracturing process to be observed. [From Muller et al. (1991).]

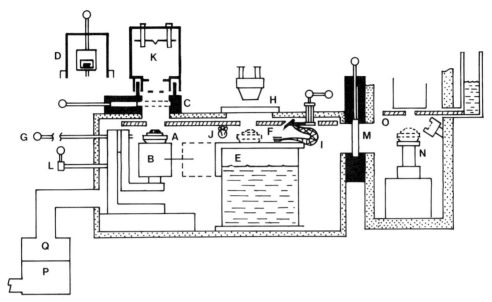

Figure 5.6. A diagram of the AMRAY Biochamber cryopreparation unit. The sample is loaded onto shuttle (A) sitting on the precooled cube (B) via the top high-vacuum gate valve (C) using the sealed specimen transfer device (D). The shuttle is moved sideways to the top of the LN_2 tank (E) below the cold shroud (F) using the shuttle transfer rod (G). The frozen sample may be observed through the top window (H) fractured using the microtome knife (I) and if necessary etched using the radiant heater (J). The shuttle with the fractured (etched) sample is moved back onto the cold cube to a position below the top high-vacuum gate valve, which is now fitted with the coating module (K). The frozen sample is rotated and tilted during coating using the lever (L). The coated sample still on the shuttle is moved back to the top of the tank where it may be inspected before being passed via the side high-vacuum gate valve (M) to the Joule Thomson cold stage (N) protected by an anticontamination plate (O) in the microscope column. The Biochamber is pumped by its own high-vacuum system (P) fitted with LN_2 traps (Q). [Redrawn from Norton and Pawley (1978).]

directions. The refrigerator is of simple design and is a reliable way of maintaining low tempertures inside an SEM. The flexibility of movement is particularly important in the examination and analysis of highly sculptured fracture faces. The refrigerator needs a supply of clean, dry nitrogen gas at a pressure no lower than 100 bar (1500 psi). When the cold stage is not being used, it is advisable to keep the gas lines at slight positive pressure 1 bar (15 psi) to prevent the 100-μm orifice from becoming blocked with dust particles.

Cooling is acheived by a large tank of LN_2 inside the Biochamber, which may be refilled from outside the instrument. The specimen shuttle, which can be moved backward and forward within the Biochamber and into the microscope column, is held in good thermal contact with a polished plate on top of the tank by a pair of leaf springs. For sample loading, the cold shuttle is moved to the top of the precooled copper cube, where it is held in good thermal contact by a second set of leaf springs. The cube is cooled by close thermal contact with a polished plate on the side of the

LN$_2$ tank. The cube is insulated and has sufficient thermal mass to hold the specimen at 100 K throughout loading. The cold cube and shuttle can be moved to a position directly below the sample loading airlock.

Samples are mounted on special holders, the upper surface of which fit into the specimen transfer module and the lower portion of which screws into the shuttle inside the Biochamber. The maximum specimen height is 5 mm. The holders can be modified in various ways to accommodate different types of specimen. Copper holders with roughened surfaces appear best for morphological studies, aluminum holders with graphite inserts are best suited for analytical studies (Echlin et al., 1982). Small specimens may be placed either in holes in the sample holder, or in metal or graphite tubes which are precooled and then placed in the specimen holder held under LN$_2$. Large samples may be cut under LN$_2$ to precise dimensions and held at known orientation in a vicelike device built into the sample holder (Bastaky et al., 1987).

The specimen transfer module consists of an insulated rod which passes through a vacuum seal in a cup-shaped piece which can be fitted to the outer seal of the airlock. The rod terminates in a partially drilled-out cylindrical brass block which has springlike fingers at the lower end to grasp the specimen stub firmly. The metal cylinder forms a cap around the sample holder and prevents ice formation on the specimen during transfer; the brass block provides sufficient thermal mass to keep the sample below 110 K during transfer from an LN$_2$ bath to the Biochamber. Frozen samples are transferred, via the top airlock, under LN$_2$ and screwed firmly into the shuttle. The transfer takes about 25 s and once the specimen is under high vacuum and on the shuttle in the Biochamber, the shuttle is moved back to its original position on top of the tank.

The frozen sample is fractured when the shuttle is on the tank where it can be observed with the binocular microscope. The knife is cooled to 100 K by a short copper braid connected to the top of the LN$_2$ tank and may be moved horizontally across the frozen specimen. The knife height is controlled by a micrometer screw concentric with the axle that holds the knife. The knife may be moved across the specimen in steps as small as 5 μm and then raised for the return strike. Fragments of frozen material invariably fall onto the large space on top of the tank, where they remain frozen thus avoiding the problem of contamination. It is usually necessary to cut one coarse fracture across the frozen sample, followed by a second or third pass with a clean part of the knife. Although the knife describes a fixed arc across the specimen, the specimen shuttle assembly can be moved slightly so that on each pass of the knife a clean, unused portion of the blade fractures the specimen.

Following fracturing, the fracture surface may be etched by moving the shuttle-specimen assembly to a new position on the tank surface approximately 10 mm below the radiant heater on the underside of the cold shroud which surrounds the whole preparative area. Talmon (1980) has shown that it is possible to calculate the rate of ice sublimation from a frozen sample heated by an overhead radiant source, provided various parameters of the system are known. Talmon based his calculations on the thermal and vacuum parameters of the Biochamber, and showed that etching for 20–30 s would remove 100–150 nm of water from the surface of a frozen sample held at 110 K. Futher details of the etching procedure may be found in the papers by Echlin et al. (1979, 1982).

The fractured sample (etched or not) is moved back to the top of the cold cube in preparation for coating. An important design feature of the copper cube is that when it is moved to a position below the air lock, it is free to be rotated and tilted using an external lever to ensure thorough coating. The evaporation coating device fits over the sample entry port and is supplied with both metal and carbon coating heads. A coating thickness of between 5–8 nm gold is usually sufficient for morphological studies and 3–5-nm layer of carbon or chromium for analytical investigations. The problems associated with coating frozen samples are considered in detail in the papers by Echlin and Moreton (1973), Echlin, (1978a, b), Echlin et al. (1982), Echlin and Taylor (1986), and Marshall (1987). The main difficulty is that unless great care is taken during the coating, the surface of the frozen hydrated sample may be damaged.

The preparative procedures are now complete and the shuttle–sample holder complex may be moved off the cube, across the cold surface of the tank, and into the microscope via the side airlock. One of the many advantages of integrated cryofracturing systems is that the sample can be quickly examined in the SEM between sequential fracturing on the cold stage. The low temperature and high vacuum ensures that the specimen remains fully hydrated at all times and that the flat and reproducible fracture faces rarely become contaminated. An example of the type of fractures that may be obtained is shown in Fig. 5.7.

Before we leave the discussion on low-temperature fracturing, we need to consider the advantages and disadvantages of sample etching. As indicated earlier, there is now much less emphasis on the need to etch samples in order to obtain improved morphological detail. In most recent studies, etching solely for the detailed enhancement of morphological features is the exception rather than the rule. There are, however, a number of circumstances where etching with a radiant heater may provide additional information about the specimen:

1. We have already discussed how etching can remove contaminating ice crystals. Beckett and Read (1986) have used the same procedure to remove environmental water surrounding aquatic microorganisms.
2. Etching can be used to localize and emphasize aqueous extracellular spaces and secretions and distinguish between different regions of perturbed water in extracellular spaces and in the fractured vacuoles of plant cells which will etch at a faster rate than water associated with the cytoplasm and organelles.
3. Etching can be used to emphasize the mass density differences that exist between organic and aqueous phases in biological systems, emulsions, and latex suspensions.
4. Beckett and Read (1986) found that etching reveals the spacing of segregation zones caused by ice-crystal damage and can be used to give qualitative information on the amount and distribution of water, the concentration of nucleation sites, and even the rate of freezing in hydrated material.
5. Surface etching of frozen fracture faces are a mixed blessing as far as x-ray microanalysis is concerned. Although the x-ray signal is vastly improved by virtue of the water being removed by sublimation, this etching process creates difficulties in quantitative x-ray microanalysis. These are matters we will return to in Chapter 11.

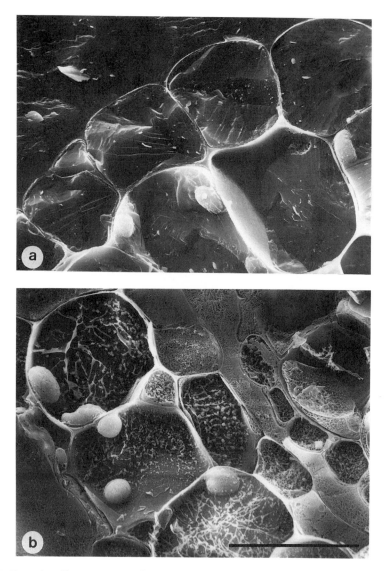

Figure 5.7. Examples of low-temperature fractures made in the Biochamber. (a) Frozen-hydrated fracture of the cortical cells of the root of *Lemma minor*. *L*. The cell walls and multiple fractures are visible in the cell vacuoles. (b) Same region of the root after deep-etching to reveal cell contents. Sample coated with ca. 10 nm Au/Pd and photographed at 110 K. Bar marker = 10.0 μm.

Marshall (1980d) considers that specimen charging is reduced in etched samples. Echlin (1978a), Walther *et al.* (1990), and Herter *et al.* (1991) found the opposite to be the case. The electrical conductivity of frozen-hydrated specimens is higher than in frozen dried specimens owing to the presence of water in the sample. The differences may also be due to variations in the operating conditions of the microscope. Low beam currents of ca. 10 pA are used for morphological studies, whereas higher beam currents of 2–300 pA are necessary for analytical studies.

The alternative to using radiant heater etching of the surface is to use conductive heating of the whole sample with the small heater incorporated into the cold stage. Beckett and Read (1986) consider that this type of etching can be carefully controlled, primarily by observing the sample surface during the etching process. Etching may also be carried out in the microscope column using the small heater incorporated into most cold stages. The process should only be observed at low voltages or using backscattered electrons to avoid the charging problem on the uncoated specimen.

The disadvantage of etching by conductive heating is that the whole sample may freeze-dry. This is not the case with the radiant heater as it has already been shown (Echlin *et al.*, 1982) that the bulk of the specimen remains fully frozen following surface etching. Even in the best designed system, the temperature at the surface of the sample will be higher than the cold stage on which it sits. If the *stage* is raised to a temperature at which appreciable sublimation will occur, e.g., 173 K at 130 μPa (10^{-6} Torr) pressure, the fracture face will dry very quickly. Theoretically, it should be possible, knowing the conductivity and temperature of the frozen sample, to calculate the temperature at which it should be held to ensure incipient drying at the surface. The presence of the cold (150 K) knife a few millimeters above the warmed fracture face will enhance the sublimation. The etching process is an interesting problem that needs further work before we can take full advantage of this procedure, which is available on most cryopreparation units.

5.3. FREEZE-FRACTURE REPLICATION

5.3.1. Introduction

In contrast to low-temperature fracturing, freeze-fracture replication is a well-established technique in which many of the preparative processes are understood and many of the problems and artifacts are recognized. The technique is well described in the literature, and it will not be necessary to give a detailed account of the procedures in this book. However, for the sake of completeness, an overview will be given of the principal features of the process and liberal reference will be made to the many excellent reviews and papers that have been written about the development and use of freeze-fracture replication. In this respect, one of the most useful and authoritative accounts is that written by Robards and Sleytr (1985), which contains a wealth of practical information on how the technique may be applied to biological and hydrated samples. An updated review of the current practice and application of freeze-fracture replication may be found in two recent issues of the *Journal of EM Technique*, 13(3) and 13(4) (1989).

The process of freeze-fracture replication, which is frequently referred to as freeze-etching, involves the production of a replica from a fracture plane made across a frozen specimen. In this respect, it is an indirect method of studying a sample as distinct from the direct methods we have been discussing earlier. The common term, freeze-etching, is slightly ambiguous, because it implies that replicas are only made from etched fracture faces, and as we will show, this is now not necessarily the case.

The process of freeze-fracture replication consists of six distinct steps, a number of which bear a clsoe similarity to some of the different steps of low-temperature fracturing:

1. Sample preparation and rapid cooling.
2. Specimen fracturing.
3. Optional etching of the freshly exposed fracture face.
4. Shadowing the fracture face at an angle with a heavy metal to create selective electron contrast.
5. Depositing a thin film of carbon over the whole fracure surface to make a continuous replica
6. Removing all the organic material from the replica after the sample has been thawed.

These six steps are shown diagramatically in Fig. 5.8. The greatest variation in the procedure is in the way the initial fracture is made and the different methods which are used for the application of the heavy metal and the carbon layers. A wide range of equipment has been devised for carrying out freeze-fracture replication, although the general principles are the same.

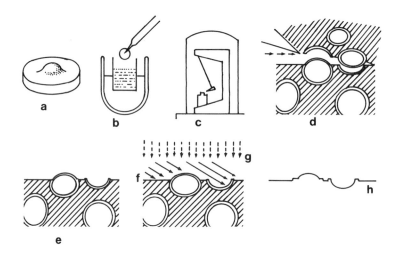

Figure 5.8. Diagram of the basic steps of freeze-fracture replication. The sample (a) is quench cooled in liquid propane (b) and transferred under LN_2 to the precooled cold stage of the freeze-fracture replication unit (c). The sample is fractured at low temperature (d), etched to remove a surface layer of water (e), and the etched fracture surface first metal shadowed (f) and then a replica made by carbon evaporation (g). After chemical removal of the remains of the thawed biological material, the cleaned replica (h) may be examined in the TEM. [Redrawn and modified from Robards and Sleytr (1985).]

5.3.2. Sample Preparation

Although a wide range of sample holders are available, they can be divided into those that are going to rely on impact devices to create the fracture plane, and those that rely on a knife edge to initiate the fracture. Different types of specimens require different types of holders. Bulk specimens need to be trimmed before being mounted in or on a holder, whereas liquid suspensions and monolayers might be mounted between thin sheets of metal, which could be subsequently pulled apart. The design of some specimen holders allows complementary replicas to be made in which both sides of a single fracturing process are retained. The actual preparation involves rapidly cooling small samples, which may or may not be infiltrated with cryoprotectants. These techniques have already been discussed in Chapter 3 (section 3.6.)

5.3.3. The Fracturing Procedure

Samples may be fractured by tensile stress (impact fracture) or by shear forces (knife cleavage) using three different conditions; under high vacuum; under a liquid cryogen; or under dry nitrogen. The three fracturing conditions have resulted in the develoment of different types of equipment.

5.3.3.1. Fracturing under High Vacuum

This is the approach to fracturing adopted in the apparatus originally described by Moor *et al.* (1961). The instrument consists of an evaporation unit containing a liquid-nitrogen-cooled microtome for fracturing the sample, which can be kept at a precise low temperature in a contamination-free environment. The basic design for this type of equipment has remained the same for the past 30 years, although there have been many significant improvements in the operational procedures. Figure 5.9 shows the interior of a modern freeze-fracture replication unit.

The most frequent problem with freeze-fracture replication is contamination, although with modern equipment and careful housekeeping, this has now been largely eliminated. All the surfaces in the immediate environment of the exposed fracture face should be at least 20 K lower than the temperature of the sample. Equipment should always be vented with dry nitrogen and checked for leaks, the component parts should never be touched by hand, and LN_2 traps should be an integral part of the vacuum pumping system and the specimen chamber. As discussed previously, one of the main sources of contamination is the frozen chips that have been removed from the sample surface during fracturing. Contamination can be virtually eliminated by using ultrahigh-vacuum conditions (5×10^{-11} Torr) (Gross *et al.*, 1978), or by enclosing the sample working area within efficient cold traps which are well below the temperature of the specimen (Steere, 1973; Sleytr and Umrath, 1976; Umrath, 1978).

The simplest device for obtaining a single fracture of a specimen under vacuum is achieved by having the sample between two holders that sit one upon the other. The bottom holder is firmly fixed to the underlying cold stage and the microtome

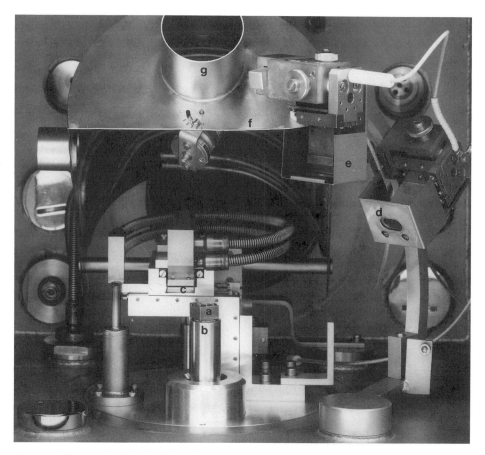

Figure 5.9. Photograph of the interior of the Balzers Union BAF 400 K freeze-etching unit. The frozen specimen (a) sits on the cold stage (b) and is fractured with the cold knife (c). The metal shadowing is made using the electron beam gun (d) and the carbon replica made with electron beam gun (e). The whole of the fracturing region is enclosed in an LN_2-cooled cold trap (f) and the fracturing process observed through port (g). The whole piece of equipment is in turn placed inside a high-vacuum chamber fitted with observation ports and electrical leads. (Picture courtesy Balzers.)

arm is used to break off the top holder to create a tensile stress fracture (Fig. 5.10). With a little ingenuity it is possible to also retain the top sample holder, which allows complementary replicas to be made. However, the most common method for producing complementary replicas is the hinged device originally developed by Steere and Mosely (1969). There are now a number of different simple and complementary fracture holders, all of which work on the impact fracture principle and have been designed to overcome the difficult problem of aligning the two complementary fracture faces and allow the precise matching of such objects as individual membrane particles on one face and the corresponding pits on the opposite face (Ting-Beall *et al.*, 1986; Fetter and Costello, 1986). Most of the commercially available pieces of

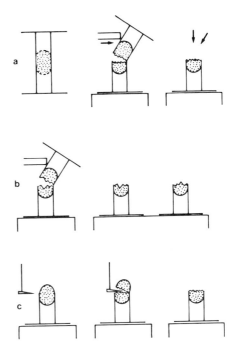

Figure 5.10. A diagram to illustrate the procedures for making simple and complementary fractures. (a) The frozen sample, enclosed between two small metal tubes, is impact fractured by a sideways blow with a cold probe. The lower half of the fractured sample is retained and may be etched, shadowed, and replicated. (b) Complementary fracture in which both halves of an impact fracture sample are retained. (c) The frozen sample may also be placed in the upper part of a single metal tube and fractured by means of a cold knife. [Modified and redrawn from Robards and Sleytr (1985).]

equipment for freeze-fracture replication will accommodate the different types of impact fracture sample holders, although each manufacturer favors their own particular holder design. The same type of sample holders will work equally well in an ordinary vacuum evaporation unit provided it is fitted with the necessary equipment for freeze-fracture replication.

Specimens may also be fractured with a knife (Fig. 5.10), although this does not allow complementary fractures to be made. The knife is usually a clean, grease-free, thick razor blade attached to the end of a microtome arm, which can be cooled to 80 K with LN_2. It may be advanced either mechanically or thermally and the cutting action automatically or manually controlled. The fracturing takes place as follows. The cold stage is first precooled, under vacuum, to ca. 123 K. The unit is then vented to atmospheric pressure using dry nitrogen, the precooled sample quickly transferred to the cold stage, and the chamber reevacuated to the working pressure. The microtome is cooled to 80 K and the sample raised to the fracturing temperature. Robards and Sleytr (1985) suggest that a sample temperature of between 153 and 163 K should be used for specimens that are not to be etched, while samples that are to be etched should be fractured at about 173 K. The knife temperature remains at 80 K and the pressure should be in the range of $1-2 \times 10^{-6}$.Torr. The cutting (fracturing) process is critical, because it can easily give rise to artifacts. Ideally, the fracture should be relatively flat and one would hope to achieve this with one pass of the knife. Such fractures are more readily replicated to reveal the fine surface topography of the sample. In practice, it is usually necessary to make several passes with the knife in order to produce a relatively smooth fracture face. A close examination of the fracturing process shows that the knife creates what Heuser (1989) refers to as "scallops"—shallow depressions (chonchoidal fractures?) that travel out in front of the knife and slightly below its actual path. Considerable experience is needed to know when to stop trimming the fracture and start the process of etching and/or replication. After the final cut is made, the microtome arm is raised slightly and placed over the freshly exposed fracture face as quickly as possible to diminish the chance of contamination. If the sample is to remain unetched it should be replicated as quickly as possible. Figures 5.11 and 5.12 show examples of single and complementary replicas produced after fracturing under vacuum. A much simpler procedure is

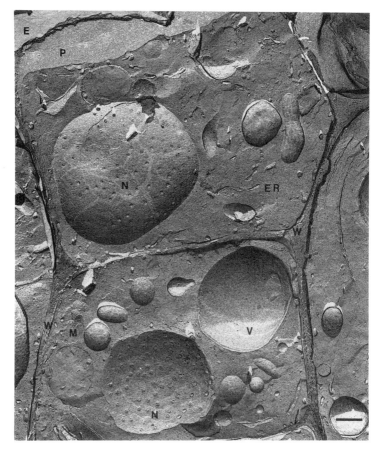

Figure 5.11. Low magnification electron micrograph of a freeze-fracture replica of meristematic tissue from a cauliflower bud. W, Cell wall; ER, endoplasmic reticulum; M, mitochondrion; N, nucleus; P & E faces of the cell membrane; V, Vacuole. Bar marker = 1.0 μm. [Picture courtesy Platt-Aloia and Thomson (1989).]

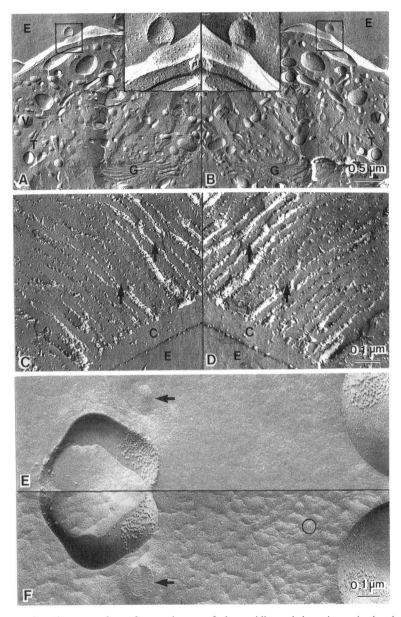

Figure 5.12. Complementary freeze-fracture images of ultrarapidly cooled specimens in the absence of chemical pretreatment. (A), (B) Images of leucotytes from human blood with a vesicle (V) attached to the plasma membrane shown in the insert. E, extracellular space; G, Golgi apparatus; T, tubular membrane organelle; V, vesicular membrane. (C), (D) Images from isolated frog retinal rod outer segment showing ice crystals (arrows) which have distorted the parallel pattern of disk membranes. C, cytoplasm; E, extracellular space. (E), (F) images of membrane vesicles isolated from bovine lens. The membrane profiles in (E) and (F) are complementary but the background ice is not because the sample in (F) has been etched slightly. The circle in (F) has a diameter equivalent to 60 nm and can be used to estimate sizes of the various features. [Pictures courtesy Costello and Fetter (1986).]

used to produce the tension fractures; the knife edge is used to break off the top holder.

The fine and gross topography of the fracture plane has been considered in some detail by Sleytr and Robards (1977, 1982) and Robards and Sleytr (1985). It is known that considerable energy is released during the fracturing process as the shear forces involved in fracturing are dissipated. Gruijters and Bullivant (1986) found a 50 K temperature rise in yeast fractured at 223 K, and Sleytr and Robards (1977, 1982) suggest that the temperature rise may be as high as 200 K in material fractured at 4 K. Molecular rearrangements are considered to take place and plastic deformation has been seen in synthetic and biological polymers fractured at temperatures as low as 4 K. These subtle changes in the fine topography of the fracture faces can lead to considerable difficulties in interpreting the images obtained by freeze-fracture replication. Robards and Sleytr (1985) show that the fracturing process can give rise to artifacts in the gross topography of the fracture in the same way as the sectioning process gives rise to problems with frozen sections. Friction between the knife and the frozen sample can cause localized melting, and such smooth areas contain little or no information. Knife marks can be recognized and the fragmented remains of the specimen can easily litter the fracture's surface. Single fractures produced by impact rather than cutting have fewer of these problems.

Further refinements to the process of fracturing under high vacuum include improving the vacuum conditions and seeking ways of further reducing contamination. Ultrahigh-vacuum technology has been applied successfully to freeze-fracture replication by a number of research groups. Escaig and Nicholas (1976) and Gross *et al.* (1978) provide specific details of the procedures adopted and Robards and Sleytr (1985) give a good general review of the problems involved. The ultrahigh-vacuum (UHV) equipment is specialized and necessarily more complex, but it does allow fracturing and replication to occur at temperatures as low as 12 K and at a presure of 7×10^{-9} Torr (Escaig *et al.*, 1980) or at 8 K and 1×10^{-9} Torr with the commercially available equipment that has been developed as a result of the work of Gross and his colleagues. The contamination rates are reduced to very low levels.

Contamination may also be reduced significantly by enclosing the sample within cold shrouds during the critical stages of fracturing and replication. The idea of the cold shroud goes back to the original work of Steere (1969), and it has been successfully developed by Sleytr and Umrath (1976), Umrath (1978), and Niedermeyer (1982). The shrouds, which can be cooled to LN_2 (or LHe_2) temperatures, reduce the total vacuum inside the shrouds to between 10^{-8} and 10^{-10} Torr. A combination of UHV and low-temperature shrouds would appear to be the ultimate refinement in the design of freeze-fracture equipment. Fracturing under vacuum is the most popular approach to freeze-fracture replication and a number of instruments are available. The principal manufacturers are Balzers Union AG, (BAF 500 K Ultra High Vacuum Freeze Etching System, BAF 400D and BAF 080T High Vacuum Freeze Etching Unit); Cressington Scientific Instruments (CFE 40 Freeze Fracture and Etching System); JEOL Ltd. (JFD 9000 Freeze Etching Equipment); Oxford Instruments (CF 4000 Cryoetch Freeze Replication); and Reichert-Jung (Leica Ltd) (Cryofract 190 Freeze Fracture Etch System).

5.3.3.2. Fracturing under liquid Gas

Bullivant and Ames (1966) developed a simple piece of equipment that very effectively produced fracture faces under liquid nitrogen (Fig. 5.13.). Frozen samples are loaded into a holder and fractured under LN_2, either by cutting or by impact. The fractured sample is now covered with an ingeniously shaped metal block which has two tunnels cut out to give a line of sight at 45° and 90° to the general plane of the fracture surface. The metal block is covered with a metal lid, and the whole assembly, which remains immersed in a dish of LN_2, is transferred to a vacuum evaporator. The vacuum evaporator is pumped out and the LN_2 first solidifies under the reduced pressure and sublimes away. When the pressure reaches 1×10^{-5} Torr, the lid is removed and the fracture face is first metal coated via the 45° tunnel and then carbon coated via the 90° tunnel. The thermal mass of the metal block maintains the fracture at ca. 133 K during coating so no etching takes place. The long, cold metal tunnels provide protection against contamination. This simple, inexpensive device, which is available from Hitachi Ltd. (Akahori *et al.*, 1972), can be used in virtually any vacuum evaporater and produces effective freeze-fracture replicas. Gruijters and Bullivant (1986) described a small modification, which allows freeze fracturing at defined temperatures followed by immersion of the specimen in LN_2. It is important to use clean liquid nitrogen with this type of equipment, because any moisture that condenses in the liquid gas as ice crystals can become deposited on the fracture face. Such contaminants cannot be removed by sublimation because of their low temperature and the relatively poor vacuum of the environment. Sleytr and Umrath (1974) have produced a system for fracturing specimens under liquid helium, which are then transferred to LN_2 prior to loading into the freeze-fracture equipment.

5.3.3.3. Fracturing under Dry Gas

This may be effectively achieved in a cryoultramicrotome, in which case the sections are discarded and the block is retained. Dempsey and Bullivant (1976b) have made controlled cleavage fractures of specimens maintained at 93 K. The best method involves precise fracturing of the first few micrometers of the sample surface which has been frozen by impact cooling, which were then transferred under LN_2 to a Bullivant Ames device for replication. Replicas produced by this technique were free of ice-crystal damage, although there was plastic deformation of some of the cellular structures.

5.3.4. Surface Etching

The fracture face may be etched by controlled sublimation of volatile components (primarily water) from the sample surface to reveal additional surfaces and structures deeper in the body of heterogeneous samples such as biological materials (Fig. 5.14). Etching is a process of freeze-drying, and although the physical conditions for the process will be discussed in more detail in Chapter 6 (section 6.6.2.), it will be briefly considered here as it forms an important part of the freeze-fracture replication

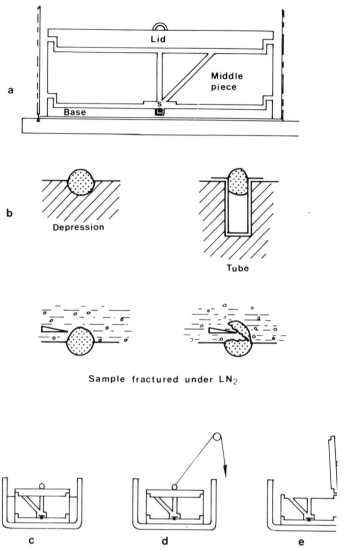

Figure 5.13. Bullivant simple freeze-fracture device. (a) The device consisting of a copper lid, a copper middle piece containing two holes, one vertical and the other at 45°, a specimen area (S), and a copper base piece. The whole device sits inside a metal container, which may be filled with LN_2. (b) Enlarged specimen area showing a specimen either in a shallow depression or in a small metal tube. The lid and the middle piece are removed and the container filled with LN_2 and the sample is fractured under LN_2. (c) The middle piece and lid are replaced and the container topped up with LN_2 and the device places ina high-vacuum evaporator. (d) Once the LN_2 has boiled away the lid is removed by an external control and (e) the fracture face metal shadowed through the 45° hole and carbon-coated through the vertical hole. The large metal mass of the middle piece stays at 83 K through pumpdown, shadowing, and replication and prevents sample contamination; the shadowing, and replication and prevents sample contamination; the base piece is slightly warmer at 133 K. [Modified and redrawn from Robards and Sleytr (1985).]

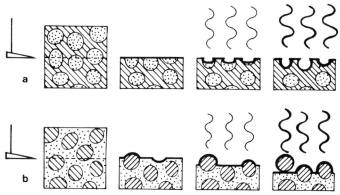

Figure 5.14. A diagram to illustrate the difference between shallow and deep etching. (a) Etchable structures (dotted) in a nonetchable matrix. (b) Nonetchable structures (shaded) in an etchable matrix. Note that deep-etched particles and deep depressions are not completely replicated. [Redrawn from Robards and Sleytr (1985).]

procedure. A good general account of the physicochemical parameters of etching is given by Moor (1973).

Frozen water will sublime if the saturated vapor pressure of ice in the specimen is higher than the partial pressure of water vapor in the immediate environment surrounding the specimen. If the reverse is the case, contamination will occur. The removal of water (sublimation) is a function of the sample temperature and the partial pressure of water vapor in the vacuum surrounding the specimen as well as the etchability of the sample, i.e., the amount and state of water in the sample. The predicted sublimation rates can be expressed by the Knudsen equation:

$$J_s = nP_s \left(\frac{M}{2\pi QT} \right)^{0.5} \tag{5.1}$$

where J_s is the sublimation rate, n is the coefficient of evaporation, P_s is the saturation vapor pressure of water, Q is the universal gas constant, T is the absolute temperature of the sample, and M is the molecular weight of water.

Kouchi (1989) and Livesey *et al.* (1991) consider that the sublimation rate is also strongly influenced by the phase of ice being sublimed. Amorphous ice (I_A) sublimes much faster than I_C or I_H. This difference in sublimation rate could have a significant effect on the etching depth achieved in well quench-cooled samples. Umrath (1983) calculates an etching rate of 2 nm s^{-1} for pure ice at 173 K and a pressure of 1×10^{-5} Torr assuming that all the water is trapped. It is not sufficient to remove the water from the sample; this water must also be effectively trapped so that it does not return to the fracture face. The microtome arm and knife assembly, which are normally kept at LN$_2$ temperatures, make an effective trap for the sublimed water when placed immediately above the fractured surface. Alternatively, any cold surface a few millimeters from the fracture face will achieve the same trapping efficiency. Lupu and Constantinescu (1989) have constructed a close-fitting copper cap which covers the specimen and is kept cold by a flexible copper braid connected to the microtome arm kept at 77 K. The cap is used to protect the sample before and after fracturing and allows long, clean etching. The cap can only be used in conjunction with tensile stress fracturing methods. Provided all the sublimed water molecules are trapped, the etching rate will be determined by the sample temperature, vacuum conditions, and time, all of which may be controlled with the freeze-fracture replication equipment. Thus the lower the sample temperature, the higher the vacuum must be and/or the longer it takes to remove a given layer of water.

Robards and Sleytr (1985) show that, for a variety of reasons, the actual etching rate is lower than that of pure water and for all practical purposes may be considered to be closer to 100 nm min^{-1}. This rate of water sublimation provides adequate etching for most samples. The rate of sublimation gradually slows down and eventually stops because only the water molecules sublime, leaving behind the solute, which builds up on the surface and gradually blocks the escape of further water molecules. Paradoxically, the better the initial cooling, i.e., the smaller the ice crystallites, the sooner the sublimation progress slows down. Poorly cooled samples can be more deeply etched. Specimens that have been infiltrated with cryoprotectants before rapid cooling are etched at a much slower rate and usually only reveal the frozen eutectic even after prolonged sublimation. It would seem prudent either to avoid the use of penetrating cryoprotectants, or to use the nonpenetrating cryoprotectants discussed in Chapter 2.

Many freeze-fracture replication studies dispense with etching entirely because there is sufficient morphological detail in the fracture face without having to resort to the further complication of subliming a layer of water from the surface of the sample. On the basis of being able to calculate precisely the etching rate of pure ice sublimed under controlled conditions, there is a mistaken belief that the etching process in heterogeneous samples can be finely controlled. This is unlikely, because it assumes that all the water molecules at the surface of such samples have an equal chance to leave the surface. We showed in Chapter 2 that water in heterogeneous systems generally, and in biological tissue in particular, is combined in various ways with the hydrophilic components of the sample. Water would be more readily sublimed from some parts of the specimen than from others, which would make it more difficult to interpret the resulting topographic images. Elgsaeter *et al.* (1984), in a study of freeze-etching of macromolecules, found that long, flexible molecules such as spectrin, alginate, and mucin would be moved and distorted during the sublimation of the ice.

5.3.5. Deep Etching

Deep etching in combination with rotary shadowing (see section 5.3.7 below) is a useful method of revealing the three-dimensional structure of cellular components

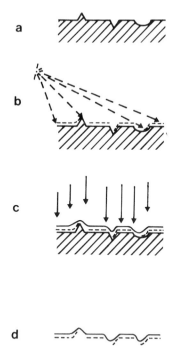

Figure 5.15. The formation of a freeze-fracture replica. The clean frozen fracture face (a) is first metal shadowed (b) from one side at an angle of about 45°. The shadowing deposits a thin discontinuous electron dense metal layer on those faces of the fracture face that are in line of slight with the evaporation source. The replica is formed (c) by evaporating a thin continuous electron transparent layer of carbon at right angles to the shadowed fracture face. The replica (d) is revealed after the underlying material is carefully removed.

by freeze-drying the surface regions (Fig. 5.15). The process is not, however, simply a matter of letting the sublimation proceed deeper and deeper into the sample, because the structures that are revealed become progressively more difficult to replicate and recognize. While the magnesium oxide surface labeling of Favard *et al.* (1990) provides a simple means for recognition, other more elaboate procedures are needed to ensure successful replication. The structural features of deep etched cytoplasm become covered with a deposit of salts and other normally solubilized components, which have to be removed before the fine structure and interconnections of macromolecules are revealed. The impressive deep-etched images (see Fig 5.16) produced by Heuser and colleagues, Heuser and Saltpeter (1979), Heuser (1981), Hirokawa and Heuser (1981), and Heuser (1989) are not solely a result of freeze-drying the frozen surface; they also involve chemical fixation, freeze-fracture replication, and freeze-drying.

Different protocols have been devised to remove the soluble components of the cytoplasm to allow deep etching into the fibrous network that makes up the cytoskeleton. Heuser and Kirscher (1980) permeabilized the cell membrane with dilute detergent solutions before quench cooling so that the soluble components leaked out. The remains were lightly fixed, rapidly cooled, and then deep etched for several minutes. The absence of cryoprotectants allows deep etching to take place, and the success of the procedure depends critically on there being good cryofixation of the samples.

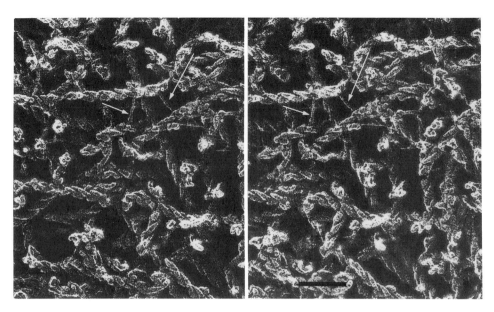

Figure 5.16. Freeze replica stereopair of the cytoplasm of hair cells from the chick ear that were detergent extracted and decorated with S1 for 10 min at room temperature, fixed with glutaraldehyde and tannic acid, washed with distilled water, rapidly frozen, deep etched, and rotary shadowed. The double-stranded ropelike structures are decorated actin filaments which are joined to each other by 3-nm connecting links (arrows). The combination of deep etching with stereophotography reveals the three-dimensional quality of the image. Bar marker = 100 nm. [Picture courtesy Hirokawa and Tilney (1982).]

The use of impact cooling has been important in the development of deep etching, although more recent work by Haggis (1985, 1986) and Haggis and Pawley (1988) has shown that it is possible to obtain good deep-etched images from specimens that had been pretreated, either by brief fixation in glutaraldehyde followed by suspension in 15% methanol (a volatile cryoprotecting agent) or by being infiltrated with 5% glycerol of DMSO prior to cooling with liquid propane.

Heuser (1983) found that macromolecules became clumped or collapsed during deep etching owing to the surface tension forces that developed during transfer of the specimens from ice to vacuum. Heuser overcame this particular problem by mixing the macromolecules with an aqueous suspension of minute mica flakes, which are then rapidly cooled, fractured, and deep etched. The molecules become attached to the surface of the mica flakes and the fracturing process strikes the surface on many mica flakes and passes through the adsorbed macromolecules cleanly enough to reveal their interior. The deep etching also exposes large areas of unfractured mica and reveals intact molecules. The molecules are not obscured by salt deposits and are revealed after 3 min etching at 175 K and 130 μPa (10^{-6}.Torr). The paper by Heuser (1989) provides an excellent review of this technique together with practical protocols for its implementation.

Deep etching is a remarkably versatile technique, for not only does it reveal the fine structural detail of the cytoskeleton and macromolecules, it also can be used in conjunction with specific labeling techniques such as antibody labeling and myosin decoration (Heuser and Kirschner, 1980; Hirokawa and Heuser, 1986). Deep etching has been used to follow a wide range of dynamic processes, and Menco (1986) and Chandler (1986) provide a summary of the structures that have been examined by the deep etch and more conventional freeze-fracture replication methods.

Etching, deep etching, and freeze drying are really the same process; they differ primarily in respect of how far the sublimation of water proceeds into the bulk of the sample. Deep etching differs from freeze drying and surface etching in that it usually involves some form of chemical treatment to remove salts and soluble components that would otherwise obscure the structures being investigated. In this respect deep etching is similar to the freeze-fracture thaw fixation procedures (section 5.5.2) that have been devised for examining the fine structures of specimens by SEM.

5.3.6. Replication

Up to this point, low-temperature fracturing and freeze-fracture replication have essentially followed the same experimental protocols, but now the two procedures show considerable divergence. In low-temperature fracturing we would continue to deal with the natural frozen fracture surface; in freeze-fracture replication we make a careful copy or mould of this surface and dispense with the underlying material.

There are five main types of replica. The *simple replica* is made from one of the two faces exposed by the fracturing process, while *complementary replicas* are made from both faces and allow (in principle) precise matching of features on the two opposing fracture faces. Anderson Forsman and Pinto da Silva (1988) developed a new method to observe the surface of cells. In the method, referred to as *fracture-flip replicas*, the carbon stabilized replica is inverted (flipped) and repeatedly washed with distilled water to expose the remaining biological material attached to the underside of replica for further study. A similar procedure is used in the production of *composite replicas* (Coleman and Wade, 1989). Membranes are fractured, replicated, and then inverted and the underlying surface deep etched and replicated. This composite preparation of two replicas with an intervening half-membrane and associated surface elements may be examined in stereo to reveal the relationship between structures on the surface and within the membrane. Fracture-flip replicas and composite replicas play an important role in the cytochemical analysis of membranes.

The fracture-flip method for membranes has been further extended and improved by the introduction of the *simulcast replica* technique recently reported by Ru-Long and Pinto de Silva (1990). The membrane fracture faces are first metal shadowed, followed by carbon stabilization, thawing, and flipping. After careful azimuthal reorientation of the replicas, the membrane surfaces are shadowed from the same direction used in the first shadowing procedure. The method can be combined with immunogold labeling and a single image of a single membrane can be used to give morphological, topological, and chemical information about dynamic membrane phenomena such as viral budding. The interpretation of some of the images is not always straightforward as the membrane-transversing proteins may partition in an unpredictable manner.

Surface replication is a two-part process. In the first stage, a thin discontinuous layer (1–2 nm) of a heavy metal such as Pt or Pt/C is evaporated onto the specimen from an oblique angle, usually 45°, to provide contrast enhancement of the topographic features of the fracture surface. The shadow coating will not be continuous over the whole sample, so it is necessary to hold all these patches of material together by a second, somewhat thicker continuous layer (5–10 nm) of an electron transparent material such as carbon or silica. These two layers form the replica; the carbon layer faithfully replicates the features of the fracture surface, while the heavy metal provides the necessary contrast for visualizing the replica in the transmission electron microscope. A diagrammatic representation of what occurs is shown in Fig. 5.15.

In a computer simulation study of metal evaporation shadowing, Colquhoun *et al.* (1985) suggest that shadowing at angles higher than 45° would give improved resolution of certain regions although the total information content of the replica may be diminished. The initial layer of metal produces a shadowing effect because the material travels in straight lines from the evaporation source. The sides of those features that face the source and are above or below an imaginary planar surface across the specimen will receive more of the metal coating than features that face away. The same effect can be seen in the way snow piles up against an object during a blizzard.

The formation of thin films is complex, but much is known about the processes, which involve thermal accommodation at the surface, surface diffusion, nucleation, crystal growth, and coalescence. The standard works of Holland (1956), Maisell and Glang (1970), and Venables (1984) describe these processes in great detail. Chapter 7 in the book by Newbury *et al.* (1986) will provide a good introduction to the process of specimen coating for electron microscopy, and additional information may be found in the references by Abermann *et al.* (1972) Willison and Rowe (1980), Slayter (1980), Peters (1986), Gross (1987), and Wepf (1991). The appropriate chapter in the book by Robards and Sleytr (1985) provides a good practical guide to the coating procedures used in freeze-fracture replication.

The aim of replication is to produce a high-fidelity copy of the fracture surface relief. The individual metal particles, or grains, which make up the thin film must be smaller than the features that one hopes to replicate (a layer of glass marbles would more faithfully replicate the surface of a cobbled road than a layer of footballs!). Small grain size and consequently increased spatial resolution is obtained by lowering the specimen temperature, decreasing the deposition rate, using high-melting-point metals, and either evaporating heavy metal alloys or simultaneously evaporating heavy metals with carbon. The use of alloys and metal/carbon mixtures minimizes epitaxial growth and crystal formation. Slayter (1980) describes techniques for producing very thin films in which the grain size is considerably reduced. He finds that the number of crystallites per unit area of platinum and tungsten is at a maximum in films 1.0–0.5 nm thick. High-melting-point metals form more crystallites per unit area, and twice as many crystallites are formed per unit area on a sample held at 77 K than at 300 K.

The sample surface also has a considerable influence on the way the films are deposited. There will be regions of different binding strength, and local topographic features will influence the amount of lateral movement of the condensing atoms. This leads to metal deposits on very small structures, which would not normally be expected to be visualized on the basis of simple geometric shadowing by the metal. These "decoration artifacts" generate their own set of problems and may make it very difficult to interpret accurately whether a feature on a replica is a true representation of the fine topography of the fracture surface, a visual manifestation of the physicochemical properties of the fracture surface, or a consequence of incipient contamination. Bachmann *et al.* (1985) have taken advantage of the physicochemical affinity of the shadowing metal for specific sites on the sample surface (decoration) as the basis of an analytical procedure to study the ultrastructure of frozen-hydrated surfaces. The information revealed by shadowing and by decoration is quite different in nature. Figure 5.17 shows the way ice crystals may decorate hydrophillic sites on freeze-fracture replicas of model membranes.

A mixture of platinum and carbon is the most commonly used material for producing the shadowing effects on replicas. The carbon is included because it will supress the mobility of the platinum atoms as they arive on the surface and diminish the formation of crystalline structures. Typically, Pt/C evaporation consists of 95% Pt and 5% C by weight, which gives approximately equal numbers of Pt and C atoms at the sample surface. It is usual to apply a layer a few nanometers thick, although ultrathin layers of 1 nm have been used in high-resolution studies of single molecules (Ruben, 1989). Fine-grain shadowing can also be obtained with a tungsten–tantalum alloy, which is superior to Pt/C for replicating the surface geometry of the sample (Bachmann *et al.*, 1985; Gross *et al.*, 1987; Menco *et al.*, 1988; Wepf *et al.*, 1991); and also superior to tantalum alone (Gross *et al.*, 1984). The backing layer of carbon should be applied as soon as the metal shadowing is completed.

Thin films may be formed by a number of different methods, and the choice of technique can have a considerable effect on the quality of the coated surface. Evaporative methods are most commonly used to make freeze-fracture replicas because the evaporant travels in straight lines and the technique is comparatively simple to use. Most metals may be evaporated as a consequence of resistive heating, which involves passing a heavy electric current through the material, or a refractory container holding the material, so that it heats up to a point at which it will evaporate in a vacuum of ca. 500 μPa. Carbon rods may be heated and evaporated in a similar fashion. Peters (1984) describes a simple method of obtaining reproducible thin films with a minimum of photon radiation onto the sample, by using flash evaporation with carbon fiber as the source material. Electron beam bombardment has to be used to heat up high-melting-point metals such as tungsten or tantalum. Evaporation of heavy metals and carbon only occurs at high temperatures and the high thermal loading may structurally alter frozen specimens. Energy can be dissipated into the

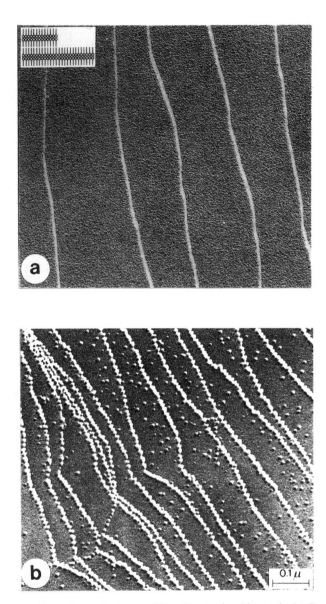

Figure 5.17. Ice-crystal decoration artifacts on multilamellar stearic acid crystals. (a) Control experiments showing crystal steps devoid of ice crystals. (b) Crystal steps are preferentially decorated with ice crystals after water vapor deposition. [Picture courtesy Walzthony *et al.* (1981).]

sample as a consequence of condensation of the vaporized material, thermal (photon) radiation from the evaporation source, and high-energy particle bombardment from poorly designed electron guns.

The formation of suitable metal–carbon replicas is crtical to the success of the freeze-fracture replication procedure but it is also one of the places where errors can occur and artifacts are generated. For this reason it is appropriate to examine the shadowing and replication processes more closely to appreciate their limitations. The success of thin-film formation by high-vacuum evporation depends on saturating the substate surface with diffusible, mobile metal atoms. This is achieved by high deposition rates of high-melting-point metals, typically 60 nm min^{-1}. In freeze-fracture replication, the low temperature of the substrate favors rapid condensation and produces a high nucleation density of metal aggregates. The arriving metal atoms diffuse freely and the chemical and physical properties of the surface have a strong influence on the sites of initial nucleation. However, secondary energy sources, such as heat of condensation and photon radiation from the evaporant surface, can cause uneven agglomeration of the metal layers as the thin film grows. Belous and Wayman (1967) calculated that thermal photon radiation can raise the surface temperature as much as 100 K even at temperatures as low as 100 K. Very low substrate temperatures (100 K) will reduce the high surface diffusion of the metal atoms, but such low temperatures require ultrahigh vacuum and/or cold shrouds to reduce surface contamination (Gross, 1984; Ruben and Marx, 1984). All these processes lead to uneven films, which limit spatial resolution.

Some of the same effects also apply during the deposition of the backing layer of carbon. Carbon evaporates at 3500 K and the photon radiation can damage thermolabile surfaces, but the presence of the previously applied metal shadow coating would be expected to help conduct away any heat. The presence of the ultraclean layer of heavy metal will tend to limit the lateral diffusion of the carbon atoms, and in any event, carbon films have a very small particle size.

A number of devices have been built which significantly decrease the thermal loading on the sample, and calculations by a number of workers have shown that the heat load is no more than 5% of the total thermal output from the evaporant source (Robards and Sleytr, 1985). Nevertheless, there is still some uncertainty as to how much damage evaporative coating can cause to organic samples. The momentum of an atom evaporated at between 2300 and 4800 K is quite considerable, and this energy is dissipated into the sample. Colquhoun (1984) calculated that the energy of a vaporized platinum atom is about 55 kJ mol^{-1}, which would be enough to disrupt intermolecular forces but not enough to break bonds. The C–C bond energy is 348 kJ mol^{-1} and the C–H bond energy is 414 kJ mol^{-1}. Colquhoun considers 55 kJ mol^{-1} to be a median value and that some of the evaporated atoms will be traveling at much higher speeds with energies approaching $150–200 \text{ kJ mol}^{-1}$, which would be sufficient to break bonds in some molecules. In this respect the H–H bond energy of 42 kJ mol^{-1} in bulk ice might be overcome. A substantial proportion of the radiant flux would consist of UV radiation, which would also damage chemical bonds. Although some of these problems will be overcome by cooling the sample, it must remain a matter of some concern that metal evaporation may cause artifacts on frozen fracture faces.

The potentially damaging effects of metal evaporation are unlikely to be alieviated by using alternative methods of coating such as the high-energy sputter deposition methods developed by Peters (1986), which produce very thin continuous films of 0.5–1.0 nm average mass thickness and have been used to resolve features as small as 1 nm in the SEM. The kinetic energy of the metal atoms produced by Pening sputtering is 5–10 times higher than the energy of evaporated metal atoms, although the *deposition* rate is 2–3 orders of magnitude slower than electron beam evaporation. Theoretically, the specimen damage should be less because of the absence of additional photon radiation, but the long coating times would increase the problems of surface contamination. The omnidirectional nature of Penning sputtering and ion beam sputtering (Clay and Peace, 1981) would make this a difficult technique to use instead of unidirectional evaporative shadowing, but it might prove to be an interesting alternative to rotary shadowing (see section 5.3.7). Wildhaber *et al.* (1985) have made a comparative study of very thin film shadowing produced by ion beam sputtering and electron beam evaporation. They find that with the same amount of material, much thinner continuous films are produced by sputtering, although there was no improvement in the granularity compared to evaporated films. From an examination of optically filtered micrographs, they concluded that images of electron beam shadowed material contained more information than similarly processed images from ion beam sputtered material. It is interesting, however, that they found that ion beam sputtered material has a tendency to shadow in a purely goemetric fashion, whereas evaporated material tended to decorate the surface. These differences are a consequence of the difference in the kinetic energy of the arriving metal atoms.

The same limitations that apply to Penning and ion beam sputtering would also apply to the sputter shadowing technique introduced by Colquhoun (1984) and Colquhoun and Cassimeris (1985). This procedure combines the omnidirectional advantage of diode sputter coating with the shadowing effects that may be achieved by carefully adjusting the position of the sample in relation to the main direction of the metal coating. More work needs to be done to see if these alternative coating methods could be modified to produce the high-resolution metal films needed for making freeze-fracture replicas. In this respect, the series of papers edited by Baumeister and Zeitler (1985) on high-resolution shadowing and the recent paper by Wepf *et al.* (1991) on coating methods for TEM, STM, and SEM should be consulted.

5.3.7. Rotary Shadowing

The uni-directional evaporation method of applying the initial heavy metal coat will only produce shadows in one direction. For most freeze-fracture replication studies this is quite satisfactory, but there may be occasions when the shadow cast by one structure may obscure a smaller structure in its lee. This may be overcome either by bidirectional shadowing or by rotary shadowing. In rotary shadowing, the sample is turned at between 50 and 200 rpm during the initial metal evaporation with the evaporation source placed at a low angle (5°–20°) relative to the specimen. This angle appears to be critical and will depend on the geometry of the object being studied. Colquhoun and Sokol (1986) used computer simulation techniques to investigate rotary shadowing at high and low angles and showed that steep angles of shadowing drastically reduced resolution. The rotary shadowing is followed by the usual application of evaporated carbon at an angle of 75°–90° to form the replica. Figure 5.18 shows the effects of rotary shadowing in comparison to unidirectional shadowing.

Rotary shadowing offers the following advantages over unidirectional shadowing:

1. It avoids the pile up of material on one side of an object and a total lack of material on the opposite side with the consequent loss of information.
2. It avoids the problem of a structure casting shadows on a neighboring structure and obscuring information.
3. When the shadowing is carried out at a low angle, it has the effect of producing a three-dimensional replica, which can enhance the contrast of small surface protrusions. This is particularly useful when studying single macromolecules on otherwise flat surfaces.

Although rotary shadowing has provided additional information about linear macromolecules such as DNA (Kleinschmidt 1960), there are some doubts about its general usefulness in shadowing fracture faces. For example, Neugebauer and Zingsheim (1979) found that the apparent diameter of ferritin molcules varied between 12 and 26 nm as the rotary shadowing parameters were changed. Care must obviously be taken in interpreting images produced by these techniques. However, before dismissing the technique entirely it should be remembered that rotary shadowing has been used to great effect in the production of deep-etched replicas. The paper by Chandler (1986) reviews the methods and applications of rotary shadowing.

5.3.8. Thin-Film Thickness Measurements

It is usually necessary to measure the thickness of the deposited films, because ultrathin layers do not give sufficient contrast and thick layers of metal and carbon will obliterate all the surface features. The simplest methods, although wildly inaccurate, are sometimes the most effective and involve either calculating the amount of material to be evaporated from the geometry of the system, or comparing the density of a deposit on a white tile or the interference colors on a glass slide. Thus a 10-nm carbon layer is a light grey-brown color on a white surface and has red to blue interference colors (Robards and Sleytr, 1985).

Film thicknesses may also be measured using quartz thin-film monitors, although doubts have been expressed about the accuracy of this method, particulary when very thin films are being measured (Flood, 1980; Slayter, 1980; Peters, 1980, 1986). Quartz-crystal monitors measure the *average mass thickness*, which in the case of the discontinuous films usd to shadow fracture faces is much less than the *average metric thickness*. These matters are discussed at length by Peters (1986) and in Chapter 7 of the book by Newbury *et al.* (1986), which also suggests other ways of measuring the thickness of thin films.

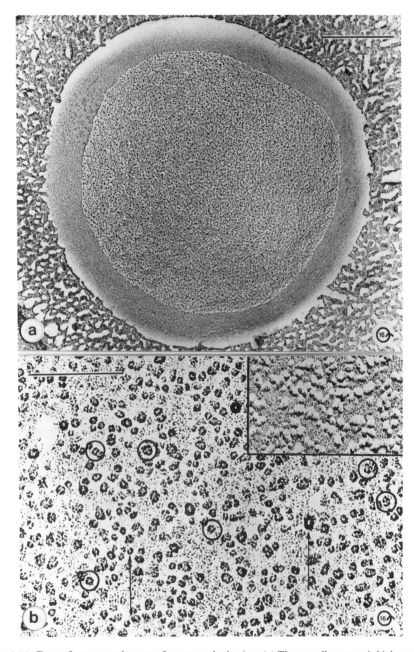

Figure 5.18. Freeze-fracture erythrocyte after rotary shadowing. (a) The overall contrast is high compared to that of a conventionally undirectionally shadowed membranes. Bar = 0.5 μm. (b) At high magnification many of the intramembrane particles show a tetrameric subunit structure (circles), which is not usually seen after unidirectional shadowing (insert). The arrows point to variations in size and structure of the subunits. Bar = 0.1 μm. [Picture courtesy Margaritis *et al.* (1977).]

5.3.9. Cleaning the Replicas

Once the replica has been produced, all that remains is to remove it from the surface of the frozen sample. The replicas are at most 10–30 nm thick and are very delicate objects. Some people consider it prudent to apply an additional, more robust, but temporary backing material, which may be removed once the replica is safely seated on the electron microscope grid. A thicker carbon coating is of little help because it will limit the resolution of the replica. Robards and Umrath (1978) applied an evaporated layer of silver to the surface of the replica, which can later be removed either with a cyanide solution or with nitric acid. Various organic polymers such as cellulose nitrate or polystyrene have also been used, although these are not always easy to remove completely and there may be a loss of image quality.

The actual removal of the replica from the fracture surface is a delicate operation, which involves first thawing the underlying frozen material and then removing all traces of the organic material by gently, very gently, washing the replica in a series of cleaning solutions, which include alcoholic KOH, chromic acid, sulfuric acid, and sodium hypochlorite. All these procedures conspire to disintegrate the replica; the initial thawing can bring about dimensional changes and different parts of the fracture face adhere more closely to the replica than others. Adachi *et al.* (1987) have described a method that permits the recovery of large electron microscope grid size replicas with comparative ease. A thin sheet of dental wax covered with a piece of rayon fabric is floated on a bath of household bleach. The tissue, covered with the Pt/C replica is placed on the floating rayon fabric and after 1–2 h the tissue is dissolved and the replica sits on the rayon. The supported replica is now washed several times in distilled water before the support is allowed to sink below the surface leaving the replica floating on the top. This can now be picked up onto the electron microscope grid with a minimum of rolling and fragmentation of the replica. The book by Robards and Sleytr (1985) contains an excellent practical account how the replica may be cleaned and mounted on the specimen grid.

5.3.10. Artifacts and Interpretation

Thirty-five years ago at the time freeze-fracture replication was being developed, it was hoped, in retrospect somewhat piously, that the method would provide artifact-free images of hydrated samples. In some respects this has occurred, for the method has played on a major role in understanding the structure of biological membranes. However, the method, like any interventionist technique, will introduce its own catalogue of artifacts. Most can be related to a specific stage of the preparation procedure and many can be avoided. Other artifacts that have been characterized, such as plastic deformation and macromolecular collapse, appear to have no remedy and their potential presence must be taken into account when interpreting the replica images. Fortunately, nearly all the artifacts are now recognized and have been fully described (Stolinski and Breathnatch, 1975; Bohler, 1975; Bullivant, 1977; Sleytr and Robards, 1977a, 1982; Kistler and Kellenberger, 1977; Rash and Hudson, 1979; Lepault and Dubochet, 1980; and Winkler, 1985.

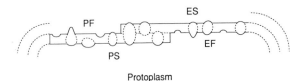

Figure 5.19. Nomemclature used for labeling the different freeze-fracture faces of biological membranes exposed by fracturing and for the membrane surfaces exposed by etching. PF, Fracture face of that half of the membrane associated with the protoplasm. EF, Fracture face of that half of the membrane associated with the extracellular space. PS, The protoplasmic surface of the membrane. ES, The extracellular surface of the membrane. [Redrawn from Branton *et al.* (1975).]

The micrographs we obtain from the transmission electron microscope examination of freeze-fracture replicas are two-dimensional projections of differences in electron scattering brought about by variations in the mass density of the replica. In order to relate the images back to the three-dimensional structures from which they were derived, we need to have some information about the direction and angle of shadowing. The generally accepted convention is for the shadows to appear *white* on the micrographs, even though this is the complete reversal of the way shadows appear in everyday life. An arrow on the micrographs indicates the direction of shadowing. If possible, the micrographs should be observed with the shadows going away from the observer, i.e., from the bottom to the top of the page. Because the fracturing process frequently splits and runs along each side of biological membranes to expose extended membrane faces, a standard nomenclature has been adopted to describe the different membrane faces (Fig. 5.19).

Freeze-fracture replicas may also be examined by scanning tunneling microscopy (STM) and atomic force microscopy (AFM), which for the first time can produce images of surfaces at atomic resolution. High-resolution images have been obtained directly from dried and hydrated material with or without metal coating, and in the context of this chapter, the techniques can be used in combination with the stripped replicas derived from planar monolayer technique (see section 5.3.1.). Fisher (1989) and Fisher *et al.* (1990a, b) and Hansma *et al.* (1988) provide good reviews of the preparative and imaging procedures and their application to nonconducting organic and biological samples. Figure 5.20 shows a scanning tunneling micrograph of a membrane surface.

5.4. *FREEZE-FRACTURE REPLICATION AS AN ANALYTICAL PROCEDURE*

Although the replicas are inert they can be used indirectly to give chemical information about the specimen. The methods that have been devised fall into two

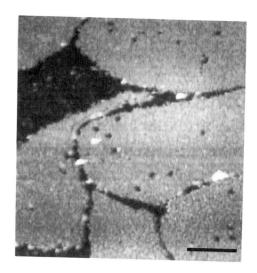

Figure 5.20. Scanning tunneling electron micrograph of purple membrane attached to polylysine-coated glass, rapidly cooled, and deep etched. The micrograph shows six membrane fragments. The platinum–iridium-coated sample was scanned *in situ* in air at 293 K using a platinum–rhodium tip operating at a tunneling current of 3 nA and a bias voltage of 1.0 V, sample negative. The scanning tunneling image was computer processed to remove drift in the *z*-axis and presented as a top-view gray scale image where black = mica substrate and white = highest point above the substrate. Bar = 200 nm. [Picture courtesy Fisher *et al.* (1990).]

general groups: those that involve labeling or chemical intervention before rapid cooling and those in which the labeling is carried out on the frozen fracture face before or after replication. The first group would include freeze-fracture and autoradiography and the use of morphologically recognizable or electron dense markers; the second group essentially correlates freeze-fracture replication with information obtained by other low-temperature techniques.

5.4.1. Freeze-Fracture Autoradiography

Freeze-fracture autoradiography (FARG), developed by Fisher and Branton (1976), involves applying a very thin layer of photographic emulsion at 143 K over a replica on the surface of a sample that had been labeled with an appropriate radioactive isotope while it was physiologically active. The entire sample, adherent replica, and the thin emulsion layer is stored in the dark at 193 K before the emulsion is developed and the replica is recovered and cleaned (Fig. 5.21). The emulsion-coated replica is examined in the electron microscope. There are a number of variations to this basic theme, which allows the emulsion to be applied in the vacuum (Fisher and Branton, 1976; Rix, 1976) or outside the vacuum (Schiller *et al.*, 1978; Fisher, 1978).

Fisher (1974, 1975, 1976) and Edwards *et al.* (1986) modified the method to incorporate monolayer freeze fracturing with electron microscopy autoradiography (MONOFARG). The isotope-labeled specimen is first bound to a support such as glass or mica, which is then cooled rapidly, fractured, and replicated and treated with an emulsion layer as described above. The thin replicas are removed from the underlying support and examined in the TEM. The monolayer technique has the advantage of increasing the spatial resolution of the autoradiography technique. The technique has been refined further to include double-labeled membrane splitting (DBLAMS) and single membrane monolayer splitting (SMMS), which allows an

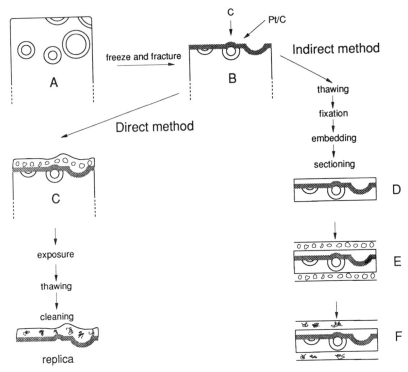

Figure 5.21. Direct and indirect freeze-fracture autoradiography. The radioactively labeled specimen (A) is frozen, fractured, shadowed, and replicated in the usual way (B). In the direct method, the replica, while still in place over the frozen sample, is covered in the dark with a thin layer (C) of photographic emulsion. The emulsion is exposed for several weeks at 173 K and atmospheric pressure to allow the sites of radioactivity to be recorded in the emulsion. The underlying material is removed by thawing and the emulsion photographically processed. The replica and attached emulsion layer is cleaned and the developed silver grains may be related to specific morphological features on the replica. In the indirect method, the replica and underlying frozen material are thawed, chemically fixed, and plastic embedded, and then sectioned parallel to the replica. The sectioned replica (D) is coated on both sides with a thin layer of photographic emulsion (E) and left at room temperature for several weeks to allow the sites of radioactivity to be recorded. The emulsion-coated section is photographically processed and the developed silver grains related to specific sites in the sectioned replica (F). [Modified and redrawn from Robards and Sleytr (1985).]

analysis of the transmembrane distribution of native and radiolabeled proteins and lipids. Details of, and multiple references to, these techniques may be found in the paper by Fisher (1989). Outlines of the FARG and the MONOFARG processes are shown in Fig. 5.22, and a diagram of the range of procedures is shown in Fig. 5.23.

5.4.2. Freeze-Fracture Labeling and Cytochemistry

Freeze-fracture labeling combines a variety of cytochemical techniques with freeze-fracture replication and makes it possible to relate the distribution of

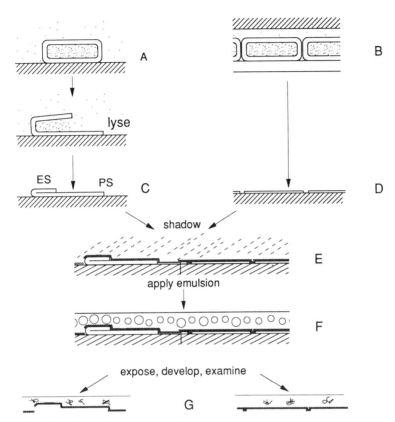

Figure 5.22. Monolayer freeze-fracture autoradiography (MONOFARG). Cells, labeled with a radioisotope, are attached to a planar cationic surface either as single cells (A) or a thick slurry of cells (B). Single cells may be lysed and freeze-fractured (C) to produce intact membranes with ES and PS faces. Alternatively, the cell slurry (B) may be fractured (D) to produce half-membranes with EF faces. The fractures are metal shadowed (E) and coated with a thin layer of photographic emulsion (F). After suitable exposure, the emulsion is developed and the silver grains related to specific sites on the replica (G). [Redrawn from Fisher (1981).]

membrane particles to surface receptors, antigens, and other chemical groups. The cytochemical labeling may be integrated into the freeze-fracture replication procedure (1) before rapid cooling (Fig. 5.24), (2) after fracturing and thawing (Fig. 5.25), (3) after metal shadowing (Fig. 5.26), and (4) after carbon replication (Fig. 5.27). The cytochemcial procedures center on colloidal gold immunolabeling of cryofractures, which has been pioneered by Pinto da Silva and co-workers (1973, 1981, 1984, 1989), who combined it with the freeze-fracture flip method (see Fig. 5.26), and by Boonstra *et al.*, (1985, 1987, 1989, 1991) and Verkleij and Leunissen (1989). Although there are many variants to the immunocytochemical procedure, the basic technique is as follows. Cells, immunolabeled with colloidal gold, are quench frozen, fractured, etched, and replicated. The 5–20-nm gold particles remain trapped in the replica

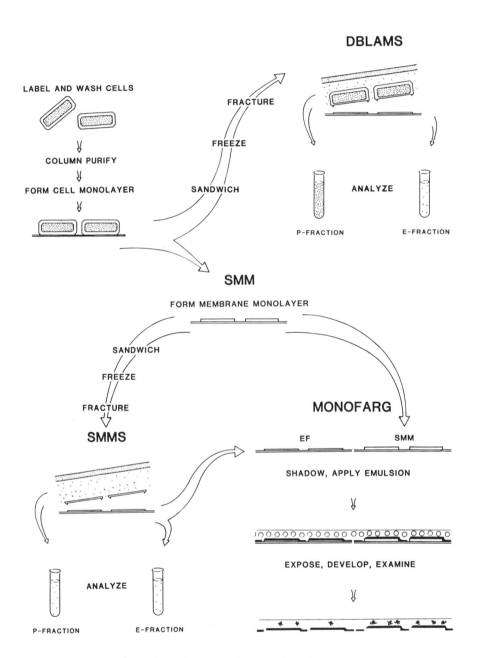

Figure 5.23. A summary of three freeze-fracture methods based on planar cell and membrane monolayer preparation. DBLAMS, double-labeled membrane splitting; SMMS, single-membrane monolayer splitting; and MONOFARG, mono-layer freeze fracture autoradiography. [Illustration courtesy Fisher (1989).]

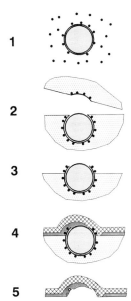

Figure 5.24. Principle of the label-fracture-etch technique. Cells in suspension are immunolabeled with colloidal gold particles (1) and after washing are frozen and fractured (2). The exposed fracture face is etched (3) to reveal structures labeled with gold particles. The etched fracture face is shadowed and replicated (4), and following cleaning (5) the gold marker particles are embedded in the replica. [Illustration courtesy Severs (1991).]

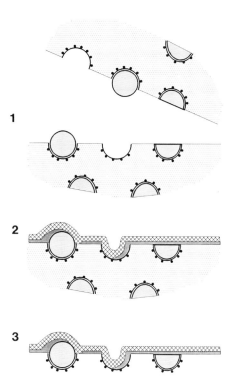

Figure 5.25. Principle of the label-fracture technique. A cell suspension is gold immunolabeled, washed, frozen, and fractured (1). A standard shadowed replica is made of the unetched fracture surface (2), and instead of cleaning the replica with chemicals, it is rinsed in distilled water leaving remnants of the fractured cells attached to the replica (3). Where a cell has been "concavely" fractured to reveal the E-face, the E-half of the membrane with its label remains attached (see middle cell). Replica of the fracture face plus label on the surface can be viewed in the TEM. Where cells are "convexly" fractured (left cell) or cross fractured (right cell) remnants of the biological material obscure the electron beam and the labeled replica is not visible. [Illustration courtesy Severs (1991).]

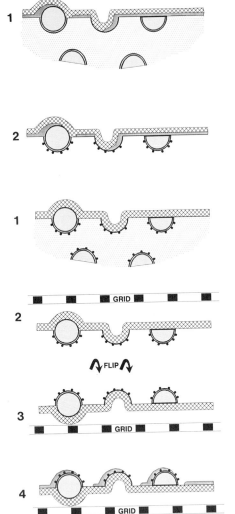

Figure 5.26. Immunogold labeling of fragments attached to replicas *after* the replica has been made. The material is freeze-fractured and shadow-replicated in the usual way (1). After thawing and washing (no chemical cleaning) the remnants of material attached to the replica may be labeled by immunogold cytochemistry (2). [Illustration courtesy Severs (1991).]

Figure 5.27. Principle of the freeze-fracture flip technique. Suitably labeled cells are freeze-fractured and carbon coated (1), and after thawing and washing the carbon replica with attached remnants are mounted on an EM grid from above (2). The EM grid is then turned over, i.e., "flipped," so that the remnants of material are now on top (3) and the replica dried and metal shadowed (4). Where a cell is initially fractured concavely (center) the labeled E surface can be examined. Cells which are convexly fractured (left) or cross fractured (right) do not provide useful images. In an alternative process, the cell remnants may be labeled after freeze-fracturing at point (2) in the diagram. [Illustration courtesy Severs (1991).]

following removal of the biological material and may be examined in the TEM (Fig. 5.28.) or in conjunction with BSE imaging in the SEM (Kan and Nanci, 1988). An excellent summary of the different procedures used in freeze-fracture cytochemistry may be found in the two review paper by Severs (1989, 1991).

The colloidal gold labeling is not the sole marker that has been used in freeze-fracture cytochemistry. Chandra *et al.* (1986) combined monolayer freeze-fracture replication with elemental analysis by ion microscopy and the subcellular distribution of a number of elements (Na to Ca) could be related to structural features revealed by freeze-fracture replication. One of the other main postfreezing labeling techniques

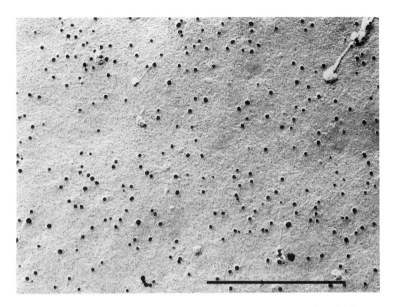

Figure 5.28. A431 human carcinoma cells fixed and labeled with a monoclonal anti-EGF receptor anti-body, rabbit antimouse, and protein A-gold conjugate before freeze-fracture replication. The 12-nm gold particles can be clearly distinguished on the platinum–carbon replicas. Bar = 0.5 μm. [Picture courtesy Boonstra *et al.* (1991).]

is the method devised by Rash (1979), which combines freeze-fracture replication and thin sectioning. Samples are cooled rapidly, fractured, and given a light shadowing of heavy metal. The fractured surface is then thawed and labeled with an appropriate marker before being chemically fixed, dehydrated, and embedded. Sections are cut parallel to the original replica surface so that information can be obtained simultaneously from the replica and the adjacent tissue sections. This method can be modified in various ways to include critical point and freeze-drying techniques.

5.4.3. Quantitative Freeze-Fracture Replication

The morphological information may be subjected to quantitative analysis. Wherever possible, stereo-pair microphotographs with a 6°–8° separation angle should be taken of the regions of interest. Stereo viewing of such pairs of images will reveal interconnections at the fracture surface as well as providing a clearer understanding of the form and nature of any artifacts. Digital image-processing techniques and stereological analysis can provide the basis for reducing some topographical features to statistically malleable terms. The frequency and distribution of intramembranous particles (IMP) have been subjected to such analyses. Appleyard *et al.* (1985) have developed statistical methods for the analysis of two-dimensional points which will unambiguously distinguish random, clustered, and dispersed patterns on replicas containing several hundred points. The method is more accurate than the earlier overlapping quadrant method developed by Niedermeyer and Wilkie (1982). Fassel and Hui (1988) have developed a reliable nonbiased method of measuring the size and density of IMP by using an automatic image analysis system. It is also possible to determine particle concentrations and molecular weight using freeze-fracture replication procedures (Bachmann, 1987). Particle concentrations in fluid specimens may be calculated from the number of particles per unit area on a freeze-fractured specimen provided the distribution is random and the particles are smaller than the etching depth. Solutions of bovine serum albumin are used to standardize the procedures and the method is suitable for measuring particle weights ranging from 10^4 to 10^8 Da. Albumin and fibrinogen molecules have been counted in this way. Bachmann has measured the size and shape of macromolecules from their appearance in rotary shadowed replicas produced by freeze-fracture replication. The relative influences of shadowing and "hydration" layers must be taken into account when calculating the true particle size dimension.

5.5. FREEZE FRACTURING AND CHEMICAL FIXATION

5.5.1. Introduction

So far we have taken a purist approach to freeze fracturing, in which we have eschewed the use of chemicals in the process. Fixatives, bulk conductors, or selective stabilizing and extraction agents may be used at various stages during the fracturing preparative procedure for preparing samples for morphological analysis in the SEM. There are a large number of different procedures that come under this heading and they are arranged into a number of main categories. It should be emphasized that these methods have been devised primarily to examine the internal structure of biological material by the SEM. They are not the sort of technique one would normally use with analytical investigations or with investigations of the ultrastructure of macromolecules.

5.5.2. Freeze-Fracture Thaw Fixation

Haggis and Bond (1979) showed that if fixation was delayed until after freeze-fracture and thawing, it was possible to wash out the soluble hemoglobin from the cytoplasm of chicken erythrocytes and reveal components of the cytoplasmic matrix. Haggis and Bond used 10% solutions of either glycerol or DMSO as a cryoprotective before rapid cooling and achieved good preservation in HeLa cells and lymphocytes. Haggis et al. (1983) and Haggis and Pawley (1988) extended freeze-fracture thaw fixation to high-resolution SEM. Cryoprotected cells, rapidly cooled and fractured under LN_2, are thawed in carefully designed fixatives, which removes a large proportion of the soluble material in the cell and exposes components of the cytoskeleton and nuclear chromatin. These structures may be preserved and observed after critical-point dryng (Fig. 5.29). The technique allows visualization of microtubules and intermediate filaments with the added advantage of seeing them dispersed within the three-dimensional space of the cell. Blackmore *et al.* (1984) have used the same sort of procedures to study the cytoplasm of plant cells.

Maruyama and Okuda (1982, 1985) have devised a variant of the Haggis and Bond technique in which the tissues are not pretreated with a cryoprotective, but after fracturing are thawed into a mixture of fixative and cryoprotective. After fixation and extraction, the remaining structures were subjected to conductive staining by tannic-acid–osmium ligation. The freeze-fracture thaw fixation procedure has many similarities to the deep etching process discussed earlier. The deep etching process is usually confined to small samples which are replicated and then examined in the transmission electron microscope, whereas freeze-fracture thaw fixation samples are destined for examination by scanning electron microscopy. The advantage of the freeze-fracture thaw fixation procedure is that it can provide either a low-power overview of the organization of the membraneous component of the cell or high-resolution images of organelles, macromolecules, or membrane surfaces.

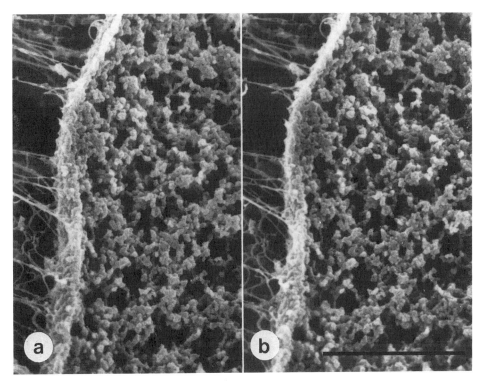

Figure 5.29. Stereopair picture of a 3T3 cell in interphase prepared by the freeze-fracture thaw technique. The fractured nucleus is compact and cytoplasmic filaments remain attached to the nuclear lamina after Triton extraction. Stereopair at (a) 0° and (b) 8°. Hitachi S-900 FESEM. Acc. voltage 2.5 kV. Bar = 10.0 μm. [Picture courtesy Haggis and Pawley (1988).]

5.5.3. Freeze Cracking

This is a derivative of the fracturing process and is one of the most effective ways of revealing the internal structure of specimens to be examined by SEM. Samples pass through a conventional ambient temperature fixation and dehydration schedule and once in pure organic solvent, e.g. acetone or ethanol, they may be solidified by plunging in LN_2. The cooling process results in the formation of either a glass or a microcrystalline solid which may be easily fractured to reveal the internal contents. The solvent may be removed by sublimation or substitution followed by critical-point drying. The same process can also be carried out on samples that have been infiltrated in unpolymerized resin. The resin can be removed by dissolving it in acetone or 1.2. epoxy propane. A general description of these methods is given in the book by Goldstein *et al.* (1981).

5.5.4. Freeze Polishing

The method has been developed by Inoue and Osatake (1984) and Inoue (1986) to observe the intracellular structure of thin cells, which are difficult to fracture or crack along their long axis. Frozen samples that have been pretreated with DMSO are polished with a thin film covered with fine (300-nm) particles of aluminum oxide. The polishing film is mounted on a metal plate cooled with LN_2 and the frozen sample polished with successively finer grain size films. After thawing in a DMSO solution at room temperature, the samples may be subjected to osmium maceration (Shiozaki and Shimada, 1990), which removes selected components of the cytoplasmic matrix.

5.5.5. Freeze-Fracture Thaw Digestion

Tanaka (1980) and Tanaka and Naguro (1981) have developed a digestion extraction procedure based on the well-known tissue extraction properties of solutions of osmium tetroxide. The mechanism by which this occurs has been summarized by Hayat (1981) and Behrman (1984). The technique involves freeze-fracturing osmium-fixed (1%) material cryoprotected with DMSO, to expose the interior. The frozen samples are thawed in a dilute (0.1%) solution of osmium tetroxide over a 3–4 day period to remove all the fibrous, proteinaceous, and soluble materials in the cytosol. The membrane systems of the cytoplasm are exposed for viewing in the SEM. An alternative procedure uses glycerol extraction after glutaraldehyde fixation rather than the prolonged osmium digestion. Under these circumstances, it should be possible to use antibody labeling on some of the components of the cytoskeleton. Inoue (1986) has reviewed a number of these freeze-fracture digestion procedures and considers that osmic digestion is the most effective method to remove excess cytoplasmic matrix components. Quite surprisingly, Inoue (1986) also found that distilled water rinsing after the initial osmium fixation revealed intracellular structures.

5.5.6. Freeze-Fracture Permeability

Heuser and Kirschner (1980) treated cells (fibroblasts) with a dilute detergent solution before they were frozen. The detergent solution partially or totally solubilizes the cell membranes, and the soluble cytoplasmic components leak out, leaving the cytoskeleton more or less intact. This may then either be lightly fixed, rapidly cooled and deep etched for examination in the SEM, or freeze-dried and rotary coated for examination in the TEM.

All these procedures produce images with a wealth of detail when examined in the SEM or, where appropriate, in the TEM. The use of stereo pair photographs enhances the three-dimensional aspects of the deep view into the cytoplasm. As Haggis (1982) has shown, these combined chemical and mechanical dissection procedures have proved to be particularly useful in revealing the interconnections of the membranes components of the cell and of the constituent parts of the cytoskeleton. It is difficult to say how the use of chemicals in association with freeze-fracture procedures affects the quality of information available from the sample. The chemicals used to extract, digest, solubilize, and remove some or most of the components of the cell are only selective insofar as they do not appear to remove the macromolecules, which are an important part of the cellular structural integrity. Differential extraction must occur and potential users of these methods must realize that numerous artifacts will appear. The process may be best compared with the way a small bulldozer can knock down the walls of a building while still leaving the structural framework intact.

5.6. APPLICATIONS OF LOW-TEMPERATURE FRACTURING AND FREEZE-FRACTURE REPLICATION

Freeze-fracture replication, and to a lesser extent, low-temperature fracturing, are applied widely to structural studies of membranes and lipid dispersions (Hope *et al.* 1989), from the very large replicas, which may be viewed at relatively low spatial resolution in the SEM, to the fine replicas, which are examined at ultrahigh resolution in the TEM. Deep etching procedures can be used to produce replicas that reveal information about the structural composition of the cytoskeleton. Unetched replicas from membranes fractured at liquid helium temperatures reveal the chemically distinct categories and domains that exist within the membrane matrix. The paper by Menco (1986) contains a comprehensive list of the biological structures and phenomena that have been studied by freeze-fracture replication. Platt-Aloia and Thomson (1989) show how these methods may be applied to plant material.

Food substances, which are generally, although now alas not exclusively, derived from biological materials, have been studied by both freeze fracturing and freeze-fracture replication techniques (Humphreys, 1989). The distribution of the fat and nonfat components in dairy products has been widely studied, and the technique has been applied to the study of the formation and growth of ice crystals in ice cream and frozen foods. The techniques have applications in the study of polymers (Sawyer and Grubb, 1987), where it has been used to provide information about the composition of a wide range of plastics. It has been used to study the structure and composition of colloidal particles in water–oil mixtures which are stabilized by polymer emulsifyers (Zasadzinski and Bailey, 1989). It has been used to study the structure of gels and emulsions, paints and oil products, fiber and paper products, cosmetic and pharmaceutical preparations.

Although low-temperature fracturing and freeze-fracture replication are similar in many respects, it is the differences in detail of the two procedures that best demonstrate their advantages and disadvantages. Low-temperature fracturing produces large fracture faces of real surfaces for examination and analysis as distinct from the replicas of surfaces produced by the freeze-fracture replication method. These real surfaces may be examined in the SEM, thus taking advantage of the increased depth of focus of this mode of electron beam imaging. This allows very rough fracture faces to be examined, a feature that is difficult by the freeze-fracture replication method. In addition, the low-temperature fractures may be examined during preparation and if unsatisfactory may be rejected and another fracture made. This is not the case with freeze-fracture replication where the information in the replica is only available at the end of the preparation procedure.

The images produced by low-temperature fracture methods generally have a lower spatial resolution than the images produced by the freeze-fracture replication technique. However, with the appearance of new high-resolution scanning electron microscope and the greatly improved coating methods, such difference are beginning to disappear. These matters will be discussed in Chapter 10 (section 10.7.8). The modes of image formation used in examining the products of the two preparative techniques are quite different and there is a need for a more critical evaluation of the interconnections between the two methods. In a few cases the two methods have been used together. The way forward would be to prepare fracture faces in equipment with high-vacuum, low-contamination characteristics such as the AMRAY Biochamber and the Balzers SCU 020, and examine them frozen and uncoated using the newly developed techniques of low-voltage microscopy (Joy, 1985; Pawley, 1988; Pawley *et al.*, 1991). The preliminary results have been shown to give a spatial resolution of 4 nm at 1.5 kV compared to 0.8–1.5 nm at 20–30 kV. Regions of general interest could be noted and the frozen sample taken back into the cryochamber for further modes of processing. If the fracture face is to be studied by x-ray microanalysis it would have been coated with a thin layer of a metal that would not interfere with the specific x-ray spectra that might be expected. If the fracture face is to be studied for morphological purposes only, then either it would have to be coated for direct viewing in the SEM or a replica would have to be made of the surface for viewing in the TEM. There is a need to develop coating techniques that will provide a coating layer that will serve the purposes of both SEM and TEM, and the approach being followed by Peters (1986) and Wepf *et al.* (1991) appears promising.

5.7. SUMMARY

Low-temperature fracturing and freeze-fracture replication are two procedures that may be used to expose the interior of a frozen sample. Samples may be fractured either by tensile stress or by the shear forces that are generated as a knife cleaves the sample. The fracture pathway–unlike sectioning, which follows a more or less predictable pathway–is very much a "hit-and-miss" affair and yields unpredictable variations within the exposed surfaces. Although low-temperature fracturing and freeze-fracture etching are similar in many respects, there are important differences. Low-temperature fracturing reveals surfaces that may be examined directly at low temperatures in the SEM, whereas freeze-fracture replication produces high-fidelity copies of surfaces that are examined at ambient temperatures in the TEM and STM. Recent advances in the instrumentation used for the preparation and examination of fracture surfaces now make these differences less distinct.

The low-temperature fracturing process, which involves fracturing together with optional etching and/or surface coating, may be carried out either in simple home-made devices or in more sophisticated pieces of equipment where much attention has been paid to the problems of sample contamination and incipient melting. The equipment is either dedicated to a particular microscope or free standing, relying on efficient sample transfer to the cold stage of the microscope. Freeze-fracture replication devices are all free standing and take the sample through the separate stage of fracturing, optimal deep or shallow etching, metal shadowing, and finally the formation of a carbon replica.

Considerable attention has to be paid to problems of contamination and to the various artifacts that occur during the preparative process. Although freeze-fracture replication was originally developed as a ultrastructural technique, it is now being used as an analytical procedure involving autoradiography, immunocytochemistry, and other labeling methods. The two fracturing procedures have broad applications, which encompass material science, foodstuffs, polymers, and biological material, in addition to being used in connection with other low-temperature methods such as freeze solubilization, fracture polishing, and freeze cracking.

6

Freeze-Drying

6.1. INTRODUCTION

In the previous two chapters we considered methods that enable information to be obtained from inside frozen-hydrated specimens. We will now consider the ways water may be removed from such specimens using low-temperature procedures. In this chapter we will discuss the process of freeze-drying, in which ice in frozen specimens is removed by low-temperature vacuum sublimation. In the following chapter, we will consider the procedures of freeze substitution, in which ice is dissolved in organic solvents maintained at low temperature. These two controlled dehydration techniques, which are an integral part of a more extended process for the preparation of hydrated specimens, serve as a link between the advantages of cryofixation and the ease of cutting thin sections at ambient temperatures. Both techniques involve considerably more intervention into the natural hydrated state of specimens than we have discussed so far. Freeze-drying, although a physical process, aims to remove all the water from a sample; freeze substitution involves contact with organic solvents and hence the risk of removing soluble components.

Because water plays such an important part in both the properties and structure of molecules in hydrated systems, its removal must be done as gently as possible in order to minimize structural damage and distortion. In some instances such damage is unavoidable. Only the most robust samples can withstand the enormous forces (146 bar) that are generated at the air–water interface as a sample is air dried. This is a consequence of the relatively high surface tension of water (7.3×10^{-2} N m^{-1} at 291 K), although these effects can be lessened by drying from solvents with low surface tension.

Freeze-drying is one of a number of dehydration processes, which include air drying, chemical drying, and solvent drying. All effectively remove water from hydrated systems, but freeze-drying has the advantage that it minimizes the trail of havoc brought about by molecular and structural collapse and the differential extraction of components during and following the dehydration process. Freeze-drying for microscopy and analysis remains an empirical process, and although there are general guidelines that can be followed, the experimental protocols have to be designed *ab initio* around the sample concerned and the imaging system to be used to examine and analyze the specimen.

Freeze-drying, which has been used for centuries by Lapps and Inouets to preserve meat and fish, is an important industrial process, which has applications in food processing, pharmaceuticals, and agriculture. The commercial and medical applications of freeze-drying have resulted in a much better appreciation of the process, even if the finer details of the process are still not always recognized (Franks, 1990). While the procedures used to produce tons of instant coffee would not be used to freeze-dry a block of frozen tissue, the general principles remain the same. Freeze-drying must be one of the few examples where the theoretical basis for an empirical laboratory procedure has been developed as a consequence of commercial exploitation. Much has been written about freeze-drying for low-temperature microscopy and analysis, and comprehensive reviews have been produced by MacKenzie (1965, 1972, 1977), Goldblith *et al.* (1975), Nermut (1977), and Mellor (1978).

6.2. FREEZE-DRYING FOR MICROSCOPY AND ANALYSIS

6.2.1. The Consequences of Freeze-Drying

There is more to freeze-drying than just removing the water from a specimen by sublimation, and before describing the process it is instructive to briefly reconsider the effects of removing water from hydrated systems. We have already shown in Chapter 2 that most molecules in hydrated systems are surrounded by a layer of water referred to as the hydration shell, which prevents the molecules sticking together. This is a particularly important phenomenon in biological material because the removal of the hydration shell can lead to aggregation and/or collapse of the macromolecular organic matrix. These changes are a consequence of thermal movements and the natural stickiness of dehydrated biological material. In addition to water associated with the hydration shells, there is the bulk or largely unperturbed water, which makes up a substantial part of the water in many hydrated systems. Removal of the bulk water is related to shrinkage, which can be a serious problem in freeze-drying, for unlike chemical dehydration, no attempt is made to replace the water with another material. For example, Lee (1984) showed that many cells retain their shape and volume until about 70% of the cellular water is replaced by solvents, but when all the water is removed, the shrinkage may amount to between 20% and 70% of the original volume. The relative seriousness of shrinkage or collapse is of course related to the resolution of the imaging system. It is unlikely that molecular collapse and aggregation would be observed in the light microscope, although shrinkage would seriously influence the topological relationships that exist between cells and tissues. At the level of the electron microscope the consequences of the molecular changes are all too evident.

MacKenzie (1972) found that freeze-drying below 210 K does not remove all the water from sugar and protein model systems or from biological organisms and tissues, irrespective of vacuum conditions or drying times. The residual water could only be removed by warming the specimens under vacuum. Raising the temperature to remove the last vestiges of water appeared to bring about the collapse phenomenon, which only occurred above a discrete temperature. This led MacKenzie to observe directly the collapse phenomenon in frozen aqueous solutions, which occurred anywhere from 223 K in a solution of 8% sucrose and 2% NaCl, to 263 K in a 10% ovalbumin solution. Collapse in these model systems was strongly influenced by the solutes. It is difficult to relate directly what happens in model systems to what may occur in the far more complex systems that exist in nature, which makes it virtually impossible to designate a collapse temperature for such materials.

Nearly all freeze-drying protocols are a two-step process in which the temperature of the sample is raised to ambient or above at the end of the drying process to remove the residual moisture. Primary drying centers on the direct sublimation of ice formed as a consequence of freezing bulk water in the sample. About 90% of the initial water is lost during this phase, leaving behind a highly porous "ghost" of the initial hydrated structure. The second step, sometimes referred to as isothermal desorption, removes the unfreezable water associated with macromolecules. This is a tedious operation which may take as long as the primary drying phase as water is desorbed from the highly porous matrix.

Nearly all the studies and applications of freeze-drying have been carried out in situations where the solidified water is in a crystalline state. Some of the fast cooling methods that were discussed in Chapter 3 are capable of vitrifying thin layers of water, and we should enquire whether there is any difference in the sublimation rates of I_v and I_c and I_h at low temperatures and high vacuum. With most freeze-drying, any differences would only be of academic interest because the ice sublimation is usually carried out at temperatures well above the recrystallization point of ice. Umrath (1983) and Franks (1986) claim that high-vacuum sublimation of vitreous water occurs at negligible rates and that solid water must be in at least a microcrystalline state before freeze-drying can occur. However, the average water molecule on the surface of vitrified water is less strongly bound than a water molecule at the surface of an ice crystal, and on thermodynamic grounds alone, one would expect a higher sublimation rate from I_v than from I_c or I_h. It is now known that the different phases of ice exhibit different stabilities, and Kouchi (1987) has shown that at temperatures at which I_a and I_c can coexist (150 K), the saturation vapor pressure of I_a is two orders of magnitude higher than that of I_c, making ultra low-temperature freeze-drying a practical possibility.

Freeze-drying is a rate-dependent process, and in addition to considering the saturation vapor pressure of the different phases of ice, one must also consider the rates of transition of one phase to another as the temperature increases. Dowell and Rinfert (1960), on the basis of x-ray diffraction patterns, showed that the $I_a \rightarrow I_c$ transition was an irreversible process between 113 and 143 K and that the $I_c \rightarrow I_h$ was a sudden event at 143 K. Livesey et al. (1991) calculate that at 133 K, the $I_a \rightarrow I_c$ transition time is about 30 min and would allow nearly a micrometer of amorphous ice to be sublimed. This finding encouraged Livesey and colleagues to develop a process of transitional drying in which each of the phases of ice is sublimed sequentially from a quench-cooled sample.

Nearly all these calculations and predictions have been made for pure ice, and the presence of solutes and embedded macromolecules will raise the temperature range over which crystallization transitions are likely to occur and will impede the rate at which water is sublimed from the frozen material. In addition, it is very difficult to measure the temperature of the surface from which ice is subliming, and although there may be an accurate measurement from the specimen holder, the subliming ice surface is probably significantly warmer. This is clearly an area of research where more careful studies need to be carried out.

6.2.2. The Theory of Freeze-Drying

The rate of freeze-drying may be predicted by the Knudsen equation:

$$G_s = N(P_s - P_c) \times \left(\frac{M}{2\pi RT}\right)^{0.5} \tag{6.1}$$

where G_s is the sublimation rate of ice in g cm^{-2} s^{-1}, n is the coefficient of evaporation, P_s is the saturation vapor pressure of ice, P_c is the partial pressure of water in the region surrounding the specimen, M is the molecular weight of the water vapor, R is the gas constant, and T is the absolute temperature.

This equation assumes that there is a dynamic equilibrium between the surface of the ice and its vapor and that the number of water molecules that sublime depends on the temperature of the ice, whereas the number of water molecules that return to the surface depends on the pressure and temperature of the water vapor. Provided the vacuum pressure is sufficiently low or a sufficiently dry atmosphere surrounds the specimen, then the specimen temperature will determine the rate of drying. From this it follows that the rate of sublimation will depend on sample temperature coupled with the efficient trapping of the water molecules that leave the surface. In this respect, the partial pressure of water vapor in the immediate vicinity of the specimen is more important than the absolute vacuum pressure.

Umrath (1983) used the Knudsen equation to calculate the maximum theoretical sublimation rate of crystalline ice between 272 and 113 K under high-vacuum conditions. His results are summarized in Fig. 6.1, together with the saturation vapor pressure of ice and the time required to sublime away an ice layer 1 μm thick. Table 6.1 expresses some of these values in the form of a table. Thus at 273 K, the maximum sublimation rate of ice is 8.2×10^{-2} g cm^{-2} s^{-1}, whereas at 173 K the rate is 100,000 times slower at 5×10^{-7} g cm^{-2} s^{-1}. These values are the *maximum* theoretical sublimation rates for crystalline ice and are considerably faster than the rates that may be obtained when freeze-drying hydrated samples. The freeze-drying of thin layers of dilute aqueous suspensions of macromolecules is about the only time where the experimental practice meets the theoretical predictions. The reasons for the large deviation from theory of most other samples are as follows:

1. The water molecules have to be removed through the finely structured and porous network of the specimen. As drying proceeds, the layer of dried material becomes progressively thicker and the pathway through which escaping water molecules have to pass becomes progressively longer. Stephenson (1953) has calculated

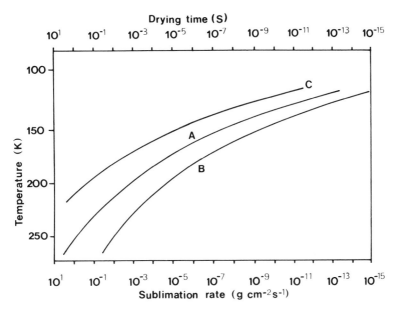

Figure 6.1. The relationship between the saturation vapor pressure and sublimation rate of ice as a function of temperature and pressure. A, Saturation vapor pressure of ice; B, sublimation rate of ice; C, time required to sublime 1 μm layer of ice. [From Umrath (1983).]

the time it would take for water vapor to diffuse through the dried shell of biological material and found that the ice in a 1-mm^3 cube of liver tissue takes 1000 times longer to sublime than it takes for a 1-mm^3 cube of crystalline ice to sublime. Stumpf and Rosas (1967) found that the drying times of rat liver : kidney : brain were in the ratio of 1 : 1.7 : 3.3. Different tissues dry at different rates. The difference between the drying time of a sample compared to the drying time of a similar size and shaped piece of crystalline ice is known as the *prolongation factor*, which remains relatively constant between 263 and 213 K. Umrath (1983) has collected the prolongation factors for different biological tissues, and the wide scatter of value (20–15,000) shows why it is so difficult to calculate with any sense of accuracy just how long it would take to freeze-dry a given sample.

   2. The sublimation of water needs energy, which must be transferred to the drying boundary. If the source of heat is radiant energy, as is usually the case with the etching procedures used on freeze-fractured samples (Section 5.1.3 in Chapter 5), then this heat transfer will take place through the progressively thicker dried layer of material on top of the frozen-hydrated sample. The rate of drying becomes slower the deeper the drying front proceeds into the sample. Alternatively, the heat transfer may occur from below and through the frozen sample, in which case the rate of drying will be primarily influenced by the mass transfer of water out of the sample. In practice, all combinations of heat transfer and mass transfer occur and the drying rate depends on the size and structure of the sample, the thermal contact the sample

Table 6.1. The Time Taken to Sublime Ice at Different Temperatures

Temperature (K)	Vapor pressure	Etching rate (nm s^{-1})	Drying time (s μm^{-1})	Drying time for 1 μm crystalline ice
213	1.08 Pa	1.48×10^3	6.76×10^{-1}	0.68 s
203	259 mPa	3.64×10^2	2.75	3.0 s
193	53.6 mPa	7.70×10^1	1.30×10^1	13 s
188	22.9 mPa	3.33×10^1	3.00×10^1	30 s
183	9.32 mPa	1.37×10^1	7.82×10^1	1 min 13 s
178	3.61 mPa	5.39	1.85×10^2	3 min 5 s
173	1.32 mPa	2.00	4.99×10^2	8 min 19 s
168	457 μPa	7.02×10^{-1}	1.42×10^3	23 min 40 s
163	148 μPa	2.30×10^{-1}	4.34×10^3	1 h 12 min 20 s
158	44.5 μPa	7.04×10^{-2}	1.42×10^4	3 h 56 min 40 s
153	12.4 μPa	1.99×10^{-2}	5.02×10^4	13 h 56 min 40 s
143	73.9 nPa	1.22×10^{-3}	8.17×10^5	9.5 days
133	28.8 nPa	4.95×10^{-3}	2.02×10^7	31 weeks
123	668 pPa	1.19×10^{-4}	8.40×10^8	27 years
113	8.03 pPa	1.49×10^{-5}	6.71×10^{10}	2157 years

has with the source of heat, vacuum conditions inside the freeze-dryer, and the efficiency of the system used to trap sublimed water molecules.

3. Water associated with the hydration shells surrounding macromolecules does not freeze and can only be effectively desorbed at elevated temperatures, which we have shown can lead to collapse and aggregation. Ideally, this secondary drying should only be carried out below the collapse temperatures, which would lead to very long drying times.

4. We have already indicated that the phase of ice crystals will influence the rate of sublimation.

5. The withdrawal of latent heat of evaporation will cause the sample to cool during freeze-drying with a concomitant decrease in the ice sublimation rate. Thus every time the temperature of the sample holder is raised, the temperature of the drying sample first decreases as water sublimes and then slowly rises to reach a new and higher equilibrium temperature.

Freeze-drying of hydrated organic samples takes much longer than the sublimation of ice alone. The data provided by Umrath (1983), although only applicable to crystalline ice, do have a practical value as they provide a rough guide to the times, temperatures, and vacuum conditions one would need to consider using to freeze-dry a hydrated sample. On this basis, the general design of a freeze-drying schedule can be assembled.

At best, only the outer 15–20 μm region of a sample is properly quench cooled and would be likely to contain either vitreous or microcrystalline ice. It is expedient to first dry this outer layer at as low a temperature as practicable, i.e., 173–178 K to diminish recrystallization and then raise the temperature to 203–213 K and dry the less-well-frozen inner core. An examination of Fig. 6.1 will show that at 173 K, 10 μm of ice will sublime in about 30 min, whereas at 213 K the same amount of ice will sublime in under 10 s. The difficulty comes in calculating the prolongation factor associated with hydrated organic samples, and this can only be determined by experimentation. If we take a prolongation factor of 50, and assume the sample is a 1-mm^3 cube of frozen biological material, then the drying times would be as follows. At a temperature of 178 K and a pressure of 400 mPa (3×10^{-3} Torr) it would take about three days to dry the outermost 30-μm layer. If we now raise the temperature to 213 K, the remaining 850 μm^3 would be freeze-dried in just under 5 h, assuming water sublimes from five faces of the cube of material. It is most important to resist the temptation to raise the sublimation temperature too high; otherwise the core of frozen material will melt and/or the sublimation rate may be so high that water vapor may condense on the cold, dried-out material. The consequences would be disastrous. If one is concerned about temperature collapse phenomena and lowers the drying temperature to 203 K, the bulk of the frozen material would be dried in about 30 h.

These rough and ready calculations assume that all the sublimed water is trapped and that the sample temperature is the same as the stage temperature. As we will show, although the former is certainly true, the sample temperature is invariably higher than the stage temperature, so that the drying proceeds at a faster rate.

6.2.3. The Practical Procedures of Freeze-Drying

Freeze-drying of hydrated material for low-temperature microscopy and analysis requires an apparatus that will permit drying to occur at between 173 and 193 K. There are a large number of freeze-dryers available, but their basic features are as follows:

1. The drying chamber should contain a platform that can be cooled low enough (173–193 K) to avoid extensive recrystallization in the sample. Liquid nitrogen is an effective cryogen for this purpose. The cold platform should be fitted with a heater, an accurate temperature controller, and a thermostat. Facilities should be available to backfill the chamber with dry nitrogen gas.

2. The vacuum system should be capable of maintaining a pressure of 1–10 mPa (5×10^{-6} to 1×10^{-4} Torr), which will allow maximum access of the subliming water molecules to the water vapor trapping system. The vacuum conditions are usually achieved with a diffusion pump or a turbomolecular pump backed by a rotary pump. If freeze-drying is to be carried out at higher temperatures, e.g., 213 K, then only a rotary pump is necessary. The vacuum lines should contain a chemical (e.g., activated alumina) or a low-temperature trapping system to trap both water vapor and backstreaming hydrocarbon vapors.

3. Optimal freeze-drying depends on trapping all the water molecules that are derived both from the drying chamber and from the subliming specimen. The partial pressure of water vapor in the sample area should be 100–1000 times lower than the saturation vapor pressure for ice at that temperature. At a pressure of 1.3 mPa (10^{-5} Torr), 70% of the residual gas is water vapor, which has a partial pressure of 1 mPa (7×10^{-6} Torr). As a rule of thumb, the saturation vapor pressure of ice on

the trapping condensor should be three orders of magnitude lower than the saturation vapor pressure of ice at the drying temperature. Thus, if freeze-drying is to be carried out at 178 K, where the partial pressure of water vapor is ca. 4 mPa (3×10^{-5} Torr), the vapor pressure at the adjacent condensor surface should be 4 μPa (3×10^{-8} Torr), which corresponds to a condensor temperature of 148 K. This is an awkward temperature to achieve in practice, and the easiest solution is to use a metal plate cooled with LN_2 (77 K). At this temperature, the partial pressure of water vapor would be 1 aPa (10^{-18}) or 1×10^{-21} Torr.

It is important to distinguish between the partial pressure of water vapor and the total pressure in the drying chamber. The partial pressure of water vapor is a measure of how much water there is in the environment. Thus at 173 K, the partial pressure of ice is 1.3 mPa or approximately 10^{-5} Torr. At 77 K, i.e., the temperature of the LN_2-cooled trap, the partial pressure is 1 aPa (10^{-18} Torr) and this 15 orders of magnitude difference between the partial pressure of ice at the subliming and condensing surfaces means that any water molecules that escape from the shell of water vapor surrounding the sample are going to be trapped on the LN_2 surface. This trapping efficiency is increased by the closeness of the 77 K condensing surface to the specimen subliming, and/or the length of the mean free path of the water molecule. At 103 μPa (10^{-6} Torr), the mean free path would be about a meter, i.e., a water molecule would, on average, travel a meter before bumping into another water molecule. At 1.3 mPa (10^{-2} Torr), the mean free path is much shorter, which means that there is a greater chance of a water molecule that leaves the ice surface colliding with another water molecule and returning to the ice surface rather than becoming trapped on the condenser. It is for this reason that a sample at 173 K in an environment with a total pressure of 1 Pa would take a long time to dry, even though it is close to a very cold condenser with a high trapping efficiency. By working at high vacuum we simply speed up the drying process. This avoids the alternative of raising the sample temperature and running the risk of recrystallization.

Chemical desiccants can also be used to adsorb the water vapor released by the specimens. Compounds such as phosphorus pentoxide, calcium chloride, and sulfuric acid undergo chemical reactions as they react with water vapor. The advantage of these materials is that they are effective at ambient temperatures and will reduce the partial pressure of water vapor to low values ca. 1 mPa (10^{-5} Torr). The chemical desiccants are particularly useful during the secondary drying procedures, which are carried out at higher temperatures. Molecular sieves (Zeolites) and copper sulfate are also effective water vapor trapping agents and can easily be regenerated by heating in an oven. Molecular sieves are more effective at low temperatures, e.g., at 77 K they can produce a total pressure of 7 nPa (10^{-11} Torr).

4. In addition to the components necessary for drying, it is necessary to have facilities to maintain the extremely hygroscopic freeze-dried material, bone dry following freeze-drying. It is useful to include facilities to enable bulk samples to be embedded in resin and for thin samples to be coated with a protective layer of evaporated carbon without breaking the high vacuum of the drying chamber. An air lock system on the freeze-dryer will facilitate the transfer of dried samples to other pieces of equipment without exposure to air.

6.3. EQUIPMENT FOR FREEZE-DRYING

There are basically four types of freeze-dryers: high vacuum (1 mPa or 8×10^{-6} Torr); low vacuum (1 Pa or 8×10^{-3} Torr); molecular distillation or transitional drying (1 μPa or 8×10^{-9} Torr); and cold molecular sieve (7 nPa or 10^{-11} Torr). Although most freeze-drying for low-temperature microscopy and analysis is carried out using the first and second types of equipment, the other two procedures have their applications. Before discussing the different approaches to freeze-drying, we should first briefly consider some of the ways the freeze-drying procedure may be monitored and measured. Monitoring freeze-drying is important, as it provides a guide not only to whether the process has been completed, but also to whether the pressures and temperatures that are being used are ideally suited to the sample being dried.

6.3.1. Monitoring Freeze-Drying

The drying process may be followed by measuring the partial pressure of water vapor in the drying chamber using a mass spectrometer. Coulter and Terracio (1977) carried out a partial pressure analysis of 250-μm^3 blocks of kidney tissue dried at a total pressure of 13 μPa (10^{-7} Torr). Figure 6.2 shows the course of water sublimation from the sample. At the start, the stage (sample?) temperature was 153 K and the partial pressure of water vapor was 780 nPa (10^{-9} Torr). As the temperature of the stage was increased, the partial pressure of the water vapor rose sharply and then declined. The dotted line that connects the top of each peak shows there is a gradual decline in the drying rate and that the drying was essentially completed at 193 K, although the final traces of water were only removed by the secondary drying process at 263 K Linner *et al.* (1986b) used the same procedure to follow the drying of 1-mm^3 cubes of rat liver, kidney, and brain tissue at a total pressure of 1 μPa (10^{-8} Torr) and found that most of the water had left the specimen at a specimen support temperature of 145 K. The secondary drying process was only completed at 298 K.

Wildhaber *et al.* (1982) have followed the course of freeze-drying in bacterial membranes under high-vacuum conditions (100 μPa or 10^{-7} Torr) in which the water in a 10–20-nm-thick sample had been exchanged with heavy water (D_2O) (Fig. 6.3). As the temperature of the frozen sample was raised, D_2O was first detected at 178 K with the maximum sublimation occurring at 193 K. Secondary drying occurred at between 233 and 223 K, where another peak for D_2O was measured. Wildhaber and colleagues also made the interesting observation that there is a characteristic H_2O peak in every freeze-drying spectrum. This peak appears as the sample is warmed up from the loading temperature, ca. 130 K, to the drying temperature, and its appearance is unrelated to the sublimation of water from the sample. This H_2O peak (also measured by mass spectrometry) is due to contamination on the cold stage and desorption of water from the walls and equipment in the drying chamber.

While it is important to monitor the drying, it would be of much greater value if this information could be applied to a feedback loop to actually control the process. Flenniken *et al.* (1990) have interfaced a computer to a freeze dryer, which can be

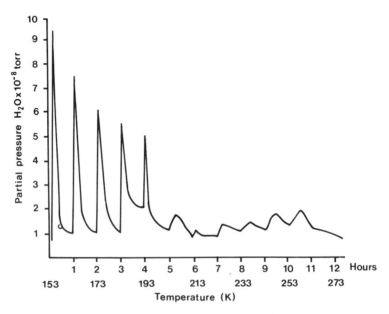

Figure 6.2. Changes in the partial pressure of water as a function of time and temperature during the freeze drying of 250-μm^3 cubes of kidney tissue. Drying was started at 143 K and a pressure of 6×10^{-9} Torr and the temperature raised 10 K h^{-1}. Each time the temperature is increased there is a substantial increase in the partial pressure of water in the system which quickly falls back as this water is trapped on the liquid-nitrogen condensor. These progressive rises and falls become less marked as the water is sublimed from the sample. At 200 K the sharp peaks have virtually disappeared and the primary drying is complete. [Redrawn from Coulter and Terracio (1977).]

used to control the warming rate of a sample as it passes through critical regions of drying.

If the freeze-drying is being carried out in a piece of equipment directly attached to an SEM or a TEM, the simplest way of checking whether the drying is complete is to examine the specimen. The surfaces of bulk specimens have a characteristic lavalike appearance when examined in the SEM, and there is a dramatic increase in the contrast of freeze-dried sections when they are examined in the TEM. In addition, it is possible to measure changes in the mass of sections by x-ray microanalysis as they lose water during the drying process. This can be done either by measuring changes in the peak height of oxygen using a windowless energy dispersive detector or by measuring changes in the continuum radiation in regions of the spectrum away from the characteristic peaks of elements. These are procedures we will return to in Chapter 11, where we discuss the analysis of frozen specimens.

6.3.2. High-Vacuum Freeze-Drying

This form of freeze-drying is performed at pressures between 1.3 mPa and 130 μPa (10^{-5}–10^{-6} Torr) and at temperatures between 173 and 193 K. At these

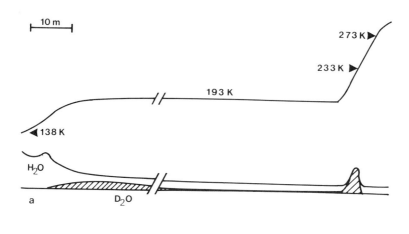

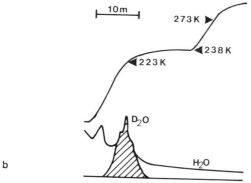

Figure 6.3. Changes in the partial pressure of water and deuterium oxide as a function of temperature during freeze-drying of bacterial membranes containing heavy water. (a) Specimen loaded at 138 K and dried at 193 K. After 2 h a trace of D_2O remains which sublimes as the temperature is raised to 233 K. (b) Specimen loaded at 138 K and the temperature slowly raised to 293 K and then to 238 K. Maximum sublimation of D_2O occurred at 223 K and at the end of primary drying at 233 K. [Redrawn from Wildhaber *et al.* (1982).]

temperatures and pressures, the sublimation rate of pure ice would be between 77 and 2 nm s^{-1} and the risk of damage due to recrystallization damage would be minimal. It is important that the samples are as small as possible in order not to prolong the drying time. The data by Umrath (1983) and Fig. 6.1 indicate that a 1-mm^3 block of pure ice at a pressure of between 1.3 nPa and 130 μPa would take 2.8 days to sublime at 173 K, 10.1 h at 183 K, and 108 min at 193 K. This assumes that the water sublimes from five of the six faces of the cube and that it is all trapped on an LN$_2$-cooled condensor. The prolongation factors for hydrated organic samples would increase these times by between one and three orders of magnitude.

At the outset of this chapter it was stated that it would be impossible to give a set formula that could be followed to freeze-dry *all* hydrated samples. The following protocols are some representative examples of high-vacuum freeze-drying of real samples:

1. Slices of mouse kidney tissue 250 μm thick were rapidly impact cooled at 77 K (Fig. 6.2). The initial temperature of the freeze-dryer was 153 K and the total pressure was 13 μPa (10^{-7} Torr). The temperature was raised 10 K per hour to 293 K and the primary drying was completed within 5 h and the secondary drying within 14 h. The sublimed water was trapped on an LN_2-cooled condensor (Coulter and Terracio, 1977).

2. A 1-mm^3 block of rat pancreas was impact cooled on a copper mirror at 4 K. The initial temperature of the freeze-dryer was 138 K and the total pressure 1 mPa (10^{-2} Torr). The drying schedule was 1 h at 138 K, 2h at 138–203 K, 30 h at 203 K, 24 h at 203–233 K, 6 h at 233–297 K, and finally, 17 h at 297 K. The condensor temperature was kept at 77 K (Roos and Barnard, 1985). An examination of this protocol suggests that very little drying would occur during the first hour, that primary drying would begin toward the end of 3 h, and that the secondary drying would be complete after about 3 days.

3. The heads of rat femurs were quench cooled in subcooled nitrogen slush at 70 K. The initial temperature of the freeze-dryer was 133 K and the total pressure was 130 μPa (10^{-6} Torr). The sample temperature was raised to 183 K over 24 h and was then raised 10 K every 12 h to 213 K and then raised to 293 K over 24 h. The condensor temperature was at 77 K (Draenert and Draenert, 1982). The primary drying took place during the first 24-h period and the secondary drying only occurred as the sample approached room temperature. This sample had a high (65%–80%) water content in contrast to the 40%–50% water content of the tissues in the other two examples.

Although a significant amount of freeze-drying for low-temperature microscopy and analysis can be carried out faster at higher temperatures and a lower vacuum, high-vacuum freeze-drying does have a special place in preparing materials for ultra-structural and analytical investigations. Provided the outermost layers (15–25 μm?) of the sample have been properly quench cooled and are mechanically undamaged, and provided this face of the sample can be identified either in the SEM or in a resin block in which it is subsequently embedded and sectioned, then this procedure is worth the extra effort.

High-vacuum freeze-drying can also be used to effectively sublime the water from thick or thin frozen-hydrated sections and from frozen suspensions. This type of sample can also be dried inside an electron microscope fitted with a cold stage. Hagler and Buja (1984) were able to freeze-dry thin (50–500-nm) frozen-hydrated sections by placing them on the microscope cold stage at 153 K and a pressure of 150 μPa (2×10^{-6} Torr). The drying was complete in 30 min. Rachel *et al.* (1985) have used a similar process to freeze-dry very thin ca. 100-nm, films of biological specimens. At 180 K and at an estimated pressure of 100 μPa (8×10^{-7} Torr) inside the microscope, the sample was dried within 1–2 h as judged by changes in contrast. At these temperatures and pressures, a similar thickness of pure ice would sublime in 13 s. Rachel *et al.* consider that the sublimation rate of ice is much reduced in the vicinity of biological material. On the basis of the theoretical value for ice sublimation calculated by Umrath (1983) and the experimental values obtained by Rachel *et al.* (1985), the prolongation factor for biological material is somewhere between 275 and 550. High-vacuum freeze-drying combines the advantages of optimal cooling with a nonchemical dehydration procedure, which virtually avoids the problems associated with the recrystallization of ice. High-vacuum freeze-dryers are available from Balzers Union AG (FDU 010 Freeze drying unit) and VG Microtech, Polaron (E7480 Low Temperature Freeze dryer) (Fig. 6.4).

With a little ingenuity, a high-vacuum evaporation unit can be converted to an effective freeze-dryer. The simplest conversions involve the use of the type of module suggested by Roberts and Duncan (1981), which do not need continuous cooling. The method is only really suitable for thin films or sections and works by surrounding the sample by a large mass of metal cooled to 77 K. The whole assembly is placed in the vacuum chamber, which is evacuated to 65 μPa (5×10^{-6} Torr) and the thin samples warm up more rapidly than the surrounding mass of cold metal, which acts as an effective trap for the sublimed water molecules. When this trapped water eventually sublimes, it in turn is trapped by a tray of molecular sieve in the drying chamber. More elaborate conversions have been made involving plumbing in LN^2 lines and electrical leads. Flenniken *et al.* (1990) have built a low-temperature, high-vacuum freeze-dryer inside a conventional evaporator which allows the sample to be coated with a protective carbon layer immediately following freeze-drying. Lindroth

Figure 6.4. Low-temperature, high-vacuum freeze-dryer. The system has a diffusion pump backed by a high-volume rotary pump and works in the 10^{-5}–10^{-6} Torr range at between 103 and 293 K with a liquid-nitrogen-cooled condensor. Accurate temperature control is achieved by a programmable proportional controller. An airlock (a) is fitted to enable quench-cooled samples to be loaded onto the precooled cold stage and freeze-dried samples to be transferred from the dryer to other pieces of equipment while still under high vacuum. Evaporation and sputtering heads (h) can also be fitted to the unit. (Picture courtesy VG Microtech, Polaron.)

et al. (1991) have modified a freeze-dryer to accommodate a magnetron sputtering head in order to sputter coat dried samples with very thin layers of tungsten while still in the drying chamber and at low temperatures.

6.3.3. Low-Vacuum Freeze-Drying

This is the form of freeze-drying that is most frequently used in low-temperature microscopy and analysis, and is performed at pressures in the range from 1.3 Pa to 130 mPa (10^{-2}–10^{-3} Torr), which can be obtained using a rotary pump. At these pressures, although the drying process may be initiated at about 193 K, there is no substantial water sublimation until 203–223 K, where the rate of water loss would be of the order of 2 μm s^{-1}. There is an increased risk of ice recrystallization at these higher temperatures, although the drying rate is going to be much faster. A 1-mm^3 block of pure ice at a pressure of between 1.3 Pa and 130 mPa is going to take 25 min to sublime at 203 K and only 5.7 min at 213 K. The prolongation factor associated with hydrated organic samples will extend these times. Low-vacuum freeze-drying can be carried out in high-vacuum equipment with an LN^2-cooled cold table (Lyon *et al.*, 1985) or in dedicated units that use multistage Peltier thermoelectric modules, which can be cooled to 213 K if the warm side of the module is cooled with chilled water. Still lower temperatures (203 K) can be obtained by cooling the module with chilled brine. The only advantage of this type of equipment is that it avoids having to use LN^2. The stage cool-down times are much longer than may be achieved with LN^2. Peltier devices can be easily incorporated into existing vacuum evaporation equipment and chemical desiccants such as P_2O_5 or molecular sieve substituted for an LN^2-cooled trap for the sublimed water. This type of equipment is easy to construct in a laboratory workshop and is commercially available from a number of different suppliers. The following protocols are some representative examples of how low-vacuum freeze-drying may be carried out.

1. Droplets of a suspension of rat hepatocytes that were 200 μm in diameter were quench cooled in melting halocarbon at 113 K. The initial temperature of the freeze-dryer was 223 K and the total pressure was 1.34 Pa (10^{-2} Torr). The sample was kept at these conditions for 24 h, after which the temperature was raised slowly over a 12-h period to 293 K. The sublimed water was trapped on P_2O_5 (Nordestgaard and Rostgaard, 1985).

2. Blocks of rat mandibular cartilage, 0.5 mm^3, were quench cooled in liquid propane at 83 K. The initial temperature of the freeze-dryer was 150 K and the total pressure was 130 μPa (10^{-3} Torr). The sample was dried at 163 K for 24 h and then raised 10 K h^{-1} until the temperature reached 218 K. The system was allowed to rise to 293 K over a further 24-h period. The sublimed water was trapped on an LN^2 cold finger placed 20–50 mm from the drying sample. Although the authors considered that the outermost shell of water sublimed at between 163 and 173 K, the rate would have been very low at these temperatures and relatively high pressures. Most of the water would have been sublimed in the third and fourth hour following the initial 24-h drying period (Lyon *et al.*, 1985).

Low-vacuum freeze-drying is a particularly useful technique for large tissue blocks, where the quality of freezing below the outer 50-μm layer is going to be less than optimal. The success of the technique rests on the hope that the rate of recrystallization is lower than the rate of sublimation. The fragmentary data from pure ice suggests that this may just be the case. In the outer parts of the sample, recrystallization is slower than sublimation, but in the inner reaches the position will be reversed because of the increased length of the diffusion pathway for the sublimed water.

6.3.4. Molecular Distillation

This is a term, originally used by Carmen (1948), which Linner and colleagues (Linner *et al.*, 1986a, b) have used to describe the ultra-high-vacuum sublimation of amorphous water from frozen samples without devitrification or rehydration. Such a claim needs to be carefully examined. 1-mm^3 cubes of biological tissue are impact cooled on a cold metal mirror at 77 K, and transferred under LN^2 to 12-mm-deep cylinders in the specimen block of the drying unit, which has been precooled with LN^2. At no time is the temperature allowed to rise above 133 K. The freeze-drying apparatus (Fig. 6.5) consists of a stainless steel, ultrahigh-vacuum assembly evacuated by a turbomolecular pump backed by a vane pump. The system achieved an ultimate vacuum of 1.3 μPa (10^{-8} Torr). The high-vacuum assembly is fitted with metal seals, high-vacuum valves, and traps for hydrocarbons and water vapor and may be baked to 523 K before use. The sample block, which can be temperature controlled with a small heater, fits inside a vacuum-tight cylinder, which is immersed in a Dewar of LN^2. The walls of the cylinder are at 77 K and act as a cold trap for the sublimed water. The apparatus is bolted together, rough pumped to 130 mPa (10^{-3} Torr) to remove the LN^2 surrounding the samples, and then evacuated to 1.3 μPa with the turbomolecular pump. The equilibrium temperature in the sample chamber is at 80 K before the drying cycle begins.

A programmable, feedback-controlled heating circuit allowed the specimens to be heated at precisely measured rates. The drying schedule first raises the temperature of the sample block to 123 K over a 10-h period and then from 123 K to 213 K over a 70-h period and from 213 to 298 K during a further 10 h. The temperature remained at 298 K for about 2 days, during which time the LN^2 in the Dewar evaporated and the water vapor that had sublimed from the sample and condensed onto the walls of the inner cylinder was either pumped away by the turbomolecular pump or trapped by a helium cryopump (Livesey and Linner, 1987). The total drying time was just under 6 days. The partial pressure of water vapor was monitored during the drying process by residual gas analysis and the drying process was considered to be completed once the partial pressure of water vapor in the drying chamber had reached 1.3 fPa (10^{-15} Torr). On the basis of the residual gas analysis, Linner and colleagues consider that primary drying occurs between 143 and 152 K, which is 2–10 K above the point at which I_a/I_v would begin to transform to I_c. Dubochet and Lepault (1982) have demonstrated with x-ray diffraction that the I_a to I_c transformation in

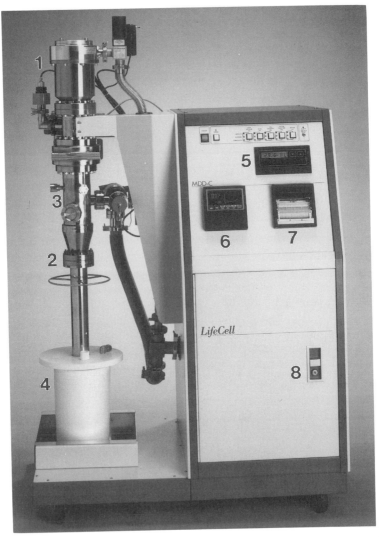

Figure 6.5. Molecular distillation dryer MDD-C by Life Cell. 1, Turbomolecular pump; 2, stainless steel column with metal-to-metal seals; 3, access ports for vapor and resin input; 4, automatic liquid-nitrogen level control; 5, low- and high-vacuum pressure gauge; 6, programmable temperature controller; 7, strip chart recorded; 8, roughing pump in cabinet. (Picture courtesy LifeCell Corporation.)

very thin biological systems begins at 153 K, so it would appear that the molecular distillation system is capable of freeze-drying amorphous ice.

At first glance it may be difficult to equate these sublimation temperatures and drying rates in biological material with the values that Umrath (1983) has calculated as the maximum sublimation rates for crystalline ice. At 143 K with a drying rate of 227 h μm^{-1}, it would take just under 13 years to sublime away a 1-mm^3 block of crystalline ice. At 148 K, the drying rate is 28 h μm^{-1} and the little block of ice would disappear in about 20 months; and at 153 K, although the drying rate increases to 14 h μm^{-1}, the total drying time is still 291 days. Even at a total pressure in the system of 1.3 μPa (10^{-8} Torr) at 163 K the sublimation rate is 72 min μm^{-1}, which would give a total drying time of 25 days, and at 173 K the comparable figures are 8 min μm^{-1} and 70 h. All these figures are exclusive of any prolongation factors.

Livesey *et al.* (1991) provide two pieces of experimental evidence that support the claim that vitreous ice may be sublimed in a high vacuum without phase transition to the crystalline state. In the first experiment, samples of deuterium oxide (heavy water) were either vapor condensed at 98 K and 163 K under high vacuum or the liquid state was slow cooled from 293 K under nitrogen gas, in the ultraclean molecular distillation unit. Following deposition, the sample chamber was immersed in LN_2 and the pressure reduced to 130 μPa (10^{-8} Torr). Each of the three heavy water samples, all now at 98 K, were separately slowly heated at 10 K h^{-1}, during which time residual gas analysis was carried out. As the original 98 K sample was heated, the D_2O started to sublime at 113 K, with a second peak at 148 K. The 168 K sample started to subline at 148 K and the slow-cooled sample at 183 K. Livesey and colleagues were able to show that the different peaks could be removed sequentially in a specimen warmed to a specific temperature and then held there.

In a second experiment, sublimation of D_2O was followed from 1-mm^3 samples of impact-cooled rat liver from animals that had been fed heavy water. The initial drying temperatures and heating rates were the same as in the first experiment and a small but reproducible level of D_2O was first detected at 113 K. In subsequent experiments (personal communication) Livesey and his colleagues are careful to show that the subliming D_2O molecules came from the sample and not from warmer parts of the equipment. The only remaining uncertainty is the existence of a temperature differential between the sample holder, where the temperature is *measured*, and the specimen, where the D_2O *sublimes*. Such differentials are likely to be very small as the samples are surrounded by cold surfaces. The evidence strongly supports the view that phasic transitional drying is a practicable proposition for small samples and can be achieved within a reasonable time to produce samples suitable for examination in the TEM (Fig. 6.6). The Molecular Distillation Dyer MDD-C is manufactured and marketed by the LifeCell Corporation.

Pawley and Ris (1987) have constructed a freeze-drying apparatus that operates at 25 μPa (2×10^{-7} Torr) and has been used to dry thin (1–10 μm) aqueous suspensions of muscle actin. The primary drying was carried out at 150 K and 25 μPa for two days followed by slow heating to 300 K over a 24-h period during which the secondary drying took place. On the basis of Umrath's calculations for crystalline ice, it would appear that about 2 μm of ice would sublime from the frozen surface of aqueous films kept at 150 K. This is well within the drying schedule suggested by Pawley and Ris and supports the earlier contention that the ultra-high-vacuum freeze-drying methods should be used only for aqueous films and thin frozen sections.

6.3.5. Ultra-Low-Temperature Freeze-Drying

These are simple techniques, which take advantage of the cryosorption pumping of molecular sieve material kept at 77 K. The device built by Edelmann (1978, 1979)

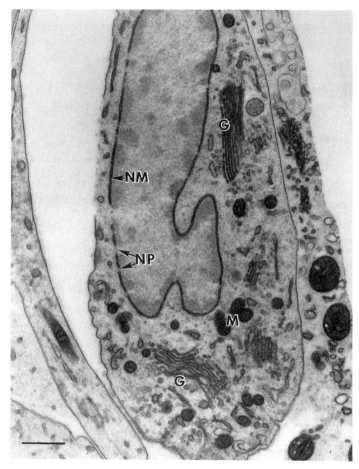

Figure 6.6. Transmission electron micrograph of rat kidney epithelial cell prepared by impact cooling, molecular distillation drying, osmium vapor stabilization, and embedding in Spurr's resin. Note the very dense cristae in the mitochondria (M) and the well-defined Golgi bodies (G). NM, nuclear membrane; NP, nuclear pores. Bar = 0.5 μm. (Picture courtesy LifeCell Corporation.)

is shown in Fig. 6.7. The lower part of the drying chamber is filled with 50 g of the molecular sieve, Zeolite 13. The precooled (77 K) specimen support is loaded with small (100 μm^3) quench-cooled samples together with a small piece of solidified low-viscosity resin. The vessel is sealed and plunged into a large Dewar of LN^2, where the cooled Zeolite produces a pressure of 130 mPa (10^{-3} Torr) without the need of any mechanical vacuum evacuation. Much higher vacuum pressures have been obtained under special conditions (Read, 1965). The specimen holder is slowly heated at 213 K for a further 6 days to complete the drying. The system is vented with dry nitrogen and warmed over 16 h to 258 K, at which point the dried sample is infiltrated with resin. The temperature is raised to 293 K and the resin-infiltrated material

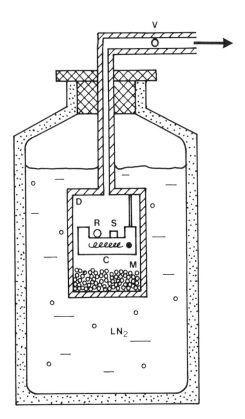

Figure 6.7. Ultra-low-temperature, molecular sieve drying device in which freeze-drying takes place by cryoabsorption. The pre-quench-cooled sample (S) and a piece of frozen embedding medium (R) are placed on the precooled specimen support (C), which is fitted with a heater and a thermostat. The cold specimen support is sealed onto the drying chamber (D), which is filled with molecular sieve (M) and immersed in a Dewar of liquid nitrogen (N). Any residual air in the system may be removed by a rotary pump, and the whole drying chamber is then sealed by valve (V). The temperature of the specimen support is raised to the primary drying temperature ca. 183–193 K and maintained there for several days. The cold molecular sieve and the cold walls of the cooling chamber act as an effective cryosorption pump, and a pressure of 130 mPa may be obtained. The system is allowed to warm up slowly to 260 K and then vented with dry nitrogen gas. At this temperature, the piece of frozen resin melts and infiltrates the now dried sample. The whole system is slowly allowed to reach room temperature and the infiltrated sample removed and passed through fresh resin mixtures prior to polymerization. (Modified and drawn from Edelmann (1979).]

removed for polymerization. The quality of the results obtained by this simple technique are impressive, although the method seems limited to small samples.

6.3.6. Cryostat Freeze-Drying of Frozen Hydrated Sections

These procedures have been considered in Chapter 4 and involve drying the frozen-hydrated sections either in the cryochamber of the cryomicrotome or on the cold stage of the electron beam instrument used to examine and analyze the sections. Freeze-drying in the cryoultramicrotome involves slowly heating the sections mounted on their supports on a temperature-controlled platform in the microtome chamber. A gentle stream of dry nitrogen will assist the drying process. Thin hydrated films and thin sections that are no thicker than 100 nm can be dried in 2–3 h in an electron microscope at 180 K and a pressure of 130 μPa (10^{-6} Torr).

6.4. HANDLING SPECIMENS BEFORE AND AFTER FREEZE-DRYING

It remains an open question whether samples should be chemically fixed before freeze-drying. Although thermal collapse and aggregation are usually decreased by prior cross-linking with chemical fixatives, this stabilization is only achieved at the expense of the soluble constituents one may wish to analyze. It may be necessary to wash samples in a salt-free buffer to avoid the formation of salt-containing eutectic lines during freezing. If the surface of the sample is to be examined by SEM, it is useful to rinse the sample in an ammonium acetate buffer before rapid cooling. Any crystals of ammonium acetate on the sample surface can be removed by sublimation. The size and form of the specimen will to a large extent determine the way it will be cooled initially, and these are matters we considered in Chapter 3. Loesser and Franzinini-Armstrong (1990) have devised a simple way of obtaining thin suspension films suitable for freeze-drying. Drops of liquid suspensions are squeezed between two glass cover slips, which are rapidly cooled. The frozen sandwich is split, and following freeze-drying, metal shadowing, and replication, the underlying coverslip is dissolved in hydrofluoric acid.

Special care has to be taken with bulk samples and to some extent with thin sections to make sure they are in good thermal contact with the underlying support. Unlike samples that are to be sectioned or fractured, it is inappropriate to clamp them into position, as this will seal off escape routes for the subliming water vapor. Wildhaber *et al.* (1982) have designed a magnetic cold table which holds electron microscope grids supporting frozen sections in close contact with the underlying support. Maintaining good thermal contact between the *sample* and the cold table of the freeze-dryer is difficult to achieve in practice, especially with pieces of material that have been prefrozen and then placed on the cold plate. Contaminating ice crystals on the specimen and the stage provide a very effective insulating layer. Most of the sample cooling and heating is by conduction through the underlying metal support; radiative and convective cooling play only a small part in keeping the sample cold. One must always assume that any temperatures that are given in freeze-drying schedules refer to the temperature of the underlying cold table, and it would be prudent to accept that the sample is several degrees warmer.

Specimens that have been freeze-dried are very hygroscopic and they must be kept under anhydrous conditions. This author has seen fluffy mounds of carefully freeze-dried 1% agar solutions quickly turn to drops of liquid as they are exposed to air with a relative humidity even as low as 40%. The freeze-dryer should always be vented with dry gas (N_2) and the samples adequately protected if they are to be transferred to another piece of equipment. The simple act of venting the freeze-dryer with dry nitrogen and quickly transferring the freeze-dried sample in air to a predried desiccator may still allow rehydration to occur. This is particularly important for samples whose freeze-dried surfaces are to be examined in the SEM. Such samples should be sputter coated or evaporative coated with a conductive layer while still in the drying chamber. If freeze-dried samples are to be shadowed, replicated, or metal coated, this is best carried out in the same equipment in which they have been freeze-dried. This is an important advantage of modifying a vacuum evaporation unit as freeze-dryer. Echlin and Van den Berg (unpublished data) have made small metal caps that fit on to Cambridge-type SEM stubs. A small O ring, inserted in the groove at the top of the specimen holder, hermetically seals the cap into position. This simple device permits labile freeze-dried proteins and carbohydrates to be transferred between the freeze-dryer, sputter coater, and SEM without any evidence of specimen collapse.

6.5. FREEZE-DRYING AND RESIN EMBEDDING

Freeze-drying is frequently used in conjunction with resin embedding and subsequent sectioning as it is argued that the finished product more closely resembles the natural state as it diminishes extraction and displacement of labile components (Fig. 6.8). The combination of freeze-drying followed by resin embedding has a wider application in analytical investigations than in ultrastructural studies, although in some circumstances the freeze-dried, resin-embedded material yields structural information that cannot be produced by more conventional means. Richardson *et al.* (1988) were only able to preserve proteoglycans in rabbit arteries by a combination of rapid cooling, freeze-drying, and plastic embedding, although the success of the approach lay less in the freeze-drying and plastic embedding than in the quality of

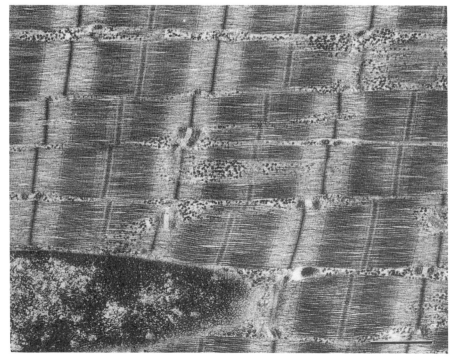

Figure 6.8. Chemically unfixed frog sartorious muscle after cryofixation, freeze-drying, and embedding in Spurr's resin; section stained with uranyl acetate and lead citrate. Bar = 1.0 μm. [Picture courtesy Edelmann(1986).]

the initial cooling. There is now a much greater emphasis on using freeze substitution as a means of preserving the ultrastructural details of hydrated samples. The resin-embedding procedures, and in particular the low-temperature embedding procedures and freeze substitution, will be discussed in the following chapter, but the general features of the methods used in connection with freeze-dried specimens are as follows.

There are two general approaches. The frozen sample and solidified but unpolymerized pieces of a resin are placed together in the specimen holder at 193 K. The sample is freeze-dried, and as it warms up so does the resin, which melts and infiltrates the dried sample either at low pressure or at ambient pressure. Alternatively, the sample is freeze-dried and the resin poured over the sample via a suitably arranged side arm and stop cock. It is also possible to vapor fix the sample at any appropriate temperature by using the side arm arrangement. The resin embedding is quite straightforward, although it is important to ensure that the resin mixtures are degassed before infiltration under vacuum to avoid bubbling. As a general rule, the embedding should take place at as low a temperature as is compatible with the viscosity of the resin mixture. Similarly, polymerization should take place at 313 K rather than 343 K (Edelman, 1986). Sjostrand and Kretzer (1975) have infiltrated freeze-dried material with hydroxy propyl methacrylate at 243 K. Polymerization took place at 246 K using a UV source. Edelman (1986) has used Lowicryl KIIM at 213 K for both infiltration and UV mediated polymerization.

6.6. FREEZE-DRYING FOR MICROSCOPY AND ANALYSIS

The applications fall into two general classes. Freeze-drying is used either to produce a finished end product, e.g., freeze-dried sections for TEM and/or micro-analysis, freeze-dried macromolecules, or freeze-dried bulk specimens which are to be examined in the SEM; or the drying process is a step in a procedure such as the resin embedding we have just been discussing.

6.6.1. Structural Investigations

Much of the success of using freeze-drying as a preparative technique for SEM centers on the way the sample is treated prior to rapid cooling and freeze-drying. These are matters that have occupied the attention of Alan Boyde and his colleagues for a number of years (Boyde, 1974, 1975, 1976, 1978; Boyde and Echlin, 1973; Boyde and Franc, 1981). Prefixation and thorough washing of the sample surface certainly enhances both the quality of preservation and the structural integrity of freeze-dried samples. Samples for SEM can be treated in a number of different ways. Natural surfaces may be freeze-dried or interior surfaces may be exposed by fracturing and sectioning and then freeze-dried.

The problems associated with drying bulk specimens, which have poor mechanical and thermal contact with the underlying cold stage of the freeze-dryer can be overcome by using the device designed by Boyde and Echlin (1973). This cryocastle

Figure 6.9. Cryocastle for freeze-drying large objects for SEM. Quench-cooled samples are placed in the dish with a crennelated edge containing liquid nitrogen and covered with the lid. The unit is placed on a precooled cold stage of a vacuum evaporator and the liquid nitrogen pumped away. The castellations at the edge of the dish allow the water vapor to escape during drying and the lid itself provides an effective shield against radiant heating. [Redrawn from Boyde and Echlin (1973).]

is shown in Fig. 6.9 and consists of a cylindrical metal dish with crenellated walls and a lid that has a larger diameter than the dish. The metal dish can hold a small amount of LN_2 and allows frozen samples to be transported from a storage Dewar to the freeze-dryer without warming. The whole assembly is placed on the precooled cold stage of the freeze-dryer and the sample remains cold as it is surrounded by cold surfaces. The cutouts on the walls of the dish allow the subliming water molecules to escape during freeze-drying, which is carried out between 173 and 203 K. Boyde (1980) and Boyde and Franc (1981) have devised experimental protocols that allow drying to proceed at a much faster rate. They argue on the basis that in SEM it is usually only the surface of the specimen that is important, and provided the outer parts are properly quench cooled, it would seem sensible to concentrate the effort on drying this part of the sample. Boyde (1980) suggests initiating the drying at 188 K and a pressure of 25 mPa (2×10^{-4} Torr), which will allow the water to sublime from the surface regions in about 10 min. The temperature is raised and the drying allowed to proceed into the specimen at a much faster rate.

6.6.2. Analytical Investigations

Freeze-drying followed by plastic embedding has been found to be a useful technique for analytical studies such as autoradiography (Williams, 1977), histochemistry (Arnold and Von Mayersback, 1975), laser microprobe mass analysis (Kaufmann, 1982), and electrolyte localization by x-ray microanalysis (Ingram and Ingram, 1984).

Zs-Nagy and colleagues (Zs-Nagy *et al.*, 1977; Zs-Nagy, 1988, 1989) have developed a technique of freeze-fracture–freeze-drying to prepare bulk specimens of biopsy material for low-resolution energy dispersive x-ray microanalysis. Quench-frozen samples are fractured and then freeze-dried carefully according to special time and temperature schedules, which are related to the sample being studied and which, it is claimed, excludes the possibility of elemental redistribution. Samples are kept under vacuum at 10^{-3} Torr to avoid rehydration and are examined and analyzed uncoated at 10 keV. We will return to the relative merits of different low-temperature techniques as applied to analysis in Chapter 11, but sections of freeze-dried resin-embedded material have been a standard preparative technique for x-ray microanalysis for more than 20 years. The results obtained from such material are inevitably

Table 6.2. Elemental Ratios Obtained by X-Ray Microanalysis from Different Regions of Rat Pancreatic Tissue Which Had Been Either Freeze Substituted, Freeze-Dried and Resin Embedded, or Cryosectioned[a]

Method	Element	Zymogen granule	Apical cytoplasm	Endoplasmic reticulum	Mitochondria
Freeze	Na	40	65	116	74
substituted	Mg	10	34	39	7
	P	45	166	48	202
	K	36	52	87	57
	Ca	6	3	7	5
Freeze-dried	Na	16	46	100	41
resin	Mg	5	23	50	16
embed	P	19	209	531	216
	K	31	121	288	142
	CA	7	4	5	2
cryosectioned	Na	15	89	168	49
	Mg	6	41	102	25
	P	18	480	1324	294
	K	18	304	732	198
	Ca	8	10	9	4

[a]There is a very wide variation in the results, and if one assumes that the data from the cryosections more closely resemble the true elemental ratios, there can be little confidence in the data obtained by the more invasive methods of low-temperature preparation. Data from Roos and Banard (1986).

compared with those obtained from either freeze-dried cryosections or frozen-hydrated sections (Table 6.2). As one might expect, the results are extremely variable, and although significant quantitative analyses have been carried out on freeze-dried embedded material, it is not possible to give an experimental protocol that could be applied to all hydrated samples. It is inappropriate to discuss here the advantages and disadvantages of the different cryomethods ued for analysis, but the papers by Ingram and Ingram (1984) and Edelmann (1986) provide a balanced account of the way sections from freeze-dried embedded material may be used to measure the local concentration of chemicals.

6.7. FREEZE-DRYING FROM NONAQUEOUS SOLVENTS

So far in our discussions we have only been considering the freeze-drying process in which the natural water content of a specimen is first transformed to ice and then removed by sublimation at low temperatures and high vacuum. We indicated earlier that there are a number of other routes for sample dehydration, and one of these involves low temperatures. In the solvent drying process, water is first exchanged for a nonphysiological organic liquid, which is then solidified and subsequently removed by sublimation. This is a far more rigorous dehydration procedure than freeze-drying from ice, and although ice crystal damage is avoided entirely, the organic liquids that are used are powerful solvents, which may selectively dissolve parts of the sample. It is usually necessary to first stabilize biological specimens with some form of mild fixation using one of the organic aldehydes such as glutaraldehyde. This fixation must be carried out under controlled conditions, and the standard texts by Glauert (1974) and Hayat (1981, 1986) should be consulted for appropriate recipes. The fixed sample is gradually dehydrated with an organic liquid such as acetone, methanol, or ethanol, which in turn may be gradually replaced with suitable organic solvents. A number of different materials have been used and a partial list is given together with some of their properties in Table 6.3. Following infiltration, the samples are cooled in liquid nitrogen. Fast cooling is not necessary for there is now no water to form ice crystals. Depending on the material, the cooled organic liquid will form either a glass or a microcrystalline solid.

Samples may either be fractured or sectioned in this low-temperature state (see Chapter 5, Section 5.5) or left untouched before being dried by sublimation at temperatures and pressures that depend on the nature of the solvent. Most of the organic liquids that are used have a sublimation point very close to their melting point, and in order to have a margin of safety, the sublimation should be carried out 10 K below the melting point. Table 6.3 shows the vapor pressure of a number of solvents at low temperatures, and this together with the data presented in Fig. 6.10 shows that some organic liquids are easier to sublime than others. Thus ethanol, diethyl ether, acetone, and 1.2.epoxy propane should be sublimed at temperatures around 173 K and at vacuum pressure between 1 Pa and 1 mPa (10^{-5} Torr). Amyl acetate and tertiary butyl alcohol can be sublimed at somewhat higher temperatures and lower vacuum pressures. Boyde and Wood (1969) freeze-dried bulk samples from amyl acetate in a few hours at 192 K and a pressure of 650 mPa (5×10^{-3} Torr). Camphene and acetonitrile are of interest because the sublimation temperatures are

Table 6.3. Some Properties of Materials That Have Been Used for Solvent Drying and or Sublimation

Material	Melting point (K)	Temperature at which vapor pressure = 130 Pa (K)	Surface tension (dyn/cm)
Acetonitrile	227	226	29.3
Acetone	177	214	26.2
Amyl acetate	202	—	—
Camphene	318	—	Sublimes
Chloroform	268	210	27.1
Diethyl ether	157	199	13.8
Ethanol	159	242	24.1
Methanol	175	229	24.5
Propylene oxide	161	198	—
T-Butyl alcohol	284	198	20.7
Water	273	253	73.1

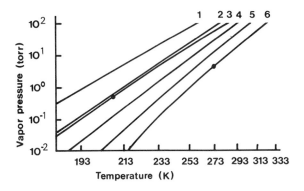

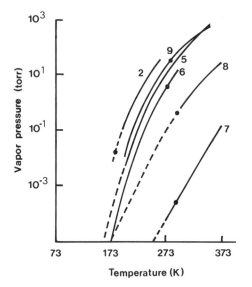

Figure 6.10. Vapor pressure of solvents used in low-temperature drying. The black dot on two of the curves marks the melting point. 1, Diethyl ether; 2, acetone; 3, chloroform; 4, methanol; 5, ethanol; 6, water; 7, glycerine; 8, DMSO; and 9, benzene. [Modified and redrawn from Robards and Sleytr (1985).]

much higher. Camphene will sublime at room temperature and acetonitrile will sublime at 130 Pa (1 Torr) at 1 K below its melting point. Katoh and Matsumoto (1980) have used a 50% aqueous solution of acetonitrile and find that the drying process is completed in under an hour with only a 3%–5% linear shrinkage of the rat liver samples they were using in their experiments. The acetonitrile has the added advantage that it has a low lipid solubility.

The low-temperature solvent-drying process may be carried out either in a freeze-dryer or in a modified vacuum evaporator. It is necessary to have a condensor to trap the sublimed solvents and LN_2 is most effective in this respect. The low-temperature solvent-drying procedure has potential hazards of toxicity (acetonitrile, amyl acetate and chloroform) and flammability (acetone, diethyl ether, 1.2.epoxy propane). In addition, many of the organic solvents will dissolve in the vacuum pump oil and diminish the effectiveness of the pumping system.

6.8. ARTIFACTS ASSOCIATED WITH FREEZE-DRYING PROCEDURES

Although freeze-drying from aqueous solutions is now generally considered to be the most gentle way to dry hydrated samples, specimen damage and artifacts can occur. Some of the damage may have occurred during the stages prior to the drying processes. Faulty or inadequate chemical fixation, failure to remove soluble and particulate surface contaminants, and poor quench cooling all create their own artifacts. The actual drying process can give rise to problems that are important to understand if not to readily recognize.

6.8.1. Molecular Artifacts

At the molecular level there are problems associated with collapse and aggregation. These two phenomena are frequently exacerbated by the strong electrostatic interactions that may develop between anhydrous samples and their supports and by thermal agitation of macromolecules freed of their hydrophilic exclusion zones. The electrostatic forces between charged groups are increased by a factor of 80 owing to the drop in the dielectric constant as water is removed by sublimation. In a discussion on the protection of proteins against dehydration, Franks (1985) stresses the importance of maintaining the hydration shell of the protein under conditions of lowered water activity. In industrial spray-drying and freeze-drying operations, the addition of sugars appears to favor the retention of the hydration shell. We should seek ways of incorporating such procedures into the preparative processes for the freeze-drying of macromolecules. In Chapter 2 we discussed the deleterious effect freeze concentration may have during cooling of dilute solutions. One way around this problem would be to start with more concentrated solutions (Franks, 1985). This would be physiologically incompatible for many biological materials, and while it is an option for biologically derived products such as in the food and pharmaceutical industry, such increases in solute concentration would have important influences on freeze-drying. The solutes in such concentrated solutions will be precipitated during the drying process and may obscure vital structural detail.

6.8.2. Structural Artifacts

Because water is such an important structural component of hydrated samples, its removal can lead to mechanical instability. As the sublimation front proceeds through the sample, and during the desorption of the last traces of water, cracking, collapse, shrinkage, wrinkling, and folding can all occur. Boyde (1978) considers that the shrinkage effects due to desorption of water during the secondary drying phase could be avoided if the incompletely freeze-dried sample could be maintained at 100% relative humidity with respect to the temperature at which the drying is taking place. This is an elegant theoretical answer to this problem which would be difficult to apply in practice to samples that are only to be freeze-dried. If the freeze-drying is to be followed by resin embedding, it should be possible to choose a resin mixture that could tolerate a small amount of water, e.g., some of the water-soluble acrylates.

There is a large variation in the structural artifacts that are principally associated with the secondary drying phase, and their extent seems to be related to the inherent lack of mechanical strength of the materials involved and local water concentrations. An extensive study by Boyde and Maconnachie (1979) and Boyde and Franc (1981) revealed that chemically fixed liver showed a 20% volume shrinkage after freeze-drying, a 65% volume shrinkage after critical point drying (a popular method of sample drying for SEM), and a 90% volume shrinkage after air-drying blocks of the tissue. Nordestgaard and Rostgaard (1985) found the opposite effect with isolated hepatocytes. With freeze-drying there was a 51% volume shrinkage; with critical point drying the figure was only 38%. Although not measured, air-dried hepatocytes would shrink to a few percent of their original volume. Although sections of resin-embedded freeze-dried material are used for ultrastructural and morphometric studies, the superior preservation obtained by freeze substitution has meant that the former type of preparations are less favored than the latter. Many sections from freeze-dried embedded material show empty spaces or "haloes" around organelles such as mitochondria. This can be avoided by extended freeze-drying prior to embedding in resin.

Artifacts due to recrystallization phenomena during sublimation are extremely difficult to separate from crystallization artifacts that may have occurred during the initial cooling. A good example of the ice-crystal artifact can be seen in some of the early work by Wolosewick and Porter (1979) on the structure of the microtrabecular network in plant and animal cytoplasm. In chemically fixed material, the network appeared finely distributed. In freeze-dried material, the solid components of the network were thicker and the spaces between them were much larger. Ice crystal growth, during either the initial cooling or subsequent recrystallization, created these spaces, forcing the network components together. Large samples are going to have both types of artifacts and the specter of recrystallization in small samples will only disappear if drying takes place at low temperatures and vacuum. As we have shown elsewhere in this chapter, the sublimation–recrystallization phenomenon is a struggle between two competing thermodynamic processes. Recrystallization reduces the entropy of the system, sublimation increases it, and in a given sample it is simply a race to see which set of conditions will prevail.

In our zeal to complete the drying process as quickly as possible we should not allow the temperature to rise to the point where the eutectic mixture surrounding the ice crystals begins to melt. The temperature at which this is likely to occur has to be experimentally determined for a given sample, although Robards and Sleytr (1985) consider that 238 K is the critical temperature for eutectic melting in biological materials. The temperature during the primary drying phase should remain below 233 K to avoid such melting. Rehydration during the drying process will result in severe surface tension mediated collapse and redistribution of soluble components.

6.8.3. Analytical Artifacts

Although freeze-drying avoids the use of chemical fixatives and dehydration, the resin monomers used in embedding are organic solvents. The effects that these monomers will have depend on the resin concerned; the Lowicryl resins appear to be much less damaging than the epoxy resins. Water may be produced during some of the polymerization processes (Marshall, 1980), and this could cause further redistribution of soluble materials, which will have been moved from their original position in the sample during the initial quench cooling. Sections should be cut without the benefit of trough liquids as water soluble materials can still be lost from resin-embedded material (Marshall, 1980a, b; Harvey, 1982). The vexed question of the relocation of electrolytes within the cells as distinct from their redistribution within and loss from tissues remains unsolved. The success of the procedure seems to depend as much on the element and tissue as on the doubtlessly cleverly devised experimental methods. Freeze-dried resin-embedded materials have been used in connection with immunocytochemical studies and the antigenicity of some protein molecules is well preserved. Although a wider range of antigens are preserved by this technique in comparison to conventional embedding procedures, the freeze-dried embedding routine is now being discarded in favor of freeze-substitution techniques, which preserve a wide range of antigens with a high biological activity. These are matters we will consider in the following chapter.

6.9. SUMMARY

Freeze-drying is a dehydration process in which the sample water is first converted to ice, which is then sublimed away at low temperatures and high vacuum. The drying process has two distinct phases. The primary drying, which usually takes place at between 173 and 183 K, removes the bulk, unperturbed water from the sample. Secondary drying takes place at more elevated temperatures and desorbs the unfreezable, perturbed water associated with hydrophilic surfaces. The different polymorphs of ice sublime at different rates and this finding has been used to advantage in the design of a freeze-dryer that sequentially sublimes the ice as the deep frozen specimen is warmed through the different phase transitions. More conventional freeze-drying can be carried out at high (10^{-6} Torr) and low (10^{-2} Torr) vacuum and at different temperatures. Other freeze-drying processes depend solely on the high affinity for water of molecular sieves maintained at liquid-nitrogen temperatures.

Although much is known about the sublimation of crystalline ice, freeze-drying of hydrated samples remains an empirical procedure. Each protocol has to be designed around the specimen concerned, and the drying process should be monitored and controlled. There are many artifacts associated with freeze-drying ranging from molecular collapse and aggregation to gross shrinkage of the sample. Freeze-drying has a wide range of applications, primarily in high-resolution structural studies of macromolecules and the analysis of sectioned material. It may be used directly to produce specimens for examination and analysis, although the extreme hygroscopic nature of such samples presents their own problems of sample handling. Freeze-drying may also be used as part of a more extended process during which the water removed from the sample is replaced by resins enabling such samples to be sectioned and fractured.

Freeze Substitution and Low-Temperature Embedding

7.1. INTRODUCTION

Although freeze substitution and low-temperature embedding are two distinct processes, they are considered together because they complement each other and are frequently used sequentially during sample preparation. Freeze substitution is a chemical dehydration process in which ice in frozen-hydrated specimens is removed and replaced by an organic solvent. This procedure is in sharp contrast to the process of freeze-drying discussed in the previous chapter, where the water is removed by a purely physical process and is not replaced. Low-temperature embedding seeks to replace either the spaces once occupied by water in a freeze-dried sample or the organic fluid that has been used during freeze substitution, by infiltrating the sample at low temperatures with resins, which are, in turn, polymerized at low temperatures. The temperatures at which these processes are carried out are critical. If it is too high, the rate of ice recrystallization will damage the sample. If it is too low, the organic solutions become so viscous that they will not adequately penetrate the specimen. The compromise is to use lower temperature (193–173 K) for substitution and somewhat higher temperatures (253–238 K) for embedding.*

The two techniques serve as a convenient link between the idealistic approach of purely physical cryofixation techniques and the pragmatic convenience of thin sectioning at room temperatures. Such a combination sounds too good to be true, and would at first glance seem to be the best approach to the preservation of ultrastructure and functionality. In many respects this is correct, but an uncritical and ready acceptance of the rationale behind this approach would be unwise. Freeze substitution and low-temperature embedding have their own catalogue of artifacts, which in many respects are more insidious and less easily recognized than the now familiar artifacts associated with rapid cooling, cryomicrotomy, and freeze-drying.

The study of hydrated material generally, and biological material in particular, by means of electron microscopy of thin sections of embedded material, is limited

---

*The low-temperature resins we will be discussing are quite distinct from water-soluble resins such as some of the glycol-methacrylates, hydroxy-propyl methacrylates, polyethylene-glycol, poly-vinyl alcohol, etc., which are all used at temperatures well above 273 K. For details, see Hayat (1986).

by many factors. The replacement of sample water with organic solvents is one of the most crucial steps in the preparation process. The aqueous environment determines the organization of many macromolecules, e.g., proteins and carbohydrates, and replacement of this water with organic liquids can seriously alter their natural configuration. Water also acts as the solvent for most of the small molecules and all the electrolytes in the cell. The distribution and local concentration of these highly diffusible materials will be significantly affected unless special attention is paid either to the stabilization or removal of the water matrix. Although chemical fixation will stabilize macromolecules, it is unlikely that it will make them any less vulnerable to the rearranging effects of dehydration. The fixation may also induce sufficient alteration in macromolecules to impair their identification by analytical techniques such as immunolabeling.

A hundred years ago, Altmann (1890) recognized that a nonchemical embedding technique, based on freeze-drying followed by embedding in wax, would be more likely to preserve the natural order of structures and processes in hydrated biological material. In the ensuing years, this essentially nonchemical embedding procedure was occasionally used by a few selected workers. (Gersh, 1932; Hoerr, 1936; Feder and Sidman, 1958) to prepare sections for examination by light microscopy. Simpson (1941) carefully evaluated the early work on low-temperature dehydration and embedding techniques and attempted to distinguish between the artifacts that had been introduced during freezing and those that occurred during freeze-drying. In an attempt to overcome some of these problems, Simpson dehydrated frozen specimens in different cooled organic liquids and was the first person to refer to this process as freeze substitution.

Low-temperature embedding procedures have also long been the concern of electron microscopists. This concern was born in the notion that the high resolution afforded by the electron microscope brings us closer to understanding the structure of macromolecules which form the framework in which vital processes occur. There was also a realization that the ephemeral and labile nature of most living systems demanded more gentle means of sample preparation than oxidizing chemicals, organic solvents, and potentially carcinogenic resins. Fernandez-Moran (1957, 1959, 1960) was the first to apply freeze-substitution techniques to ultrastructural studies. Glycinerated tissues that had been quench cooled using liquid helium at 2 K were substituted in alcohol/acetone/ethyl chloride mixtures at temperatures between 143 and 193 K. The substituted specimens were infiltrated with methacrylate mixtures at 173–193 K and polymerized at 193–253 K using UV radiation coupled with chemical catalysts. Bullivant (1970) modified the procedures pioneered by Fernandez-Moran to include methanol and ethanol as substitution media, and Pease (1966, 1967, 1968) attempted to use low-molecular-weight glycols as substitution fluids. Other workers, including Rebhun (1961) and Van Harreveld and Crowell (1964), further developed the use of ethanol/acetone mixtures as a means of substituting animal cells. Details of much of this earlier work may be found in the reviews of Rebhun (1972), Pease (1973), and Glauert (1974).

In spite of these advances toward what now appears to be one of the better ways of preserving the ultrastructure and functionality of hydrated tissues, the methodology appeared to stagnate. In an early review of freeze substitution, Pease (1973) provided two main reasons why freeze substitution and low-temperature embedding did not attract very much attention at the time. There was then, as now, difficulty in obtaining good initial cryofixation of hydrated samples. There was confusion over which was the best substitution fluid, experimental procedure, and which resin mixture to use for ultrastructure and/or analytical studies.

In addition to this uncertainty and lack of a sense of direction, one must also appreciate the general trend of microscopical investigations in the 1960s and early 1970s. With a few notable exceptions, most microscopy was concerned with obtaining structural information, a goal that was easily obtained using heavy metal fixatives, room-temperature dehydration with polar organic liquids, and embedding in epoxy and acrylate resins. By the mid-1970s, there was a perceptible change in the direction and resolve of electron microscopists as we became more interested in the chemical reactivity of the structures we were examining with increasingly better spatial resolution. There was a conscious attempt to link structure and function, and it quickly became apparent that it would be necessary to seek new ways of avoiding conventional fixation, staining, dehydration, and embedding. The existing procedures all conspired to limit spatial resolution to 3–5 nm and cause a loss of upwards to 95% of the soluble constituents of cells and tissues. In addition, the conventional techniques abolished biological characteristics such as enzymatic activity and antigenicity.

In the last ten years there has been a resurgence of interest in developing techniques to improve the potential for structural preservation. Much of the impetus for reviving and further developing the appropriate preparative technology has come from the research group at the Biozentrum, University of Basel in Switzerland (Kellenberger and Kistler, 1979; Kellenberger *et al.*, 1980; Carlemalm *et al.*, 1982; Armbruster *et al.*, 1982). These workers considered that three steps in the preparative procedure needed closer examination to ensure that dehydration and embedding resulted in minimal loss of structure and maximum retention of biological activity:

1. The macromolecules should be immobilized in an inert manner to minimize the likelihood of massive supramolecular rearrangement and subsequent denaturation of cellular components. This would preserve cell architecture in a more lifelike form and, provided the integrity of the membranes remained intact, should retain soluble components in their correct subcellular compartments.

2. The solvent environment should remain as "waterlike" or as polar as possible throughout the entire dehydration and embedding. This would maintain the hydration shell surrounding molecules and prevent components of the embedding medium from disrupting the chemical groups and bonds involved in inter- and intramolecular contact. (This factor, although well suited for preserving macromolecular structure, now seems less than optimal for the smaller molecular weight soluble components of the cell.)

3. The use of low temperatures would enhance the stability of biological materials during solvent exchange by reducing the amplitude of molecular thermal vibration.

Having established the general procedures that have most influence during sample preparation for electron microscopy, the Basel group sought ways of separating these factors into a number of variables that might be studied in more detail. They consider freeze substitution and low-temperature embedding as an integrated whole, made up of fixation (where applicable), dehydration, resin infiltration, and resin curing, within which the following variables are important:

1. The chemical interactions between fixatives and specimen.
2. The polarity of the dehydrating and embedding materials.
3. The temperature of dehydration and embedding.
4. The amount of water maintained in the sample during tissue processing.

This bold new approach to the preservation of ultrastructure and molecular integrity needed certain criteria of excellence against which one could judge the success of a given preparative procedure. The most obvious criterion would be an equivalency of structure as seen by existing technologies. Carlemalm *et al.* (1980, 1982) chose protein crystals as model systems because their structural integrity could be monitored quantitatively by x-ray diffraction techniques. They found that many of the newly developed solvents, fixatives, and embedding media caused only very small changes in the order of the lattice configuration. About 30 different proteins have now been studied at subzero temperatures in mixed cryosolvents. The freeze-fracture replication images of biological membranes are considered to more closely resemble their natural structure than the trilamellar unit membrane shown by conventional fixation. Sjostrand (1982) found that images similar to those seen after freeze-fracture replication could be seen in mitochondria after freeze substitution and low-temperature embedding. In the course of their studies, Sjostrand and his co-workers (Sjostrand and Barajas, 1968; Sjostrand, 1977) demonstrated that dehydration may alter the tertiary and quaternary structures of proteins, particularly those associated with protein-rich biological membranes. Petsko (1975) and Franks (1985) have shown that low-temperature methods make it possible to obtain information on the spatial fluctuations that occur within macromolecules. Such fluctuations are obviously significant in relation to the biological functionality of the proteins. Studies by Roth *et al.* (1980, 1981) showed by sensitive immunocytochemical techniques that freeze substitution and low-temperature embedding preserved much of the three-dimensional structure of proteins, which was normally lost using conventional methods.

Numerous workers have been able to show that freeze substitution, in combination with conventional heavy metal fixatives and stains, produced images equivalent to, and frequently better than those produced by existing chemical techniques. In some instances the techniques revealed new structural information. Concurrent with this exploratory work to seek validation of freeze substitution and low-temperature embedding as a more sensible approach to the preservation of cell ultrastructure, low-temperature studies were being carried out on enzyme kinetics. Cryoenzymology studies by Douzou (1973, 1977) showed that at 223–173 K, enzymes resist denaturation in 79%–80% wt/wt organic solvent solutions provided the pH, dielectric constant, and polarity of the solutions are adjusted to the optimal conditions of the enzyme system. The role of low temperatures in this technique is to reduce the amplitude of thermal vibration and thus minimize the effects of solvent exchange

(Frauenfelder, 1979; Artymink *et al.*, 1979). Low temperatures alone do not slow down protein denaturation.

The renaissance of these two techniques, in which both the structure and functional integrity of biological and hydrated organic materials is preserved, offers a viable alternative to frozen-hydrated sections in which unfixed structures are embedded in a matrix of amorphous or microcrystalline ice. Although this latter approach will probably remain the goal of low-temperature microscopy and analysis, the technology for achieving it is expensive, complex, and of limited availability. Freeze substitution and low-temperature embedding are now an excellent second best sample preparation procedure.

7.2. GENERAL OUTLINE OF THE PROCEDURES

In terms of practical usage, the various steps of freeze substitution and low-temperature embedding are integrated into a series of processes that are continuous through dehydration, embedding, and polymerization. Within this general scheme there are, however, a number of variations depending on the ultimate destination of the samples—i.e., whether they are to be used for ultrastructural examination, analysis of soluble components, or analysis of functional integrity. There are very few "standard" procedures, and Fig. 7.1 provides a generalized account of the different options that may be followed. Within this overall picture it is possible to identify four main areas of activity, each of which is sufficiently distinct to warrant separate more detailed discussion:

1. Progressive lowering of temperature (PLT). These are procedures that ensure that chemically fixed, hydrated samples are dehydrated slowly and at progressively lower temperatures. The PLT technique is particularly applicable in studies on the immunocytochemical localization of antigens in biological material.

2. Freeze substitution. These are the processes whereby the ice crystals in a frozen sample are dissolved in organic liquids maintained at subzero temperatures.

3. Low-temperature embedding, where the spaces once occupied by water in the samples are infiltrated at low temperatures with low-viscosity resins, which are subsequently polymerized at low temperatures.

4. Isothermal freeze fixation. A more specialized technique, which aims at preserving the frozen state of cells and tissues close to the melting point of ice in order to study ice crystal damage.

Before considering these four main topics it is appropriate first to consider briefly sample preparation and the circumstances under which chemical fixatives, cryoprotectants, and stains may be applied to the sample either prior to the low-temperature procedures or during the procedures.

7.3. SAMPLE PREPARATION

The samples should be as small as possible for optimal rapid cooling and to present a short diffusion path for the substitution and embedding chemicals. Large

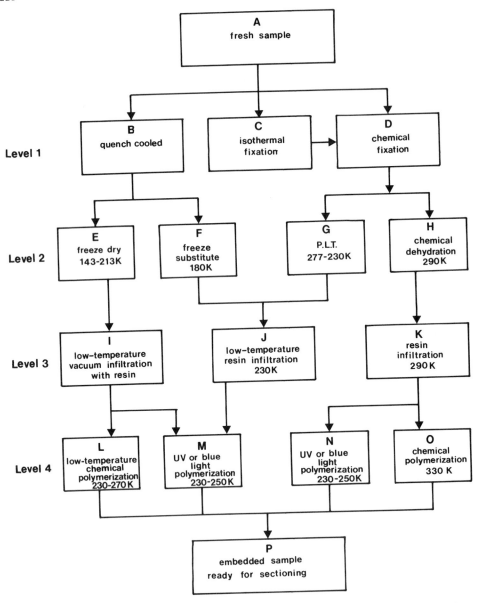

Figure 7.1. Flow chart to illustrate the different options available with freeze substitution and/or low-temperature embedding. The different options may be exercised at four stages during processing. There will be variations within each option depending on the material and the chemical being used.

specimens should be sliced thinly and liquid suspensions presented either as small droplets or thin-film suspensions. Lancelle *et al.* (1986) placed small samples on a copper loop coated with a Formvar film prior to quench cooling. The thin films remain intact and the copper wire forms a convenient handle for sample transfer through the various substitution and embedding liquids. All the exhortations that have already been made about quench cooling apply equally well to the freeze substitution and low-temperature embedding procedures.

7.4. CHEMICAL ADDITIVES

7.4.1. Fixatives

Fixatives (and other forms of chemical intervention) can either be applied to the sample while it is in the aqueous phase prior to quench cooling or, as is more usual, incorporated into the substitution fluids which are used immediately following quench cooling. It is tacitly accepted that there are no chemical fixatives that do not cause some alteration to macromolecules, and that none of the various chemicals that have been used are capable of rendering structures unaffected by the influence of dehydration.

So-called mild fixation, i.e., low concentrations for short periods of time, with organic aldehydes such as glutaraldehyde or acrolein, appears to stabilize molecular components by inter- and intramolecular cross-linking without significantly affecting their structure. There is both indirect and direct evidence to support the use of mild chemical fixatives. The direct evidence reveals that there is a significantly greater retention of antigenicity of samples fixed with glutaraldehyde rather than osmium tetroxide, and in the absence of either fixative, the structural preservation is generally poor. In addition there is frequently greater extraction of soluble materials in unfixed samples during freeze substitution. This is not, however, a universal truth, as some proteins retain their antigenicity after fixation at ambient temperature with both glutaraldehyde and osmium tetroxide. On the other hand, Monaghan and Robertson (1990) found that it was not necessary to fix material prior to freeze substitution in methanol and embedding in a low-temperature resin (Lowicryl) in order to get good structural preservation and positive immunolocalization of antigens. Clearly some proteins are structurally more robust than others. It is also known that many biological substances—e.g., nucleic acids, lipids, and polysaccharides—have a very low reactivity towards aldehydes and that very few cross linkages are established. Carlemalm *et al.* (1986) have shown that ambient temperature embedding of aldehyde fixed cells such as phage-infected *E. coli*, which contain substantial amounts of nucleic acid, results in aggregation of the DNA. The DNA pool with its low protein content, does not form a stable cross-linked structure with glutaraldehyde, although at sub-zero temperatures the same fixation schedule coupled with low-temperature embedding in Lowicryl resulted in good structural preservation.

Indirect evidence for the efficacy of mild fixation comes from x-ray diffraction studies of aldehyde-treated protein crystals in which minor changes only amounting to a few angstrom units have been observed (Steitz *et al.*, 1967; Langen *et al.*, 1975;

Wiskner *et al.*, 1975). This is not to suggest that these fixatives have no effect, although this has never been tested experimentally. Immunocytochemical techniques are effective largely on the basis that it is only necessary to retain a significant amount of the antigenicity of a molecule, not necessarily all the structural integrity.

There is considerable variation (and uncertainty) over the correct length of time required for fixation prior to freeze substitution. Weibull *et al.* (1980) prefixed isolated pea chloroplasts for 20 min with a dilute aldehyde solution, but found that there was little variation in the appearance of the chloroplasts if this time varied between 5 and 60 min, although very short fixation times, i.e., less than 5 min, resulted in disrupted chloroplasts. Mersey and McCully (1978) found that plant cytoplasm would not gel satisfactorily after short periods of fixation. The early freeze substitution work on animal tissues by Sjostrand and his colleagues (Sjostrand and Barajas, 1968; Kretzer, 1973; Sjostrand and Kretzer, 1975; Sjostrand, 1977), was all based on very short (2–15 min) fixation times at quite high temperatures (273 K). They found that longer fixation caused membranes to appear more like the unit membrane configuration of conventionally osmium-fixed material. Later studies by Kellenberger *et al.* (1980, 1982) showed that these fixation times were probably too brief to provide adequate protection to the sample during dehydration.

For ultrastructural studies alone, the best results are obtained if the fixative is used as an integral part of the freeze-substitution schedule. Steinbrecht (1980) found that if samples were fixed with osmium tetroxide *before* rapid cooling and freeze substitution, the resultant images were the same as those obtained by ambient temperature procedures. Fixatives such as glutaraldehyde and acrolein, or heavy metal salts such as osmium tetroxide or uranyl acetate, should be mixed with the organic substitution media. It is presumed that the dissolved fixatives penetrate the specimen at the same speed as the substitution media. Very little is known about the chemical reactivity of fixatives at subzero temperatures, and consequently many substitution schedules include a gradual temperature rise at the end of the substitution period. Humbel and Muller (1986) have investigated the action of fixatives on a 2% aqueous solution of bovine serum albumin (BSA), which was freeze substituted at 183 K for a 8 h in methanol containing 3% glutaraldehyde. After 8 h the temperature of the freeze-substitution medium was gradually raised and samples of BSA removed at various times during the warming period. The relative amounts of soluble and cross-linked protein were determined at the different substitution temperatures. Figure 7.2 shows that the glutaraldehyde in methanol exhibits characteristic temperature-dependent fixation curves and that an extrapolation of the data would show that, at 223 K, about half the protein would be cross linked in 16 h. No such studies have been carried out on other fixatives, although Van Harreveld *et al.* (1964) found that osmium tetroxide would not blacken tissues below 248 K. Work by White *et al.* (1976) and Humbel *et al.* (1983) suggests that both uranyl salts and osmium tetroxide will react with lipids and the double bonds of unsaturated fatty acids at temperatures as low as 183–200 K. The reports on the action of chemical fixatives on biological samples are largely anecdotal, and all that is certain is that fixation schedules that work for one specimen may not necessarily work for another.

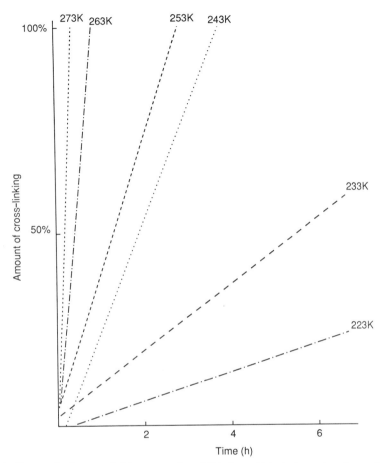

Figure 7.2. Changes in the cross-linking properties of 3% glutaraldehyde in methanol as a function of temperature. The vertical axis is a measure of the relative amounts of cross-linked protein. The horizontal axis is time of fixation. [Redrawn from Humbel and Muller (1986).]

Heavy metal fixatives such as uranyl and osmium salts should not be used in conjunction with low-temperature embedding procedures which rely on UV polymerization of the embedding media. The darkened specimens do not absorb the radiation so strongly and give rise to uneven polymerization. The same codicil applies to heavily pigmented specimens.

If freeze substitution is to be used to prepare samples for x-ray microanalysis of diffusible ions or autoradiography of soluble molecules, no fixatives should be used either before or during the substitution process. Humbel *et al.* (1983) have shown that biological material freeze substituted in the absence of fixation is more strongly influenced by polar than nonpolar embedding media. In these circumstances it is advisable to use an apolar solvent such as diethyl ether as the substitution liquid.

7.4.2. Stains

In addition to using heavy metal fixatives in the substitution fluid, a number of investigators have also included heavy metal salts, which it is hoped will increase the electron scattering of the organic material being substituted. Fernandez-Moran (1957, 1959) used platinum and gold chlorides in the substitution fluids. Bridgeman and Reese (1984) have examined a whole range of substitution media and found that the addition of tannic acid in the primary substitution media enhances the staining effect of both osmium tetroxide and uranyl acetate. Heuser and Kirschner (1980) have used hafnium chloride as a stain and Tosuyama *et al.* (1984) have used a complex of periodic acid thiocarbohydrazide silver proteinate as a staining protocol during freeze substitution.

7.4.3. Cryoprotectants

The situation with regard to the use of cryoprotectants is equally uncertain. We have already considered how the use of these materials imposes restraints on the total information that may be obtained from the specimen. There seems to be no *a priori* reason why cryoprotectants should not be used prior to quench cooling for specimens that are going to be used for only morphological studies and possibly for immunocytochemical investigations, particularly if chemical fixation is also going to be included in the preparative protocol.

7.5. THE REMOVAL OF WATER FROM SAMPLES

Water may be removed from the sample by two processes. The cooled, but *unfrozen*, sample is passed through increasing concentrations of dehydrating agents, which are kept at progressively lower temperature. This is referred to as low-temperature dehydration. In the second procedure, the *frozen* sample is kept in a liquid dehydrating agent at low temperatures and then slowly warmed up. This is referred to as freeze substitution.

7.5.1. Low-Temperature Dehydration

Pease (1973) and Sjostrand (1975) chose ethylene glycol (one of the weakest denaturing organic solvents) at below 273 K as a highly polar dehydrating agent on the basis that very polar solvents are relatively inert to biological material. The later studies by Carlemalm *et al.* (1982) contradicted this generally accepted hypothesis as their investigations made a distinction between the *disruption* of macromolecular structure and the *denaturation* of its molecular components. They found that although ethylene glycol was minimally denaturing, it was nevertheless causing disruption of the macromolecular structure of a number of proteins.

Care must be taken in choosing the correct dehydrating agent, particularly if the aim of the experiment is to retain the functional integrity of the components of

Table 7.1. The Effect of Different Dehydrating and Embedding Agents on the Crystal Lattice Dimensions of Catalase and AAT (Aminacid Transferase)[a]

Polarity of whole process	Protein	Dehydration agent	Resin	Embedding temperature (K)	X-ray resolution (nm)
High	AAT	Ethylene glycol	K4M	238	0.6–0.7
High	AAT	OH-Ethyl acrylate	K4M	238	0.6–0.7
Medium	AAT	Methanol	K4M	238	0.8–1.0
High	AAT	[glycol methacrylate]		273	0.8–1.0
Low	AAT	Ethanol	Epon	293–343	
Low	Catalase	Acetone	Epon	343	0.8–1.0
Low	Catalase	Acetone	HM20	293	0.8–1.0
Medium/Low	Catalase	Ethanol	HM20	223	0.8–1.0
Medium/Low	Catalase	Methanol	HM20	223	1.4
Medium	Catalase	Ethanol	K4M	238	1.6
High	Catalase	Ethylene glycol	K4M	238	2.2
Medium/Low	Catalase	Ethanol	Epon	293–343	2.9
Medium/Low	Catalase	Methanol	Epon	293–343	No difference
High	Catalase	Ethylene glycol	K4M	273	No difference

[a]For AAT, a polar environment and low temperatures more effectively maintain molecular order. For catalase, an apolar environment is preferred, but low temperatures mitigate some of the disruptive effects of a polar environment. The temperature is that at which the dehydration, embedding, and polymerization take place. All x-ray diffraction carried out at 293 K. [Data from Carleman *et al.* (1982a).]

the biological matrix. Table 7.1 shows how different dehydrating agents (and resins) can influence the activity of two proteins as measured by changes in their crystal lattice. These data show that in order to preserve protein functionality, polar solvents such as ethylene glycol could be used to dehydrate the enzyme aspartate aminotransferase, whereas relatively apolar solvents such as acetone should be used to dehydrate the enzyne catalase. Retention of molecular integrity and functionality are not the sole criteria for choosing a particular dehydrating agent. The solvent must also be compatible and miscible with the resin(s) that are to be used to embed the sample. If polar resins such as Lowicryl K4M or K11M are to be used, then methanol, ethanol, ethylene glycol, glycerol, dimethylformamide, and acetone can be used. Glycerol and ethylene glycol should not be used in conjunction with nonpolar resins such as Lowicryl HM20 and HM23.

7.5.2. Progressive Lowering of Temperature

This low-temperature dehydration and embedding procedure was developed by Carlemalm *et al.* (1982) and starts with unfrozen aldehyde-fixed samples at 273 K. The consecutive dehydration steps are done at progressively lower temperatures to prevent the tissues freezing. Care is taken to ensure that the interior of the sample

Table 7.2. An Experimental Schedule for Progressive Lowering of Temperature Dehydration Routines for Ethanol and Acetone[a]

Ethanol	Acetone
30% for 30 min at 273 K	30% for 30 min at 273 K
50% for 60 min at 258 K	50% for 60 min at 258 K
70% for 60 min at 238 K	70% for 60 min at 238 K
95% for 60 min at 238 K	100% for 60 min at 243 K
100% for 60 min at 238 K	100% for 60 min at 243 K
100% for 60 min at 238 K	

[a]Data from Carlemalm *et al.* (1985) and Acetarin *et al.* (1986).

equilibrates with the dehydrating agent at each step before it is transferred to the next step. The temperature in a subsequent dehydration step is not allowed to be lower than the previous step. By observing these simple rules, Carlemalm and colleagues found that it was possible to design dehydration and resin infiltration schedules for a variety of chemical agents. Table 7.2 shows a typical progressive lowering of temperature procedure for ethanol and acetone. It must be stressed that the dehydration is an integral part of the total preparation procedure (see Fig. 7.1) and that the solvents and procedures used must be compatible with the ensuing processes. Thus the dehydration solvents must either be miscible with the resins to be used for embedding or exchangeable with solvents that are miscible with these materials. Similarly, if the low-temperature dehydration process is to be used to prepare the sample for scanning electron microscopy, the dehydrating solvents must be miscible with the fluids used in solvent drying, critical-point drying, or freeze-drying.

7.6. FREEZE SUBSTITUTION

7.6.1. General Principles

Organic solvents are used to dissolve the ice in frozen specimens at temperatures low enough to avoid ice recrystallization. Once the substitution is complete, the sample may either be freeze-dried or critical-point dried for scanning electron microscopy or infiltrated with embedding resin in order that thin sections may be cut for transmission electron microscopy. The resin infiltration and polymerization can either take place at low temperatures or at room temperature and above. The freeze substitution is always carried out at temperatures below 273 K.

7.6.2. Substitution Fluids

Ideally, one should only use organic liquids that can dissolve sample ice at or below the recrystallization temperature of ice, i.e., ca. 140 K. There are very few organic liquids that would dissolve ice at these temperatures, and the process would take far too long to be of practical use. However, as we discussed earlier, substantial

Table 7.3. Melting Point and Water Containing Capacity at 193 K of Solvents Used in Freeze Substitution

Compound	Melting point (K)	Percent water dissolved at 193 K
Acetone	177.6	2.5
Diethyl ether	156.5	0.8
Dimethyl ether	134.5	—
Ethanol	155.7	16
Methane	179.1	32
Propane	83.3	0.1
Dimethoxypropane	226.0	(Reacts chemically)

recrystallization does not occur in hydrated samples until much higher temperatures are reached, ca. 190 K, and there are a number of organic liquids that will dissolve ice at these temperatures in a reasonable period of time.

Table 7.3 gives the properties of the organic liquids that are now most commonly used as substitution media. Robards and Sleytr (1985) give a much larger selection, although many of the chemicals included on their list are now no longer used as substitution fluids. In addition to the chemicals listed in Table 7.3, it is also possible to use eutectic mixtures such as 70% ethylene glycol and 30% water, which has a MP of 223 K (Pease, 1967), or 40% ethylene glycol and 50% methanol, MP 193 K (Barlow and Sleigh, 1979). Such mixtures only achieve partial substitution, and dehydration has to be completed in another fluid such as methanol or ethanol. The chemicals listed in Table 7.3 have their own advantages and disadvantages.

7.6.2.1. Acetone

This is one of the most widely used substitution media and when used as a freshly made solution containing 0.4%–4% osmium tetroxide gives excellent structural preservation in a wide range of samples. A typical substitution procedure would be as follows. Small pieces of rapidly cooled tissue are placed in an open mesh wire basket immersed in liquid nitrogen. The basket is quickly transferred to anhydrous acetone at 193 K, with or without fixatives, and kept at this temperature for 48–72 h. The cold anhydrous acetone may be changed during this time. The most convenient drying agents for the substitution fluid are pellets of Linde Molecular Sieve or anhydrous silica gel, but care must be taken not to get pieces of the desiccant in the specimen as it makes it difficult to section properly. A study by Ross et al. (1983) showed that molecular sieves release varying amounts of sodium, potassium, and calcium into the substitution fluid, which would compromise any analytical studies. Silica gel was found to be an excellent solvent desiccant which did not have these problems. Following substitution at 193 K, the samples are allowed to slowly warm to 233 K over a 12–24-h period. At this stage the specimens may either be processed further for low-temperature embedding, or allowed to warm up to room temperature, rinsed several times in dry acetone, and either dried or embedded by more conventional methods. Other fixative mixtures include 5%–10% acrolein (Ornberg and

Reese, 1981; Murata, 1985) and 0.3% glutaraldehyde (Ichikawa *et al.*, 1989). Steinbrecht and Muller (1987) caution that uranyl acetate should not be used in conjunction with acetone because of its low solubility, and that the relatively high water content of glutaraldehyde, even 50%–60% solutions, would introduce too much water into the substitution mixture. Kaeser (1989) and Lehmann *et al.* (1990) found that water can be removed from aqueous glutaraldehyde solutions by vigorously shaking it with ten times the volume of dimethoxy propane (DMP). The DMP reacts chemically with water to produce acetone and methanol. In spite of the warning by Steinbrecht and Muller (1987), Kiss *et al.* (1990) have successfully used a substitution mixture containing 0.1% uranyl acetate on high-pressure frozen roots.

7.6.2.2. Diethyl Ether

This highly apolar solvent is used mainly in experiments where the principal objective is to retain the original location of water soluble substances rather than preserve ultrastructural detail. Harvey (1982) and McCully and Canny (1985) recommend using diethyl ether containing acrolein. Other investigators (Marshall, 1980a; Wroblewski and Wroblewski, 1984) have used anhydrous diethyl ether without any chemical additives. A typical substitution schedule would be to keep samples at 193 K for 20 days followed by 1 day at 233 K before low-temperature resin infiltration or a slow warm up to room temperature. Because water has a very low solubility in diethyl ether it is important to use this substitution fluid in conjunction with a drying agent, and to very slowly rewarm the mixture after keeping it at 193 K for several days. We have had some success using dimethyl ether as a substitution fluid (unpublished results). It has a lower melting point (134 K) than diethyl ether (157 K) and has a slightly higher affinity for water. The advantage of dimethyl ether is that it can be used at very low temperatures, although the drying times run into several weeks. It also has a low boiling point (250 K), which makes it less easy to use than diethyl ether.

7.6.2.3. Ethyl Alcohol

This has been used by Nagele *et al.* (1985) in the preparation of samples for both immunocytochemical and structural studies. Frozen samples were substituted in a mixture of 3.5% glutaraldehyde in 95% ethanol for two days at 183 K and following embedding were used for immunocytochemical studies. Samples for ultrastructural studies were postfixed in 1% osmium tetroxide in acetone at 183 K.

7.6.2.4. Methanol

This has the advantage that it can be used to substitute samples even in the presence of quite high amounts of water. It substitutes faster than acetone and has been used alone (Monaghan and Robertson, 1990), in combination with both acrolein, (Zalokar, 1966) and osmium tetroxide (Ryan *et al.*, 1990), and as part of a complex mixture containing uranyl acetate, osmium, and glutaraldehyde (Muller *et al.*, 1980). Muller *et al.* (1991) have successfully combined methanol substitution with 0.5% hafnium chloride and 0.5% uranly acetate. A typical substitution schedule using methanol would start at 178 K for 24–28 h, after which the mixture is slowly warmed to 273 K over a 24-h period to allow the fixatives to react with sample.

7.6.2.5. Propane

Pogorelov *et al.* (1991) have freeze-substituted specimens in propane at 163 K for up to 21 days using molecular sieve to scavenge water from the propane as the slow dehydration takes place. The weight of the molecular sieve was calculated so that its capacity to absorb water was an order of magnitude greater than the total water in the frozen sample.

7.6.2.6. Dimethoxypropane

Kaeser (1989) added DMP to an acetone–methanol substitution mixture containing both glutaraldehyde and uranyl acetate. The DMP effectively combines with any traces of water and gives better preservation than in methanol alone. The substitution mixture consists of 6 ml acidified DMP, 3 ml acetone, 2.25 ml methanol, 0.75 ml of 10% uranyl acetate in methanol, and 0.6 ml of 25% aqueous glutaraldehyde. The substitution is carried out over a 24–48-h period at $193 > 213 > 243$ K.

7.6.3. Temperature of Substitution

The lowest temperature at which freeze substitution can be performed is the melting point of the solvent (see Table 7.3). In practice, substitution is started at 173–193 K and samples remain at this temperature for several days. The temperature is then slowly increased, using a continuous temperature rise rather than stepwise increase, either to the temperature at which the low-temperature embedding processes are used (190–240 K) or allowed to rise to room temperature for conventional embedding and/or drying for scanning electron microscopy. The substitution temperatures are also influenced by the identity of the substitution fluid, and to some extent by the water content and size of the sample.

7.6.4. Rate of Substitution

The rate of substitution is a function of the temperature, water-dissolving capacity, viscosity, and relative diffusion rate of the substitution liquid. Robards and Sleytr (1985) consider that freeze substitution rates will also be affected by the following:

1. The water content of the organic liquid during the time course of the substitution process.
2. The temperature and water content of the substituting liquid, which is at equilibrium with the ice in the specimen.

3. The presence of drying agents such as molecular sieve, which will continually remove water from the substituting fluid.
4. The final warming rate of the sample and substituting fluid.

The rate of dehydration can be checked qualitatively by the simple method devised by Zalokar (1966), who followed the loss of methylene blue from frozen blocks of stained agar and/or pieces of frozen filter paper soaked in the dye. The frozen dyed substrates were placed in different substitution liquids at different temperatures and the completion of the substitution was determined by the decolorization of the dyed material. Zalokar found that acetone did not dissolve ice (dye ?) after several days at 195 K, whereas ethanol and methanol are much more efficient substitution liquids.

More quantitative measurements on the rate of freeze substitution have been made by Ornberg and Reese (1981) and Humbel *et al.* (1983), who have measured the ^3H content in the substitution liquids during the freeze substitution of tritiated ice. Figure 7.3 shows the rate of ice substitution by three different solvents and reveals that the fastest substitution occurs in methanol followed by acetone and diethyl ether. The substitution velocities are also strongly temperature dependent. Methanol is an efficient substitution fluid even at temperatures as low as 183 K, which for biological systems is considered sufficiently safe to prevent any substantial recrystallization. Acetone and diethyl ether will very slowly dissolve ice at these temperatures provided they remain absolutely dry. Figure 7.3 also shows that the freeze substitution properties of acetone and methanol are influenced by their water content. At 183 K, the substitution rate is significantly reduced even by 1% water, whereas methanol containing 10% water will substitute water completely in 3 h at the same temperature.

Acetone is a less effective substituting agent at low temperatures, because at 193 K it only takes between 1% and 2% water to form an equilibrium mixture with acetone. At these low temperatures, this water has to be continually removed from the acetone by an additional drying agent in order that the organic liquid can function as an effective substituting agent. It is thus important that the substituting fluid is kept as dry as possible throughout the drying process and that its volume should be a thousand times greater than the volume of the sample (Steinbrecht and Muller, 1987). The sample vials should be gently agitated from time to time during the substitution process. Too much agitation is inadvisable as this may cause pieces of the molecular sieve to break off and these may eventually become incorporated into the specimen and surrounding resin.

Experiments by Ornberg and Reese (1981) using acetone at 193 K suggest that the initial rate of sample cooling may affect the subsequent rate of freeze substitution. Samples that were impact cooled at 4 K substituted faster than samples plunge cooled at 160 K. It was first thought that this showed that smaller ice crystals (impact cooling) substituted more rapidly than larger ice crystals (plunge cooling). A more detailed analysis of the results suggested, however, that the higher cooling rate gave rise to a continuous network of ice throughout the sample (a single ice crystal ?), through which the substituting fluid could more easily diffuse. The lower cooling rate

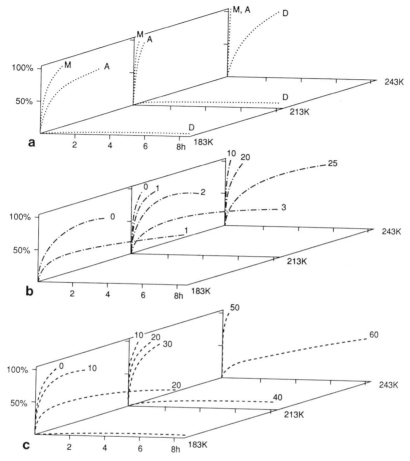

Figure 7.3. Substitution of ³H-crystalline ice by various solvents as a function of time and temperature. (a) Pure solvents: M, methanol; A, acetone; D, diethyl ether. (b) Acetone containing different percentages of water (n%). (c) Methanol containing different percentages of water (n%). [Redrawn from Humbel *et al.* (1983).]

gave rise to a discontinuous network of ice crystals, which would slow down the diffusion.

Steinbrecht (1982) has studied the damage caused by recrystallization, as very small, ca. 100 μm specimens were warmed up in a substitution medium of acetone containing 2% osmium tetroxide. The substitution time at 194 K could be reduced from 7 days to 5 min without any recrystallization artifacts provided the subsequent rate of warming was slow (half-time 3 h). Severe damage occurred if the warm-up time was fast, e.g., half-time 30 s, and no damage was observed in samples that were substituted for 7 days at 194 K and then warmed rapidly, e.g., half-time 30 s. Steinbrecht's work on biological systems and the earlier work by MacKenzie (1977)

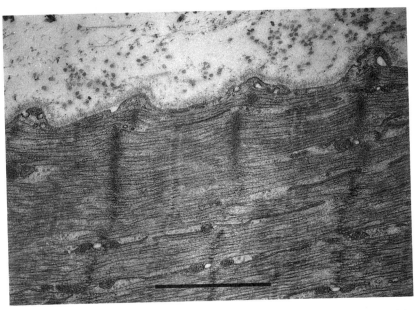

Figure 7.4. Freeze-substituted contracting frog sartorious muscle. Substitution in acetone containing 2.5% osmium tetroxide and 0.2% uranyl acetate for 2 days at 193 K. Temperature then raised at 3 K h^{-1} to 213 K and kept at 213 K for 1 day; raised at 3 K h^{-1} and kept at 223 K for 4 days. Embedding in Lowicryl HM23 and polymerized at 223 K. Bar = 1.0 μm. [Picture courtesy Edelmann (1989).]

on model systems suggests that no significant recrystallization occurs, provided the samples remain at low temperatures for a long time and/or are slowly warmed up. Unfortunately very few detailed studies have been made on the freeze-substitution process and in particular on the behavior of different substitution media at varying temperatures and with specimens with varying water content. There is also a wide variation in the recrystallization temperature in different parts of individual cells, e.g., the vacuoles and cytoplasm of mature plant cells. Under these circumstances it would be prudent to allow substitution to proceed as slow as is practicable, bearing in mind that the longer substituting times increase the chance of solubilizing cell components. It is impossible to show an example of material that has only been freeze substituted, because as we have already mentioned, the process is part of a complete preparative package which includes fixation and embedding. Nevertheless, Fig. 7.4 shows an example of material that has been freeze substituted as part of the preparative procedure.

7.7. LOW-TEMPERATURE EMBEDDING

7.7.1. General Procedures

Low-temperature embedding is most frequently used as a continuation of the freeze-substitution process and the same equipment can be used for both procedures.

The slow rise in temperature, which is initiated at the end of the substitution process, is stopped at a predetermined point depending on the embedding medium to be used. The sample is then slowly infiltrated with a resin mixture, which first replaces the substitution liquid and is then polymerized at low temperature. Low-temperature embedding can also be used in connection with fixed samples which have been dehydrated using the progressive lowering of temperature discussed earlier. It can also be used to embed freeze-dried specimens. The common feature of all three approaches is that the entire embedding and polymerization process occurs at temperature substantially below 273 K. It is quite distinct from situations where samples are either freeze-dried or freeze substituted and then embedded at temperatures above 273 K.

Much of the current success of low-temperature embedding centers on the Lowicryl resins developed by the group in Basel in conjunction with Chemische Werke Lowi in Germany, and the new LR acrylic resins developed by the London Resin Company. For the sake of completeness, this section will first briefly discuss some of the early studies before considering in some detail the use of the Lowicryl and LR resins. It is important to appreciate that many of the components of low-temperature embedding media are potential health hazards. The paper by Tobler and Freiburghaus (1990) reviews these hazards and provides information on how to handle these chemicals.

7.7.2. Early Low-Temperature Embedding Studies

This section will not serve as a historical introduction to low-temperature embedding but will briefly review some of the earlier procedures that may still be used. Low-viscosity epoxy resins have been used at 223 K to infiltrate tissue samples (Lauchli *et al.*, 1970; Harvey *et al.*, 1976), but these materials will only polymerize at temperatures well above the freezing point of water and as a consequence may damage the specimens. They are not considered, *sensu stricta*, low-temperature embedding media. Some later unpublished studies by Carlemalm (Carlemalm *et al.*, 1980) showed that epoxy resins have only a limited application to low-temperature studies and that all attempts to alter their physicochemical properties has only yielded inferior results. The epoxy resins do not polymerize at subzero temperatures.

Sjostrand and Kretzer (1975) have used the apolar polyester resin Vestopal for low-temperature embedding on the basis that it is a weak lipid solvent. Because the resin is highly viscous, Sjostrand and Halma (1978) had to use high-speed centrifugation to speed up the infiltration process. Vestopal can be polymerized at between 253 and 293 K with UV light and produces specimens with high contrast and a low background structure. Carlemalm *et al.* (1980) and Kellenberger *et al.* (1980) have since altered the Vestopal formulation to give it improved cutting characteristics, and the more extensive work of Acetarin (1981) has shown that the class of polyester resins, although highly viscous, have promising possibilities as a general purpose embedding agent.

Table 7.4. An Experimental Schedule for Low-Temperature Substitution and Embedding Using Aqueous Glycol Methacrylate

5% GMA for 30 min at 273 K
10% GMA for 30 min at 273 K
20% GMA for 30 min at 273 K
40% GMA for 30 min at 273 K
60% GMA for 30 min at 273 K
80% GMA for 2 h at 273 K
90% GMA for 2 h at 253 K
95% GMA for 2 h at 253 K
100% GMA for 16 h at 253 K
100% GMA for 2 h at 253 K
100% GMA for 30 min at 253 K

Bartl (1962) used glycol methacrylate both as a low-temperature substitution solvent and as an embedding medium, and Table 7.4 shows a typical embedding schedule for this resin mixture. The monomer mixture consists of 95% glycol metha crylate, 5% polyethylene methacrylate, and 0.15% azobisisobutyronitrile (Hayat, 1986) mixed with water in increasing concentrations. The embedded samples are polymerized for 36 h at 253 K using UV radiation. Pease [see Pease (1973) for a review of this work] used 70% ethylene glycol both at embient temperatures and at 223 K, or pure propylene glycol at 233 K, as a drying agent in a process referred to as "inert dehydration." The dehydrated samples were embedded either in hydroxy-ethyl or hydroxypropyl methacrylate, which was polymerized at room temperature. Cope (1967) successfully embedded specimens at low temperatures in water-miscible methacrylates after treatment with cryoprotectants. The polar groups of some of these acrylate resins disappear during polymerization, resulting in the expulsion of any water remaining in solution in the unpolymerized resin, which, in turn, could combine with hydrophilic components of the biological matrix. There are fears, as yet unquantified, that this water could also cause the relocation of water-soluble substances in cells and tissues. No water is produced during the polymerization of methacrylate, although many of the methacrylate resin mixtures exhibit severe prob-lems of shrinkage during their exothermic polymerization.

These earlier studies on the methacrylate group of resins revealed that these chemicals showed the most promise as low-temperature embedding agents, and a number of papers were published attesting to its usefulness (Sjostrand and Barajas, 1968; Bullivant, 1970; Kretzer, 1973). These earlier studies provided the impetus to improve the low-temperature embedding capability of acrylic resins.

7.7.3. Low-Temperature Embedding with Lowicryl

Kellenberger and Kistler (1979), Kellenberger et al. (1980), and Kellenberger (1991) consider that perturbation of the hydration shell that surrounds proteins leads to conformational changes and denaturation. Their theoretical predictions and subsequent experiments showed that the water in hydration shells is strongly "bound" at subzero temperatures and that there will be competition for this water by highly polar substitution liquids. Highly polar resins would have the same effect and remove the hydration shell from the biological material. Nonpolar resins have no affinity for water, and provided the hydration shell remains intact during dehydration, such resins will preserve hydrated specimens in a more natural state. On the basis of extensive studies on the influence of temperature, water content, and resin polarity, two highly cross-linked resins, one polar (K4M) and the other apolar (HM20), have been developed for low-temperature embedding. The two resins do not change their polarity during polymerization and are usable at temperatures between 223 and 243 K.

Details of the two resins Lowicryl K4M and Lowicryl HM20, are given in a paper by Carlemalm (1982). Both resins are based on a mixed methacrylate–acrylate system in which the resulting resin backbone remains unchanged after polymerization. The physical nature of the resin is altered by modifying the side groups, and the chemical constitution of the two resin mixtures is shown in Table 7.5. Within certain limits, the amount of cross linker can be varied to give blocks of different hardness. The resin K4M is a low-viscosity polar resin which remains fluid down to 238 K and is soluble in most polar dehydrating agents, including hydroxyethyl methacrylate. It can absorb up to 5% water, which means that specimens can be kept in a partially hydrated state during dehydration and infiltration with resin. The resin HM20 is a low-viscosity apolar resin which remains liquid down to 193 K and is miscible in most polar dehydrating agents except ethylene glycol and dimethylformamide. It can absorb up to 0.5% water.

Both resins can be polymerized by long-wave (360 nm) indirect UV radiation at temperatures between 223 and 273 K and by the initiator benzoyl methyl ether at temperatures between 273 and 333 K. When these resins are used at low temperatures, Carlemalm et al. (1982a, 1986) and Armbruster (1984) have shown that there is a minimal loss of molecular structure and that embedded protein crystals retain molecular order to between 6 and 8 Å. The resins have proved to be especially useful for immunocytochemical studies and preserve antigenicity to levels greater than that which can be usually achieved in conventionally embedded material. The newly

Table 7.5. The Chemical Composition of Lowicryl K4M (Polar) and HM20 (Apolar) Resins[a]

Lowicryl K4M (polar)		Lowicryl HM20 (apolar)	
Wt %	Monomer	Wt %	Monomer
48.4	OH-Propyl methacrylate	68.5	Ethyl methacrylate
23.7	OH-Ethyl methacrylate	16.6	N-Hexyl methacrylate
9.0	N-Hexyl methacrylate	14.9	Cross linker
13.9	Cross linker		

[a]The cross linker in both resins is triethylene glycol dimethacrylate. The resins are available commercially as separate packs of the premixed monomer mixture and cross linker. (Data Lowi Chemische Werke.)

formulated resins show much less shrinkage and are well suited for general electron microscopy. A special formulation has been developed for light microscopy (Franklin and Martin, 1980).

One drawback to the polar resin K4M is that it tends to be hydrophilic and as such will tend to absorb water. This can lead to softening of the polymerized resin mixture and will create sectioning difficulties. The polymerized blocks should be stored under anhydrous conditions. The problem can also be overcome by making the resin slightly less polar by adding two flexibilizing resin monomers, one with hydrophilic side groups the other hydrophobic side groups. K4M-embedded micro-organisms have a tendency to separate at the cell-surface–resin interface. Pretreatment of cells with 0.5% aqueous uranyl acetate for 30 min before dehydration appears to cure this problem and the treatment does not seem to affect the immunolabeling of a number of different antigens (Benichou et al., 1990).

Another disadvantage of the Lowicryl resins is that both formulations may extract a substantial amount of lipid from aldehyde fixed samples (Weibull et al., 1983). This loss of lipid material is substantially lowered when freeze substitution and resin infiltration are carried out at lower temperatures (Weibull et al., 1984). Weibull and Christiansson (1986) have shown that up to 99% of the protein and 88% of the lipid was retained after dehydration in ethanol or acetone at between 223 and 277 K and embedding in Lowicryl at 224 K. Much less lipid was retained when methanol was used for dehydration. Glutaraldehyde-fixed erythrocytes and spinach chloroplasts retained substantial amounts of phospholipids when prepared under the same conditions. These studies show that although low temperatures greatly increase the retention of lipid material during embedding, the amount varies considerably between different biological systems.

Electron beam exposure can cause mass loss in both resins, which will create problems where these resins are being used in connection with quantitative analytical studies. Wroblewski and Wroblewski (1984) found that although visualization of Lowicryl-embedded material was improved after 10–20 s irradiation by the electron beam, subsequent mass loss measurements showed that the material suffered a greater mass loss than Araldite-embedded material. The resin can be made somewhat more beam, subsequent mass-loss measurements showed that the material suffered a greater mass loss than Araldite-embedded material. The resin can be made somewhat more beam resistant by the addition of cross-linking agents such as styrene or divinyl Table 7.6. There are three principal routes by which specimens may be processed before they are embedded with these resins: (1) Ambient temperature fixation followed by progressive lowering of temperature dehydration; (2) freeze substitution (in both these cases, the samples remain in a liquid environment throughout the whole process): (3) freeze-drying, in which the specimens alternate between the liquid environment and the dried state. A good review of the interrelationships of these procedures is given by Chiovetti et al. (1987).

In all three cases, the temperature of the specimen is never allowed to rise above the freezing point of water until it has been fully infiltrated with resin and polymerized. Polymerization is initiated at 223 K using UV radiation and remains at this temperature for between 12 and 24 h. The polymerization may be completed for a

Table 7.6. An Experimental Schedule for Lowicryl K4M and HM20 Low-Temperature Resins[a]

Lowicryl K4M	Lowicryl HM20
30% Ethanol for 30 min at 273 K	30% Ethanol for 30 min at 273 K
50% Ethanol for 1 h at 253 K	50% Ethanol for 1 h at 253 K
70% Ethanol for 1 h at 238 K	70% Ethanol for 1 h at 223 K
90% Ethanol for 2 h at 238 K	90% Ethanol for 2 h at 203 K
1 pt resin: 1 pt 90% ethanol for 1 h at 238 K	1 pt resin: 1 pt 90% ethanol for 1 h at 203 K
2 pt resin: 1 pt 90% ethanol for 1 h at 238 K	2 pt resin: 1 pt 90% ethanol for 1 h at 203 K
100% Resin for 1 h at 238 K	100% Resin for 1 h at 203 K
100% Resin for 12 h at 238 K	100% Resin for 12 h at 203 K
Encapsulate in fresh resin	Encapsulate in fresh resin

[a]At temperatures below 273 K care must be taken to maintain anhydrous conditions and to assure that any residual water in the samples does not form ice crystals.

further 2 days at 293 K if necessary. The UV should only be used to polymerize colorless samples and should not be used with samples that have been freeze substituted with heavy metal salts. Care should be taken that the atmosphere around the blocks contains little or no oxygen as this affects the polymerization process. In practice it is convenient to mix the various components of the resin mixture by means of bubbling nitrogen and to have a slightly positive pressure of nitrogen in the low-temperature UV polymerization chamber.

The polymerization process is best carried out in a chamber that will fit into a low-temperature refrigerator. Details of suitable equipment will be discussed later. Tissue samples may be loaded into BEEM capsules or gelatine capsules which are placed in a suitable holder so they can receive UV radiation from all sides. Bou-Gharios *et al.* (1988) have devised a simple procedure for the *in situ* embedding of monolayers of cells. They used a Lab Tek Flaskette for growing, fixing, processing, and embedding adherent cells maintained in tissue culture. The plastic flask has a glass base, which is sealed to the main body by a gasket that allows the glass slide on which the cells are growing to be easily separated from the plastic body. The container is also fitted with a screw cap so it may be flushed with CO_2 or N_2 prior to resin polymerization. The embedded cells may be easily separated from the glass base.

It is important that the appropriate low temperatures are maintained throughout the polymerization process. A low-temperature bath filled with an appropriate anti-freeze agent is more efficient at keeping the capsules cool than suspending them in air. A pair of 15-W UV fluorescent tubes are a suitable light source and a right-angled baffle should be placed just below the tubes so that the samples only receive indirect UV light to ensure a more even polymerization of the resin. A certain amount of trial and error is necessary to judge the distance between the UV source and the sample. Specimen shrinkage and deformed blocks indicate that polymerization has been too fast. The polymerized blocks should be stored in a dry, dark container.

The apolar HM20 resin is highly cross linked and can be sectioned with a glass or diamond knife by the usual procedures; see Reid (1974) for details of ultramicrotomy. The polar K4M resin requires a little more attention during sectioning. The level of trough liquid should be slightly below normal to ensure that the block face is not wetted, but should not be so low that the sections are cut dry, although this may be necessary when cutting sections for elemental analysis. The sections should be picked up from the trough liquid as soon as possible.

The two types of resin show significant differences in their staining properties, although they both seem to stain reasonably well with lead and uranyl salts. Robards and Sleytr (1985) recommend staining with saturated aqueous uranyl acetate (35 min for HM20; 5–10 min for K4M) followed by lead acetate for 1–3 min. As a general guideline, shorter staining times should be used with the hydrophilic K4M resin (Fig. 7.5). Semithick sections of material embedded in K4M stain poorly with many of the traditional stains used for light microscopy. Some enzyme localization procedures such as the diaminobenzidine method for peroxidases are less effective in K4M-embedded material. Monaghan and Robertson (1990) found better structural preservation in HM20 than in K4M.

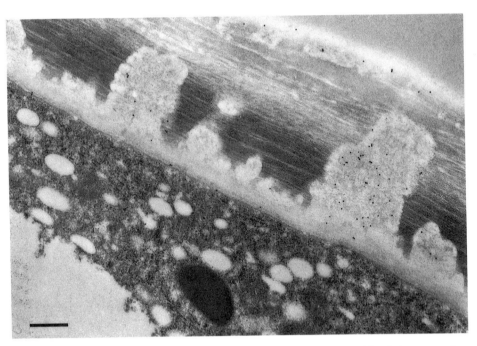

Figure 7.5. Immunocytochemical localization of a-amylase with the protein–A gold technique on sections of gibberelic-acid-treated barley aleurone embedded in Lowicryl K4M at 233 K. The gold labeling occurs specifically in areas of the wall where material has been lost, indicating that a-amylase like other enzymes is released via digested wall channels. Bar = 0.5 μm. [Picture courtesy Ashford et al. (1986).]

7.7.4. Improvements and Modifications to Lowicryl Embedding

There have been a number of alterations to the two basic Lowicryl mixtures since their introduction in the early 1980s. Before discussing these, we should first briefly consider a potential problem with the Lowicryl resins that was not fully appreciated in the earlier studies. The whole aim of low-temperature embedding is to preserve the ultrastructure and biological activity of the living cell, and it is thus important that the temperature be kept as low as possible throughout the whole embedding procedure. Ashford *et al.* (1986) monitored the temperature of both glycol methacrylate and Lowicryl K4M throughout the embedding procedure and discovered that the heat associated with the exothermic polymerization process was not rapidly dissipated, even in a liquid-cooled bath. As a consequence there was a significant rise in the temperature of both the resin and the embedded sample. Kellenberger (1987) showed that it was particularly important to avoid any temperature increases while the resin was still liquid and the structural components of the specimen were still mobile. With glycol methacrylate, the temperature rise was smallest at the lower polymerization temperatures but still amounted to between 2 and 4 K at 240 K. At room temperature, the temperature rose by as much as 14 K. Lowicryl behaved in a similar fashion, except that there were two peaks of temperature increase and in both cases the temperature rises were higher—i.e., as much as 40 K above ambient—than that seen in the glycol methacrylate resin. In both glycol methacrylate and Lowicryl K4M the onset of the temperature rise coincided with the time the polymers were exposed to the UV source. In order to achieve maximum hardness in K4M resin it is customary to complete the polymerization process at room temperature. At these temperatures, Ashford *et al.* (1986) found the resin temperature would rise by about 20 K, although these effects are reduced in smaller samples.

Weibull (1986) has investigated the same phenomenon in samples embedded in small volumes (0.7–1.7 ml) of four different Lowicryl resins. There were significant temperature rises when the samples were air cooled but these temperature excursions were reduced to 2–4 K when the samples were kept immersed in a low-temperature liquid bath. Weibull noted that there was a transient rise in temperature (5 K) when previously cooled sample were brought up to room temperature. Glauert and Young (1989) found that even a 2 K temperature rise in a 0.6-ml K4M sample resulted in uneven polymerization and, as a consequence, gave rise to problems during sectioning and deformation of ultrastructure. Glauert and Young (1989) showed that the temperature rise could be suppressed by using diffuse UV irradiation and heat sinks for the embedding containers. Flat embedding containers should be placed in recesses in a cooled aluminum block and capsules suspended in a cold ethanol bath.

Care should be taken to absorb the heat that is generated when Lowicryl sample are polymerized, even at low temperatures. It is not known whether this temperature rise is a general phenomenon throughout the whole resin block as it polymerizes, or whether it is confined to local "hot spots." The volume of the resin should be kept as small as possible, and during the final stages of polymerization it would be prudent to use extended times at lower temperatures rather than shorter times at higher temperatures. No experimental data are available that would suggest that these relatively small temperature rises may in any way be deleterious to the molecular integrity of the embedded material. Finally, an additional potential hazard has been recognized by Ashford *et al.* (1986), who found that even the process of sawing a sample from an embedded block of resin can cause a rise in specimen temperature.

Carlemalm *et al.* (1985) and Acetarin *et al.* (1986) introduced two additional resins that can be used at lower temperatures. Lowicryl K11M is a polar resin, which can be used down to 210 K, and Lowicryl HM23, an apolar resin, can be used at temperatures as low as 190 K. Photopolymerization is achieved by indirect UV radiation although catalysts are necessary to initiate the process at temperatures lower than 223 K. These two resins, which are complex mixtures of different methacrylates, can be used in conjunction with appropriate freeze-substitution procedures, and with the progressive lowering of temperature dehydration method. Weibull *et al.* (1983, 1984) found that there was a significant decrease in the amount of lipid lost from specimens embedded in the two resins. Another advantage of the two new Lowicryl resins is that they give good morphological and immunocytochemical information with unfixed specimens.

Acetarin *et al.* (1986) discuss the use of the two new Lowicryl resins in light of the theoretical prediction that the hydration shell surrounding many biological molecules will not be removed at lower temperatures. While this may be a limitation to obtaining high-resolution ultrastructural information, Acetarin and colleagues give three reasons in favor of using resins at the lower temperatures.

1. There appears to be better preservation of antigenicity and therefore a greater immunochemical efficiency.
2. Some solvent-induced aggregates appear much smaller at lower temperatures.
3. There is a decreased solubility of lipids.

Simon *et al.* (1987) have developed a rapid method of embedding tissues in Lowicryl K4M for immunocytochemical studies. The chemical initiator has been changed and the polymerization temperature has been raised to 263 K. It is claimed that this procedure gives superior structural preservation and more specific immunolabeling, although the authors appear unaware of the problems associated with the rise in temperature during polymerization.

7.7.5. Other Low-Temperature Embedding Resins

In spite of the apparent success of the Lowicryl family of low-temperature resins, a number of other resins that can be used at low temperatures have appeared in the last two or three years. One of these resins, LR gold, is based on a low-viscosity acrylic embedding material; others (Nanoplast MUV 116 and FB 101) are aminoplastics. Acetarin *et al.* (1986) have developed a styrene-based, low-viscosity, negative staining resin for use with unstained biological material. Methacrylate molecules are incorporated as comonomers and cross-linkers. The resin contains 3 at. % tin and can be used at 243 K. Polymerization, which is relatively slow, is initiated by UV irradiation (360 nm) at 243 K for 6 days. All dehydration and infiltration stages have to be carried out using acetone as the styrene and related monomers are not completely soluble in alcohol. The resin provides sufficient contrast with bright field CTEM and excellent contrast in the STEM mode of imaging. The tin-based resin provides a further opportunity of studying unstained biological material in the electron microscope by complementing the scattering properties of material in the specimen with the surrounding media.

LR gold is a very low-viscosity monomer polyhydroxylated aromatic acrylic hydrophilic polar resin, which can be used and polymerized at low temperatures. This resin is attracting a lot of interest as it is easy to use and gives very good results. Samples may either be dehydrated through the progressive lowering of temperature schedule or freeze substituted in ethanol or methanol. A typical schedule for uncooled 2–3-mm diameter tissue blocks would be to dehydrate fixed or unfixed samples, by slow rotation in increasing concentrations of ethanol (with or without added aldehydes) initially at 277 K and eventually at 253 K, followed by gradual infiltration with increasing concentrations of resin. Acetone and osmium tetroxide should not be used as they both interfere with the polymerization process. Embedding should be carried out at temperatures no lower than 248 K because some of the components begin to precipitate out of the resin mixture at about 243 K. The resin can be used on unfixed tissues, although under these conditions it is recommended that for animal tissue, polyvinylpyrrolidone (PVP mol. wt. 40,00) is added to the dehydration schedule to protect the unfixed material from osmotic changes during processing. Studies in my own laboratory have found that it is not necessary to add PVP to plant tissue. An embedding schedule for LR gold in shown in Table 7.7.

A light-sensitive initiator, an alpha ketone, is added to polymerize the resin. The polymerization is carried out at ca. 253 K with a blue light source rather than UV and oxygen must be excluded from the mixture. The samples must be in contact with a good heat sink as the polymerization is an exothermic reaction and may cause undue temperature rises in the sample. McPhail *et al.* (1987) have reported excellent results with immunolabeling of nonosmicated tissues, and in my own laboratory we

Table 7.7. An Experimental Dehydration and Embedding Schedule for LR Gold Low-Temperature Resin for Use with Animal Tissues[a]

50% Methanol + 20% PVP 15 min at 273 K
70% Methanol + 20% PVP 45 min at 248 K
90% Methanol + 20% PVP 45 min at 248 K
50% LR Gold monomer + 50% methanol with 10% PVP 30 min at 248 K
70% LR Gold monomer + 30% methanol with 10% PVP for 1 h at 248 K
100% LR Gold monomer for 1 h at 248 K
100% LR Gold monomer + initiator for 1 h at 248 K
100% LR Gold monomer + initiator for 16 h at 248 K
100% LR Gold monomer + initiator for 248 h at 248 K
Encapsulate in fresh resin

[a]The PVP is added to prevent osmotic damage during dehydration of unfixed tissue. With plant material much less PVP is needed. The amounts of PVP should be varied according to the nature of the specimen and whether they have or have not been fixed prior to dehydration.

are using it to prepare tobacco tissue for the immunocytochemical localization of nicotine. More details on the use of this resin are available from the London Resin Company, who also make a closely related resin, LR White, which can be used at ambient temperatures.

Bachhuber *et al.* (1987) have published details of Nanoplast MUV 116, a melamine–formaldehyde water-soluble resin for use at low temperatures. Very small samples, spread as thin layers on electron microscope grids, are quench cooled in LN_2 and then transferred under LN_2 to a sealed container containing a small amount of the solidified resin/substitution mixture. This mixture consists of 50% w/v hexamethyl–melamine–ether in anhydrous acetone. The samples are freeze substituted in this mixture at 190 K for 24 h. The container and grids are transferred to a precooled cold stage of a low-temperature vacuum unit and the acetone removed by sublimation at 190 K and a pressure of 13 mPa (10^{-4} Torr). The pressure is brought up to ambient conditions with dry nitrogen and polymerization is initiated at 190 K by means of UV radiation. During the 2-h polymerization process the temperature was slowly raised to 273 K. The blocks are posthardened at 333 K.

The images obtained by this procedure (see Fig. 7.6) are comparable with those obtained by conventional embedding methods. This resin has only recently appeared

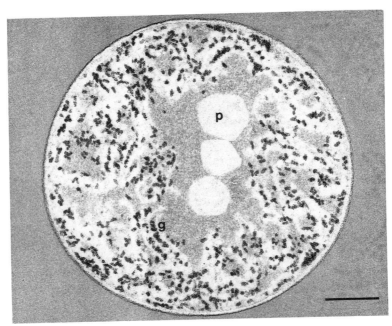

Figure 7.6. Cross section of the cyanobacteria *Anabaena variabilis* fixed with glutaraldehyde and embedded in the low-temperature resin Nanoplast FB 101 and stained with 1% silver proteinate at room temperature for polysaccharides. The structure of glycogen granules (g) and their distribution along thylakoid membranes is clearly visible; p, polyhedral body. Bar market = 1.0 μm. [Picture courtesy Bachhuber *et al.* (1987).]

on the market and clearly more studies are needed in order to judge the breadth of its applications.

7.8. FREEZE SUBSTITUTION AND LOW-TEMPERATURE EMBEDDING EQUIPMENT

7.8.1. Low-Temperature Refrigerators

The refrigerators needed for freeze substitution and low-temperature embedding must be capable of providing more than just low temperatures. The following specifications should be considered, when designing and building pieces of equipment, and/or buying commercially available equipment for these procedures:

1. The refrigerator must cover the temperature range being used, maintaining it, if necessary, for several weeks.
2. It should be possible to raise and lower the temperature of the samples in a controlled and reproducible manner.
3. Facilities must be available to store and exchange all substitution fluid under strictly anhydrous conditions.
4. It should be possible to create an oxygen-free and anhydrous environment during the period of resin polymerization.
5. The equipment should be fitted with the appropriate UV sources and controls for use with most of the embedding media.
6. It should be possible to easily manipulate small samples and pieces of equipment kept at low temperatures.

It is relatively easy to achieve a whole range of temperatures from 273 K (melting ice) to 77 K (boiling nitrogen). Domestic refrigerators, deep freezes, and various mixtures of salts and ice and organic liquids and dry ice (solid CO_2) can provide intermediate temperatures. These are, however, fixed temperatures, and many of the advantages of both freeze substitution and low-temperature embedding are derived from being able to accurately vary the temperature during tissue processing. From the technical point of view, the simplest and most effective way of achieving, maintaining, and varying low temperatures in the 273–173 K range is to use LN_2 as a coolant and electric power as a source of heat.

The samples are best kept in airtight transparent containers with lids that are easy to remove at low temperatures. The containers should fit snugly into holes in a metal block which can be cooled by LN_2 and heated by a small electrical element. The cooling and heating components should be linked to a programmable temperature controller and thermocouple. The sample cooling block should be placed at the bottom of a well-insulated box, the top of which is either fitted as a glove box or fitted with a transparent, tight-fitting lid. There should be sufficient space in the insulated box to hold a slow rotary stirrer and an indirect, low-wattage UV source. The interior of the box should be lined with shiny aluminum foil and continually flushed with a gentle flow of dry nitrogen to create an anhydrous anoxic environment.

The various substitution and embedding liquids should, where appropriate, be kept over molecular sieve in sealed dark glass bottles in the cooling cabinet.

A number of pieces of equipment, ranging from the simple homemade to the rather complex commercial, have been designed either for specific freeze substitution and low-temperature embedding routines, or for more general use and a wider range of applications. Sitte (1981) designed a relatively simple piece of equipment for freeze substitution. It consists of a container sitting inside a larger metal block, which is initially cooled with LN$_2$ before being placed in an outer insulated container. The quench-cooled samples are placed on the surface of the frozen substitution medium and the whole apparatus covered with an insulated lid. The frozen samples slowly warm up and sink into the now melting freeze-substitution medium. There is no temperature control, but the whole insulated box may be placed inside a deep freeze. The substitution medium may be exchanged periodically with fresh cold fluids.

A slightly more complicated device is also proposed by Sitte (1981), in which the inner metal container is placed in a Dewar vessel over LN$_2$ (Fig. 7.7). This device is fitted with a heater and temperature sensor so that a whole range of temperatures can be readily obtained. A similar device is proposed by Marshall and Kent (1991), with the additional refinement of a reciprocating cam shaker to gently agitate the samples. None of these devices properly address the problems associated with the exchange of cold, hygroscopic fluids under anhydrous conditions, although a simple answer would be to quickly remove the metal container holding the samples and place it in a large plastic bag, which may be filled with dry nitrogen. The exchange

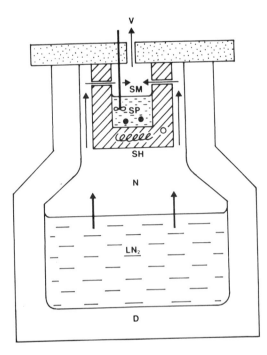

Figure 7.7. Freeze-substitution equipment consisting of a sample chamber sitting inside a liquid-nitrogen container. The metal specimen holder (SH), which contains an electric heater and thermostat, is placed in the neck of a large liquid-nitrogen Dewar (D). The sample holder is cooled by cold nitrogen gas (N), which slowly boils off from the large resevoir of liquid nitrogen. The specimens (SP) are placed in the substitution medium (SM), which may be slowly stirred at a preset controlled temperature. The substitution medium is kept anhydrous by the constant flow of dry nitrogen gas, which is vented through a hole (V) in the insulated cover. [Modified and redrawn from Sitte (1986).]

of fluids could be made by handling the containers through the wall of the bag, and the large metal mass of the container will keep the samples cold.

Wells (1985) has designed a relatively simple low-temperature box which will handle all the procedures associated with freeze substitution, low-temperature embedding, and UV polymerization in the temperature range 273–233 K. Cold nitrogen gas acts both as a coolant and as the means of obtaining anoxic anhydrous conditions inside the chamber. A more sophisticated device, which automatically controls the time and temperature of substitution, has been described by Muller *et al.* (1980) and Humbel and Muller (1986). An interesting feature of this equipment is that it incorporates a Beckman Microfuge B surrounded by a LN_2-cooled copper cylinder, in order to handle cell suspensions. The Balzers Union AG FSU 010 Freeze substitution Unit, which can also be used for low-temperature embedding and polymerization, is based on the Muller design. The Balzers Union AG LTE 010 Low-Temperature embedding cabinet has been designed solely for embedding. Samples may be infiltrated with resin at various temperatures to as low as 223 K using a Peltier cooling device in a moisture-free environment. An accessory UV light source allows the resin-infiltrated samples to be polymerized at low temperatures.

Sitte *et al.* (1986), give details of an elegant but complex piece of equipment that is now commercially available as the Reichert–Jung (Leica) CS Auto cryosubstitution apparatus. The unit, which has sophisticated stirring, transfer, and temperature control facilities, can handle a large number of different samples. The problem of transferring substitution fluids under anhydrous condition has been overcome by introducing the liquids at room temperature. These are cooled to predetermined temperatures under anhydrous conditions inside the equipment before being added to the specimen. It is also possible to infiltrate samples with resin polymers and polymerize the material under a UV source.

7.8.2. Specimen Handling Devices

One of the major difficulties with freeze substitution and low-temperature embedding is to be able to handle small transparent samples at low temperatures and under anhydrous conditions. The principal operations are to change the substitution and embedding liquids and to ensure that these substances flow freely around the sample. Wire baskets are one of the most popular containers, which may be free or attached to the bottom of a plastic tube (Ridge, 1990). Wells (1985) describes an ingenious device for handling the samples through the substitution and embedding process. The device, which is shown in Fig. 7.8, is constructed of pieces of plastic transfer pipette, vinyl tubing, and nylon filters. It allows liquids to be dispelled and fresh liquids drawn up without losing the tissue pieces and under anhydrous conditions. From personal experience, this simple device is one of the most effective ways of sample transfer.

Volker *et al.* (1985) has also constructed a simple device which ensures that polymerization goes on under anhydrous and anoxic conditions. It is based on an outer 20-ml plastic syringe, which acts as a mini environmental chamber and an inner 5-ml plastic syringe, which has been modified to hold five plastic embedding capsules.

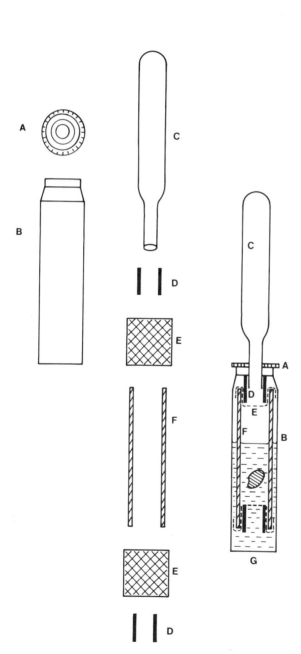

Figure 7.8. Tissue handling device for use during low-temperature substitution and embedding. The plastic cap (A) of a glass vial (B) is drilled with a 3-mm hole. The teat of a plastic transfer pipette (C), cut off about 15 mm from the bulb base, is forced into a 8-mm piece of vinyl tubing, o.d. 4.1 mm (D), which in turn is forced into a larger piece of vinyl tubing (F) 30 mm long and with an o.d. of 5.5 mm. A 10-mm square of 100-μm mesh nylon filter (E) is squeezed in between the two pieces of vinyl tubing and acts as a permeable stopper. A similar stopper, less the plastic teat, is assembled at the other end of the large piece of vinyl tubing, which now acts as a container for the sample. The device is assembled and used as follows. The nylon mesh filter is assembled at the lower end of the vinyl tube container, placed in the bottle, two thirds filled with the substitution fluid and cooled to the working temperature. The quench-cooled sample is placed in the substitution fluid in the vinyl tube container and the upper nylon filter/pipette/cap assembly forced into position. The whole assembly (G) should be placed in a hole in a large precooled metal block and kept in the refrigerator. The substitution fluid may be agitated by periodically gently squeezing the plastic bulb, which remains pliable even at quite low temperatures. If it is necessary to change the substitution or embedding liquids, the cap (A) is carefully loosened from the bottle and the whole assembly quickly transferred to another bottle containing the appropriate precooled liquid. This operation may be repeated several times without changes in the sample temperature and with substitution and embedding liquids remaining anhydrous. All the operations can be done with one hand inside the refrigerator. [Modified and redrawn from Wells (1985).]

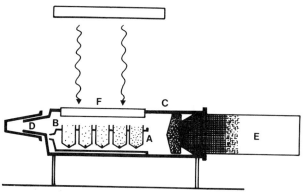

Figure 7.9. A simple device for the low-temperature polymerization of resin-embedded specimens. The samples are placed in capsules (A) which are held in holes drilled in the barrel of a small plastic syringe (B). The capsules are filled at low temperature under anhydrous and anoxic conditions, with the appropriate resin mixture and placed inside the barrel of a larger syringe (C), which is flushed with dry nitrogen through the needle attachment nipple (D). After a few minutes, the nitrogen is turned off and the barrel of the large syringe sealed at both ends with the plunger (E) and the plastic cap (D). The sealed container is placed on a small stand. If the samples are to be polymerized under UV (Lowicryl) or blue light (LR Gold) a narrow strip of masking tape (F) should be placed along the top of the barrel to shade the specimens from direct light. Polymerization should be carried out in a refrigerator lined with aluminum foil to reflect the light onto the capsules. [Modified and redrawn from Voker *et al.* (1985).]

The device is shown in Fig. 7.9. The capsules are filled with the final Lowicryl resin mixture together with the sample and the 5-ml syringe–capsule holder placed inside the barrel of the 20-ml syringe, which is then purged with dry nitrogen gas and sealed at both ends. The plastic walls of the syringe barrel are transparent to UV radiation and the nitrogen provides a protective atmosphere. If direct UV light is to be used, the top of the syringe barrel should be covered with a strip of opaque tape, and the whole sealed device is then placed on a piece of aluminum below the light source.

7.9. LOW-TEMPERATURE EMBEDDING OF FREEZE-DRIED MATERIAL

Although considerable attention now focuses on low-temperature embedding in conjunction with freeze-substitution processes, it should not be forgotten that low-temperature embedding can also be used to complete the preparation of samples which have been freeze-dried, preferably without exposing them to the atmosphere. Wroblewski and Wroblewski (1984) and Wroblewski *et al.* (1985) overcame this problem by transferring freeze-dried samples into Lowicryl HM20 under vacuum at 243 K. The freeze-drying was performed in a conventional freeze-dryer, which had been redesigned to allow freeze-dried samples to be introduced into cold resin mixtures under vacuum. Tissue samples were dried on a Parafilm layer stretched across the surface of a sample tube containing cold Lowicryl. Once the drying was completed, the tissue samples were plunged into the embedding media by breaking the

Parafilm layer (Fig. 7.10). The partial pressure of water vapor and oxygen was very low in the freeze-drier and provided ideal conditions for the anoxic polymerization of the resin. Polymerization was accomplished by indirect UV radiation for 20 h at 243 K and then at 293 K for 2 days. Edelmann (1986) has embedded freeze-dried material in Lowicryl K11M at 213 K.

Dry plant cell walls can form a considerable barrier against the penetration of apolar and/or high-viscosity resins. Fritz (1980, 1989) found that vacuum infiltration at 10^{-2}–10^{-3} Torr of freeze-dried plant tissue with diethyl ether facilitated the subsequent infiltration with apolar resin mixtures such as styrene–methacrylate. Apolar resins are used to avoid displacement of diffusible, water-soluble elements.

Chiovetti et al. (1986) have compared the structure of freeze-dried rabbit renal arteries in material that has been embedded at low temperatures in Lowicryl resin, with material that has been embedded in low-viscosity eposy resins at ambient temperature. Samples were taken with and without prior treatment with osmium tetroxide and/or glutaraldehyde. Only Lowicryl K4M embedded material that had been freeze-dried with or without glutaraldehyde gave good structural preservation.

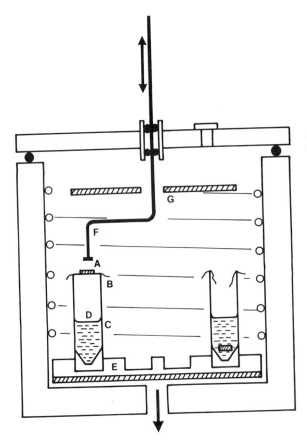

Figure 7.10. A combined freeze-dryer and low-temperature embedding device for tissue sections. Frozen tissue sections (A) are placed on a thin Parafilm layer (B) stretched across the top of plastic tubes (C) filled with the appropriate embedding medium (D), all at low temperature on the cold stage (E) of a freeze-dryer with light-tight walls. The tissue sections are freeze-dried and while observed through transparent top of the freeze-dryer, are pushed into the embedding medium using the rod (F), which breaks the Parafilm layer. Once the sample is infiltrated with resin, the freeze-dryer is brought up to ambient pressure, flushed with dry nitrogen, and the resin polymerized at the appropriate low temperature, using the indirect light source (G). [Modified and redrawn from Wroblewski and Wroblewski (1984).]

7.10. ISOTHERMAL FREEZE FIXATION

Isothermal freeze fixation has proved useful in investigations into the formation and distribution of intra- and extracellular ice, and the effects of freezing damage in hydrated specimens at relatively high subzero temperatures. The amount and distribution of ice in frozen biological systems is an important factor in understanding the processes of low-temperature injury and for low-temperature preservation. Both phenomena are associated with slow, near equilibrium cooling conditions. Isothermal freeze fixation, which was first introduced by McKenzie *et al.* (1975) involves chemical fixation of frozen tissue in the presence of cryoprotectants at subzero temperatures under conditions of constant temperature and water activity. The composition of the fixative solution is carefully adjusted so that its freezing point corresponds to the temperature to which the sample has been cooled to in the first place. This ensures that the fixative diffuses through the tissue under equilibrium conditions without disturbing either the configuration of the ice crystals or the state of tissue hydration. Once fixation is completed, the tissue is thawed and processed at ambient temperatures for study in the microscope. The technique has been used at temperatures between 252 and 273 K by incorporating different solutes into the fixation liquid to ensure that it is isotonic with the concentrated solute phase in the tissue.

Isothermal freeze fixation has the advantage that it preserves the structure of ice at the temperatures at which it was formed. Techniques such as freeze-drying, freeze substitution, and freeze-fracture replication, which are based on nonequilibrium cooling, all require the sample to be cooled to low temperatures. This can seriously affect the configuration and morphology of ice formed under near equilibrium conditions at higher temperatures, which will in turn alter the structural appearance of the tissue.

Hunt (1984) has reviewed the procedures of isothermal freeze fixation and shows how it is possible to design freeze fixatives for a wide range of subzero temperatures. A given freeze-fixation schedule will depend on the temperature and nature of the tissue concerned, the post thawing processing, and the choice of co-solutes used to lower the freezing point of the fixative. Thermal equilibrium is ensured by maintaining both the sample and the fixative at a constant temperature during fixation. Hunt (1984) recommends that a small amount of finely divided ice be added to the freeze-fixation solution. These ice crystals will act as foci for any ice-crystal growth and will ensure that minor fluctuations in temperature will not affect the ice phase in the sample. Table 7.8 gives a list of some freeze-fixation formulations that have been used by different investigators.

Table 7.8. Fixatives and Co-solutes Used for Isothermal Fixation

Temperature (K)	Fixature	Co-solute
271	2% Osmium tetroxide	3.4% NACL
263	53.5% Glutaraldehyde	3.4% NACL
263	2% Osmium tetroxide	18% NACL
253	3% Glutaraldehyde	23% NACL
252	5% Glutaraldehyde	34% DMSO/TRIS

Hunt (1984) has shown that isothermal freeze fixation at high subzero temperatures has a number of distinct advantages over freeze substitution carried out at the same temperature. No solvents are used and the ice matrix remains essentially undisturbed during fixation of the tissue matrix. In freeze substitution the stability of the matrix phase is critically dependent of the polarity and ice solubility properties of the substituting fluids. This is not a problem with isothermal freeze fixation, which eschews the use of solvents that affect ice crystals.

All the work that has been carried out on isothermal freeze fixation has been confined to the initial fixation phase of tissue processing. A logical extension of these studies would be to complete the dehydration and embedding at subzero temperatures and to take advantage of the low-temperature resins that are now available.

7.11. APPLICATIONS OF FREEZE SUBSTITUTION AND LOW-TEMPERATURE EMBEDDING

We will first examine the notion enunciated at the beginning of this chapter, namely, that the two procedures are one of the best ways of preserving the functional ultrastructure of biological material, before discussing the disadvantages of the two procedures and the breadth of their application.

7.11.1. The Effectiveness of Freeze Substitution and Low-Temperature Embedding as a Means of Tissue Preservation

It is difficult to analyze objectively how well the ultrastructure has been preserved for there are no good criteria against which this may be judged. The direct approach of examining the effect these two procedures have on living cells by light microscope observation suffers from limited spatial resolution. If, however we accept the premise that structure is a manifestation of the revolving phases of metabolism, it will be realized that it is probably impossible to ever "see" the true structure of any biological polymer *in situ* in the living cell. The uncertainty principle must apply, and in addition we must accept that any chemical or physical intervention with the natural state will, even in some small way, alter the structural integrity of most biological macromolecules. There is, however, reasonable circumstantial evidence suggesting that freeze substitution and low-temperature embedding are well suited for preserving cellular ultrastructure. The evidence is as follows:

1. There is good structural correspondence between living and freeze-substituted material in a wide range of biological and hydrated organic specimens.
2. There is a close morphological similarity between the structure of material prepared by freeze substitution and low-temperature embedding and samples prepared by other low-temperature techniques.
3. Novel (nonartifactual) structural features have been found in a number of tissues prepared by freeze substitution and low-temperature embedding which have not been seen using other closely related techniques.

The most commonly used objective and quantifiable criteria for judging the quality of material prepared by the two methods are as follows. Cell membranes should appear smooth, continuous, and sharply contrasted within the biological matrix, even though the familiar contrast of the unit membrane may be reversed. The size and distribution of cellular components and compartments remain unchanged. The cytoplasm should appear dense and slightly granular, indicating that little extraction has taken place. There should be no large "ice crystal ghosts," which would indicate poor initial cooling or subsequent recrystallization. There should be no transparent halos around organelles such as mitochondria or chloroplasts, which would indicate incipient shrinkage. There should be no separation between the specimen surface and the embedding resin.

There is more direct evidence that the functional integrity of macromolecules is preserved by the two preparative procedures. The evidence of retention of antigenicity will be considered later, although uncertainty remains whether immunosensitivity is truely indicative of the complete integrity of the molecule. The evidence from x-ray diffraction studies that some of the procedures do not significantly alter the molecular order of some proteins is more convincing. The most convincing evidence comes from studies like those undertaken by Weibull *et al.* (1980), who found that Photosystem I and II were not irreversibly damaged in chloroplasts that had been freeze substituted and infiltrated in Lowicryl. Other studies have shown that different enzymes retained their activity after low-temperature embedding. The studies of Weibull and colleagues suggest that some integrated enzyme complexes may also be unaffected.

7.11.2. Disadvantages of Freeze Substitution and Low-Temperature Embedding

The use of organic solvents to dissolve ice also presents the possibility of dissolving lipids. This loss is significantly decreased at low temperature and reduced further in the presence of heavy metal fixatives, these materials will, in turn, adversely affect other components of the biological matrix. Investigations by Menco (1984) revealed sample shrinkage, but there are insufficient studies to know whether this is a general problem. For the moment there is still considerable uncertainty whether freeze substitution and low-temperature embedding retain *all* the soluble constituents at their natural location in the cell.

7.11.3. The Application of Freeze Substitution and Low-Temperature Embedding to Structural Studies

Both Menco (1984) and Steinbrecht and Muller (1987) provide extensive citations on the use of freeze substitution and low-temperature embedding in structural investigation in the light and electron microscope. Nearly all the investigations show that the two techniques, particularly when used in connection with heavy metal staining of thin sections, provide crisp, clearly defined images in the transmission electron microscope. In many instances there is a substantial improvement in the quality of preservation of intracellular structures in comparison to the results obtained by ambient temperature fixation. Unfortunately, there have been few studies where a careful comparison has been made of the differing effects of temperature, dehydration, solvent, degree of dehydration, and embedding medium on a set specimen. One such study is the work of Horowitz *et al.* (1990) on avian chromatin and nuclei, which showed that the best results were obtained with spray frozen samples prefixed in 3% glutaraldehyde, substituted in methanol at 183 K, and embedded in K11M at 218 K.

Freeze substitution and low-temperature embedding have also been used to prepare samples for light microscopy and in some instances have revealed structures not seen in the electron microscope. McCulley and Canny (1985) found that freeze substitution was the only means by which it was possible to preserve the elongate vacuolar structure called the "pleiomorphic canalicular ssytem," which had only been seen previously in living plant cells. Osumi *et al.* (1988) used freeze substitution in connection with low-voltage scanning electron microscopy and have for the first time revealed the three-dimensional arrangement of fine filaments in yeast cell walls. An earlier paper by Baba and Osumi (1987) showed similar structures in freeze-substituted material prepared for the transmission electron microscope.

7.11.4. The Application of Freeze Substitution and Low-Temperature Embedding to Analytical Studies of Diffusible Substances

A considerable amount of data on the concentration of electrolytes in plant cells has come from the x-ray microanalysis and, latterly, electron spectrosopic imaging of thin sections of freeze-substituted and, occasionally, low-temperature embedded material. The relative effectiveness of freeze substitution and low-temperature embedding in comparison to other low-temperature techniques will be considered in Chapter 11, but the review papers by Harvey (1980, 1982, 1986) and Harvey *et al.* (1985) should be consulted for the application of freeze substitution to the elemental analysis of plant specimens. Thin sections of freeze-substituted material have been used to measure the distribution of Na^+, K^+, and Ca^{2+} in tomato leaves (Ross *et al.*, 1982), and for the co-localization of P and Ca^{2+} in mouse cartilage by electron spectroscopy (Arsenaut *et al.*, 1988). Lehmann *et al.* (1990) and Michel *et al.* (1991) have used freeze substitution to prepare thin sections for study by electron spectroscopic imaging. Papers by Marshall (1980a, b), Chandler (1985), Sumner (1988), and Wroblewski *et al.* (1987) give a number of examples where freeze substitution has been used to prepare animal tissue for x-ray microanalysis of electrolytes. All the studies show that varying amounts of different electrolytes are retained in cells and tissues. None of the studies show that all of a particular element is retained in a given cell. These are matters we will return to in Chapter 11.

Freeze substitution *and* low-temperature embedding have been combined to prepare samples for x-ray microanalysis. Wroblewski and Wroblewski (1984) used the two procedures for the x-ray microanalysis of muscle tissue. Although significant ion shifts characteristic of muscle tenotomy were found, there were a number of anomalous results that can only be related to the redistribution of ions with the cells.

Very few critical studies have been made on what appears to be a promising approach, presumably because it is believed that freeze-dried and frozen-hydrated sections provide more accurate answers. More analytical studies need to be carried out using freeze-substituted material, particularly with unfixed specimens.

One of the few comparative studies has been reported recently by Condron and Marshall (1990), who carried out quantitative x-ray microanalysis of avian tissue prepared either as bulk frozen-hydrated samples, freeze-dried, or freeze-substituted material. Changes in elemental concentration were found in the freeze-substituted and freeze-dried material but not in the untreated frozen-hydrated material. It is uncertain whether these changes were due to the substitution or drying *per se*, or to the ensuing resin infiltration.

Freeze-substituted material has been used for laser probe analysis, autoradiography, and electron energy loss spectroscopy—see Echlin (1984) for relevant details of these procedures. Hippe (1987, 1989) elegantly demonstrates how freeze substitution and low-temperature embedding may be used for the intracellular localization of a tritiated fungicide by autoradiography and the immunochemical localization of a foreign protein in transgenic tobacco plant. A recent paper by Marshall and Wright (1991) combines the low-temperature procedures for use with confocal scanning microscopy.

7.11.5. The Applications of Freeze Substitution and Low-Temperature Embedding to Immunocytochemical and Enzymatic Studies

A substantial amount of the evidence presented in this chapter suggests that the structural configuration of many proteins are well preserved by a combination of freeze substitution and low-temperature embedding. We have already shown that enzyme activity may be retained following freeze substitution, and immunocytochemical methods have been applied to a wide range of tissue. Monaghan and Robertson (1990) have used unfixed, methanol-substituted tissue for immunocytochemical studies (Fig. 7.11) and equally good preservation has been found in tissues fixed with osmium (Fig. 7.12). The papers by Roth *et al.* (1981), Bendayan and Shore (1982), Armbruster *et al.* (1984), Nagele *et al.* (1985), Abrahamson (1986), Hobot *et al.* (1987), Durrenberger *et al.* (1988), Monaghan and Robertson (1990), and Muller *et al.* (1991) are but a few of the examples where Lowicryl has been used. The studies by Monaghan and Robertson are particularly interesting as they have obtained excellent structural and immunochemical preservation with embedded but unfixed material. The papers by McPhail (1987) and Herken *et al.* (1988) give similar applications for the resin LR Gold. Sautter (1986) has used the progressive lowering of temperature dehydration technique and Lowicryl embedding in combination with protein A gold for the immunolocalization of enzymes in water melons. Finally, Kann and Fouquet (1989) and Ichikawa *et al.* (1989) provide a good up-to-date introduction to the different freeze-substitution and low-temperature methods used in connection with immunocytochemical studies. Immunocytochemical labeling is currently an area of active research and wide application in the biological sciences;

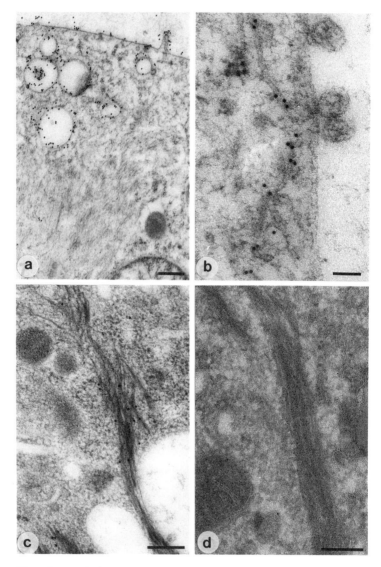

Figure 7.11. Tissue freeze substituted in pure methanol without aldehyde or osmium fixation, embedded in a low-temperature resin and used for immunocytochemical labeling. (a) ZR-75-1 cells immunolabeled for the presence of epithelial membrane antigen detected with 10 nm colloidal gold. The gold is localized both on the cell surface and in association with cytoplasmic vesicles. Bar = 0.25 μm. (b) ZR-75-1 cells labeled with antitubulin antibody. The microtubule is labeled with 10 nm gold. Bar = 0.1 μm. (c) ZR-75-1 cells showing labeling of keratin filaments with antikeratin antibody LE61. (d) The keratin filaments are not detected by the antibody in sections treated with pre-immune serum. Bar = 0.2 μm. [Pictures courtesy Monaghan and Robertson (1990).]

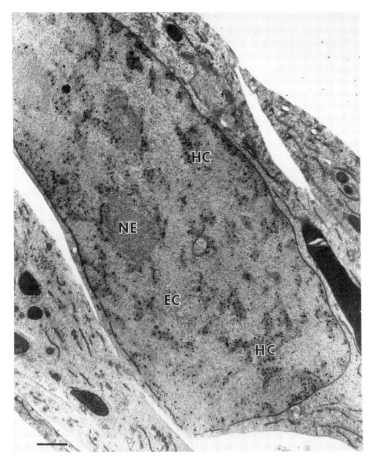

Figure 7.12. Cultured cell prepared by impact cooling, molecular distillation drying, and osmium vapor stabilization, and embedded in Spurr's resin. Postembedding labeling with anti-DNA (ANA) antibodies and 17 nm colloidal gold. Localization is specific to the heterochromatin areas (HC) in the nucleus and absent in areas of the euchromatin (EC) and nucleolus (NE). Bar = 1.0 μm. (Picture courtesy LifeCell Corporation.)

and many different approaches are being adopted for tissue preservation. Nearly all the techniques use low temperatures at one stage or another in the preparative process. Freeze substitution, with or without low-temperature embedding, features prominently as one of the methods that are giving consistently good results.

7.12. SUMMARY

Freeze substitution and low-temperature embedding are two distinct processes which are frequently conjoined, first to dehydrate frozen specimens by dissolving the ice in organic liquids and then infiltrating the vacant spaces with liquid resins, which are subsequently polymerized. Both procedures are invasive and make extensive use of organic polar and apolar solvents and chemicals. Freeze substitution is carried out at between 173 and 193 K in order to minimize the disruptive influence of ice recrystallization. Low-temperature embedding is best done at higher temperatures (223–253 K), which favor higher diffusion rates of the high-viscosity resins. Both procedures must be carried out under strictly anhydrous conditions.

By careful choice of solvents and resins, and with close control of the times and temperatures of infiltration and embedding, it is possible to dehydrate and embed samples of biological and hydrated organic materials without seriously disturbing the supramolecular arrangement and functionality of their constituent hydrophilic macromolecules. The quality of preservation in sections cut from such samples is excellent and may be used for immunocytochemical analysis, autoradiography, and enzyme studies. The ultrastructural preservation is thought to most closely resemble that found in the natural state. Although some new techniques are beginning to appear, there is still some uncertainty whether these methods will permit the *in situ* retention of small molecular weight species and dissolved elements and compounds in the aqueous phases of hydrated specimens. The equipment and techniques needed to carry out all these low-temperature procedures are neither complicated nor expensive, although the actual processes are time consuming and require great attention to detail. Two additional processes, progressive lowering of temperature and isothermal freeze fixation, which work at somewhat higher subzero temperatures, are designed to remove water while still retaining the form and location of any ice crystals in near-frozen specimens.

8

Low-Temperature Light Microscopy

8.1. INTRODUCTION

Low-temperature light microscopy, or cryomicroscopy as it is sometimes referred to, is a powerful technique for investigating the wide range of phenomena associated with freezing and thawing in hydrated systems. Its singular advantage over all other methods is that it allows dynamic observations and analysis to be made under thermally controlled conditions. The technique provides a readily visible and sometimes dramatic view of phase change processes and their effect on cells and tissues. An impressive literature exists on the subject, some of it dating back to the late 19th century. Most of the earlier studies were of an entirely qualitative nature and centered on observing and recording the processes of freezing and thawing in a range of biological tissues. These were heroic times, because the only way cryomicroscopy could be performed was to ensure that the sample, stage assembly, microscope, *and* observer were all in a cold environment. Cohn (1871) and Kunisch (1880) went outdoors in winter to study the process of freezing in cells of the alga *Nitella*. Weigand (1906) did the same thing to study freezing in fruit tree buds. The German botanist Sachs (1892) was a little more cautious and put his microscope on an open window sill at 267 K in order to follow the formation of ice crystal formation in pumpkin tissue.

A significant advance in cryomicroscopy came with publication in 1897 of the now classical study by Hans Molisch, *Das erfrieren der Pflanzen*. Molisch gave details of the first cryomicroscope, which consisted of a light microscope inside a box packed with ice and other freezing mixtures. The apparatus, which cooled to about 260 K, lacked any thermal controls, although the optics could be adjusted and the specimens moved by controls outside the ice box. Although Molisch devised a cryomicroscope that essentially relieved the observer from hypothermia, he did advise that it was best used in an unheated room. The problems studied by Molisch nearly a century ago are the same ones that engage the attention of cryobiologists today; only the technology for studying these problems has improved. An English translation of this remarkable study was published a few years ago (Molisch, 1982), and any aspiring cryomicroscopist should read the lucid description Molisch gives of these early attempts at low-temperature microscopy.

Cryomicroscopy has shown considerable progress since these early beginnings and these advances will be discussed in this chapter. It is intended to confine our discussions to what is seen as the prime role of cryomicroscopy: an experimental system that can provide a quantitative description of freezing and thawing processes. With this in mind we will concentrate on the equipment and procedures that have been developed in the past 20 years and give a general coverage of the applications. References to the early work may be found in the reviews by Diller (1982, 1985) and McGrath (1985, 1987).

8.2. SIMPLE COLD STAGES FOR QUANTITATIVE LIGHT MICROSCOPY

A significant landmark in the development of cryomicroscopy came with the development of a separate cold chamber, which could be mounted on the stage of the microscope. The chamber was cooled by carbon dioxide gas. For the first time, the microscope (and the observer!) could be kept at ambient temperatures while only the cold stage and sample were cooled to subzero temperatures. There then followed a series of cold stages, which were designed principally for specific applications. One of the most sophisticated systems was that devised by Mason and Rochow (1934) for the study of freezing processes in plants. The stage was cooled by circulating refrigerant, had accurate temperature control down to 248 K, and was capable of reasonably fast cooling rates.

Although it was possible to obtain quite low temperatures with these early cryomicroscopes, the thermal control was limited. The range of sample temperatures was governed solely by the range of environmental temperatures. In order to raise or lower the sample temperature, it was only necessary to vary the environmental temperature above or below that of the sample. Such an arrangement, which is shown diagrammatically in Fig. 8.1, would be of limited usefulness in any studies that required precise temperature control. The temperature of the sample and the rate of temperature change were subject to a number of variables. These included the size and thermal characteristics of the sample and the cold stage, and the temperature difference and thermal conductivity between the cold stage and the sample. Such cryomicroscopes had limited and low rates of temperature change: a serious limitation for studies involving precise measurements of phase changes in cooled hydrated systems. However, not all light microscope studies require such precise temperature

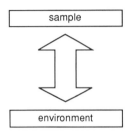

Figure 8.1. Diagrammatic representation of a simple cold stage in which heat is transferred between the sample and the environment with little or no measure of control. [Redrawn from McGrath (1987).]

control, and in some circumstances, a simple cold stage can be used to solve a particular problem.

A number of these simple cold stages will be discussed here as they are the type of cold stage that can easily be constructed in a laboratory workshop. Gallup *et al.* (1975) constructed a cold stage which consisted of a transparent plastic box fitted over the microscope stage and cooled by a continuous flow of cold dry nitrogen gas. The base of the box was cut away to accommodate a standard glass microscope slide, and the lid of the box had a hole through which the microscope objective could be lowered. A plastic washer around the objective effectively sealed the inside of the box from the outside air. The constant gas flow prevented misting on the objective lens and some degree of temperature control was provided by varying the rate of gas flow. The main part of the objective lens is warmed by a few coils of electrically heated resistance wire wound around the length of the lens barrel. This prevents any condensation *inside* the objective lens. A transparent glass plate made of a pair of standard microscope slides is fitted into a space cut out in the floor of the plastic box. A pair of nichrome resistance wires are placed between the two glass slides at their edges and the whole assembly is held together by epoxy glue. A low voltage applied to the two wires warms the glass slides just enough to prevent condensation forming on the front lens of the condensor.

Strips of glass, cemented to the upper surface of the glass floor, support the specimen slide and allow the cold gas to flow above and below the sample. A thin wire thermocouple may be placed in the immediate vicinity of the specimen in order to measure the local temperature. This simple, inexpensive cold stage, which is illustrated in Fig. 8.2, has been used with oil immersion objectives and at temperatures as low as 233 K. The temperature control is minimal, although with a little practice it is possible to get low and high cooling rates.

Mersey *et al.* (1978) give the specifications for an inexpensive controlled temperature stage which allows observation of living cells down to 268 K. The stage, shown in Fig. 8.3, consists of two aluminum plates, one of which has been milled out in the form of a circuitous channel, which acts as a passage for refrigerated liquid. Holes, in line with the optical axis of the microscope, are cut out at the center of each plate. The two plates are welded together at their edges and around the center hole. Hose connections are made with the circulating refrigerated liquid channel and the cold stage is fitted to the microscope by means of suitably spaced screws. The stage is cooled by circulating a variety of liquids such as ethanol or a water–glycerol mixture, which are cooled in a low-temperature water bath. The temperature is measured using ultrathin thermistors mounted close to the specimen. Cooling rates of between 0.3 and 3.0 K min^{-1} have been obtained by varying the flow of the coolant. Holding temperatures of 273 K have been obtained with oil immersion lenses and 268 K if a dry objective is used. This stage has been successfully used with phase and Nomarski optics to observe living cells at temperatures between 295 and 268 K. The surface of the stage tends to frost up, but this could be alleviated by enclosing the stage area with a plastic bag through which dry nitrogen is slowly circulated. The performance could be improved further by using insulating nylon offsets between the cold stage and the substage assembly of the microscope.

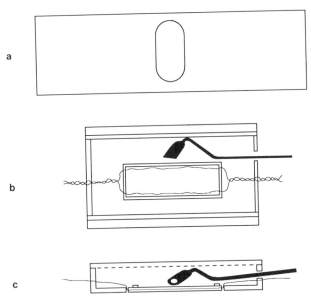

Figure 8.2. A cryostage cooled by cold dry nitrogen gas. The stage is mounted inside a plastic box, 70 × 140 mm and 22 mm deep, covered by a fenestrated lid (a) for viewing. In plan view (b) the dry cold nitrogen is directed onto the specimen area from the flattened end of a copper pipe. The stage is electrically heated but the temperature is controlled by the rate of gas flow. The side view (c) shows how the gas flow is directed close to the sample area. [Redrawn from Gallup *et al.* (1974).]

Makita *et al.* (1980) have devised a very simple system for observing frozen hydrated sections immersed in a shallow bath of liquid nitrogen, using an incident light fluorescent microscope equipped with a water immersion lens. Cryostat sections cut at between 248 and 243 K were mounted at low temperature onto glass microscope slides using double-sided adhesive tape. A large circular cover slip was placed over each section, which was gently flattened with a precooled glass rod. This cold sandwichlike preparation was held together by two clips and attached to a metal stage as shown in Fig. 8.4. The whole assembly is immersed to a depth of 5 mm in a shallow bath of liquid nitrogen, which in turn sits in another bath of LN_2 to prevent bubbling and movement. The sections are observed directly using low power (×10 and ×20) water immersion objective lenses. This simple system is surprisingly effective, particularly as the fluorescent intensity increases by an order of magnitude at the low temperatures. It is not known what effect LN_2 has on the objective lenses.

Although these simple stages, and many others like them, are adequate for the purpose for which they were designed, they lack any independent control over the various thermal variables such as temperature and the cooling and warming rates. The cooling rate of the specimen is dictated only by the sink temperature and the thermal properties of the stage and the sample. The cooling rate would tend to decrease asymptotically to zero as the sample temperature approached that of the sink, and the driving potential for the cooling flux would decrease accordingly. Under these circumstances, precise temperature control is impossible.

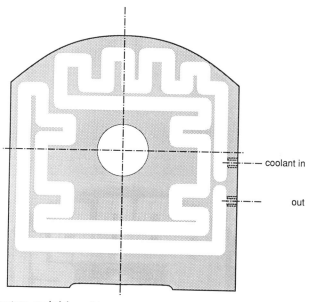

Figure 8.3. A cryostage cooled by refrigerants at preset temperatures, which circulate through sealed channels in the metal cold stage. There is no heating and samples with closely apposed thermocouples are placed on glass slides on the cold surface. [Redrawn from Mersey *et al.* (1978).]

This problem was elegantly solved in a cold stage designed by Harmer (1953), which introduced an additional degree of thermal freedom into the heat transfer path. Regulation of the cooling rate, independently of the sample temperature, may be achieved by adding an ancillary heat flux, which functions separately from the low-temperature sink. The balance between the heating and cooling flux may be used to set the rate at which the heat is removed from the sample, independently of the magnitude of the temperature. This principle, which is illustrated diagrammatically in Fig. 8.5, is the basis on which most cryomicroscope controllers are constructed.

In Harmer's stage design, a heater is introduced into the heat transfer path between the sample and the low-temperature sink to buffer the sample from the low temperature. The sample temperature, although still governed by the same factors

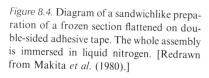

Figure 8.4. Diagram of a sandwichlike preparation of a frozen section flattened on double-sided adhesive tape. The whole assembly is immersed in liquid nitrogen. [Redrawn from Makita *et al.* (1980).]

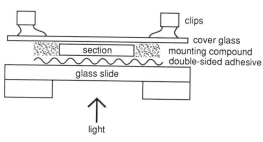

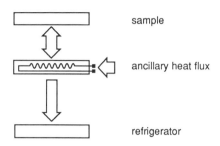

sample

ancillary heat flux

refrigerator

Figure 8.5. Diagram illustrating the principles of the Harmer cryomicroscope system. Sample cooling is regulated by introducing an electrically powered ancillary heat flux between the specimen and refrigerated heat sink. [Redrawn from McGrath (1987).]

we saw in the simple stages, is now also affected by the presence of a heater. The thermal capacitance of the specimen area should be made as small as possible in order to minimize inertial effects in the system and obtain the highest thermal transients for a given heat flux. In practice, the sample area and the heater are quite small and are placed in good thermal contact with the low-temperature refrigerator. When the heater is turned off, the maximum rate of cooling is achieved. Lower cooling rates and higher ultimate temperatures are achieved by simply varying the power input into the heater. In addition, the temperature range over which the stage may be operated is greatly extended in comparison to the thermally unbuffered systems. Because the power input is electrical, the whole system may be linked directly to a microprocessor and computer. Thus the specimen temperature may be continually monitored and its value compared to a preprogrammed signal. The difference in the signal can be used to operate a closed loop proportional controller to automatically adjust the input heating and cooling flux. The development of the thermally buffered cold stage lead cryomicroscopy from purely qualitative studies to quantitative investigations, which require both well-defined cooling and heating rates and control of the spatial distribution of temperature in addition to its change with time.

8.3. COMPREHENSIVE COLD STAGES FOR QUANTITATIVE LIGHT MICROSCOPY

The fundamental components of a cold stage for light cryomicroscopy are shown in Fig. 8.6. They consist of a sample viewing area, heater, and heat sink. This heat transfer assembly is usually quite small and is attached to the x–y substage assembly of the microscope. Stage design has followed two main routes depending on the way the heat transfer occurs in the immediate vicinity of the sample.

Convective heat transfer stages have been developed by Diller and Cravalho (1971), Diller (1982), Schwartz and Diller (1982), Schweiwe and Korber (1982), Korber *et al.* (1986), Hayes and Stein (1989), and Walcerz and Diller (1991). *Conductive heat transfer stages* are favored by McGrath *et al.* (1975), McGrath (1987), and Kochs *et al.* (1989). Before discussing the design of these two types of stage and considering the advantages and disadvantages of the two heat transfer systems, it might be useful to first consider the basic operating and performance characteristics that one would expect to find in a modern cryomicroscope.

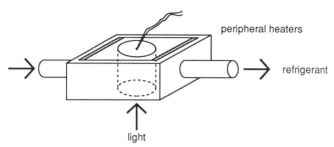

Figure 8.6. Functional characteristics of a light microscope low-temperature stage. Refrigerant flows below the transparent sample viewing area, the temperature of which is controlled by peripherally located resistive electrical heating pads. Sample temperature measurements are made with a closely apposed thermocouple. [Redrawn from Diller (1982).]

8.3.1. Optimal Features of a Cryomicroscope

McGrath (1987) considers that the following features should be included in the design of a low-temperature microscope system:

1. It should be capable of operating over a wide temperature range, i.e., 77–373 K. The lower temperatures are easily achieved because of the ready availability of LN_2 as a cryogen. The upper temperature limit will be determined by the amount of energy that may be put into the system.

2. It should be capable of a wide range of cooling and heating rates of 10^{-2}–10^5 K min^{-1}. The actual rates that will be achieved are going to depend on the physical size and thermal properties of the specimen and its thermal contact with the stage.

3. If needed, it should be possible to use arbitrary, nonlinear temperature changes.

4. The system should be easy to operate, and it should be easy to program the temperature changes. The introduction of microprocessors and microcomputers with associated software now makes this a relatively trivial task.

5. The system should be relatively robust and easily fitted to most light microscopes.

6. The control and measurement of temperature should be accurate, stable, and reproducible. Most modern cryomicroscopes have a temperature resolution in the range ±0.1–1.0 °C.

7. It should be possible to use all standard optical techniques such as bright and dark field, reflected and transmitted light, interference contrast, fluorescence, and polarizing optics.

8. The microscope optics should be capable of being linked to photomultiplyer, video, and cine recording systems and image analyzer devices.

9. The optical and thermal data should be displayed and recorded in a convenient and accessible form.

In addition, it would also be useful if electrodes could be inserted into sample solutions to measure freezing potentials.

8.3.2. Convective Heat Transfer Cryomicroscope Stages

The salient feature of a convective cryomicroscope stage is that the sample is cooled by convection using a refrigerated liquid or cold gas, which circulates below a thin transparent window on which the sample is mounted. One of the first convection cryostages was that designed by Diller and Cravalho (1971) and shown diagrammatically in Fig. 8.7. The system consists of a chamber 11 mm square by 7 mm deep made of 0.5-mm-thick stainless steel with refrigerant inlet and discharge tubes. Viewing windows approximately 2.5×6.5 mm are cut out from the top and bottom of the chamber. These openings are fitted with 165-μm-thick quartz cover slips sealed to the surface with epoxy glue. A small deflector placed to one side of the bottom chamber directs the inlet refrigerant against the underside of the top window to enhance the convective component of heat transfer. The cryostage is mounted on a low thermal conductivity phenolic plastic block to isolate it from the main part of the microscope. The complete unit, mounted on the substage assembly of the microscope and kept inside an airtight box, can be used in conjunction with a wide range of optical viewing and recording systems.

The dimensions of the cryostage are kept as small as possible to minimize the thermal mass and can achieve high cooling warming rates. The total thickness of the stage is approximately 10 mm, in order not to exceed the working distance of the light microscope condensor. A variety of objective lenses can be used, although dry objectives are used in preference to oil immersion lenses. The heater consists of a thin, transparent tin oxide coating layer deposited on the underside of the top quartz window. Any desired heater resistance can be achieved by varying the thickness of

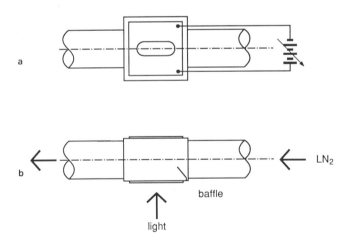

Figure 8.7. A diagram of a convective heat transfer cold stage in top (a) and side (b) view. A thin quartz window, coated on the lower side with electrically conductive material, is bonded to the upper side of a small hole, which passes through a stainless steel tube. The lower hole in the tube is sealed with a second quartz window without heating elements. Cooling is provided by liquid nitrogen or refrigerated gas, which passes through the stainless steel tube, and a small baffle ensures convective heat transfer. The temperature is controlled by the electric heating circuit in the upper quartz window and monitored by thin wire thermocouples placed close to the sample viewing area. [Redrawn from Diller and Cravhalo (1970).]

this electrically conductive layer. Electrical leads are attached to the heater by conductive paint and connected to a suitable dc power supply. The refrigerant is dry nitrogen or helium gas at a pressure of 99 kPa (1 bar), which is passed through an LN_2 heat exchanger kept at 77 K. Both the heating and cooling sources are in thermal communication with the specimen via conduction through the quartz window.

Specimens may be mounted for examination by two methods. They may either be placed directly onto the upper quartz window and covered with thin cover slips, or mounted between two cover slips, sealed with petroleum jelly, and the sandwich placed directly onto the top window of the cryochamber. The first method, although more difficult to assemble, is preferable since it minimizes the thermal mass and resistance between the refrigerator and the specimen. Irrespective of which method is used, the upper and lower cover slips contain pockets etched out using hydrofluoric acid into which a 20-μm copper–constantan thermocouple is fitted. It is important to have a small probe in order to minimize the thermal mass and, in turn, the response time of the thermocouple. The thermocouple output is connected to a suitable recording device. The microscope objective and the cryochamber are enclosed in a transparent plastic housing maintained at a slight positive pressure with dry gas at 293 K. This prevents water vapor condensation on the specimen and lens and enhances the contact between the specimen and refrigerator.

When in use, the stage is in continual balance between the magnitude of the heating and cooling fluxes so that at any given moment the measured temperature matches a predetermined temperature stored in the controller memory. The heater has a much shorter thermal time constant compared to the refrigerant flow. In practice the refrigerant flow is kept constant with time and the heater–cooling balance is achieved by changing the voltage across the resistive tin oxide layer. Because the thermal mass of the heater window and the specimen is so small, it is possible to achieve very rapid temperature changes of up to 10^3 K min^{-1} in both the heating and cooling phases. Although the original stage of Diller and Cravalho (1971) worked very well, it suffered a number of drawbacks. The thin heater window was very delicate and easily broken. It also took considerable time to mount the specimen and to ensure that the thermocouple was located correctly between the cover slips.

The original cryostage design of Diller and Cravalho (1971) has been improved by Diller and his colleagues to give better thermal performance and ease of use (Diller *et al.*, 1976; Diller, 1982; Evans and Diller, 1982; Schwartz and Diller, 1982, 1983). The result of these improvements is a more robust stage with much higher cooling rates of up to 10^5 K min^{-1}. However, this created difficulties since the necessary increased current density in the metal oxide resistive heater could cause the window to fracture. The thermocouple was embedded in a shallow groove etched into a thin cover slip, which was permanently attached to the upper window of the cryochamber. Specimens were placed on top of this cover slip and careful measurements showed that there was less than 1 K difference between the sensor and the specimen for any achievable cooling rate. Sample preparation time was much reduced with this improved stage.

A much simpler (and less expensive) convective stage is described by Diller (1982) and further modified by Hayes and Stein (1989). The convective heat transfer is more uniform, and although the cooling rates are of the order of 10^2 K min^{-1} over a 120–373 K temperature range, this is quite adequate for a number of studies where the water permeability rate is low. The modifications make it possible to study the effects low temperature have on the microcirculation in animal tissues. In these circumstances the principal concern is to retain the electrical integrity of the cold stage during the time large sheets of living tissue are bathed with physiological fluids. The main bulk of the stage is made from a block of phenolic plastic with windows at the top and bottom. The cooling and heating arrangements are similar to those described earlier except that in the Hayes and Stein modification, the heater slide is sealed into position with an O ring rather than being bonded with epoxy resin. This makes it much easier to replace a broken heater slide. Thin foil (5 μm) thermocouples are used and the stage parameters are controlled by a personal computer drive temperature controller. Further developments in convective stages may be found in the paper by McCaa and Diller (1987), which provides a useful practical guide to this type of cryostage. In the latest variant of the convective cold stage (Walcerz and Diller, 1991), the specimen is in a sealed chamber, which has inlet and outlet ports for admitting and collecting perfusate solutions during freezing and thawing. The entire volume can be exchanged in a few seconds and enables the samples to be viewed continuously during a cryopreparation protocol.

8.3.3. Conductive Heat Transfer Cryomicroscope Stages

One of the disadvantages of the convective heat transfer cryostage is that it is too thick to use with very short working distance objectives. To overcome some of these problems, McGrath *et al.* (1975) produced a conductive stage, the principal components of which are shown in Fig. 8.8. The basic design is based on the system developed by Diller and Cravalho (1971) except that the transparent heater is bonded to a copper plate, which conducts heat away from the specimen to a refrigerant stream flowing around the edge of the copper plate. The heat flux into and from the specimen is balanced by varying the power into the transparent quartz heater while constantly cooling the copper block with LN$_2$. The temperature is measured with a

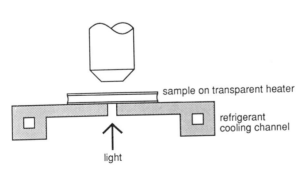

Figure 8.8. Diagram of a conductive heat transfer stage seen in side view. The sample sits on a transparent heater and thermal buffering is provided by heat flux from the sample area to the metal plate cooled at its periphery with circulating refrigerant. The measurement and control of the stage temperature is by a thermocouple embedded at the edge of the central viewing window. [Redrawn from McGrath (1987).]

5-μm copper–constantan thermocouple placed close to the specimen area, which is regulated and monitored by a control unit. It is possible to select temperatures between 77 and 310 K and cooling rates up to 7×10^3 K min^{-1}. In addition, the cooling or heating may be stopped at any intermediate point between the maximum and minimum temperature and maintained at this temperature indefinitely. The stage is 5–10 times thinner than the typical convection stage and allows oil immersion optics to be used with a condensor working distance reduced to 1.5 mm.

The principal modifications that have been made to this type of stage have centered on further reducing the size of the specimen chamber and on ensuring that it is in good thermal contact with the underlying cold stage window. Schweiwe (1981), Schweiwe and Korber (1984), and Schweiwe *et al.* (1978), have integrated a 5-μm foil thermocouple into the heating element to give a total thickness of 250 μm. This sandwichlike arrangement for heating and measuring is clamped or cemented tightly to the underlying copper block cooled by LN$_2$. A small sample droplet of a few microliters is placed on top of the sandwich, covered with a cover slip, and sealed with silicone grease, to give a sample layer of 10–12 μm. With this arrangement, the sample is within 50 μm of the thermocouple and the temperature variation between the sample and the temperature probe is no more than 0.1 K. This type of cryostage, shown in Fig. 8.9, produces cooling and warming rates of up to 10^4 K min^{-1} over a temperature range of 123–293 K.

Reid (1978) has constructed a simple program-controlled stage capable of attaining temperatures between 160 and 373 K and providing heating and cooling rates of up to 100 K min^{-1}. A brass substage, through which a refrigerant passes, provides the cooling capacity and the heater is a standard glass microscope slide coated on one side with a thin layer of chromium. Copper contacts are vapor deposited at each end of the slide leaving a 12-mm uncoated chromium strip in the middle. Electrical contacts are made to the copper layers using conductive adhesive. A thin foil (5 μm) copper–constantan thermocouple is glued into position over the central region on the opposite side to the chromium layer. The slide is placed chromium coated side down onto a brass substage, so that light passes through the 12-mm-wide central strip. A thin plastic film on the brass substage provides electrical insulation. The heater-sensor stage is held in good electrical contact with the cold substage by means of spring clips. The thermocouple acts both as a temperature indicator and as a sensor for the temperature controller.

It is possible to choose between two stage geometries to follow the propogation velocity of an ice front through a specimen. In situations where the heating and measuring assembly is placed over a hole cut in the copper plate used for cooling, this gives centripetal cooling and the ice front moves inward from the periphery. These configurations represents conditions similar to those found in the cylindrical vials used in many freezing experiments (Korber *et al.*, 1986). If the cut out is in the form of a long slit, this gives parallel cooling and the ice front moves inward from the two long sides. The slit geometry is used mainly for unidirectional freezing experiments where a planar ice front proceeds through the sample. Schweiwe and Korber (1982a) provide the mathematical basis for the analysis of the temperature fields across the two different stage geometries.

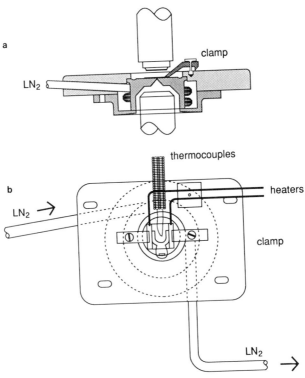

Figure 8.9. Drawing of a conductive heat transfer stage seen in side (a) and face (b) view. The thin sample under a cover glass is firmly clamped to the upper surface of the central transparent portion of the stage. Cooling is by liquid nitrogen circulating through a metal coil at the periphery of the central window of the stage. The stage temperature is controlled by electrical resistive heaters and measured by thin wire thermocouples embedded in the upper transparent window. The whole stage sits on a sheet of low-conductivity material, which is mounted on the microscope stage. The stage is cut away on the under surface to allow the use of short working distance condensor lenses. [Redrawn from Schweiwe and Korber (1982).]

Kourosh and Diller (1984) give details of a unidirectional conductive heat transfer stage. The heat source and heat sink are placed at opposite ends of the sample viewing area and the temperature is monitored by a linear array of thermocouples arranged along the primary heat flux axis. The specimen chamber is designed for freezing thin films no more than 25 μm thick and is essentially concerned with studying phase change effects in two dimensions. The temperature variation in the vertical direction is negligible. Boundary cooling rates up to 60 K min^{-1} have been followed and the operating temperature range is between 298 and 153 K.

Kochs *et al.* (1989, 1991) have devised a conductive cryostage which allows dynamic observation of freezing and freeze-drying in solutions and suspensions. Samples, no more than 30 μm thick, are placed at the edge of the cold stage, which is mounted inside a small vacuum chamber on the microscope substage. The specimen is thus only cooled from one side, with the far side remaining open to the high-vacuum atmosphere. Although there is a linear decrease in temperature from the cooling plate to the far edge of the sample which is being observed, there is adequate control of the temperature during solidification and sublimation. During freeze-drying, the sublimation interface proceeds from the outer edge of the sample toward the cooling block. The stage has been used to study freezing and drying of aqueous polymers (Fig. 8.10) and has provided further insight into the way the pathway for vapor transport is established in a sample during the drying process.

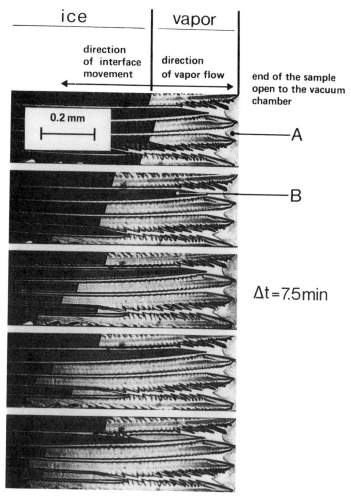

Figure 8.10. A series of five light microphotographs made at 7.5-min intervals, of the process of freeze-drying in a 20% hydroxyethyl starch solution. The ice sublimation proceeds from right to left. In some parts (A) regions of high solute composition form a significant barrier to vapor transport during drying. In other regions (B) the ice crystals sublime very slowly. [Picture courtesy of Koch *et al.* (1989).]

8.3.4. Comparison of Conductive and Convective Heat Transfer Cryomicroscope Stages

The convective cryostages have a very fast thermal response and there are negligible vertical temperature gradients in the sample even at high cooling rates. This allows large areas of the sample to be examined with the knowledge that there is little variation in the measured temperature. These advantages must be offset by a number of disadvantages, which include the inability to use optical techniques which rely on short working distance objectives and condensors, and the extreme fragility of the transparent window on the stage. Schweiwe and Korber (1982a) suggest that additional optical problems arise as the coolant streams across the light path. Both of these problems have to some extent been overcome in the convective stage designed by Cosman *et al.* (1989), which incorporates a thin flow passage to give improved optical performance and a sapphire window laminate to reduce thermal gradients.

In contrast, the conductive stages are much thinner and permit the use of high-resolution and high-numerical-aperture (N.A.) objective lenses. The thin sample sandwich is not in direct mechanical contact with the refrigerant and is thus less prone to mechanical damage due to the shear stress and the pressure variation of the refrigerant stream. If LN_2 is used as a refrigerant, vibration due to bubbling may present a problem at high-resolution imaging. There may be considerable temperature gradients across the viewing area because of the large heat flux in the lateral direction. This can lead to errors in temperature measurement and it is necessary to have thermocouple(s) in close contact with the specimen.

Misting, due to atmospheric condensation, is a problem with both types of stage and it is necessary to enclose the sample viewing area in a transparent box kept at positive pressure with a dry, inert gas. McCaa and Diller (1987) provide a simple and effective solution to this problem and recommend that after the sample is placed on the cryostage, the whole region around the objective and condensor is packed lightly with cotton wool balls. Such insulation is inexpensive and can be easily removed or replaced.

8.3.5. Thermoelectric Cold Stages

A Peltier thermoelectric element can be used as a cold stage operating at a relatively high subambient temperatures. When an electric current is passed through a suitable bimetallic junction, the temperature at one side of the junction decreases while the temperature on the other side increases. The hot and cold sides of the junction can be reversed by simply changing the polarity of the applied voltage. If heat is removed from the warm side of the junction, either by connecting it to another thermoelectric junction or by connecting it to a large heat sink, the temperature of the cool sides decreases further. Thermoelectric devices are usually stacked one upon another and the final heat sink is usually a large block of metal, which is either air or water cooled.

Most Peltier devices that have been used as cryomicroscope stages suffer from the serious disadvantage that the cooling element(s) and heat sink are located below the sample. This limits the illumination to condensors of long focal length and hence low aperture, which means that high-resolution objectives cannot be used.

Troyer (1974) fitted a thermoelectric ring coupled to a water-cooled heat sink to a standard rotating microscope stage. After setting the temperature control, the thermoelectric elements either heat or cool until the temperature sensor, a bead thermistor, reaches the nominal temperature. The temperature regulation is ±0.5 K in the range 270–313 K and the stage can be used with oil immersion objective lenses with a 1.4 N.A. condensor. Wharton and Rowland (1984) have improved the cooling performance of the Peltier stage by using a refrigerated circulator as a heat sink. Temperatures as low as 213 K have been achieved with the Peltier refrigerator set at 233 K. Temperature measurements are accurate to ±0.5 K and cooling rates of 1–2 K min^{-1} have been obtained.

8.3.6. Cryostage Temperature Measurement and Control

Accurate temperature measurement and control of rates of cooling and heating are an essential part of any cryostage assembly. It is usually only necessary to use a single sensor to monitor the sample temperature, which is used as an input to control the rate of temperature change in the specimen. In early models of cold stages, bead thermistors or thin wire copper–constantan thermocouples were used to measure the temperature. These have now been largely superseded by thin foil (5 μm) copper–constantan thermocouples, which can either be embedded within the thin glass heating and observation window, or placed in close proximity to the specimen during the final stages of sample preparation. It is important that there should be a minimal temperature difference between the sample and the measuring sensor in order to obtain accurate temperature control of the sample. Kourosh and Diller (1984) used a linear thermocouple array arranged along the primary direction of the heat flow. This array is embedded permanently in the specimen chamber of the unidirectional temperature gradient stage they have developed for solidification studies of aqueous solutions. The accuracy of temperature measurement is 0.5 K for each thermocouple and to within ±1 mm for position.

McGrath (1987) clearly delineates the various components involved in temperature control and considers they may be divided into a number of interactive subsystems. These systems are shown in diagrammatic form in Fig. 8.11. The *sample, heater, heat sink*, and *temperature sensor* have already been discussed. The *reference signal generator* creates the desired sample temperature and is part of the system that allows the operator to program a sequence of temperature changes. Signals from the temperature sensor and the reference signal generator are matched in *comparator*, which sends a signal to the *temperature controller*. The controller operation is based on a closed loop proportional feedback system and allows precise regulation of both the sample temperature and its rate of change with time. A number of methods have

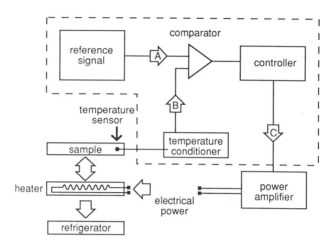

Figure 8.11. A diagram showing the basic elements of a temperature controller for a modern cryomicroscope. A, desired temperature; B, actual temperature; C, error signal. [Redrawn from McGrath (1987).]

been used to control the temperature, but a detailed discussion of the circuitry involved is beyond the scope of this book. Only an outline of the principles involved will be considered. Diller and Cravalho (1971) used an analogue control system, which generates a voltage representative of the desired linear temperature time profile. Reid (1978) used the voltage drop across an electrical–mechanical programmer to obtain the desired temperature changes. These two methods have now largely been abandoned because they are prone to long-term electrical drift. Most controllers are now based on digital electronics, which are inherently more stable and less expensive than the older systems. Such systems are described by Morris and McGrath (1981a), Schweiwe and Korber (1982a), and Diller (1982). It has also been possible to make use of microprocessor technology in the system by incorporating an ADC interface between the output of the thermocouple and the temperature controller of the cryomicroscope, which can be programmed to produce quite complex patterns of thermal changes. Most modern cryomicroscopes are now interfaced to microcomputers, which has the advantages that all the thermal functions and many of the other functions of the microscope including light levels, exposure times, and numerical data recording can be rapidly monitored and changed. Specific software may be written to control the thermal parameters and extend the versatility of the system. Morris and McGrath (1981a) obtained a temperature resolution of ± 0.05 K over a temperature range of 123–323 K, together with a wide range of reproducible non linear and discontinuous cooling and heating rates. Microcomputers have also been used by Diller (1982) and Korber *et al.* (1986), and a paper by Shah (1987) shows how an IBM PC/AT personal computer can be used to control thermal protocols in a cryomicroscope.

    The controller is at the heart of the cryomicroscope, as it compares the actual temperature–time course with that set up for a particular experiment. The output of the controller is used to drive a power amplifier, which in term sends electrical energy to the cryostage heater. As the electrical energy is transduced to thermal energy, the temperature in the stage rises, and the closed feedback loop is completed.

8.3.7. Directional Solidification

Although most cryostages have been designed for general use some have been designed for specific tasks. This applies in particular to directional solidification in which two of the governing parameters of freezing effects, the propagation velocity of the ice–water interface, and the temperature gradient can be varied and adjusted independently. In early systems (Brower *et al.*, 1981; Korber *et al.*, 1983), which are all based on conductive cold stages, the sample is placed on a stationary glass slide and freezing is initiated by decreasing the temperature at the edge of the slide. Time-dependent temperature gradients appear on the substrate and directional solidification occurs in the sample. Unfortunately, the freezing rates were variable and uncontrolled because the temperature history of the frozen sample is dependent on the thermal conditions at the glass substrate and its thermal properties. Many of these problems have been overcome by Rubinsky and Ikeda (1985), who designed a new cryomicroscope stage in which discrete samples, as distinct from thin aqueous films, are frozen in a predetermined direction and at a constant cooling rate. They made use of the Bridgeman effect, which is characterized by the creation of a well-defined, nonvanishing thermal gradient throughout the sample by placing it between two heat reservoirs of different temperatures.

This new design, which is shown in Fig. 8.12, gave independant control over the thermal gradient and ice front velocity. The sample solution or cell suspension is contained in a flat glass capillary, which is pushed at constant velocity from a warm to a cold reservoir, both of which are kept at constant temperature. After an initial

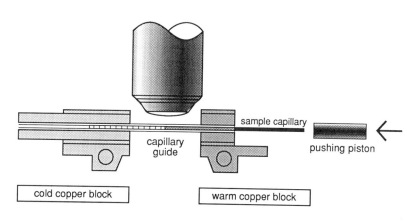

Figure 8.12. A schematic diagram of a directional solidification cryostage. [Redrawn from Beckmann *et al.* (1990).]

transient phase, the ice front grows at the same velocity as the capillary is being pushed, but in an opposite direction. The ice–liquid interface thus remains at a fixed position somewhat in the middle of the gap between the two heat reservoirs. In a later paper, Rubinsky *et al.* (1990) applied the technique to thin (1–3-mm) slices of tissue, which are placed on a microscope slide and moved at constant velocity between two heat reservoirs. The frozen samples were immediately plunged in LN_2 and subsequently examined in more detail by low-temperature scanning electron microscopy. The results from directional solidification studies clearly demonstrate that thermal gradients and ice front velocity are equally important as the cooling rate in determining the viability of frozen and thawed cells and tissues. In a recent paper, Kochs *et al.* (1991) have shown how directional solidification can be used to follow the process of vapor transport during freeze-drying of solutions of aqueous polymers.

8.3.8. Low-Temperature Microspectrofluorometry

If microspectrofluorometry is carried out at low temperatures, there is a significant improvement in both the spatial resolution and intensity of fluorescence and a decrease in the rate at which fluorescence fades. Giordano *et al.* (1978) constructed a cold stage to make high spatial resolution microspectrofluorometric measurements on organic samples at temperatures as low as 77 K. The cold stage, shown in Fig. 8.13 is a metal block cooled with liquid nitrogen inside a sealed container. The sample is deposited on a small cover glass and placed on the cold stage using zinc oxide cement to achieve good thermal contact. A second cover glass is sealed in position above the specimen and acts as a viewing window. The sample is cooled down to 80 K on a cold stage, which is thermally isolated by means of low vacuum (60 mPa/ 5×10^{-3} Torr), and is viewed by reflective optics. Figure 8.14 shows that there is a significant improvement in the excitation and emission spectra of the fluorescent probe, quinacrine mustard, when examined at 77 K and 293 K. A more elaborate cold stage has been constructed by Tiffe *et al.* (1979), which allows measurements to be made between 4 and 300 K in both reflected and transmitted light.

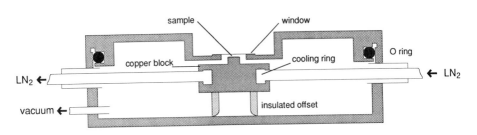

Figure 8.13. Diagram of a cold stage used for low-temperature microspectrofluorometric studies. The sample, which is examined by reflection optics, sits on the copper block below the window and next to a thermocouple. [Redrawn from Giordano *et al.* (1978).]

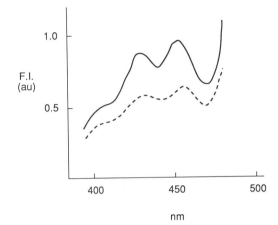

Figure 8.14. Uncorrected excitation spectra (nm) of human lymphocyte nucleus stained with quinacrine mustard. Solid line at 77 K; dotted line at 293 K. F.I.(au) = fluorescent intensity in arbitrary units. [Redrawn from Giordano *et al.* (1978).]

8.3.9. Specimen Holders

Unlike specimen preparation for low-temperature electron beam microscopy, sample preparation for low-temperature light microscopy is simple and straight-forward. The ideal samples for light cryomicroscopy are thin flat transparent sheets of cells or tissue, or thin layers of liquid suspensions with a low thermal mass. The maximum thickness should not exceed 25 μm; otherwise considerable thermal variation may develop within the thickness of the sample. A few microliters of sample may be placed directly on the thin transparent window, which incorporates the heater and temperature sensor of the cryomicroscope. The sample is covered with a second thin cover glass and sealed in position with paraffin wax or silicon grease. Alternatively, the sample is prepared as a sealed microliter size sandwich between two thin cover glasses, which is placed on to the transparent heater window. In both cases, this type of specimen produces uniformly thin (10–20 μm) aqueous layers which are ameniable for examination by this type of microscopy. Roberts *et al.* (1987) have made microliter-size sample holders by boring 0.5-mm-diameter holes in thin glass cover slips using a laser beam. The fenestrated cover slip is fixed to an intact cover slip using glass adhesive. A few microliters of sample may be placed in these microliter containers and covered with another thin glass cover slip. Samples may be examined by cryomicroscopy and the container is sufficiently small to fit onto the cold stage of a scanning electron microscope.

8.4. IMAGING AND RECORDING SYSTEMS

One of the advantages of cryomicroscopy is that it allows dynamic events such as freezing and thawing to be observed. Many of the events associated with these processes occur quite rapidly, and much effort has been put into devising suitable imaging and recording systems. Not only is it necessary to record the fast optical events that may be observed in the light microscope, but it is also necessary to store the multiplicity of thermal data that are associated with the visual images.

The thermal information may be stored conveniently on a strip recorder as this provides a continuous record as a function of time. Alternatively, the temperature signal may be digitally processed and displayed and stored via a computer terminal or as a head-up display on a videotaped image of the optical signal. This latter procedure is probably the most convenient as it allows the simultaneous recording of thermal data with the observed image at a given point in a time sequence.

There are few constraints to the variety of optical imaging techniques that may be used with modern cryomicroscopes. The optical limitations of cryomicroscopes are no different from those that occur at ambient temperature light microscopy. A survey of the literature will show that bright and dark field, phase, interference, and Nomarski optics and polarizing light microscopy have all been used with low-temperature stages. If the cryomicroscope assembly is too thick, it may prevent the use of short-working-distance condensors and objectives in association with transmitted illumination. As McGrath (1987) points out, this is primarily a problem with the condensor rather than the objectives lenses. In many instances, epi-illumination can be used as an alternative. Immersion lenses may be used, but care has to be taken over the choice of the immersion fluid. Solidification of the immersion medium may degrade its optical properties and repeated thermal stress may also damage the optical system.

A number of different ways are used to record optical images. Most cryomicroscopes have a beam splitter to allow simultaneous observation and recording. Although images continue to be recorded by still or motion picture photography, there is a move to using video recording devices and/or to inserting a photomultiplier into the optical pathway for use in association with quantitative techniques such as fluorescent microscopy. The use of electronic imaging systems allows the signal output to be fed, either to an image storage system such as a frame store or to a computerized image analysis system.

The image analysis system developed by Korber and Schweiwe and colleagues (Korber *et al.*, 1984; Schweiwe and Korber, 1983, 1984, 1987) is based on the Intellect 100 (Microconsultants, Kenley UK) coupled to an LS 11/12 image processor (Digital Equipment Corporation). Up to 50 digitized pictures (512×512 pixels) with eight grey levels may be stored and subjected to further manipulation. A detailed consideration of image analysis is outside the scope of this book, but it would include such features such as image enhancement, filtering and noise reduction, and the detection of cell outlines for area and volume measurements. The application of this type of analysis is elegantly illustrated in a recent paper by Korber (1988).

Diller and colleagues (Dietz *et al.*, 1982, 1984; Diller, 1982; Diller and Knox, 1983; Diller and Aggarwal, 1987) have developed computer-mediated techniques for the quantitative analysis of size and shape changes in cells frozen on a cryomicroscope. Fully automatic computerized image analysis is performed in which images are digitized from film negatives using a video camera and then processed off-line using a VAX 11/780 computer. In the integrated cryomicroscopy system described by Cosman *et al.* (1989), the image analysis is computer-aided using an IBM compatible PC. It is argued that this latter approach gives greater flexibility in analysis because of the high level of interaction between the computer, video system, and the operator. The analytical routines involve digitization, thresholding, and enhancement processing followed by edge detection and perimeter identification. Diller and Aggarwal (1987) have evaluated the effectiveness of different microscope systems in producing images compatible with automated computer image analysis. They found that when ice forms in the system, there is a wide variety in image size depending on the type of microscopy that is used. Cells appeared larger when imaged by differential interference contrast, smaller when examined by phase contrast, and virtually unaffected by bright field imaging. Green and Diller (1981) and Evans and Diller (1982) have developed digital computer systems for the quantitative analysis of the effects of freezing on the microcirculation in animals in which the blood volume had been tagged with a fluorescent dye.

8.5. APPLICATIONS OF CRYOMICROSCOPY

Applications of cryomicroscopy center largely on examining the responses of living tissues to near and just below subzero environments. Direct observations are coupled with an analysis of the physiological effects measured as samples are subjected to a range of thermal parameters. The main areas of interest include the following:

1. The effect of temperature and cooling rates on cell volume.
2. The effect of cooling rates on the formation of intracellular ice.
3. Cell survival after freezing and thawing.

In addition to these biological applications, cryomicroscopy is used to study the dynamics of solidification and segregation in salt solutions and hydrated organic systems, the growth of ice, the behavior of gaseous solutes, and the formation of gas bubbles and behavior of suspended particles at the freezing interface. The paper by Korber (1988) provides an excellent theoretical background to these phenomena together with a wide range of experimental results and observations. Space will not permit a detailed discussion of all the applications, but the following summary of representative examples will serve to show how widely cryomicroscopy is used to study these low-temperature phenomena.

8.5.1. Intracellular Ice Formation

Mazur *et al.* (1972) had shown that for a given cell type, high cooling rates increased the chance of intracellular ice formation. McGrath *et al.* (1975), Schweiwe and Korber (1983), Leibo *et al.* (1978), and Diller *et al.* (1976) are among the many people who have observed the formation of intracellular ice under different cooling rates. Beckmann *et al.* (1990) have used directional solidification to study the dynamics of the process as observed by cryomicroscopy. There appears to be a direct correlation between visually detected ice and cell survival, although McGrath (1987) makes the important point that such studies are limited only to those ice crystals that may be observed at the optical resolution of the light microscope.

8.5.2. Nucleation Phenomena

Cryomicroscopes have been used to study this phenomenon in a number of systems (Leibo *et al.*, 1978; Morris and McGrath, 1981; Rall *et al.*, 1983; Lipp *et al.*, 1987). The available evidence suggests that the intracellular nucleation temperature is only weakly dependent on cooling rate. Diller (1975) observed that if a cell and its surrounding media is subcooled prior to nucleation of the external ice, then slow rates of cooling must be used to avoid intracellular ice formation during freezing.

8.5.3. Gas Bubble Formation

Although small amounts of gases may become incorporated into the ice crystal lattice, they more often form bubbles as solutions are cooled. The appearance of such bubbles is shown in Fig. 8.15. Bubble formation is a consequence of gas supersaturation at the freezing interface during cooling, and there is some confusion as to the mechanism by which they are formed. Steponkus and Dowgert (1981) and Morris and McGrath (1981a) consider the bubbles to be a direct consequence of intracellular ice formation. Schweiwe and Korber (1982b, c) consider that gas bubbles may occur even when no intracellular ice is visible, and McGrath (1987) points out that small gas bubbles and ice crystals are at about the limit of resolution of the light microscope and that this may be the principal source of confusion with the problem. Korber *et al.* (1986), Lipp *et al.* (1987), and Korber (1988) had made a detailed cryomicroscopic study of the process of bubble nucleation in solutions containing known amounts of dissolved gas.

8.5.4. Solute Polarization and Redistribution During Freezing

The progressive increase in electrolyte concentration during freezing is a major cause of osmotic damage in biological tissue. Korber and colleagues—see for example Korber *et al.* (1986) and Korber (1983)—have studied solute concentration profiles formed under controlled thermal conditions. They used solutions of sodium permanganate, which Korber (1981) has shown has a phase diagram and mass diffusion properties very close to NaCl. The pink color of sodium permanganate is easily seen and may be measured spectroscopically. Figure 8.16 shows a sequence of micrographs in which the ice front advances through a solution of sodium permanganate. The

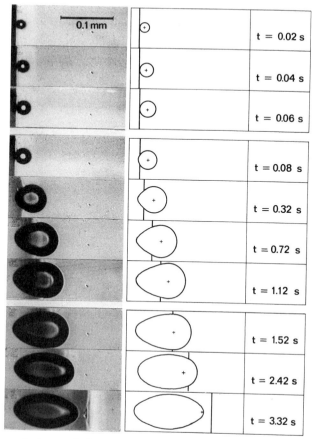

Figure 8.15. Sequence of micrographs illustrating the initial rapid growth of an air bubble ahead of an ice front, its subsequent partial entrapment and deformation, and finally encapsulation of the distorted bubble which is left behind the ice front (velocity = 37.5 μm s^{-1}). [Picture courtesy Lipp *et al.* (1987).]

progressive blackening in advance of the interface is a concentration region of solute that is not incorporated into the solid ice. A detailed analysis of the micrographs shows that solute enrichment occurs in the space between ice crystals until eutectic solidification takes place. Such studies on model systems have given valuable information on the process of solute enrichment in biological solutions.

8.5.5. Electrical Charge Separation

The differential incorporation of minute amounts of ions such as Na$^+$ and Cl$^-$ into the ice lattice, produces a charge separation and an electric potential difference between the solid and liquid phases. Steponkus *et al.* (1984) have suggested that this phenomenon may be important in cryobiology as local changes in transmembrane

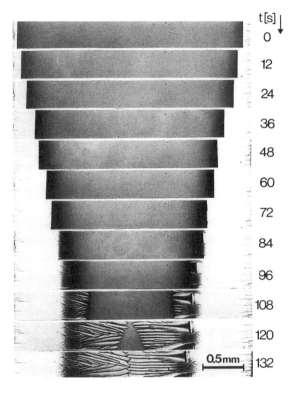

t[s]

0

12

24

36

48

60

72

84

96

108

120

0.5mm   132

Figure 8.16. Sequence of micrographs taken at uniform intervals of 12 s showing the freezing process in a 2% wt. % sodium permanganate solution. Ice crystallization starts at the edge and then advances to the center. (Photograph courtesy Ch. Korber.)

potential could affect the permeability of cell membranes. Hubel (1985) has used a cryomicroscope fitted with electrodes to measure changes in the voltage of a solution while simultaneously observing the advancing ice interface. The peak voltage appears to be a function of both cooling rate and solute concentration. During freezing and thawing, membranes are exposed to shear and bending stresses, which arise directly or indirectly from the ice matrix. Thom and Matthes (1988) have examined this phenomenon by utilizing high-frequency electric fields, which have been shown to cause cell stretching and polarization (Engelhardt, 1984). A pair of 20-μm gold electrodes are placed in a cell suspension on a microscope slide on a programmable cold stage. The samples are cooled to the measuring temperature and photographed before and after exposure to different electric field strengths. The distinct change in cell morphology can be related to variations in the thermal parameters and the presence of additives in the suspending medium.

8.5.6. Interface Interactions with Particles and Cells

During freezing, suspended particles and cells are pushed ahead of the advancing ice front. As the velocity of the solidification front increases, the distance between the suspended particles and the phase boundary decreases. At fast freezing rates, the particles or cells become engulfed by ice. Korber *et al.* (1985) and Korber (1988) have measured the critical solidification velocity at which the majority of particles are left behind the freezing front and have shown that this value is inversely proportional to the radius of the particle. Information on the velocity of solidification has an important bearing on the rate at which cells should be cooled for the maximum preservation of vitality.

8.5.7. Membrane Permeability and Exosmosis

Mazur (1963) shows that the permeability of membranes is strongly influenced by temperature. Schwartz and Diller (1983) have used the cryomicroscope to obtain information about the low-temperature permeability of cell membranes. Water transport models feature prominently in any understanding of the way biological cells react to ice formation. Shabana and McGrath (1988) have investigated the cell membrane transport characteristics of single-celled hamster ova and have calculated the mean osmotically inactive cell volume. The extracellular enrichment of impermeable solutes changes the osmotic equilibrium between the cell and its environment and the cells lose water. This in turn affects the solute concentration within the cell. The visible manifestations of these dehydration kinetics are changes in the cell volume, which may be easily observed in the cryomicroscope. A large number of cell types have been studied to determine the kinetics of cell volume reduction during freezing. The review articles by McGrath (1987) and Korber (1988) contain a useful summary of the work in this area.

8.5.8. Direct Chilling Injury or Cold Shock

Many tissues and cell types are damaged in the absence of freezing following rapid exposure to low temperatures. This phenomenon, referred to as direct chilling injury or cold shock, is reviewed by Morris and Watson (1984) and Morris (1987). Cryomicroscopy is one of the tools that have been used to study this problem. McLellan *et al.* (1984) found that the velocity of cytoplasmic streaming in higher plants was reduced as the temperature was lowered, and stopped at 283 K. No further changes were observed when the cells were subcooled to 263 K. On warming, the cytoplasm contracted and the plasmodesmata were grossly deformed. Subsequent ultrastructural studies showed severe aggregation of the actin microfilaments, which form an important component of the cellular cytoskeleton.

8.5.9. Other Applications of Cryomicroscopy

There are a number of other examples that may be used to demonstrate the useful application of cryomicroscopy to the study of low-temperature processes and phenomena. The effects of cryoprotective additives on various aspects of cell freezing have been studied by Diller (1979), Rall *et al.* (1980), and Shabana (1983). The osmotic response of cells subjected to low freezing rates have been investigated by

Knox *et al.* (1980) and Callow and McGrath (1982). Liposomes have been studied in the cryomicroscope by a number of workers as they offer a promising model system to Morris and McGrath (1981b) and McGrath (1983). Viability assays on cells that have been subjected to freeze–thaw cycles using fluorescent dyes such as fluorescine diacetate or ethidium bromide have been carried out using a cryomicroscope (Dankberg and Perdisky, 1976; Rotman and Papermaster, 1966; and Korber *et al.*, 1986). Summaries of the many applications of cryomicroscopy may be found in the papers and review article of McGrath (1987), Korber (1988), Korber *et al.* (1986), and Diller (1982).

8.6. INTERPRETATION OF CRYOMICROSCOPE IMAGES

As discussed earlier in this chapter, the optical aspects of cryomicroscopy are no different from those at ambient temperatures. There are obvious limitations of low spatial resolution in addition to variations in the quality of the picture due to different imaging modes of the optical system. Individual planar images may well contain some out-of-focus components, which may degrade the image quality and make it more difficult to determine cell boundaries clearly. The halo around images of suspended particles can make it difficult to obtain an accurate measure of the particle diameter and thence, by stereological techniques, its volume. Many of these imaging problems can be avoided by linking the output of the cryomicroscope to a computer-driven image analyzer. Cryomicroscopy of dynamic events does, however, have its own set of image peculiarities.

8.6.1. Intracellular Darkening or Flashing

Optical darkening or flashing, due to changes in the image gray level, during intracellular ice formation have been known for some time; see, for example, Luyet and Gibbs (1936). The darkening phenomenon progresses through the cell in 0.2–0.5 s, and it is generally considered that it arises from the *diffraction* of light due to the formation of many small ice crystals (Fig. 8.17). The size of the ice crystals appears to be important, as Korber *et al.* (1986) associate the darkening process, in which cells have a coarse grain structure, with the formation of relatively large ice crystals. Following direct observation with a video cryomicroscope, there is a body of evidence that contends that the darkening is a direct result of the *scattering* of light due to the formation of air bubbles rather than the formation of ice crystals. Darkening has also been observed in cells during rewarming (McGrath *et al.*, 1975; Rall *et al.*, 1980; Schweiwe and Korber, 1983; Englich *et al.*, 1986) and it is uncertain whether this is indicative of recrystallization or devitrification. It is accepted that darkening is associated with high cooling rates and a high degree of subcooling. It is difficult to resolve this question by optical means alone as the objects in question are at or near the spatial resolution of the imaging system. Those interested in this phenomenon should consult the papers by Ashahina (1961), Diller *et al.* (1976), Leibo *et al.* (1978), Brown (1980), Steponkus and Dowert (1981), Schweiwe and

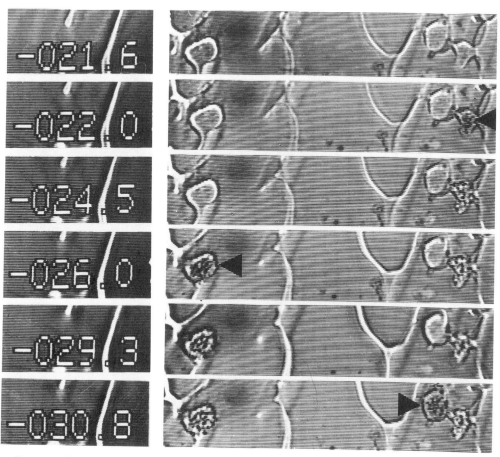

Figure 8.17. Sequence of micrographs illustrating the phenomena of darkening associated with the appearance of intracellular ice in human lymphocytes. Cooling rate 100 K min^{-1}. Arrows indicate the darkening effect associated with intracellular crystallization. Temperature ($^\circ$C) indicated at left of the picture. (Photograph courtesy Ch. Korber.)

Korber (1982b, c), and, more recently McGrath (1987). The optical darkening or flashing of cells is an important phenomenological observation because depending on the interpretation of the possible cause, different kinds of information will be available on the state of the water in the cell. These matters are discussed at some length by Steponkus and Dowert (1981).

8.6.2. Twitching

Schweiwe and Korber (1982a, b) and Englich *et al.* (1986) recognize another optical effect in cells during cooling. Twitching is believed to be associated with intracellular ice formation, although no visible ice front can be observed. It is more difficult to observe than flashing and is usually only seen in dynamic studies where the images have been recorded on movie film or by a video camera. Twitching is characterized by a brief, ca. 20-ms, movement in the whole cell without any change in either the image gray level or intracellular structure. Schweiwe and Korber (1987b) observed that in human granulocytes, twitching occurred 6–7 K above T_{HOM}, whereas darkening occurred at a much higher temperature of 23–28 K above T_{HOM}. In contrast to the darkening type of intracellular ice formation, gas bubbles have not been observed with the process of twitching. The occurrence and extent of both darkening and twitching are governed by the cooling and heating rates and by the amount of cryoprotective additives associated with the sample. In a more recent paper (Korber *et al.*, 1991), it is shown that twitching may be attributed to liquid–solid phase changes and correlates well with supplementary studies made using differential scanning calorimetry.

The only real limitation to cryomicroscopy is the lower spatial resolution of light optics. This limitation is more than balanced by the facility of the system to examine living cells and tissues and to follow dynamic events associated with freezing and thawing. If increased spatial resolution is to be the prime objective of a given cryomicroscopic investigation, it will be necessary to resort to electron optical imaging systems. We will proceed to a discussion of these methods, principally transmission and scanning electron microscopy, in the following two chapters.

8.7. SUMMARY

The sole limitation of low-temperature light microscopy, decreased spatial resolution in thin transparent specimens, is more than offset by the advantages of being able to study in real time dynamic events associated with freezing and thawing. Such processes have been observed for more than a century, initially by putting the sample, microscope, and observer in an ice box; latterly by sophisticated, computer-driven cold stages, which fit directly onto the microscope stage. There are two main types of cold stages. Convection stages are cooled by a transparent refrigerant circulating immediately below the sample, while conduction stages are cooled by heat transfer from a peripherally located low-temperature reservoir. The heating side of both types of cold stage is achieved by the electrical resistance of optically transparent metallic layers vapor deposited onto the cold stage surface. There are advantages and disadvantages to both systems. The high cooling rates that may be achieved with the convection stages must be balanced against their fragility and the difficulty of using oil immersion lenses. The thinner conductive stages present no problems to the optical pathway of the cryomicroscope, but their thermal response is slower than the rapid response time of the convective stages. Microprocessors, video recording systems, personal computers, and sophisticated software have made it much easier to control the various parameters of the cold stage and to record and analyze the images and thermal data obtained during experiments. The applications of low-temperature light microscopy fall into two main groups. Most of the studies center on the changes that may be seen during the freezing and thawing of biological samples, particularly in relation to the problems associated with cryopreservation, chilling phenomena, and ice crystal damage. A minor, but nevertheless important, group of applications is where cryomicroscopy is used to study phase changes in hydrated organic material such as emulsions and polymers.

Low-Temperature Transmission Electron Microscopy

9.1. INTRODUCTION

During the past 60 years, the transmission electron microscope (TEM) has been used to investigate the fine structure of practically all the main categories of the inorganic and organic components of our planet and—in some notable instances—a few extra-terrestrial objects. All these investigations have made use of the very high spatial resolution available in the microscope, which currently hovers around 2–3 Å. The physical events on which TEM relies are well characterized and need not concern us here except to appreciate that the processes of imaging and recording involve energy transfer to the sample and will also damage the sample, particularly those of an organic constitution. Scherzer (1949) calculated that an electron dose of $10,000 \, e^- \, nm^{-2}$ is needed to take a high-resolution image with sufficient contrast; we now know that in some cases, only $100 \, e^- \, nm^{-2}$ are needed to damage a specimen. The images of most organic and biological material examined in the TEM have a very low amplitude contrast, which can only be ameliorated by staining with heavy metals. The presence of water in almost all biological material was solved by the draconian expedient of removing it and then replacing it with plastic.

The difficulties in sample preparation paled into insignificance as a plethora of new information became available about the fine structure of ourselves and of our environment. The specters of radiation damage, chemical fixation, and dehydration were conveniently forgotten. New structural elements were discovered and the TEM had a virtual monopoly on characterizing the structural conformation of macromolecules of biological significance. Everyone must have realized that all this new information, most of which by modern standards is at medium to low resolution, was being obtained from altered and damaged material. Some people must have dreamed about the possibility of examining material in its natural and unaltered state. A few enlightened people realized that low-temperature microscopy might be a way to achieve this goal. This chapter will discuss the techniques and advantages of low-temperature transmission electron microscopy (LTTEM), from the slow beginnings in the 1960s and 1970s to the acceleration in interest that has occurred in the last ten years.

It is important to define what is meant by LTTEM. It is simply the examination of specimens maintained at subzero temperatures, almost always well below 130 K, in a TEM either as thin suspensions or sections. In many instances the samples have themselves been prepared by low-temperature techniques and retain their natural water content albeit as crystalline or vitreous ice. In other cases, such as organic crystals, the specimens have simply been cooled down in the TEM prior to examination.

The early efforts at low-temperature TEM were developed in an attempt to deal with problems of contamination, which in turn led to studies on the structure of ice crystals (Leisegang, 1954; Honjo *et al.*, 1956; Fernandez-Moran, 1960). Fernandez-Moran was an avid proponent of the direct observation of frozen hydrated samples and in the mid-1960s constructed a cryo-electron-microscope with superconducting lenses at liquid helium temperatures. A summary of these pioneering studies may be found in two papers by Fernandez-Moran (1966, 1985), in which a wide range of inorganic, organic, and biological materials were examined at 4.2 K and at a spatial resolution that approached 1 nm. Equally important, Fernandez-Moran was able to show that specimen damage was reduced at low temperatures.

By the mid-1960s everything seemed to be set for an upsurge in low-temperature TEM, but for the next ten years very little happened. Superconducting electron microscopes were expensive, unreliable, difficult to operate, and few and far between. Other less complicated liquid-helium and liquid-nitrogen stages were developed, but their use was not very widespread. Low-temperature specimen preparation was still in its infancy. The full potential of low-temperature TEM only became apparent from the work of Glaeser and his colleagues at Berkeley. Glaeser (1971, 1975) quantified the extent of beam damage in the TEM and showed that these effects could be ameliorated at low temperatures. At about the same time Unwin and Henderson (1975) obtained 1-nm spatial resolution images of an unstained bacterial membrane embedded in glucose at room temperature. Unwin and Henderson considered that the hydroxyl groups in the glucose would substitute for water and form hydrogen bonds with the protein and preserve the structural conformation. The disadvantage to this approach centered on the uncertainty about how much water is essential for the structural conformation of the protein and how much of this water remains around a sample embedded by the glucose method. The images that were obtained were of low contrast, even when strongly defocused, because there is little difference between the scattering density of glucose and protein. The problem of image contrast was overcome by using a combination of electron diffraction patterns and micrographs taken at room temperature using low dose techniques ($100 \ e^- \ nm^{-2}$). Image processing was used to recover the information from these low-contrast images.

Taylor and Glaeser (1974) obtained high-resolution electron diffraction patterns of catalase embedded in ice, and two years later (Taylor and Glaeser, 1976) obtained images of unstained catalase crystals embedded in ice in a fully hydrated state.*

---

*An alternative approach to examining aqueous specimens is to use a hydration chamber in which the water vapor can be regulated around the region of the specimen (Matricardi *et al.*, 1972; Parsons, 1974). The images which have been obtained have poor contrast because of the large amount of scattering by water vapor. In addition, there are unresolved problems of radiation damage and the perceived movement of components at high magnifications.

In the few instances where samples were being examined in a frozen-hydrated state it was assumed that the ice surrounding the samples was in a crystalline form. The conventional wisdom at this time stated that it was impossible to vitrify water from the liquid because none of the cooling techniques were fast enough. This myth was shattered, first by Brugeller and Meyer (1980), who showed that very small droplets of water could be vitrified, and then by Dubochet and McDowell (1981), who vitrified thin water layers spread on electron microscope grids by rapid immersion into liquefied ethane.

Concurrent with these important advances in specimen preparation, more efficient (and less expensive) cold stages became available for different microscopes. Although liquid helium stages were not abandoned entirely, much more emphasis was placed on liquid-nitrogen-cooled stages, which were relatively simple to construct and use. In addition, an increasing body of evidence suggested that liquid-nitrogen temperatures would provide the means of answering most of the questions being posed by LTTEM. The inherent low contrast of unstained organic samples embedded in water (ice) was remedied when it was shown by Adrian *et al.* (1984), following the much earlier work of Erickson and Klug (1971), that phase contrast images of underfocused specimens would overcome this difficulty.

Having briefly set the historical perspective and the *raison d'être* of low-temperature TEM, we will now consider the technical aspects of the procedures, the problems it can (and cannot) solve, and some of its applications. Most low-temperature TEM is concerned with the examination of hydrated biological material and organic samples, for it is here that the problems of maintaining the native state and countering the effects of beam damage are most exacting. The examination of vitrified thin-film suspensions is now a matter of routine, although the production and examination of fully hydrated thin sections still remains a specialized procedure. The advantages of cryoelectron microscopy of unfixed and unstained hydrated samples are self evident. It permits both high-resolution substructure and low-resolution suprastructural investigations to be carried out. It allows dynamic reactions to be studied and investigations to be made on the natural distribution of particles in hydrated suspension. Low-temperature TEM also has much broader applications beyond biological studies. It is being used to study the structure of solid gases, ferromagnetic phases of rare earths, including those involved in the manufacture of high-temperature superconductors, low-temperature phases in metals, organic polymers, and materials or defined chemical and hydrated states, and the products of electron beam lithography. Low-temperature TEM is a relatively new discipline and as such has not resulted in many comprehensive reviews. The noteworthy exceptions are the excellent review by Dubochet *et al.* (1988), written before the Diaspora of the thriving group of cryomicroscopists at the EMBL at Heidelberg and the earlier review by Chiu (1986).

9.2. SPECIMEN PREPARATION

9.2.1. Introduction

The two central features of sample preparation for low-temperature TEM are (1) to produce specimens thin enough—i.e., 100–200 nm, to transmit a substantial portion of the illuminating beam of electrons; (2) to vitrify any aqueous phases in order to prevent ice crystal damage. In the case of bulk samples it will be necessary to cut frozen sections that are either dried before being examined at low temperatures or examined in the frozen hydrated state. If the objects are small enough—e.g., macromolcules, membrane fragments, viruses, particles, polymers—they may be incorporated into a thin liquid suspension which is rapidly frozen prior to being examined in the TEM.

9.2.2. Frozen Sections

The process of preparing thin frozen sections has been discussed in considerable detail in Chapter 4. In spite of a few notable exceptions, it is not an easy and routine procedure to cut ultrathin cryosections, particularly if the water in the sections is to remain in a fully vitrified state throughout preparation and examination. Vitrification of uncryoprotected bulk samples still remains a problem, although the examination of frozen sections in which the fully hydrated state is assured is now much easier thanks to the work of Euseman *et al.* (1982). These workers demonstrated that it was possible to monitor the degree of hydration in a section by taking optical density readings from images at different areas off and on the regions in question, before and after drying. Using these techniques, Dubochet *et al.* (1983) were able to show that the frozen sections they obtained from *E. coli* and *Staph. aureus* were fully hydrated. The mass thickness of the sections was 200 mg m^{-2} and the mass thickness of the space between bacteria decreased by 20% during drying. Although the contrast of the biological material increases during drying, and in most cases the sections showed considerable shrinkage, these changes are subjective. Changes in mass thickness are the only reliable way to assess whether a frozen section is (was) fully hydrated. Egerton and Chang (1987) review the various methods that can be used to derive the thickness of electron microscope specimens held at ambient temperatures by using the transmitted energy loss spectrum. The methods, which use a low radiation dose, have an accuracy of ±10% in the 10–50-nm thickness range. There seems to be no reason why these techniques could not also be applied to frozen-hydrated sections.

If frozen-hydrated sections are successfully obtained it is important that they remain flat and in good thermal contact with the underlying specimen support. Large sections should be picked up on bare grids and flattened with a polished weight to ensure good contact. These procedures are not made any easier by the electrostatic buildup in the necessarily dry environment surrounding the sample and knife. Ultrathin sections are usually only satisfactorily cut from small block faces. In this case it will be necessary either to use small-mesh (400–700) grids or to use grids coated with a thin carbon or plastic support film, which remains flexible at low temperatures. The sections must be in firm mechanical contact with the grid to ensure good thermal conductance and that they do not become dislodged during transfer and examination. Sections that are freeze-dried outside the microscope are usually coated with a thin layer of evaporated carbon to enhance the thermomechanical properties of the section/grid sandwich. No satisfactory method has been found to coat frozen hydrated sections without causing some degree of surface sublimation, although Jakubowski and Mende (1991) have devised a low-temperature method to carbon coat thin layers of amorphous ice without inducing any phase changes or sublimation.

McDowall *et al.* (1983) demonstrated that it is possible to obtain images of unstained, unfixed, noncryoprotected sections of vitrified water (Fig. 9.1). There are, however, few examples of such totally unsullied material, which is in sharp contrast to the much larger amount of information from frozen-hydrated material that has been chemically treated, i.e., cryoprotected, fixed, and/or stained, prior to rapid cooling and sectioning. The images obtained from such material are virtually identical to the images that may be obtained using more conventional means, and there seems to be little advantage in including low-temperature methods in the preparation protocol. The possible exception may be found in the images from frozen-hydrated

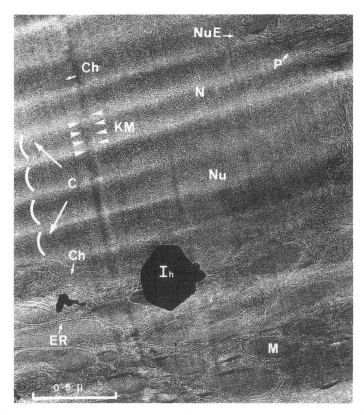

Figure 9.1. Cryo-transmission electron micrograph of a frozen-hydrated section of rat liver. The 180-nm-thick section was cut at 120 K from untreated material vitrified by impact freezing. N, nucleus; Nu, nucleolus; NuE, nuclear envelope; P, nuclear pore; Ch, chromatin; M, mitochondrion; ER, endoplasmic reticulum; V, vacuole; I_h, contaminating hexagonal ice crystal; KM, knife marks; C, chatter. Note that although the general contrast of the picture is low, the main features of the cell can be distinguished. [Picture courtesy McDowall *et al.* (1983).]

sections which have been cut from vitrified material following the addition only of cryoprotective agents such as 15% glucose or 20% sucrose (McDowall, 1986) (Fig. 9.2), even though such materials are acknowledged to give rise to problems of specimen shrinkage. The image quality is no better than that which can be obtained from properly freeze-substituted material although occasionally new structural interpretations have been possible when these images are compared with unstained frozen-hydrated images. (Dubochet *et al.*, 1983). This general lack of new information from unstained but cryoprotected frozen-hydrated sections of bulk material has discouraged many people from preparing and looking at such types of specimen in the TEM. The difficulties associated with ultracryomicrotomy, the presence of severe cutting artifacts, the low yield of good images, and the even greater difficulty of maintaining the sections in a fully frozen-hydrated state, have diminished the popularity of this type of preparation for ultrastructural studies. The reluctance to use ultrathin frozen sections for structural studies does not lie solely with the sectioning process but also with our inability to vitrify large pieces of hydrated material without the addition of cryoprotective chemicals. The high-pressure freezing method described elsewhere in this book offers the only way forward.

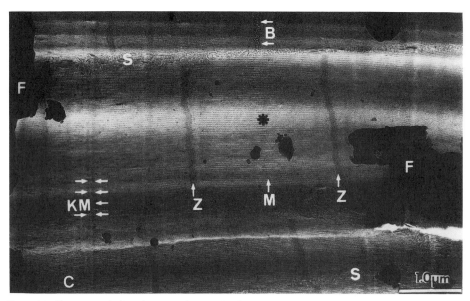

Figure 9.2. Cryo-transmission electron micrograph of a frozen-hydrated vitrified section of insect flight muscle fibrils in the relaxed state. The material is unfixed and unstained but cryoprotected with 15% sucrose. The section thickness is 80–150 nm and was cut in the vertical direction at ca. 113 K. Z, Z-band; M, M-band; S, remains of the sarcoplasmic reticulum; F, remains of an ice flake; KM, knife marks; C, crevasse; B, beam-induced bubbling. The state of ice was tested by selected area electron diffraction. [Picture courtesy McDowall *et al.* (1984).]

9.2.3. Thin Liquid Suspensions

In many ways these are the ideal samples for low-temperature TEM, and although they are much easier to prepare than frozen sections, considerable attention has been paid to the exprimental detail. Ideally, the frozen-hydrated specimen should be embedded in an ice layer thin enough to give sufficient contrast and thick enough to ensure that the samples are completely encapsulated. The effect of a thick layer of ice is analogous to having a thick support film which reduces the visibility of the specimen.

Water has a high surface tension and its natural tendency is to form spheres with a low surface-to-volume ratio rather than thin confluent layers, which have an unfavorable surface-to-volume ratio. But aqueous solutions can be persuaded to form thin layers if the substrate is sufficiently hydrophilic and the wetting angle approaches zero. Glaeser et al. (1991) have made a comprehensive study of the interfacial energies and surface tension forces involved in the preparation of thin crystals of biological macromolecules for electron microscopy. High-resolution studies require samples that are flat to at least 1° and perhaps to 0.2° over areas as large a square micrometer. Glaeser and colleagues found that the initial interface reaction involves adsorption of the specimen to the air–water interface rather than the adsorption to the substrate. Interfacial surface tension forces and an apparent repulsive interaction between the specimen and the hydrophilic carbon support film seem to be the major factors determining sample flatness. Although these findings provide a better insight into the biophysics of sample–substrate interaction, the wide variation in the composition and charge of the two components does not permit a single preparative method to be prescribed which would routinely ensure flat specimens. However, as a generalization, negatively charged surfactants provide some amelioration of the problem.

The preparation of thin vitrified layers involves a number of discrete steps, which include the formation of thin hydrophilic support films, the production of a thin aqueous layers, and rapid cooling procedures.

9.2.3.1. Thin Hydrophilic Support Films

The most popular support film is a thin ca. 10–20-nm layer of evaporated carbon although thin polyimide films have been used (Talmon and Miller, 1978). Polyimide films have both advantages and disadvantages. Falls et al. (1982) showed that such films are wetted by both lipidophilic and aqueous layers, but as we shall discuss later, the ice–plastic interface is a region of enhanced radiation damage in the TEM of frozen hydrated layers. For this reason carbon films are now considered the most reliable support film.

Carbon support films are made by depositing a thin layer of evaporated carbon onto a clean surface such as freshly cleaved mice. The evaporated film is then floated off and picked up on a TEM specimen grid with a 300–400 mesh size. Films may

also be made by evaporating a thin layer of carbon onto a collodion-coated EM grid
and then dissolving away the plastic with a suitable solvent.

Considerable success has been achieved by using holey carbon films, originally
introduced by Sjostrand (1956), which can be prepared by the methods of Fukami
and Adachi (1965), Fukami (1972), and Adrian *et al.* (1984). A thin plastic film is
first made in a very humid atmosphere, and in order to obtain clean holes the film
thickness should be much smaller that the hole size, which is typically from 100 nm
to 10 μm. The plastic film is carbon coated and the underlying plastic dissolved away.
Papers by Murray and Ward (1987), Vinson (1987b), Bellaire *et al.* (1988), and
Glaeser *et al.* (1991) give details of how these films may be prepared specifically for
cryo TEM. The advantage of the holey carbon film is that it substantially decreases
the amount of phase contrast from the support film at moderate and low resolution
(Fig. 9.3).

Freshly made carbon films are usually slightly hydrophobic and may be made
hydrophilic in a number of different ways including glow discharge in air (Dubochet
et al., 1982a,b), water vapor (Glaeser *et al.*, 1991), or an alkyamine atmosphere
(Lepault *et al.*, 1983). Other procedures which have been used include treatment with
surfactants, solvent cleaning or coating with a thin layer of silicon monoxide (Taylor

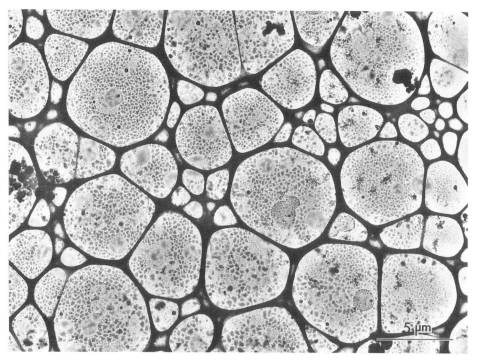

Figure 9.3. Cryo-transmission electron micrograph of purple bacteria membrane embedded in a thin
discontinuous film of I_H on a holey carbon film. Note the varying sizes of the holes. [Picture courtesy of
Lepault *et al.* (1983).]

and Glaeser, 1973). Alternatively, it may be necessary to make the support films more strongly hydrophobic and the carbon film aging techniques proposed by Ceska and Henderson (1990) and Henderson (1990) appear to change the surface properties. All types of carbon films should be used shortly after they are prepared.

9.2.3.2. Production and Maintenance of Thin Aqueous Layers

Thin frozen films may be made either by subliming away ice layers inside the TEM (Heide and Grund, 1974; Taylor *et al.*, 1976) or by first producing thin aqueous layers which are quench cooled. The first method is very difficult to control (the specimens are inevitably freeze-dried) and most studies now rely on the second approach. Immediately prior to quench cooling, a small drop of the aqueous suspension, ca. 2–5 μl is placed on the grid and the bulk of the liquid drained away by touching the edge of the grid with a point of filter paper. Although this procedure produces thin aqueous films, a large proportion of any suspended material can be sucked up into the filter paper. Toyoshima (1989) has overcome this problem by applying suspended specimens to one side of a holey carbon grid and blotting away the solution from the other side. Aqueous layers may be thinned further by evaporation under controlled conditions. The thickness of the aqueous layer can only be judged from experience and involves observing the interference pattern changes at the air–liquid interface by phase contrast microscopy, as shown in Fig. 9.4. Capillary and hydrodynamic forces and even pressure effects can produce areas of uneven thickness within films during the draining process (Falls *et al.*, 1982).

Care has to be taken that the evaporation of water from the thin suspension does not result in changes in the concentration, ionic strength, osmotic properties, pH, and even the temperature of the sample. It is well known that even quite small changes in the composition of the aqueous environment can affect the conformation and structure of proteins. Temperature and concentration gradients in the evaporating sample may affect surface tension and partial segregation of components within the aqueous suspension. Bellare *et al.* (1988) consider that there is an increased chance of artifacts forming in a sample the closer it or any of its components approach the saturation condition. This may lead to the appearance of salt and liquid crystals in liquid solutions and particle agglomeration in colloidal dispersions. Trinick and Cooper (1990) consider that the salts can become sufficiently concentrated to dissolve muscle thick filaments prior to freezing. There is insufficient information to predict what is occurring during the evaporation of thin liquid films, and unless great care is taken, the microstructure of rapidly cooled liquid suspensions may not be representative of the structure of the bulk liquid. The existence of long-range interactions with the surface and possible concentration effects brought about by drying prior to freezing are rarely investigated.

Because evaporation occurs so quickly and is very difficult to avoid, a number of workers have been sufficiently concerned to construct special equipment to ensure that the thinning process occurs under controlled environmental conditions. The amount of evaporation may be measured quantitatively by comparing the mass of the thin vitrified layer, measured as a function of the optical density of the microscope

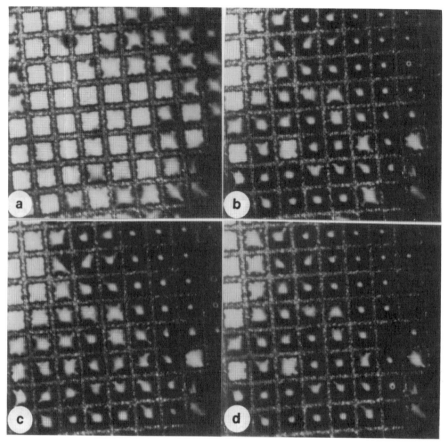

Figure 9.4. Light micrograph of the successive changes (a)–(d) seen during the draining of a water droplet on a bare grid. The thin films which span the grid openings become evident as white rings at the center of a dark background as the thinning proceeds. [Picture courtesy of Bellaire *et al.* (1988).]

image, before and after freeze-drying (Eusman *et al.*, 1982; Dubochet *et al.*, 1983). Alternatively, the amount of evaporation may be measured by comparing the contrast of known particles such as polystyrene spheres in a known concentration of high-density salt solution such as caesium chloride, or by using small lipid vesicles as sensitive osmometers (Dubochet *et al.*, 1988).

Taylor and Glaeser (1976), Rachel *et al.* (1986), Murray and Ward (1987b), and Bellaire *et al.* (1988) describe systems that ensure that a liquid or a liquid particle system is maintained in its original state while being prepared prior to rapid cooling. Chang *et al.* (1985) have suggested a novel way of obtaining thin films which avoid the problems of uncontrolled evaporation. They sandwiched the sample between a monolayer of fatty acid on a polylysine coated carbon coated grid. As the fatty acid layer beneath became attached to the specimen grid, it squeezed out excess liquid and allowed a thin film of water to form.

A number of studies have been carried out that show that phospholipids are useful in producing thin aqueous layers. Frederik *et al.* (1989) examined catalase crystals in thin films of dimyristoyl phosphatidyl choline (DMPC) and found that the water content of the film drops from 99% to 33% during the draining process needed to thin the aqueous layer. Ward *et al.* (1990) describe a technique by which paracrystals of actin/tropomyosin filaments may be incorporated into unsupported lipid monolayers spread across hydrophilic holey carbon films. Jakubowski and Mende (1991) find that the incorporation of lipid monolayers (DMPC) into water films prolongs the film thinning process carried out in a controlled environment and permits the production of ultrathin amorphous ice films for cryo-TEM.

Evaporation can also be reduced by sandwiching the sample between two support films as shown in Fig. 9.5 (Taylor and Glaeser, 1973; Talmon *et al.*, 1979; Jaffe

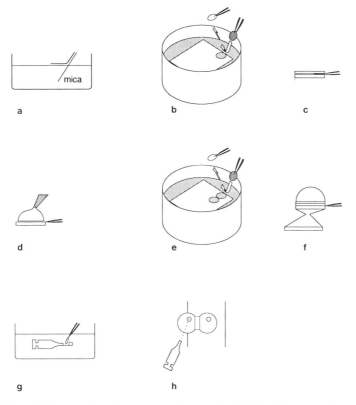

Figure 9.5. Outline of the procedure to prepare a frozen-hydrated thin film suspension encapsulated between two thin carbon films. A carbon film deposited on a cleaved mica surface is (a) floated of on water, (b) picked up on an EM grid, and (c) allowed to dry. (d) A suspension of the specimen is applied to the grid and allowed to settle for a few minutes. The grids are washed with distilled water and (e) a second carbon layer is picked up on top of the sample sitting on the first carbon film. (f) Excess water is blotted away and the aqueous specimen is now encapsulated between two carbon films. The grid is quench cooled, (g) loaded onto the specimen holder under liquid nitrogen, and (h) transferred to the cold stage of the microscope. [Redrawn from Taylor and Glaeser (1975).]

and Glaeser, 1984). This method is particularly useful when the sample adheres strongly to the carbon film. An early approach to producing thin aqueous samples was to spray fine droplets (20–50 μm) onto a carbon-coated grid as it descended into a cryogen (Dubochet et al., 1982b). If the surface support is perfectly hydrophilic, the drops spread rapidly across the surface and form thin, but not confluent, layers. In practice, the frozen sample contains large droplets and thin and thick films. The disadvantage of this approach is that the shear rates induced by passing a jet through a small aperture may be sufficient either to impose a preferred orientation on the sample or to change its microstructure.

Thin aqueous layers may also be formed over the surface of bare 2–400 mesh grids devoid of any support film—the so-called bare grid method (Adrian et al., 1984). Because surface tension effects and evaporation conspire to break these films, a small amount of surfactant is usually included in the aqueous layer to increase the stability. The excess water is removed with a fine pipette momentarily before quench cooling. Dubochet et al. (1985) consider that blotting is less critical with these types of preparation as the thin liquid layer is remarkably self-stabilizing. Although this type of suspension should provide the best specimen for cryo-TEM, it is still susceptible to concentration-induced artifacts due to partial drying. In addition, although the absence of a support film will enhance the contrast of the specimen it also makes it more difficult to detect the contrast function in the optical diffractogram of the image (Chiu, 1986). A modification to the bare grid method is seen in the procedures proposed by Milligan and Flicker (1987). These workers absorbed myofibril filaments and F-actin to carbon coated holey films which had been glow discharged in an atmosphere of alkylamine. The grids were washed with appropriate buffers and covered with 2–3 drops of a solution containing a myosin subfraction. After two minutes, the grids were blotted and quench cooled. This procedure is a satisfactory method in conditions where the specimen is attached firmly to the carbon substrate. In an earlier paper (Milligan et al., 1984) it was found that specimen absorption to the carbon film is enhanced if the films are first treated with cytochrome c (ca. 1 mg/ ml) or sodium deoxycholate (ca. 0.01%).

9.2.3.3. Time-Resolved, Stopped-Flow Procedures

We should not forget that low temperatures are an effective way of stopping dynamic processes and chemical reactions. This advantage has been capitalized on by Bellaire et al. (1988), Talmon et al. (1990), and Chestnut et al. (1991), who have devised a new technique of time-resolved cryo-TEM to study changes in microstructure which occur during phase transitions and chemical reactions. The sample is prepared on an EM grid maintained at a fixed temperature in a controlled environment. The dynamic process is induced in situ by changes in pH, temperature, irradiance, salt concentration, or by rapidly mixing different reactants. The reaction is allowed to run for a set time, terminated by quench cooling, and the products examined by cryo-TEM. An example of such a reaction is shown in Fig. 9.6. The same grid in situ method has been used by Lepault et al. (1991) to follow the induction of G-actin polymerization. While adopting the same general principle, a different

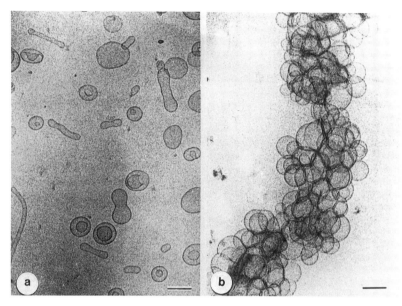

Figure 9.6. Time-resolved cryo-transmission electron micrographs of liposome dispersions at pH 9.9, (a) before and (b) after 5 s mixing with a bicine buffer at pH 8.3. The drop in pH causes the liposomes to aggregate. Bar = 100 nm. [Picture courtesy of Talmon *et al.* (1990).]

approach has been adopted by Pollard *et al.* (1990), who have built a machine capable of freezing within 5 ms the transient intermediates formed in rapid biochemical reactions. The frozen mixtures cannot be viewed directly and require further processing by freeze-fracture replication.

9.2.3.4. Rapid-Cooling Procedure

The rationale and operational procedures for rapid cooling have been discussed in Chapter 3 and the preparation of suitable thin, frozen-hydrated films is consequent on following these now well-defined methods. In practice, the carbon-coated or bare EM grid is held at the tip of fine anticapillary forceps, which are attached to some form of mechanical device which is capable of rapidly propelling the grid into the cryogen. The grid holding-propelling system should be kept in a high-humidity environment or integrated into one of the controlled environmental vitrification systems such as that proposed by Murray and Ward (1987), Jeng *et al.* (1988), Bellaire *et al.* (1988), Trinick and Cooper (1990), and Jakubowski and Mende (1991). The device designed by Bellaire and colleagues is shown in Fig. 9.7. A suitable cryogen is prepared and placed in a container below the poised EM grid. A number of different cryogens have been used, but liquefied ethane is now used almost exclusively as it gives the most satisfactory and consistent results. Although propane has a slightly lower M.P. (85 K), it becomes rather viscous at these low temperatures

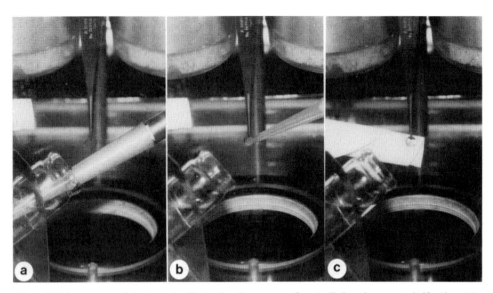

Figure 9.7. A sequence of photographs illustrating the process of controlled environment vitrification. (a) A small aliquot of the sample is withdrawn from a container in the chamber and (b) applied to the surface of an EM grid held at the end of fine forceps. (c) Excess liquid is removed from the grid by touching the back of the grid with a piece of filter paper and the thin film formed by touching the front of the grid with a second piece of filter paper. The thinning process is observed through a binocular microscope. When the film is judged to be thin enough, the grid is rapidly propelled into liquid ethane held in a container below the central aperture shown on the pictures. [Photographs courtesy of Bellaire *et al.* (1988).]

and there is an increased risk of damaging the delicate EM grid. It is most important to use dry ethane, as Dubochet *et al.* (1988) have shown that a small amount of cryogen remains on the specimen when it is transferred into the microscope and that any moisture will be evaporated as the cryogen is removed by sublimation. A fresh batch of ethane is prepared for each rapid cooling and is allowed to subcool in LN_2 to the point at which a semisolid skin begins to form on the surface. At this point, a small drop of the suspension to be examined is placed on the grid. The concentration of the suspension should be sufficiently high to ensure that enough particles are present in the small volume of the film. Any excess liquid is removed by blotting or pipetting. When the layer is considered to be thin enough—i.e., 100–300 nm—and this is probably the most critical and subjective part of the whole process—the grid is ready to be plunged into the cryogen. Various methods have been devised to judge when the film is sufficiently thin, although they are all subjective as no direct measurement of film thickness is made. The simple and pragmatic approach adopted by Adrian *et al.* (1984) is that the film is thin enough when about half the grid squares are filled with liquid as seen by low-power binocular microscopy. Jaffe and Glaeser (1984) judge the correct thickness of the aqueous layer by observing the grid surface by reflected light. A shiny surface indicates that the layer would be too thick, but that satisfactory results would be obtained from a purple-orange reflection.

Another method is to observe the grids by light microscopy and begin the quench-cooling process when some of the films begin to rupture (Heide and Zeitler, 1985). The films should be sufficiently thick to enclose the particles, but not so thick as to obscure the embedded material.

The various high-speed plunging devices have been described earlier (Chapter 3, Section 3.6.2). In the improved sample preparation techniques described by Bellaire et al. (1988), a shutter over the cryogen is opened simultaneously with the release of the plunging device. This prevents the thin liquid suspension from pre-cooling below the ice crystallization point prior to rapid cooling. The speed of entry (100 m s^{-1}), the thinness of the liquid layer (100–300 nm), and the temperature of the cryogen (90 K) results in vitrification of substantial parts of the liquid sample. Following quench cooling, the grid is raised just above the surface of the cryogen to allow any excess ethane to drain away and then immediately placed into the surrounding LN$_2$ where the remaining ethane solidifies. The solidified ethane forms a protective layer around the vitrified sample. Jakubowski and Mende (1991) describe a procedure to carbon coat both sides of ultrathin (less than 20 nm) amorphous ice films prior to examination in the TEM. The evaporated carbon films are claimed to greatly improve the stability of self-supported vitreous ice films during exposure to the electron beam. The grid may now be loaded onto the precooled microscope specimen holder under LN$_2$ for examination in the microscope.

9.3. COLD STAGE DESIGN FOR THE TRANSMISSION ELECTRON MICROSCOPE

A large number of systems have been designed to enable spcimens to be examined at low temperatures in the TEM. They range from simple modifications of standard specimen holders to the complex cold stages which form an integral part of the LHe$_2$-cooled superconducting lenses which have been designed for some microscopes. It is only possible here to give a general account of the design requirements and operation of cold stages as they effect the end user. The important design parameters of heat transfer, input and balance, physical properties of materials, and mechanical stability are best considered elsewhere, and the comprehensive review papers by Heide (1982), Turner et al. (1989), and Downing and Chiu (1990) provide a good starting point. In addition to discussing the devices that are necessary to keep the sample cold, it will also be necessary to consider the devices that enable the uncontaminated, frozen samples to be transferred into the microscope and onto the cold stage.

9.3.1. The Basic Requirements for Low-Temperature TEM Stages

The general and specific requirements for equipment have been long recognized and the major problems identified. The sample must be kept cold and in a stable position relative to the pole piece, and residual gases must be prevented from condensing onto the sample surface. Sample cooling may be achieved either by continuous flow cryostats or bath cryostats in which the coolant is carried close to the sample. Alternatively, the specimen may be cooled by a flexible heat conductor cooled by a cryostat some distance from the sample. In both cases the principal mechanism of heat transfer from the sample is by conduction, which is proportional to the temperature gradient, the contact area between the sample and the cold sink, and the thermal conductivity of the materials used in construction. Radiation from the warm regions surrounding the cold stage, condensation of residual gas molecules onto the sample, and the energy input from the electron beam are the principal sources of heat transferred into the sample. Thermal energy from warmer regions of the microscope is transferred to the cold specimen via molecular conduction of residual gases in the microscope column as they bounce from warm to cold surfaces. This form of energy transfer decreases linearly with decreasing pressure. Direct radiant heat transfer from the warm surfaces surrounding the specimen area to the cold sample is quite small provided the sample is well shielded by cold anticontamination plates. Dubochet *et al.* (1988) comment on the fact that vitrified samples have sometimes been observed to crystallize during the rapid condensation of a water layer. The water molecules are thought to have come from the warmer parts of the microscope column and their condensation contributes more energy to the sample than can be removed by conduction through the sample and the specimen holder. This phenomenon emphasizes the importance of cold anticontamination plates around the specimen area.

The increased interest and usage of low-temperature TEM prompted the Council of the Electron Microscope Society of America to establish a committee on Cryoelectron Microscope Technology to assess the requirements for equipment for this type of work. Their far-reaching report (Chiu *et al.*, 1988), which has been extended by Downing and Chiu (1990), contains a series of design specifications for cryoelctron microscopy in both the physical and biological sciences. A brief summary of their recommendations for cold stage design is given below.

For biological and chemical studies we have the following recommendations:

1. A temperature of 123 K is recommended, which should reach final stability within 15 min of loading a new sample. Variable temperatures, i.e., heating and cooling, would be useful, although temperature stability is more important. The cooling Dewar (LN_2) should only require filling every 3 h.

2. Chiu and colleagues consider the stage in relation to the overall modulation transfer function of the electron microscope, which is particularly important when considering the low contrast of biological specimens examined at high resolution. Under these conditions the maximum amount of drift and planar vibration should be less than 0.1 nm during a 2 s exposure period.

3. There should be a single tilt axis of $+60°$ to $-60°$, while still maintaining high spatial resolution.

4. The eucentricity of the stage should be limited to a movement of 1 μm throughout the entire tilt range.

5. There should be a suitable cryotransfer device to enable precooled samples to be loaded into the microscope at low (123 K) temperatures with minimal contamination.

6. Contamination should be limited to no more than a 2 nm h^{-1} ice layer. This would require cold anticontamination plates in the region of the specimen and a vacuum pressure in the microscope column of 130 nPa (10^{-9} Torr).

We also include, for completeness, the following recommendations for analytical studies of biological specimens, although these recommendations are directed towards scanning beam instruments and x-ray microanalyzers:

1. The temperature should be as close to 77 K (LN$_2$) as possible. Heating and cooling are important, particularly in situations where samples have to be outgassed before cooling and/or freeze dried inside the microscope.

2. The short-term stability (1 s) should be better than 0.5 nm; the long-term stability (1 h) should be of the order of 10 nm. For scanning electron microscope (SEM) studies the resolution should be in the 1–3 nm range.

3. There should be a 30° tilt toward the x-ray detector(s).

4. For scaning transmission electron microscope (STEM) imaging the X and Y movements should cover a 3-mm EM grid. For SEM these movements should be increased to 10 mm.

5. Cryotransfer devices are most important with sample temperatures maintained at 123 K in a clean environment.

6. Contamination should be limited to 1 nm h^{-1}.

For structural physical and material sciences applications we have the following recomendations:

1. The temperature should be as close to 4 K (LHe$_2$) as possible and certainly less than 10 K. It is important that this temperature is measured precisely and controlled heating and cooling are important. The temperature stability should be within ±0.5 K.

2. The resolution should be of the order of 0.3 nm over 1 s, although for many studies, 1 nm over 5 s is acceptable.

3. A single tilt angle of 60° is recommended.

4. The X and Y movements should be sufficient to observe 2 mm of a 3-mm EM grid.

5. It would be useful to have a cryotransfer device operating at 4 K.

6. Contamination should be limited to no more than 2 nm h^{-1}.

This brief survey shows that there are close similarities in the stage design requirements for the different applications of cryomicroscopy. The main differences center on the actual temperatures needed for the different disciplines. Liquid-helium stages continue to be important for some applications in the materials sciences, chemistry, and physics, such as the structure of solid gases, phase stability, and irradiation effects. Such low temperatures are generally of little advantage in the biological sciences and for analytical applications, where, it may be argued, they exacerbate the problems of ice contamination. We will return to the LN$_2$/LHe$_2$ radiation protection effects on beam sensitive materials later in this chapter.

9.3.2. Cold Stages for Cryo-Electron Microscopy

The types of cold stages may be conveniently divided into those that are cooled by LN$_2$ and those cooled by LHe$_2$. Both types of stages use liquefied gases as the primary cryogen. We will not consider Peltier stages, which only effectively cool down to 223 K; and we will reserve discussion on Joule–Thompson stages, which rely on the adiabatic cooling effects of a rapidly expanding high-pressure gas, to the following chapter on low-temperature SEM.

9.3.2.1. Liquid-Nitrogen-Cooled Cold Stages

Transmission electron microscopes are designed either with top-entry stages *or* side-entry stages, and any cold stages have to be accommodated in one or other of these two geometries. Figure 9.8 illustrates the two types of stages and how, in simple terms, they may be modified to take low-temperature fittings. Cold stages usually only have a single tilt axis capability, and while this is generally not a serious limitation, it has presented difficulties where dislocations are being studied under different diffraction conditions. Baker (1986) suggests a way around this problem by employing computer-simulated images.

Top-entry cold stages have a large cooling surface and a symmetrical temperature gradient, which minimizes thermal drift, but have limited tilt facilities. One simple and effective way of overcoming this problem is either to construct a set of fixed angle holders (Chiu *et al.*, 1988) or to bend the microscope grids and determine the tilt angle from the micrographs (Unwin and Zampighi, 1980). If this latter procedure is used, a substantial part of the microscope grids must remain in good thermal contact with the specimen stage if frozen-hydrated samples are to be examined. Top-entry stages are used on some JEOL and Zeiss instruments. Chiu (1986) and Jeng and Chiu (1987) give details of the way they have modified the stage on a JEOL 100CX electron microscope in order to obtain 123 K with a 0.35-nm resolution. They found that multiple copper leaflets were more effective conductors than copper braid. Figure 9.9 shows a diagrammatic view of a top-entry cold stage.

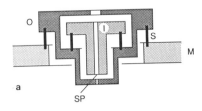

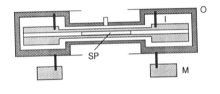

Figure 9.8. Schematic diagrams of (a) top-entry and (b) side-entry cold stages used in cryo-TEM. The beam axis is vertical and the stages are separated from the basis microscope stage (M), which is at ambient temperature, by insulated supports (S). The stages have outer (O) and inner (I) cooling regions, both of which surround the specimen area (S). [Redrawn from Heide (1982).]

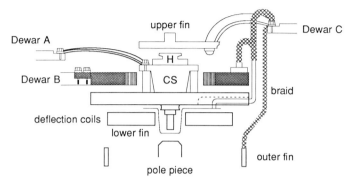

Figure 9.9. Schematic cross-sectional diagram of a top-entry cold stage with the beam axis in a vertical direction. The specimen holder (H) sits in the insulated cold stage (CS), which is cooled via flexible copper leaflets from Dewar A containing liquid nitrogen. Dewar B cools the middle fin of the cold trap and Dewar C is connected via flexible copper braids to the upper, lower, and outer fins of the anticontamination traps, all of which effectively surround the specimen area. [Redrawn from Jeng and Chiu (1987).]

The alternative to top-entry cold stages are the side-entry cold stages. Although they suffer from some constraints of space, they have been carefully designed to provide low temperatures with high mechanical stability. Side-entry cryostages and transfer devices are available on JEOL (EM-CTH10), Hitachi (H-500IC), and Zeiss (EM902 Cryo) electron microscopes, and separate single- and double-tilt cold stages are available from Oxford Instruments and Gatan Inc.

Side-entry cold stages all follow more or less the same design. The sample is placed at one end of a long insulated specimen rod made of a high-conductivity metal such as copper, which is cooled by liquid nitrogen; see, for example, Taylor *et al.* (1984), Henderson *et al.* (1991), and the diagramatic representation in Figs. 9.10 and 9.11. This basic design has a number of variations. The specimen may rely on conductive heat transfer along the length of a metal bar cooled from outside the microscope by a small Dewar of LN_2. Alternatively, liquid nitrogen may be circulated to a reservoir located to within a few millimeters of the surface area and the sample cooled by conductive transfer along a flexible metal braid. A similar system may operate in which dry nitrogen gas, precooled to LN_2 temperatures, is passed to within a few millimeters of the sample area. In all three cases, the cryogen (or cold metal rod) passes into the microscope along an insulated tube to prevent heat gains and frosting due to atmospheric condensation. In addition, the specimen support area is connected to the low-temperature source by a flexible coupling to allow the specimen to be moved and tilted under the electron beam. Although this is an effective procedure to couple the required flexibility of sample movement with the good heat conduction of a high-purity copper rod, there are heat gains into the system. These changes may be calculated, but a good rule of thumb is to assume a 10° temperature rise at each metal–metal contact along the cooling chain and between the sample and the specimen support grid. Thus a copper reservoir containing LN_2 (77 K) or circulating cold N_2 gas (85 K) may be connected to a flexible copper leaflet (+10°),

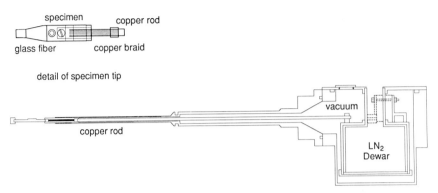

Figure 9.10. Schematic cross-sectional view of a side entry cold stage with the beam axis in a vertical direction. The EM grid is clipped into position over the central hole in the specimen tip, which is shown in detailed plan view. The specimen area, which is insulated by glass supports, is cooled via a short flexible copper braid connected to a long copper rod, which in turn is connected to a small liquid-nitrogen Dewar made of high-purity aluminum. The copper rod and flexible braid are contained in a thin stainless steel tube and the internal space, which also surrounds the Dewar, is pumped by the microscope vacuum. The cooled specimen holder replaces the regular side entry holder of the microscope. [Redrawn from Henderson *et al.* (1991).]

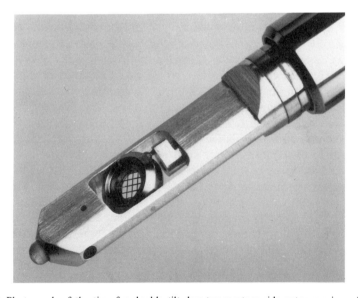

Figure 9.11. Photograph of the tip of a double tilt, low-temperature side entry specimen holder. The transmission electron microscope grid is loaded into the central hole while the tip is kept cold and frost free with dry, cold nitrogen gas. The grid is held in position with a clip. The specimen is transferred to the precooled cold stage using a frost-free transfer device. (Picture courtesy Gatan Inc.)

which in turn are connected to the cold stage ($+10°$). The specimen grid sits on the cold stage ($+10°$) and the sample sits on the specimen grid ($+10°$) and results in a *sample* temperature of ca. 120 K. This is only an estimate, and it is important to take accurate temperature measurements as close as possible to the specimen.

The biggest problem with all stage design is the vibration associated with the heat transfer involved in cooling the sample. Liquid cryogens have a tendency to boil and the vibrations from a Dewar outside the microscope or a small reservoir inside the microscope are transferred to the sample area. Boiling, and hence vibration, can be minimized to some extent by providing good insulation around the primary liquid cryogen. Gas-cooled cold stages are less prone to boiling-induced vibration problems but have an additional problem due to vibration arising from turbulent gas flow through the specimen cooling area. Many of these problems appear to have been solved in the side entry cryoholder designed by Henderson *et al.* (1991)—see Fig. 9.10. The holder is cooled by a small (50 ml), lightweight LN_2 Dewar made of aluminum in which particular care has been taken to minimize any movement in the liquid coolant that may cause vibration in the image. This cold stage, which is available from Oxford Instruments Ltd., is designed for use with the Philips 400 and CM TEM, and routinely gives 0.3 nm images.

Drift and vibration are the main factors limiting resolution that can be achieved with LN_2-cooled cold stages. Jakubowski (1985) has suggested that the drift could be minimized if the metallic heat conduction elements used in LN_2 cold stages were replaced by a heat pipe. Heat pipes work in a closed circuit and transfer heat by capillary flow of a cryogen from a cold source where the liquid is condensed from the gas phase, to the warm source where it is evaporated. The capillary forces are maintained by a wick lining the inner wall of the heat pipe. Jakubowski has calculated that if LN_2 is used as the working medium inside the heat pipe with supercooled LN_2 as the cryogen, there would be a several-hundred-fold increase in thermal conductivity and that temperatures of ca. 90 K could be obtained. If the heat pipe is constructed of low thermal expansion metal, such as Invar steel, drift would be of the order of $0.2 \, nm \, s^{-1}$. Vibration would be diminished by using supercooled nitrogen, which would melt before it boiled. These interesting cryogenic concepts have yet to be applied to functional TEM cold stages.

Nearly all the commercially available side-entry cold stages will operate at or below 123 K. The spatial resolution in most cases is not as good as that suggested by Chiu and colleagues (1988) but is in the 0.3–0.6-nm range.

9.3.2.2. Liquid-Helium-Cooled Cold Stages

Ultra-low-temperature TEM is by no means a new phenomenon. LHe_2-cooled specimen stages have been available for about 25 years, and during this time a number of different stages have been designed in different laboratories. The papers by Heide (1982) and LeFrance *et al.* (1982) provide an excellent review and source of secondary references on microscopy at LHe_2 temperatures. The design parameters that are critical for LN_2 stages are equally exacting for LHe_2 stages. The requirement for specimen stability coupled with flexibility of movement relative to the electron beam are critical, and in addition, the sample is much colder. The usual procedure is to use one or more LHe$_2$ cryostats backed by a LN$_2$ cryostat. The whole lens and stage system is surrounded by thermal shields to minimize heat flow into the system. The sample area is usually first cooled to LN$_2$ temperatures and then cooled to lower temperature (4–10 K) using the LHe$_2$ cryostat. The diagram in Fig. 9.12 gives some idea of the arrangement of the cryostats. Many LHe$_2$ stages are an integral part of a superconducting lens which has been designed for use at high resolution. The first of these instruments was built by Fernandez-Moran (1964, 1965, 1966), who obtained 1–2 nm spatial resolution at 4 K with a superconducting Nb–Zr objective lens. Other instruments have been constructed by Dietrich (1977) at the Siemens Research Laboratories in Munich and the Fritz–Haber Institute in Berlin (LeFranc *et al.*, 1982); the EMBL in Heidelberg; and at the Oak Ridge National Laboratory (Worsham *et al.*, 1974), now relocated at Duke University (Lamvik *et al.*, 1983). Prototype superconducting microscopes are available elsewhere and they are also now available commercially, e.g., the JEOL-2000Scm, which is fitted with a top-entry stage with limited tilt facilities. These later microscopes have a much improved performance, and in some cases the specimen drift has been reduced to 0.002 nm s^{-1} (Iwatsuki, 1987).

It is, however, not necessary to have a LHe$_2$-cooled superconducting lens system in order to have a LHe$_2$-cooled cold stage. A number of people have built LHe$_2$ cold stages which are not an integral part of a superconducting lens system and work in the 5–30 K temperature range; see, for example, Venables (1963), Valdre and Goringe (1965), Siegel (1972), Heide and Urban (1972), and Hobbs (1975). Gibson and McDonald (1984) describe a simple top-entry cold stage for the JEOL 200CX TEM and have obtained a resolution of 0.25 nm at 30 K. Following the earlier work of Fernandez-Moran, Aoki *et al.* (1986) give preliminary details of a superfluid helium-cooled stage which operates at very low temperatures (ca. 1.5 K) and gives a spatial resolution of 0.29 nm, Superfluid helium is a particularly suitable cryogen because the evaporation of the fluid causes no vibration. Side-entry liquid helium cold stages are available from Oxford Instruments Ltd. both as single-tilt

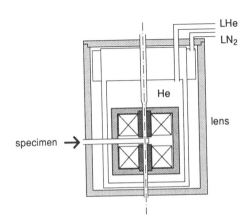

Figure 9.12. Schematic diagram of a liquid-helium-cooled superconducting lens incorporating a side-entry specimen holder. The beam axis is in the vertical direction. The lens assembly is surrounded by a liquid helium cryostat, which is inside an outer liquid-nitrogen cryostat. [Redrawn from Lefranc *et al.* (1982).]

(HCHST 3008) and double-tilt (HCHDT 3010) devices. Gatan Inc. also offer a liquid-helium-cooled stage.

9.3.3. Transfer Devices for Cryo-Transmission Electron Microscopy

Most modern cryomicroscopes are equipped with devices for transferring frozen specimens onto the cold stage in the microscope column without the sample warming above the recrystallization temperature or becoming contaminated. The early days of plastic bags taped to the side of the microscope column and flushed with dry nitrogen are fortunately a thing of the past (Milligan *et al.*, 1983). Most cryo-transfer devices are built on the principle that the sample grid is loaded onto the specimen holder outside the microscope under clean, dry, ice-free LN_2. The specimen transfer rod is screwed into the holder, which is withdrawn from the LN_2 bath into a precooled temporary storage area or frost shield. The whole system is quickly transferred to the microscope via an air lock, which may be prepumped. In the device built by Frederik and Busing (1986), sliding cold shields are used to protect the specimen during sample transfer. The shields are only withdrawn after the sample has equilibrated on the cold stage in the microscope. Other transfer devices are described by Lichtenegger and Hax (1980) and Perlov and Talmon (1983), and systems are available from Gatan and Oxford Instruments. A transfer device is shown in Fig. 9.13. In the system designed by Fotino (1986) for a high-voltage electron microscope (HVEM), the top-entry specimen holder remains under LN_2 in a mini-Dewar following rapid cooling and insertion into the airlock leading into the microscope column. After the pressure in the air lock has reached an appropriate level, the sample holder is removed from the now empty Dewar and lowered into the precooled cold stage. In all these transfer processes, speed and cleanliness are of paramount importance, for as Dubochet *et al.* (1988) have wryly observed, "the quality of transfer is inversely proportional to its duration."

Cleanliness is achieved by wiping metal parts, including the tools used to load the samples, with methanol before use and making sure the LN_2 containers are bone dry. All grid holding and transfer systems are constructed of several interlocking parts, and any moisture will form ice and prevent these parts fitting together properly. A hairdryer is an indispensible part of any cryotransfer kit. This apparent obsession with cleanliness is based on an awareness of the problems brought about by contamination. Unless care is taken, the partial pressure of water vapor in the specimen chamber will increase for a few minutes following sample transfer. The ice that forms on exposed surfaces inside the microscope that are at a temperature higher than that of the specimen is a source of contamination. This ice formation can be largely prevented by using fresh, clean, dry, LN_2, filtered through fine muslin to remove ice crystals, for each transfer. LN_2 surfaces should be covered, and it may even be necessary to war face masks and plastic gloves. If the ambient humidity is consistently high, i.e., above 50%, then refrigerated dehumidifying units should be installed in the laboratory. Jeng and Chiu (1987) recommend using dry nitrogen gas to flush the camera chamber on the microscope during film changes as a way of further reducing the amount of water vapor introduced into the microscope column.

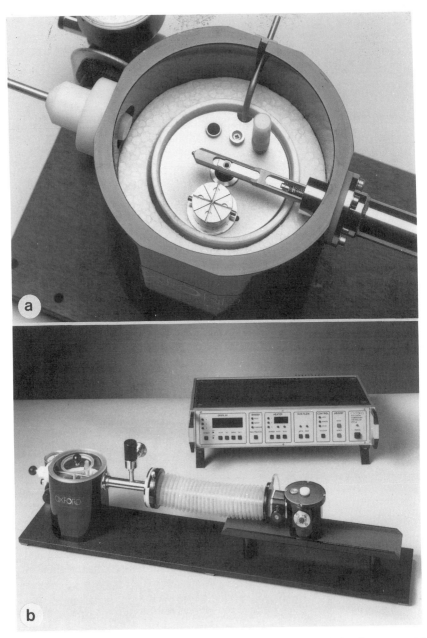

Figure 9.13. Photograph of a side entry cryo-transfer device suitable for transferring frozen material onto the cold stage of a cryo-TEM. The specimen is loaded onto the cold specimen tip, is (a) cooled by a terminal liquid-nitrogen Dewar, and (b) withdrawn into the protective plastic bellows, which allows frost-free transfer into the microscope. (Picture courtesy Oxford Instruments Ltd.)

9.3.4. Anticontamination Devices for Cryo-TEM

Although a properly designed cold stage will keep a sample well below the ice recrystallization temperature, it does nothing to protect the sample from water and organic contaminants, which abound in the warmer parts of the microscope column and the photographic and recording systems. Indeed, a cold specimen is an efficient cryopump and presents an ideal focus for condensates. For this reason the sample and cold stage must be protected by anticontamination devices which operate at temperatures below that of the specimen and have the proper geometry to effectively trap contaminants. This is usually best achieved by surrounding the specimen area with metallic shrouds, which are cooled to near LN_2 temperature by a separate LN_2—or even LHe_2—Dewar. This type of shielding can be achieved with both top-entry and side-entry stages, and the anticontamination devices should aim to shield as much of the specimen as possible, leaving only a small aperture (ca. 500 μm) at the top and bottom to allow the electron beam to traverse the sample. The solid angle between the cold specimen and the warmer parts of the microscope should be as small as possible to minimize contamination but still allow efficient removal of noncondensable gases from around the sample. The temperature on the anticontaminaters must be below the temperature at which water vapor evaporation is negligible (140 K) and preferentially well below the temperature of the specimen. Most LN_2 anticontamination devices operate at around 100 K and their design requires considerable attention to detail particularly in relation to the tilting movements of the stage. For example, Chiu and his colleagues found that they had to substantially rebuild an anticontamination device which was purchased for a particular model of microscope, because careful measurements showed that the anticontamination device was warmer than the specimen. The redesigned anticontamination device is shown in Fig. 9.14. In other instruments, the position of the anticontamination device may limit the tilt angle, which may be obtained on a eucentric goniometer stage. In systems that have been built for side-entry cold stages (see, for example, Figs. 9.15 and 9.16), the frozen sample can be withdrawn entirely into cold shrouds inside the microscope while the total pressure and partial pressure of water and other contaminants are measured. The specimen is only exposed when the contamination is judged to be negligible. Other devices are available which enclose the sample with a nonfenestrated shroud. Such systems are very effective during sample transfer and vacuum pumping but are less effective during observation as they have to be retracted from the specimen area leaving it at the mercy of invading condensates. Nearly all modern transmission electron microscopes are fitted with their own built-in liquid-nitrogen-cooled anticontamination devices, which are located close to the sample area. Separate side-entry anticontamination devices are also available from Gatan (Micropositioning LN_2 cooled anticontaminator 651N) and a liquid-helium version (651 LHe). Similar devices would be available on special order from Oxford Instruments.

9.3.5. Cold Stage Temperature Measurement and Control

Provided there is minimal contamination, most cryo-TEM applications are carried out at as low a temperature as possible, and it is only necessary to achieve good

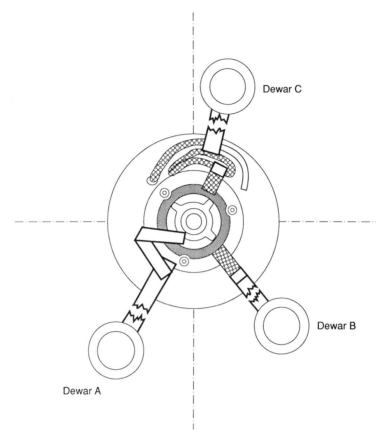

Figure 9.14. Schematic plan view along the optical axis of a top entry cold stage showing the multiple cooling pathways for the anticontamination traps which surround the specimen area. Dewar A cools the centrally located specimen stage via flexible copper leaflets. Dewar B cools the upper, lower, and outer fins of the cold traps, and Dewar C cools the middle fin of the cold trap. The side view is shown in Fig. 9.9. [Redrawn from Jeng and Chiu (1987).]

temperature stability and accurate measurement. Stage temperatures down to 90 K can be measured using thin copper-constantan (Cu-Co) or chromel-alumel thermocouples placed on the cold stage as close as possible to the sample. Care must be taken to minimize heat loses down the thermocouple leads and to ensure good thermal contact between the thermocouple and the sample holder. For temperatures below 90 K, and down to 2 K, it is more convenient to use gold + 0.07% iron versus chromel-P thermocouples (Rosenbaum, 1968). Other methods of measuring very low temperatures include silicon or gallium-arsenide semiconductor diode sensors or germanium or carbon glass resistence thermocouples (Heide, 1982a). At very low temperatures, a layer of condensed inert gas over the sample can provide a useful way of monitoring the performance of a cold stage and calibrating the accuracy of

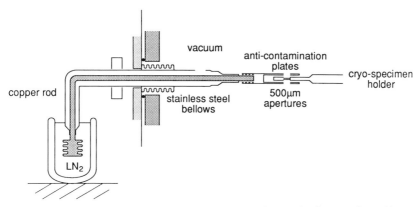

Figure 9.15. Schematic side view diagram of a double-blade anticontamination trap for a side entry cold stage. The beam axis is vertical. The cold trap consists of a pair of fenestrated copper plates, which shroud the specimen area. The copper plates are connected to a copper rod, which sits in a liquid-nitrogen Dewar. The space surrounding the Dewar and the copper rod is pumped by the microscope vacuum system. [Redrawn from Homo *et al.* (1984).]

Figure 9.16. Photograph of a double-bladed liquid-nitrogen-cooled anticontaminator. The pair of copper blades can be aligned in the *x* and *y* direction so they are positioned precisely around the cold specimen, which is inserted between the two copper anticontamination blades from the opposite side of the microscope. (Picture courtesy Gatan Inc.)

thermocouples (Tatlock, 1982). This is one of the few ways of measuring the temperature at *the level of the specimen* in a cryomicroscope. A single-point temperature
determination may be made by observing a known change of phase in the sample or
the surrounding water. It is, however, difficult to measure the local temperature of
a specimen while it is being irradiated with the electron beam. Temperatures at the
level of the sample grid may be measured using thin-film thermocouples fabricated
onto part of the TEM grid. Ota (1970) deposited 0.1-μm-thick Cu–Co thermocouples
onto a 0.1-μm-thick carbon-formvar substrate and used it to measure the temperature inside the TEM. Clark *et al.* (1976) devised methods for making thin-film
thermocouples which were used to measure the temperature at the level of the specimen inside an SEM. In both cases only small temperature rises were recorded when
the electron beam was placed over the thermocouple junction. In situations where
heating is required or where experiments have to be carried out at known rates of
heating or cooling, a small electrical heater is incorporated into the cryoholder. Care
must be taken to ensure that any electromagnetic emissions from the heater do not
interfere with the stability of the electron beam. The general principles of control
mechanisms and feedback loops have already been discussed in Chapter 8.

9.4. OBSERVING AND RECORDING IMAGES IN THE CRYO-TRANSMISSION ELECTRON MICROSCOPE

9.4.1. Introduction

This is not the place to provide a general description of image formation in the
TEM, which may be found in most standard texts of electron microscopy. However,
a brief description of the two main contrast mechanisms that can take place in the
TEM will help to explain the process of image formation in the cryo-TEM. Dubochet
et al. (1988) have provided a simple and concise account of the processes involved
in imaging "native" frozen samples, which is based on whether the electron is considered as a particle or as a wave.

The electron beam may be considered as a flux of particles impinging on and
largely passing through thin samples. A few of the electrons will be scattered
(deflected) by the material within the sample. An aperture (objective aperture) may
be used to remove all the electrons that are scattered through a given angle, with the
consequence that these electrons do not contribute to the final image. Some regions
of the specimen, i.e., those of higher atomic weight or increased mass and/or thickness, scatter electrons more than other regions and this gives rise to amplitude
contrast in the image. To a first approximation, the total scattering in an image
element in biological and hydrated samples is proportional to the mass of thickness
(density and thickness) of the image element. Amplitude contrast plays an important
part in images of positive stained, i.e., heavy metal, sections, freeze-dried cryosections, and liquid suspensions that have been negatively stained by dilute solutions of
heavy metal salts. Amplitude contrast is much less important in image formation of

unstained frozen hydrated samples where low atomic weight material are embedded in water (ice).

If the electron beam is considered as a series of waves impinging on the sample, a different contrast mechanism operates as the electron beam is diffracted by the specimen. The optical system of the microscope is so arranged that this diffracted wave interferes with the undiffracted part of the wave. The phase shift between the diffracted and undiffracted wave is used in image formation. In this mode of imaging, referred to as phase contrast imaging, all the electrons traversing the specimen are involved in image formation, and contrast results from the interference of all the electrons. Although phase contrast makes a contribution to all TEM images, it is a particularly efficient method of image formation for very thin, unstained frozen aqueous suspensions and thin unstained frozen-hydrated sections where amplitude contrast only makes a small contribution to the image.

9.4.2. Amplitude Contrast

Amplitude contrast, bright field images, although the dominant imaging mode of large stained specimens that are examined in focus with a small objective aperture, can also be used to obtain low-resolution (better than 10 nm) structural information about unstained specimens at dimensions greater than ca. 10 nm. Euseman *et al.* (1982) have shown how this mechanism plays a major role in the formation of low-resolution images of well focused amorphous specimens where the phase contrast transfer function is small. In addition, provided the electron scattering properties of the various constituents are understood, it is possible to calculate the degree of hydration between freeze-dried and frozen-hydrated specimens. Chang *et al.* (1983) and Eusemann *et al.* (1982) show that the mass density difference between vitrified ice ($0.93 \, \text{g cm}^{-3}$) and protein ($1.35 \, \text{g cm}^{-3}$) is sufficient to provide the basis for an imaging signal from unstained frozen hydrated biological material.

9.4.3. Phase Contrast

The retrieval of high-resolution (less than 10 nm) information from thin unstained frozen-hydrated specimens and thin frozen-hydrated sections relies on phase contrast because the contribution from amplitude contrast is very small (Erickson and Klug, 1971; Unwin and Henderson, 1975; Lepault and Pitt, 1984). Phase contrast is increased by substantially underfocusing the image and the optimum defocus for a particular structural detail can be estimated from transfer theory (Hanzen, 1971). In practice an underfocus image is taken to maximize the signal from the smallest space frequency that is thought to exist in the image. The phase contrast produced by defocusing varies the intensity of components of the specimen having different spacings to give the so-called phase contrast transfer function. Figure 9.17 shows a focus series from a thin vitrified suspension of the Semliki Forest virus with the corresponding contrast transfer function. There is thus a direct relation between the object and the image, and the variations in the phase contrast function can be processed by Fourier-based computation methods to give an accurate

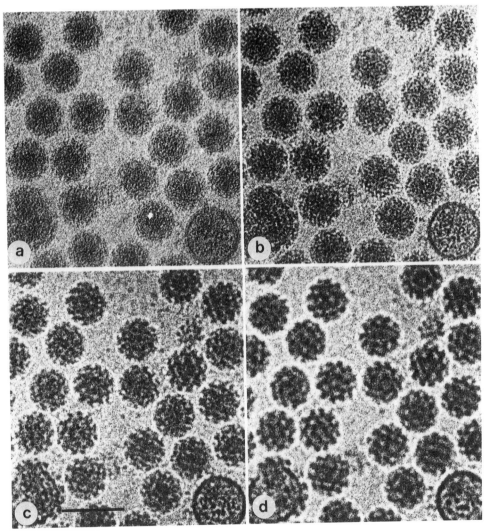

Figure 9.17. Focus series of a thin vitrified unsupported layer of unfixed and unstained Semliki Forest virus suspension. The photographs were taken at 113 K at a magnification of ×33,000 with an electron dose of 1 ke⁻ nm⁻². The micrographs (a)–(d) are underfocused by 1.5, 3.0, 6.0, and 11 μm, respectively. These values were chosen so that the first zero of the transfer function of one micrograph corresponds to the first maximum of the preceding photograph. In these conditions, image reconstructions of the four micrographs retrieve 70% of the information in the 3–12-nm spatial frequency band. [Picture courtesy of Vogel *et al.* (1986).]

representation of the densities in the specimen (Erickson and Klug, 1971). These same image analysis techniques have been applied to thin, unstained ice-embedded specimens (Taylor and Glaeser, 1976; Adrian *et al.*, 1984; Milligan *et al.*, 1984; and more recently, Toyoshima and Unwin, 1988).

9.4.4. The Appearance of Images

The biggest problem with the production of electron microscope images is the conflict between the high beam doses required for an adequate signal-to-noise ratio and the low beam doses needed to avoid radiation damage. Although electron diffraction indicates spatial resolution as good as 0.3 nm can be achieved in biological material, the low contrast of unstained frozen hydrated specimens limits resolution in images to between 2 and 3 nm. Low contrast is a particular problem with high-resolution images from thin frozen suspensions where low beam doses have to be used to avoid damaging the specimens. In these cases, the signal-to-noise ratio may be so poor that nothing can be seen in the images by direct visual observation. Image deconvolution techniques have to be employed to extract the relevent information from the specimen (Unwin and Henderson, 1975). This lack of contrast is also a feature of frozen hydrated sections, which are equally sensitive to beam damage. Although the contrast of frozen-hydrated sections is 3× times lower than that of freeze-dried specimens, it does not mean that the contrast is so poor that nothing can be seen in frozen-hydrated sections. In actual fact, the contrast in unstained frozen-hydrated cryosections is better than the contrast in similar thickness plastic-embedded unstained material. In ultrathin sections, ca. 50–80 nm, it is difficult to identify objects much smaller than 5 nm. In thin sections, ca. 100–300 nm, it is possible to identify membranous components such as endoplasmic reticulum and chloroplast thylakoids. In the thicker sections, ca. 1–2 μm, examined in the STEM, only the cell outlines and possibly the major organelles such as nuclei, mitochondria, and chloroplasts may be identified unambiguously. However, as we shall see, the pictures become much clearer when the frozen-hydrated sections are freeze-dried. The contrast in frozen-hydrated material can be improved by staining with heavy metal salts prior to imaging, although such additives may well compromise the ultrastructure and natural chemistry of the specimen.

Other modes of imaging have been examined to see if they can improve the information that is available from bright field imaging in the conventional transmission electron microscope (CTEM). McDowall *et al.* (1983) have found that the dark field imaging mode, generally used in STEM instruments, is not particularly useful for examining medium thick, i.e., ca. 100 nm, frozen-hydrated sections because the multiple inelastic scattering of the electron beam within the section causes a loss of contrast.

9.4.5. Scanning Transmission Imaging

The STEM is generally less effective than the CTEM for observing small structures embedded in a thick layer of ice. However, Bachmann and Talmon (1984) consider that provided the sections are thin enough, dark field imaging, and a mixture of dark field and bright field imaging in the STEM, can be used to enhance contrast. McDowall *et al.* (1983) found that in order to obtain the same signal-to-noise ratio, the STEM working in a dark field mode applies a higher beam dose than the CTEM working in the phase and amplitude contrast imaging mode. STEM imaging could be further improved by using the phase contrast mode (Rose, 1974). Henderson and Glaeser (1985) and Downing and Glaeser (1986) found that the spot-scan procedure reduces the blurring caused by beam-induced movement, which is particularly troublesome perpendicular to the surface of highly tilted specimens. Bullough and Henderson (1987) consider that the use of the small beam limits the effect of beam motion by minimizing the area of the specimen exposed at any given instant. Although beam damage is evident in some specimens, the use of this method on samples maintained at low temperatures may prove to be an effective way of imaging unstained organic or biological specimens.

Crewe *et al.* (1975) had shown that the STEM could be used to form an image from the ratio between the inelastic to elastic scattered electrons. Carlemalm and Kellemberger (1982) have used this imaging mode, which is sometimes referred to as *Z* contrast, on unstained sections. With *Z* contrast, the influence of local density and section thickness are eliminated and contrast is principally a function of the average atomic number in the specimen. Kellenberger and Chiu (1982) have calculated that this contrast mechanism should be able to produce well-defined images in thin sections of unstained material embedded in ice. More experimental studies, particularly in relation to the relevent beam doses, are required before it is known whether *Z* contrast imaging can be used satisfactorily on frozen-hydrated sections and suspensions.

9.4.6. Alternative Imaging Modes

Electron energy loss spectroscopy (EELS) is an imaging mode that is particularly applicable to thin frozen suspensions and ultrathin cryosections. EELS filters and simultaneously records various fractions of the electron beam, which passes through the sample. Low dose images of frozen-hydrated suspensions have a large noise component due to the inelastic scattering from the ice. Langmore and Athey (1987) found that this background noise reduces both scattering and phase contrast by a factor of about 3. They have used energy filters to eliminate this effect and to increase scattering contrast in both bright and dark field microscopy. Dunlap *et al.* (1989) have made some preliminary observations and calculations on the mean free path of electrons and plasmon losses in vitreous ice. They find that at 200 keV, the path length is about 250 nm, which is close to the theoretical estimates. A recent review article by Wolf *et al.* (1991) on the histochemical application of EELS using ultrathin cryosections shows that the technique can be used for the localization of small molecules, antigens, and elements. Michel *et al.* (1991) have successfully used the Zeiss 902 EM energy-filtering microscope to improve the poor contrast of frozen-hydrated sections by zero-loss mode imaging. Figure 9.18 shows that there is a considerable increase in image contrast.

9.4.7. The Nature of Ice and Image Quality

In addition to the influence of the various imaging modes and contrast mechanisms on the quality of the final image, the form of ice that constitutes the embedding

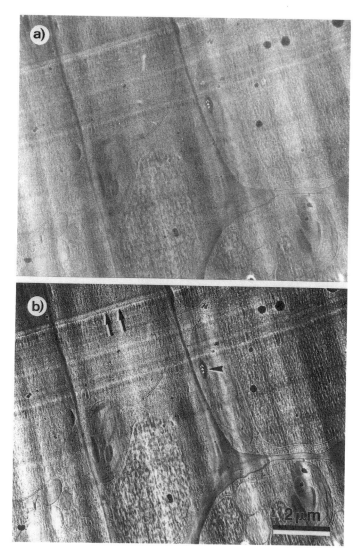

Figure 9.18. Electron energy loss images, taken in a Zeiss EM 902 transmission electron microscope fitted with an energy filter, of frozen-hydrated sections cut from high-pressure frozen apple leaf tissue. Image (a) is recorded without energy filtering and results in low contrast. Image (b) is recorded in the low-loss filtering mode and gives much higher contrast, but also intensifies artifacts like knife marks (large arrows). The small arrow indicates a region of beam damage. [Pictures courtesy Michel *et al.* (1991).]

medium in frozen-hydrated sections and suspensions also have an effect on image quality. Bachmann and Talmon (1984) consider that amplitude contrast is the dominating contrast mechanism in vitrified specimens, whereas diffraction contrast is the principal mechanism for a crystalline ice matrix (Taylor, 1978). The contrast in frozen specimens embedded in I_v is much lower than material embedded in thin layers of crystalline ice. Falls *et al.* (1982, 1983) provide details of the rather complicated mechanism of diffraction contrast. An electron beam passing through a crystalline solid splits into transmitted and diffracted parts. The intensities of the two parts vary inside the crystal depending on features such as the crystallographic orientation, lattice defects, and bend contours. The optical density in the micrograph is influenced by the structure and orientation of the ice crystals. Depending on the thickness and form of ice, the embedded specimens may appear in both positive and reverse contrast as shown in Fig. 9.19. Taylor (1978) considers that the contrast at the interface between the biological material and crystalline ice cannot be considered simply in terms of the difference in density between the material and water. The water in the immediate vicinity of the hydrated sample is unfreezable and forms an ice exclusion zone, and a hydrated sample, such as a protein, is acting as its own cryoprotectant. This means that at high resolution, the fine structural detail, which depends more on variations in the internal density of the sample, will be less affected by the diffraction contrast engendered by the crystalline ice.

Image interpretation becomes more problematic at lower resolution, particularly for unidimensional crystalline objects. The best solution is to ensure that the samples are embedded in I_v rather than crystalline ice. Specimens composed of two- or three-dimensional crystals are less affected by this problem as the edges of the crystal are less important in the imaging process (Taylor, 1978). Good contrast can be obtained in dispersions of noncrystalline objects embedded in layers of I_h. The interpretation of images of thin noncrystalline samples in a crystalline ice matrix and crystalline samples in a vitreous ice matrix is greatly assisted by applying spatial averaging techniques to the micrographs.

The contrast in both frozen-hydrated suspensions and sections can be enhanced considerably by *in situ* freeze-drying. Unfortunately, frozen sections do not dry at a uniform rate and there may be considerable shrinkage and loss of fine structural

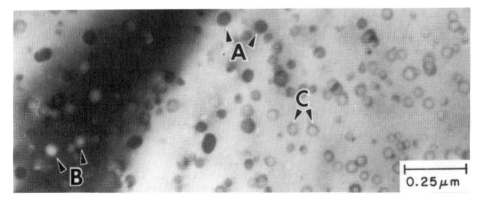

Figure 9.19. Bright field cryotransmission electron micrograph from an unstained, frozen-hydrated vesicular suspension of bovine phosphatidylcholine. Vesicle display both negative (A) and positive (B) contrast due to changes in thickness of the ice matrix. Some vesicles (C) appear surrounded with a dark ring. [Picture courtesy of Falls *et al.* (1982).]

detail as the water is sublimed from the sample. Dubochet *et al.* (1983) showed that it is possible to calculate the shrinkage in a section during freeze-drying by multiplying the mass thickness of the section by the inverse square of the linear shrinkage ratio. While this procedure goes some way to restoring the linear dimensions and gross topographic interrelationships within the sample, it will do nothing to restore the loss of fine structure. Freeze-drying may be done gradually so that images at varying degrees of sample hydration may be studied and used to check whether the sample was originally in the frozen-hydrated state. This has, however, led to problems in some earlier studies, and the only reliable test of hydration is to carry out mass and contrast determinations. *In situ* freeze-drying is a useful procedure for frozen-hydrated sections as it provides the only sure way of identifying structures in otherwise low-contrast images. This is particularly important with the thicker, ca. 500–1000-nm sections used for x-ray microanalysis, which, as we will show, are more susceptible to radiation damage in the frozen-hydrated state.

9.4.8. Procedure for Imaging Frozen Specimens

The process of recording high-resolution images of frozen samples is quite complex and involves taking more than one photograph. As we will discuss later in this chapter, most unembedded organic samples and all frozen-hydrated organic samples embedded in ice are particularly sensitive to beam damage. However, for samples observed at low temperatures the radiation damage only becomes significant after an *accumulated* electron exposure ca. $\times 5$ higher than the critical exposure at ambient temperatures. The low-dose imaging methods, which play an important part in recording images of frozen-hydrated samples, when used alone produce virtually featureless images. The image contrast of a frozen specimen depends on the inherent scattering of the sample, the defocus value of the objective lens, and the thickness of the ice layer surrounding the specimen (Chiu, 1986). The visibility of low-resolution structural detail may be enhanced by setting the defocus several micrometers away from the optimal defocus setting, although the ultimate spatial resolution will remain unaltered.

For high-resolution studies of frozen-hydrated suspensions it is necessary to take several different images of the sample, which are then analyzed and combined using computer-aided image processing such as the correlation averaging technique developed by Saxton and Baumeister (1984) and Verschoor *et al.* (1984). The theoretical basis for these imaging techniques is discussed by Lepault and Pitt (1984). Several different image analysis protocols have been devised; see, for example, Grant *et al.* (1986), Toyoshima and Unwin (1988), Jeng and Chiu (1984), Rachel *et al.* (1986), Hayward and Stroud (1981), Jaffe and Glaeser (1984, 1987), Chiu *et al.* (1988), and Schalz and van Heel (1990). The general procedure that has been adopted as a result of these studies is as follows.

The initial searching of the image is carried out at a low dose (ca. $5e^- \text{ nm}^{-2}$) and low magnification, ca. 3000–4000 at an underfocus of between 10 and 15 μm. If crystalline material is being examined, an electron diffraction pattern is recorded using low-dose conditions ($100\ e^- \text{ nm}^{-2}$). The microscope is then switched back to the magnification mode and a low-dose (200–300 e^- nm^{-2}), small underfocus picture taken at between ×15,000 and ×40,000 for retrieval of phases. This is followed by one or two higher-dose (20,000 e^- nm^{-2}) images of the same area at the same magnification in order to evaluate the contrast transfer function. The first high-dose image is taken at the same defocus setting as the low-dose image; the second high-dose image is taken at slightly greater defocus. After several sets of these 3–4 micrographs are recorded from a grid, the sample may be freeze-dried *in situ* and a final low-magnification image taken for each data set. This enhanced contrast image can be used to determine crystal orientation. All photographs and the initial diffraction pattern are taken at low temperatures, ca. 123 K. Figure 9.20, taken from the work of Toyoshima and Unwin (1988) on the crystalline acetycholine receptor tubes of the fish *Torpedo marmorata*, shows the different types of images that may be obtained.

The presence of ice in the sample can be detected from the characteristic electron diffraction patterns or the bend contours of ice. It is important to know the hydration state of the specimen when attempting to interpret the image. The contrast of a partially or completely dried specimen is better than a fully hydrated sample. Chiu (1988) points out that one of the many advantages of working with specimens suspended across a hole in the support film is that one may be reasonably certain that the sample is surrounded by ice. There is, however, no guarantee that the *sample* is fully hydrated, because it may protrude, partially or entirely, above the ice surface layer. Chiu suggests that it is advisable to examine a large number of images of specimens embedded in ice of varying thickness and with different defocus values to determine which image represents the optimal preparation and best structural representation.

The methods for recording information from frozen-hydrated sections are much less complicated, if only for the fact that high-resolution structural information cannot be obtained from such thick samples, which are usually on the top of a support film. The initial searching should be carried out at slight defocus and under low-dose conditions and the actual image taken using as low a dose as is practicable. The main use of images of frozen-hydrated sections is to provide a map for associated x-ray analytical studies. High-magnification, high-resolution images are not required and it is usually only necessary to be able to see and identify the principal components of the sample. This identification is facilitated by freeze-drying the section *in situ*, followed by recooling to the original temperature and taking a second image of the section in the freeze-dried state. This two-image procedure works equally well in the TEM and in the SEM working in the transmission signal mode.

9.5. BEAM DAMAGE

9.5.1. Introduction

The accuracy with which a structure can be analyzed in the TEM is limited by dehydration and radiation damage. The problem of dehydration has been neatly solved by using frozen-hydrated specimens. The problem of radiation damage

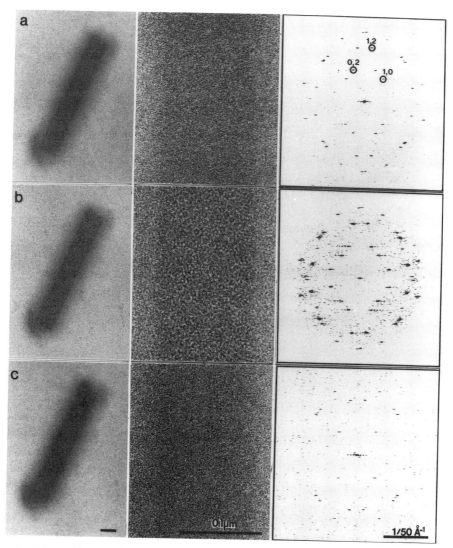

Figure 9.20. A set of images of acetyl choline receptor tubes embedded in amorphous ice at 106 K over a carbon film. The left hand column has a general view, the middle column a detailed view and the right hand column the computed Fourier transforms. Images (a) and (b) were recorded at underfocus values of 570 and 200 nm, respectively in order to determine the amplitude contrast contribution. Image (c) was recorded to evaluate the effect of electron irradiation. Although it is difficult to see any evidence of crystallinity in the underfocused images (a) and (c), the corresponding diffraction patterns show its presence. Note how the diffraction pattern in (c) has a poorer signal/noise ratio than (a) because of radiation damage. [Picture courtesy Toyoshima and Unwin (1988).]

remains an inevitable and serious consequence of using high-energy beam instruments to examine labile organic and biological material. Beam damage will always be present because the process of electron beam–specimen interaction that produces the image also leaves sufficient energy in the specimen to cause damage. The extent of the radiation damage problem is succinctly summarized in an early paper by Glaeser and Taylor (1978). At ambient temperatures a dose of only 20 e^- nm^{-2} is needed to cause blackening of the photographic emulsion in a single picture of a specimen taken at low magnification. But even at these low doses, the high-resolution features of some specimens may already be damaged. The problem of radiation damage cannot be solved simply by resorting to low-dose (100 e^- nm^{-2}) microscopy because the statistical fluctuation from one picture element to another may greatly exceed the inherent contrast in the specimen. Although low temperatures decrease the radiation sensitivity of most specimens by a factor of 5, these procedures have to be coupled with image averaging methods to realize the full potential of this type of microscopy.

Cryo-electron microscopy has added interesting new dimensions to these problems, for although low temperatures are generally accepted to slow down radiation damage, the presence of water (ice) in the frozen-hydrated sample may significantly diminish this amelioration of beam damage. The main purpose of this section is to provide a brief introduction to radiation damage and consider the special conditions that are associated with low-temperature microscopy and frozen specimens containing ice. Much has already been written on this subject, but the papers by Isaacson (1977), Glaeser and Taylor (1978), Cosslett (1978), Hobbs (1979, 1984), Glaeser (1975, 1979), Talmon (1982, 1984, 1987), Dubochet et al. (1982a, 1988), and Jeng and Chiu (19843 will provide a comprehensive introduction to the problems associated with radiation damage.

9.5.2. The Processes of Beam Damage

Beam damage falls into two broad categories, radiation damage caused by the ionizing effects of the electron beam and the thermal damage where a substantial proportion of the energy absorbed by the specimen is eventually dissipated as heat. Radiation damage is manifest by the introduction of structural and/or compositional changes into the specimen. These changes range from the directly observable damage at low resolution, e.g., bubbling, to the more insidious nonobservable but nevertheless measurable damage at higher resolution, e.g., mass loss and losses of crystallinity.

The fundamental physical processes that result in specimen damage are complex and incompletely understood. Elastic collisions, in which the electron is deflected with negligible energy loss, provides useful information about the specimen. At accelerating voltages up to 100 keV, elastic scattering events leave little energy in the specimen that might cause damage. Most of the damage is caused by inelastically scattered electrons, which lose energy as they traverse the specimen. The amount of energy deposited depends on the chemical composition and the thickness of the specimen, the speed (voltage), and number of electrons (beam current) at the level of the specimen.* Thus in a thin film of carbon, a 60-keV electron beam loses on average 1 eV per nanometer of sample traversed. This figure is equivalent to a significant proportion of the bond energies of organic compounds. Cosslett (1978) and Dubochet *et al.* (1988) calculated that the average energy loss per inelastic scattered event in water is ca. 20 eV. There are about three elastic scattered events for every inelastic scattered event in biological material stained with uranium salts; the reverse ratio is true for water and unstained organic and biological material. This fact alone is sufficient to be concerned about beam damage in biological and organic specimens. Radiation damage is a sequential set of processes, which occur on different time scales, and a detailed sequence of events is presented below:

1. Primary events (10^{-18}–10^{-12} s):
   a. Excitation of an individual atom.
   b. Excitation of a group of atoms (plasmon excitation).
   c. Ionization of an atom.
   d. Displacement of an atom (knock-on collision).

These primary events give rise to one or more secondary interactions:

2. Secondary physical interactions (10^{-12}–10^{-7} s):
   a. Electron emission.
   b. X-ray emission.
   c. Light emission.
   d. Temperature rise.
   e. Electrostatic charging (conductivity dependent).
3. Secondary chemical interactions (10^{-10}–10^{-4} s):
   a. Breaking chemical bonds (bond scission) (temperature dependent).
   b. Promoting chemical bond formation (cross-linking) (temperature dependent).
4. Tertiary consequences (10^{-4}–10^{-2} s):
   a. Production and movement of free radicals (temperature dependent) and as a consequence:
      i. Mass loss.
      ii. Loss of crystalline order.
      iii. Structural damage.

Most radiation damage in organic, biological, and hydrated specimens is caused by ionization, in which energy from about 10 eV to several kiloelectronvolts is transferred to the sample, and plasmon excitation, where the energy transfer is of the order of 10–30 eV. Ionization and plasmon excitation occur very rapidly and there is nothing we can do to prevent them other than turning off the electron beam. The secondary interactions of radiation damage, which are somewhat slower than the primary events, are due primarily to the diffusion of free radicals and other reaction

---

*Specimen damage *decreases* with increasing accelerating voltage (the faster the electrons travel the more quickly they pass through the specimen) and *increases* with increasing beam current (the greater the number of electrons the greater the chance of damage occurring).

products. Free radicals are highly reactive atomic or molecular species, which possess an unpaired electron. They are formed in nearly all materials that are irradiated with an electron beam, although there is considerable variation in the number of radicals that are formed. For example, water forms six times as many radicals per 100 eV than aromatic compounds (Henglein, 1984). A dose of $100\,e^-\,nm^{-2}$ into biological material corresponds to a free radical concentration of $12\,M$, which Henderson (1990) has calculated represents about 1000 reactions per protein molecule. There is every reason to believe that free radicals are one of the principal causes of radiation damage in frozen-hydrated material, although their effects are much reduced at low temperatures.

Some of the secondary interactions, i.e., electron, x-ray, and light emission, are an important source of alternative modes of signal aquisition in electron beam instruments; the remainder cause damage to the sample. In the context of low-temperature electron microscopy, temperature rises may cause phase changes and/or alteration in the crystalline form of ice; electrostatic charging will create problems with quantitative x-ray microanalysis; bond scission can give rise to free radicals; cross-linking leads to the production of carbonaceous contamination and mass loss will give rise to problems during quantitative x-ray microanalysis. In addition, plasmon excitations are the principal sources of beam heating in the sample. There is every reason to believe that low, i.e., ca. 90 K, temperatures will ameliorate some of these secondary effects. For example, the diffusion of free radicals is significantly reduced at low temperatures, thermal effects will be constrained, mass lossess will be diminished, and contamination will be localized to regions away from the specimen. At liquid helium temperatures these effects are reduced even further, although the amount of extra cryoprotection is extremely variable. But even at these very low temperatures, radiation damage can still occur.

9.5.3. Units of Radiation Dose

The extent to which a sample is damaged by the beam is a function of electron exposure or dose and sample temperature. In electron microscopy the dose is most conveniently expressed as the number of electrons per square nanometer ($e^-\,nm^{-2}$). Dose can also be expressed as the electric charge per unit area, i.e., coulombs per square meter ($C\,m^{-2}$). The conversion between the two units is given by $1\,e^-\,nm^{-2} = 0.160\,C\,m^{-2}$, where $C = 1$ ampere second. It must be remembered that radiation effects are *cumulative*, and it should be assumed that all the values given here and elsewhere in the literature are the total dose used. The situation is complicated by the fact that in the general radiation literature, the intensity of radiation is measured by the amount of energy deposited into the sample. In order to convert one to another it is necessary to know the stopping power (i.e., rate of energy loss) of the incident electrons. At 10 keV, a typical stopping power in organic material is of the order of 1 eV per 0.1 nm, so the actual energy deposited for a flux of $100\,e^-\,nm^{-2}$ of the order of $10^5\,J\,ml^{-1}$ because 1 eV is $1.6 \times 10^{-19}\,J$. Assuming a density of 1.0 for frozen-hydrated biological material, this would represent a dose of $10^6\,J\,kg^{-1}$. The correct S.I. unit for dose is the gray (Gy), where $1\,Gy = 1\,K\,kg^{-1}$. The conversion from charge deposited to energy deposited is energy dependent because it varies with the stopping power, which falls as the energy of the incident electrons rises. In this respect it is necessary to have a means of comparing radiation measurements made at different energies. This aspect is considered in some detail by Lamvik (1991), who recommends using the calculations proposed by Isaacson (1977) to scale the dose values by the energy dependence of the inelastic scattering cross section. The results presented by Lamvik on the radiation damage in dry and frozen-hydrated organic material are all scaled to dose values of 100 keV. For example, a dose of 2.5 $ke\,nm^{-2}$ at 30 keV into a solid biological sample corresponds to 6.1 $ke\,nm^{-2}$ at 100 keV.

These simplified calculations for radiation dose only give a measure of the amount of energy (or charge) arriving at the surface of the sample, and we really need to know how much energy is being deposited in the sample. Thus a dose of 15 $ke^-\,nm^{-2}$ at 100 keV would be expected to deposit substantially more energy into a bulk sample than into a thin section where most of the electrons would pass through without scattering.

For the sake of simplicity and consistency with most of the recent work on radiation damage in the transmission electron microscopy of biological, organic, and hydrated samples, we will use the term $e^-\,nm^{-2}$.

9.5.4. Manifestations of Beam Damage During Low-Temperature TEM

Specimen damage, which may occur either as a result of beam heating or as a result of radiation damage, can be conveniently separated into two classes: observable damage, in which there are distinct changes in the appearance of the image, and nonobservable damage, where the sample appears unaltered but where measurements such as mass loss or electron diffraction show that damage has occurred. Observable damage is usually associated with higher energy inputs into the sample, and the unobserved but measurable damage is associated with lower energy inputs.

9.5.4.1. Beam Heating

Even though practically all the energy of the electron beam passes through the sample, a proportion of the energy absorbed by the specimen (plasmon excitation) will eventually be dissipated as heat. Specimen heating used to be a major problem in electron microscopy, and in the early days of low-temperature electron microscopy it was thought that this would be one of the limiting factors to imaging frozen-hydrated material, as both ice and biological material are poor conductors of heat. These fears were unfounded because early experiments by Clark *et al.* (1976) and more precise experiments by Talmon and Thomas (1977a,b, 1978, 1979) and Talmon (1982) showed that beam heating in frozen specimens, examined in the transmission mode in both the TEM and the SEM, is probably limited to only a few degrees, provided the sample is in good thermal contact with the cold stage. It is important to ensure that specimens are flattened on the supporting grid and that there are no contaminating ice crystals betwen the specimen grid and the cold stage. Kohl *et al.* (1981) consider that a few degrees of temperature rise might present a problem with very low temperature microscopy where a LHe$_2$ stage is being used to study phenomena that are dependent on a precise temperature being maintained during examination and imaging. At higher temperatures, although moderate temperature rises of ca. 10–20 K may not normally give rise to structural damage, they may just be sufficient to raise the temperature of the sample to a point at which substantial changes may occur. It is advantageous to have prior knowledge concerning the physicochemical properties and thermal characteristics of samples to be examined by cryomicroscopy. At higher temperatures it is possible that a small amount of beam heating at the devitrification temperature of I_v could initiate ice crystal formation. But as this is a time-dependent phenomenon it is not considered to be a serious limitation to low-temperature electron microscopy.

9.5.4.2. Observable Radiation Damage

Bubbling, in which numerous minute bubbles, or cavities, appear in the sample, is one of the most readily observable phenomena in frozen specimens. This was first observed by Unwin and Muguruma (1971) and Glaeser and Taylor (1978) and later by Dubochet *et al.* (1982). With increased radiation dose these bubbles grow, coalesce, and eventually break as shown in Fig. 9.21. There is, as might be expected, considerable variation in the threshold radiation dose for bubbling. Dubochet *et al.* (1987) consider that bubbling generally starts after a dose of about 1–10 ke^- nm^{-2} and that the extent of bubbling depends on the density and size of the organic

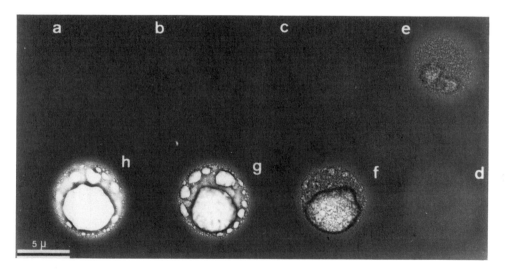

Figure 9.21. A series of cryotransmission electron micrographs taken at 110 K of a condensed layer of vitrified water on a carbon-coated Formvar film. Areas a–h are irradiated at 80 keV with electron doses of 5, 20, 40, 80, 120, 240, 340, and 450 e^- nm^{-2}, respectively, at a rate of 2 ke^- nm^{-2} s^{-1}. The ice layer shows progressive bubbling with increased dose until a hole is bored through the sample. [Picture courtesy Dubochet *et al.* (1982).]

particles. With T_4 bacteriophage, bubbling was observed at 7000 e^- nm^{-2}; in apoferritin, which is much smaller, bubbling only began to occur at a dose of 50 ke^- nm^{-2}. In a later publication, Dubochet et al. (1988) showed that smaller viruses were more resistant to bubbling than larger viruses, and in T_4 bacteriophage, the dense core of DNA showed bubbling before the less dense protein coat. Dubochet and his colleagues conclude that bubbling is marginally dependent on dose rate and that measurements in STEM instruments are about the same as those obtained from the CTEM.

There is some uncertainty about the cause of bubbling. Talmon and co-workers (Talmon, 1984, 1987; Talmon et al., 1986) consider that it is the result of a specific chemical reaction betwen organic material and the water. They consider that small cavities form in the ice either as a result of defects in the ice lattice or from small organic particles. The observation by Talmon (1987) that the holes (bubbles) grow along preferred crystallographic planes supports the suggestion that the interface betwen the ice crystal and the organic material plays an important part in the process. Dubochet et al. (1987) consider that this type of ice–specimen interaction only occurs at doses much higher, i.e., ca. 10 ke^- nm^{-2}, than those needed to initiate bubbling. Chang et al. (1983a,b) have shown that much higher radiation doses are needed to initiate bubbling in material embedded in vitrified ice than in crystalline ice. Pure I_v and I_c are 100× less sensitive to radiation damage than pure I_h. Dubochet and his colleagues consider that bubbling arises as a result of gas formation by electron beam initiated decomposition of the organic sample. The gas is initially trapped within the frozen matrix but is released as the bubbles coalesce and eventually burst. Bubbling and mass loss generally occur at about the same electron dose.

While the debate about the mechanism(s) causing bubbling will continue, the visible manifestation of the phenomenon is a tiresome reminder that the electron beam can damage frozen samples. Bubbling causes severe structural damage and can be taken as an early warning that mass loss is beginning to occur in the sample.

The other main effect that may be observed is specimen drift, which varies in both direction and intensity. This drift is distinct from any movement associated with the specimen holder and probably arises as a consequence of a buildup of electrostatic charge in poorly conductive samples. Specimen drift is even more severe at low temperatures, where there is a marked decrease in the conductivity of iced specimens, and is particularly pronounced in unsupported specimens. It is important to keep the beam at low intensity during focusing. The specimen should remain unirradiated for a period immediately prior to image recording. Even then drift may be a problem as the first electrons cross the specimen. Dubochet et al., (1987) provide a number of useful practical suggestions as to how the effect of drift may be minimized. They suggest that the illumination should be well centered and that electrons scattered from the objective lens aperture and a closely opposed anticontaminator could form a gas, which would in some way neutralize the charge on the specimen.

9.5.4.3. Nonobservable but Measurable Radiation Damage

The commonly used methods to evaluate specimen damage in the absence of any immediately observable damage in the image are to monitor changes either in the electron diffraction pattern or in the mass of the specimen. Changes in the electron diffraction pattern, which are represented either by fading of specific intensities or by loss of resolution in the diffraction pattern, are an especially useful tool for observing radiation damage in crystalline samples. The sharp spotlike electron diffraction pattern may be observed to fade away as the sample is exposed to increasing radiation dose and is replaced by the diffuse ringlike patterns, which are characteristic of amorphous (noncrystalline) materials. Figure 9.22 shows the change in electron diffraction patterns as an organic sample is irradiated. For most crystalline biological material at room temperatures, these changes occur at low radiation doses, i.e., 200–$1000\ e^-\ \mathrm{nm}^{-2}$ and in some instances are initiated at the doses used to take low-dose images. The specimen alteration due to damage is a continuous process, which may not be adequately represented by a specific fading curve of loss of particular spots on the diffraction pattern. The visual estimation of the fading of spots in a diffraction pattern is less precise than densitometric measurements or direct measurements with an appropriate detector (Chiu, 1986). The resolution limits defined by instrumental and sample parameters must be considered and the influence of any radiation damage related to such limits. The changes in electron diffraction patterns are particularly useful in low-temperature electron microscopy of vitrified specimens. Vitrified ice will give the diffuse ringlike patterns characteristic of amorphous material, whereas cubic and hexagonal ice will give the sharp spotlike diffraction patterns of crystalline materials. It is thus possible to monitor temperature changes in vitrified specimens by observing the recrystallization processes. The electron diffraction pattern is also a very simple way of checking that the water in the sample has been converted to the vitreous state. A carefully coordinated study by a number of research workers in Europe and the United States (Chiu, 1986; Chiu *et al.*, 1988) showed that the electron diffraction patterns of crystalline protein and hydrocarbon materials lasted 3–$5\times$ longer at 150 K than at 300 K. Cooling to 4 K gave an additional twofold cryoprotection, but the results were variable. The earlier studies by Knapek and Dubochet (1980), in which there appeared to be a 30–300-fold decrease in radiation sensitivity of catalase crystals at 4 K, have not been substantiated. The one limitation of electron diffraction patterns is that they only give useful information about changes between the amorphous and crystalline states.

Another aspect of radiation damage is mass loss. It has been known for more than 25 years that irradiation of nonhydrated organic specimens with an electron beam can cause massive losses of light elements (Bahr and Stenn, 1965; Stenn and Bahr, 1970). Indeed, in early electron microscope studies, before it was possible to cut ultrathin specimens, this was one of the more draconian methods of thinning sections of resin-embedded samples prior to observation. Thinning and mass loss of the specimen in the electron beam can be measured in terms of changes in optical or electron transmission, by autoradiography or by decrease in x-ray emission. The details of the physicochemical processes that lead to mass loss remain unclear. Dubochet *et al.* (1988) suggest that atoms and small molecules are lost from the specimen surface every time an inelastic scattering event occurs. Symons (1982) suggests that atoms are lost from the sample as a result of the shock wave that is produced by elastic scattering inside the sample. Lamvik *et al.* (1989) and Lamvik

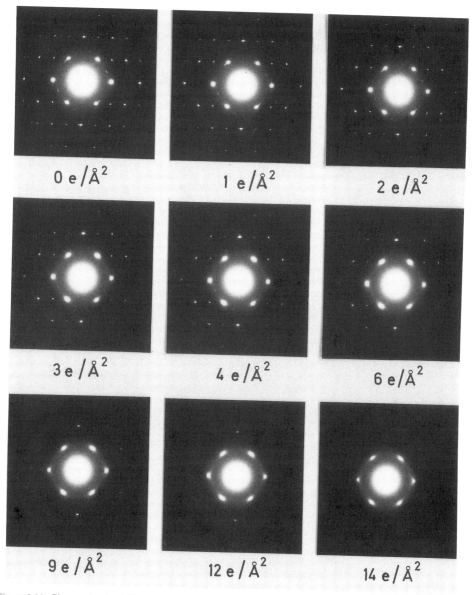

Figure 9.22. Changes in the diffraction pattern of an *n*-tetracosane crystal irradiated with increasing doses of 100-keV electrons. Even electron beam doses as low as 100–200 e^- nm^{-2} (1–2 e^- Å^2) cause visible changes in the diffraction pattern. [Picture courtesy Dorset and Zemlin (1985).]

(1991) suggest that electrical conductivity may be involved. During irradiation, a local electrostatic charge builds up that would attract ionized molecular fragments which would be etched away.

Mass loss is directly proportional to electron exposure and is generally negligible for observations made at 80 K under low dose conditions (Dubochet, 1975; Freeman et al., 1980). Egerton (1980, 1982a) showed that the loss of light elements from plastic films was considerably reduced by cooling the sample to 100 K. Figure 9.23, from the work of Lamvik (1991), shows how the mass loss from collodion supported on a carbon film is linear with electron dose and is considerably reduced at low temperatures. In a more recent study (Lamvik et al., 1989), it was found that the mass loss from collodion on a titanium support proceeded exponentially with exposure at liquid-helium temperatures. At these temperatures, the conductivity of thin titanium films is 10^6 higher than carbon films. This suggests that the differing conductivities of the substrate may play an important role in explaining the different mass loss effects. Mass loss is generally not a problem with low-temperature TEM where low-dose imaging procedures are being used, although the results are extremely variable. Lamvik (1991) found that some materials, such as albumin, show the same amount of damage at 130 K and 10 K. With phosphatidylcholine, a constituent of biological membranes, the mass loss is reduced 20-fold at 130 K compared to room temperature. With nitrocellulose there is a 40-fold reduction at 10 K compared to 130 K, and with the epoxy resin Epon there is a 70-fold reduction over the same temperature range. However, as we shall see in Chaper 11, mass loss is a serious problem in qantitative x-ray microanalysis were the electron exposures are several orders of magnitude higher than those used in low-dose TEM. The loss of material from the sample is of concern in biological and organic microanalysis, because in addition to causing changes in the mass of the analyzed microvolume by the pronounced loss of light

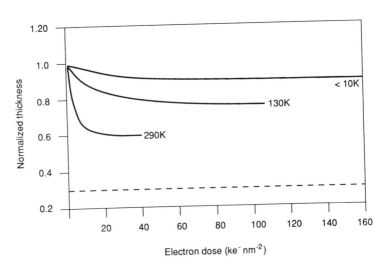

Figure 9.23. Mass loss from 60-nm-thick sections of epoxy resin as a function of electron dose at three different temperatures. The sections are coated with a thin layer of titanium and the graphs show that as the temperature is reduced the rate of mass loss is slower and more residual mass is retained. [Redrawn from Lamvik (1991).]

Table 9.1. Comparative Radiation Doses Associated with Different Phenomena in a TEM Operating at 100 keV and at a Temperature of 100 K[a]

Low-dose searching	$50 \, e^- \, nm^{-2}$
Observation of electron diffraction pattern	$100 \, e^- \, nm^{-2}$
Low-dose imaging	$200 \, e^- \, nm^{-2}$
Loss of electron diffraction pattern	$1 \, ke^- \, nm^{-2}$
Free radical formation	$1 \, ke^- \, nm^{-2}$
Ice bubbling	$2 \, ke \, nm^{-2}$
Mass loss	$2 \, ke^- \, nm^{-2}$
Removal of 1-nm layer of ice	$3 \, ke^- \, nm^{-2}$
Normal viewing procedures	$10 \, ke^- \, nm^{-2}$

[a] From Echlin (1991).

elements such as C, H, O, and N, there are differential losses of the elements that are the object of analysis. Although low temperatures *reduce* these mass losses, they do not *eliminate* the problem.

The different manifestations of beam damage occur at different accumulated doses, and while it is not possible to give precise figures for all types of material studied by low-temperature TEM, the values given in Table 9.1 from Echlin (1991) give a general impression of the radiation doses associated with different viewing phenomena and specimen damage. We have so far only considered mass loss in dry samples. The situation is different in hydrated samples, where the presence of water in one or more of its solid forms creates additional problems in the low-temperature TEM of frozen-hydrated specimens. These matters will be discussed in the next section.

9.5.5. Radiation Damage to Pure Ice

Earlier experiments outside the electron microscope showed that ionizing radiation can damage ice. The realization that frozen-hydrated samples could be examined and analyzed in electron beam instruments has resulted in considerable work being carried out in an attempt to understand the nature of the processes. Details of this work may be found in the papers by Heide (1982a,b, 1984), Heide and Zeitler (1985), Talmon *et al.* (1979, 1986), Talmon (1982, 1984), and Dubochet *et al.* (1982a,b). The review by Talmon (1987) provides an excellent introduction to the subject.

The main primary event in the radiolysis of ice is the ionization of water molecules with a consequent ejection of an electron. The products of ionization interact with water molecules to form a large number of active chemical species, including free radicals. Details of the various radiochemical reactions that occur in water may be found in the papers by Heide and Zeitler (1985) and Talmon (1987). The physicochemical changes that occur as a result of irradiating pure ice depend largely on the temperature and the form of the ice. Although there is still variation in the results, the following general phenomena begin to emerge from the various studies that have been made:

1. Mass loss in pure ice is linear with dose at 90 K and is a result of radiolysis, not beam heating and sublimation (Talmon *et al.* 1979; Talmon, 1982).
2. Pure water at 90 K sublimes under the electron beam at a rate of about 1 atom for every 100 e^-, or 1 nm per 3000 e^- nm^{-2} (Talmon *et al.*, 1979; Dubochet *et al.*, 1982a; Heide, 1982b, 1984).
3. In spite of some conflicting evidence (Heide, 1984), the etching (sublimation) rates for microcrystalline and vitreous ice are approximately the same.
4. The etching rate of ice is strongly dependent on temperature (Heide, 1984).
5. At low temperatures, i.e., 8–80 K, crystalline ice may be transformed to vitreous ice by electron beam radiation; this effect is not seen above 80 K (Lepault *et al.*, 1983a; Heide, 1984; Heide and Zeitler, 1985).

The radiation studies on *pure ice* show that it is more stable under the electron beam than many frozen-hydrated biological samples. For example, a dose of 3000 e^- nm^{-2} are needed to remove 1 nm of ice at 90 K, whereas only 100–1000 e^- nm^{-2} may be used to take a low-dose image of a frozen biological sample at the same temperature without damaging the sample (see Fig. 9.24). This removal rate of ice refers to one face only (Heide, 1984), which means that the ablation rate in a thin section would be twice as high. Although pure ice is an ideal embedding material, it is also a highly productive source of free radicals. If these radicals are produced in high enough concentration, they will both initiate the self-destruction of the ice, as well as attacking any organic material embedded in the ice. Herein lies one of the problems of using ice as an embedding material for high-resolution TEM, for electron exposures as low as 100 e^- nm^{-2} are sufficient to initiate damage.

9.5.6. Radiation Damage in Frozen-Hydrated Specimens

It has been known for a long time that water may affect the sensitivity of organic specimens to electron beam irradiation. Residual water vapor inside the electron microscope will condense onto cold organic surfaces and electron beam etching at this ice–organic surface interface has been shown to cause specimen damage (Heide, 1965); Hartman and Harman, 1971; Somlyo *et al.*, 1976). Even today, most electron

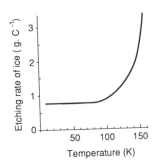

Figure 9.24. Mass loss of pure ice by electron irradiation as a function of temperature. [Redrawn from Heide (1984).]

microscopes still have a problem of residual water vapor which can condense onto cold specimens unless they are adequately protected by anticontamination devices. A small amount of water is an integral part of most biological molecules; many crystals contain water of crystallization and it is frequently difficult (if not impossible) to completely dehydrate many organic samples. These small amounts of water, which may well be in an unfrozen state at low temperatures, are probably sufficient to initiate radiation damage.

The first indication that there may be a radiolysis problem at the ice–organic interface in frozen-hydrated material comes from the work of Heide and Grund (1974), who showed bubbles and holes in their frozen specimens. Later work by Glaeser and Taylor (1978) and Talmon *et al.* (1979) showed similar effects when ice in contact with organic material was subjected to relatively high radiation doses. Bubbling has also been seen in vitrified sucrose and glycerol solutions (Talmon, 1982, 1984) and in pure vitreous ice (Dubochet *et al.*, 1982b). The radiation-induced bubbling phenomenon is similar in many respects to the radiation-induced bubbling seen in wet organic samples studied in an environmental chamber (Bourgeois *et al.*, 1980). The common factor in all these observations is that the observable effects of radiation damage occur at the interface between ice (water) and organic materials. Talmon (1987) discusses the mechanisms of radiation damage to organic inclusions in ice and considers that the effects are due primarily to the production of free radicals in the ice, which attack both the ice and the organic material. The decreased sensitivity of dry versus hydrated specimens tends to support this idea. Experiments have been carried out by a number of workers to try and understand this problem, and the following general features begin to emerge:

1. There is considerable damage at the ice–substrate interface where small ice crystals have been deliberately deposited on organic polymers films commonly used in electron microscopy, e.g., Formvar, collodion, and polyimides (Talmon *et al.*, 1979; Dubochet *et al.*, 1982b).

2. Although I_v deposited on carbon films may be etched away under the electron beam, the underlying substrate is only damaged as the last traces of the ice are removed. Dry carbon films are unaffected by the same radiation doses. Heide (1982b) and Talmon (1987) consider that this shows that a free surface is important for the radiolysis products of ice to cause any significant damage.

3. The extent of radiolysis in a model system of ice-embedded polymer latex spheres depends on the chemical nature of the polymer and whether the material is embedded in crystalline or vitreous ice. High radiation sensitivity polymers such as the polymethylmethacrylates (PMMA) are damaged at doses as low as $5 \, \mathrm{ke^- \, nm^{-2}}$, whereas low sensitivity polymers such as polystyrene show little damage even at doses as high as $40 \, \mathrm{ke^- \, nm^{-2}}$, even though damage can be observed in the ice surrounding the specimen (Talmon, 1984, 1987; Talmon *et al.*, 1986).

4. In radiation tests on beam sensitive PMMA, Talmon *et al.* (1986) found that material embedded in I_v would only show damage at a dose of $40 \, \mathrm{ke^- \, nm^{-2}}$, whereas material embedded in crystalline ice was damaged at $5 \, \mathrm{ke^- \, nm^{-2}}$.

5. Vitrified frozen-hydrated suspensions of natural and artificial surfactants are $10\times$ more radiation resistant than material embedded in a crystalline matrix (Talmon

et al., 1986; Bellaire *et al.*, 1988; Miller *et al.*, 1987). This cryoprotection effect is less pronounced in protein suspensions (Talmon, 1987).

9.5.7. Practical Consequences of Radiation Damage in Frozen-Hydrated Organic and Biological Samples

Although the general problem of radiation damage in electron microscopy and the particular problems associated with frozen-hydrated samples are being subjected to continuing scrutiny, some general information and guidelines are beginning to emerge:

1. Beam damage is reduced by a factor of 3–5 at temperatures between 90 and 150 K (LN_2 temperatures) in comparison to ambient temperatures. This effect is also seen in nonhydrated samples.

2. Although the primary ionization processes are unaffected by low temperatures, the diffusivity of free radicals is considerably slowed down at 90 K with a consequent lessening of radiation damage. In this respect, cryoprotection influences the chemical reactions that follow the initial ionization event. Henderson (1990) considers that at 90 K and below there is no reduction in the number of free radicals that are formed, but their movement away from the point of origin is slowed down. The protective effects of low temperatures are provided by a process of mechanical support for radiolytic products and for the undamaged molecules, which would otherwise be disrupted.

3. Cooling to 4–30 K may improve cryoprotection by a factor of 2 in addition to the advantage of increased electrical stability associated with superconducting lenses. Studies by Iwasaki *et al.* (1977) and Downing (1983) suggest that some free radicals are only significantly immobilized at temperatures below 30 K. The rate of mass loss is dramatically reduced at liquid helium temperatures.

4. Because ice is such a good source of free radicals when exposed to ionizing radiation, it is important to remove any excess water from inside the microscope and on the sample. This may be achieved by fitting a clean high-vacuum system and efficient anticontamination devices to the microscope, and using effective transfer devices when introducing frozen-hydrated samples into the TEM.

5. Unsupported, vitrified suspensions appear to be the most radiation resistant of all frozen-hydrated specimens.

6. Carbon or silicon monoxide should be used as support films for frozen hydrated-specimens rather than organic polymers films such as Formvar or collodion.

7. At very low temperatures, the use of conductive supports or coating with a thin conductive metal layer may help to diminish mass loss.

8. Organic material embedded in crystalline ice is more susceptible to radiation damage than material embedded in I_v.

9. Organic material partially embedded in ice is more susceptible to radiation damage than material completely embedded in ice, even though this may result in low contrast images.

10. Frozen-hydrated sections are particularly susceptible to radiation damage. They are invariably cut from material with a (micro?) crystalline ice matrix, are

usually contaminated with extraneous ice deposits, and can rarely be made to form close contact with the underlying support film.

11. Radiation damage is much less of a problem in structural studies using frozen-hydrated suspensions than with microanalytical investigation, which have to be made with thicker (0.5–$1.0\ \mu$m) frozen-hydrated sections.

12. It should always be assumed that *some* radiation damage will *always* occur in hydrated organic material examined and analyzed in electron beam instruments.

13. It should be appreciated that the resolution decreases in linear proportion to the logorithm of the accumulated dose (Glaeser and Taylor, 1978). As a rule of thumb, the resolution (in nanometers) that can be obtained from a vitrified specimen maintained at 90 K is one one thousandth the total electron dose applied to the sample (Dubochet *et al.*, 1988).

9.6. APPLICATIONS OF LOW-TEMPERATURE TEM

Although most of the applications of low-temperature TEM are from the biological sciences, the technique is beginning to be used on a wider variety of organic and inorganic samples. Rather than provide a catalogue of applications, it would seem more appropriate to consider the general categories of the types of hydrated and radiation sensitive materials that are best suited to investigation by these low-temperature procedures. If one was entirely honest, this would include *all* samples to be examined in electron beam instruments. The decrease in radiation sensitivity at LN_2 temperatures is beyond question as is the cleaner environment inside the microscope. Both factors substantially increase the quality of information that may be derived from an electron microscope image. Nearly all transmission electron microscopy would be improved by working at 150–90 K, and ambient temperature microscopy should be the exception rather than the rule. Before considering the general categories of applications, we should first briefly discuss the special cases of high-voltage cryo-TEM and low-temperature scanning tunneling microscopy.

9.6.1. Low-Temperature High-Voltage TEM

High-voltage cryo-TEM offers a number of advantages for the observation of radiation-sensitive and hydrated materials:

1. There is a lower characteristic energy loss of high-voltage electrons within the specimen resulting in significantly less beam-induced damage in the sample.
2. The higher-energy electrons have greater penetration, which enables thicker samples to be examined. Although such samples are easier to prepare, this advantage may be lost if such thicker sections are not cut from properly quench-cooled samples.
3. The decrease in spherical and chromatic abberation at higher beam voltages would be expected to result in improved spatial resolution in thin samples.

4. The large pole pieces and specimen chamber allow sturdier and potentially more stable cold stages to be used (Fotino and Giddings, 1985).

In spite of these advantages, very little low-temperature high-voltage TEM has been carried out. Fotino and his colleagues (Fotino, 1986; Fotino and Gidings, 1985) have built a top-entry cold stage and transfer device for a JEOL 1000 HVEM that operates down to 80 K. Images have been obtained of intact frozen-hydrated tissue culture cells up to 3 μm thick (O'Toole *et al.*, 1990), but in spite of the increased accelerating voltage, the intracellular ice obscures most of the cellular ultrastructure. Brink and Chiu (1991) have made use of intermediate high voltages (400 keV) in a study of a beam sensitive organic crystal, a C^{44} *n*-paraffin. They find that the contrast values are improved over those obtained at 100 keV and that a spatial resolution of 0.4 nm can be obtained routinely. It remains to be seen whether the full potential of combining low temperatures with HVEM can produce new information about radiation-sensitive hydrated material.

9.6.2. Low-Temperature Scanning Tunneling Microscopy

Scanning tunneling microscopy (STM) provides the opportunity of studying surfaces with atomic resolution (Binnig *et al.*, 1982) and is being used as a cryo-STM to investigate metal surfaces covered either with LN_2 or LHe_2 (Giambattista *et al.*, 1987). The samples are immersed in cryogenic liquids in order to keep them clean at low temperatures as well as diminishing thermal drift. The very fact that it is possible to obtain atomic resolution images from such preparations opens the way to applying this technique to study problems in catalysis, corrosion, electrochemistry, lubricants, electron beam lithography, and superconductivity (Berthe *et al.*, 1988). Cryo-STM has enormous potential in biology, although there have, so far, been very few applications (Fisher *et al.*, 1990a,b). It should, for example, be possible to examine the two faces of a membrane in vitrified samples immersed in LHe_2 at 4.2 K, provided of course, the problems of low sample conductance and high radiation sensitivity are resolved.

9.6.3. Applications of Low-Temperature TEM

A review of the literature reveals that several hundred papers have been published in which important information has been obtained using low-temperature TEM. A well-illustrated range of applications may be found in the comprehensive review paper by Dubochet *et al.* (1988). Some additional examples covering a broader range of applications are given below.

One of the most significant achievements of high-resolution cryo-TEM and image analysis is the 0.35-nm resolution atomic model of bacteriorhodopsin reported by Henderson (1990) as a result of combined studies in Cambridge, Berlin, and Berkeley. The same techniques are now being applied to the bacterial membrane pore forming protein, porin (Jap *et al.*, 1990; Sass *et al.*, 1990). At the molecular level, Taylor (1978) has studied the crystal structure of catalase, Degn (1987) has obtained high-resolution images of a neurotoxic protein, crotoxin, embedded in I_v, and Brink et al. (1989) have used cryo-TEM at two focus settings in a computer image analysis of two-dimensional crystals of NADH : ubiquinone oxidoreductase. Macromolecules such as actin filaments (Trinick et al., 1986; Lepault et al., 1991), microtubules (Mandelkow, 1987; Wade et al., 1989), and tropomyosin (Milligan and Flicker, 1987) have all been imaged at high resolution in the frozen-hydrated state. Ribosomes subunits suspended in I_v have been imaged by Heide and Zeitler (1985) and reconstructions achieved by Penczek et al. (1989); chromatin fibers have been studied by Athey et al. (1987) and eukaryotic flagella investigated by Murray (1986). An interesting artifact became apparent in the study of flagella. Flagella embedded in a very thin layer of water were always wider than those embedded in a thick water layer. This variation is thought to be due to compression of the flagella between two air–water interfaces before freezing. This suggests that very thin frozen suspensions may not provide the best embedding procedure for cryoelectron microscopy. The structures of many plant and animal viruses have been determined by electron microscopy and x-ray crystallography (Lepault, 1985; Adrian et al., 1984). Cryo-electron microscopy and image analysis is now being used to investigate the "native" three-dimensional structure of viruses such as the rhesus rotovirus (Yeager et al., 1990) and the complex that forms between intact virus and antibody (Prasad et al., 1990). A wide range of membranes have been studied by cryo-TEM including post synaptic membranes in fish (Unwin et al., 1988; Toyoshima and Unwin, 1988), the purple membrane of Halobacterium (Jaffe and Glaeser, 1987), and the tubular photosynthetic membranes of the bacterium Thiocapsa (Rachel et al.,1986). The same methods are being used to study phospholipid monolayers and bilayers, unilamellar vesicles and liposomes (Frederik et al., 1989, 1991; Lucken, 1990). All the applications considered so far have been of samples prepared as frozen suspensions. Although frozen-hydrated sections of biological material such as the ever popular rat liver/kidney have been casually imaged, there are very few studies where frozen sections have been used on a regular basis to solve a particular problem. The study on frozen-hydrated E. coli and Staph. aureus by Dubochet et al. (1983) and the high-resolution study of DNA in vitrified section (Richter and Dubochet, 1990) are two good examples of what may be achieved.

Cryo-TEM has not been the sole preserve of the biological sciences. A large amount of work has been carried out on solid gases (Klein and Venables, 1976; Tatlock, 1982). The structure and transformations of high-temperature superconductors are an obvious application of cryo-TEM. It has been used to study phase transformations in ceramics (Sugiyama et al., 1990) and for the direct observation of colloidal sols and the formation of ceramic sol-gel transformations (Bailey and Mecaetney, 1988; Bellaire et al., 1987). Tazaki et al. (1988) have examined the water-absorbing capacity of smectites which form at the edge of clay particles as fresh rocks are weathered in Arctic soils. Following on from earlier work on the study of organic crystals—see Fryer and Holland (1983) for a brief review—a number of workers including Vinson (1988), Miller et al. (1987), and Talmon (1987), have been studying frozen suspensions of surface active agents, microemulsions, and polymers.

All the nonbiological applications, like the majority of their biological counter-parts, have been carried out on frozen suspensions. In spite of a few notable excep-tions, frozen sections do not feature prominently in cryo-TEM. The uncertainty and difficulties surrounding all aspects of the preparation, transfer, and examination makes them unreliable specimens for structural studies. Although cryo-TEM of frozen-hydrated suspensions is the optimal approach for ultrahigh-resolution micros-copy, it is not the best approach for medium- to low-resolution studies on intact and fractured surfaces. Low-temperature SEM is better suited for these types of study, and this is the form of microscopy we will discuss in the following chapter.

9.7. SUMMARY

Low-temperature TEM can be used for examining either thin frozen sections or thin film suspensions up to 100 nm thick. The frozen sections are best examined either unsupported on fine mesh grids or on grids covered with a thin carbon film. Frozen suspensions, which give information far superior to that given by frozen sections, are prepared by vitrifying thin liquid films formed either on unsupported grids or on holey, carbon-coated grids in a controlled humidity environment. It is necessary to use a transfer device to bring both types of specimen from the low-temperature preparation equipment to the microscope cold stage. There are two types of cold stages that may be used for cryo-TEM: side entry and top entry. These stages, which may be single or double tilt, may be cooled either by LN_2 or by LHe, and some have been designed to give better than 0.5 nm spatial resolution. The cold stages have to be used in conjunction with low-temperature anticontaminators and in a microscope pumped to a high clean vacuum.

Because frozen-hydrated specimens are composed primarily of low-atomic-weight material, amplitude contrast only plays a small part in image formation. The principal image-forming mechanism is phase contrast, in which the phase shift between the diffracted and undiffracted wave is used to produce the image. The main problem with producing an image from thin frozen-hydrated specimens is the conflict between the high beam doses required for an adequate signal–noise ratio and the low beam doses needed to avoid beam damage. Although alternative imaging modes can be used to improve image contrast and information transfer, radiation damage still remains a major problem. Provided the sample is kept below 123 K, beam heating has a negligible effect in low-temperature TEM. Although the processes of ionizing radiation damage are still not fully understood, enough is known to recog-nize the ways by which the high-energy electrons used in imaging quickly destroy samples. Frozen-hydrated organic material is especially sensitive to these predations owing to the production of free radicals that arise from the interactions between electrons and ice. All the effects are dose dependent, and by carefully controlled low dose imaging coupled with computer-enhanced image analysis, it is possible to obtain high-resolution images from a wide range of biological macromolecules and hydrated organic material in their native state.

10

Low-Temperature Scanning Electron Microscopy

10.1. INTRODUCTION

The scanning electron microscope (SEM) is primarily associated with the examination of the surfaces of specimens. In this chapter we will be concerned with the ways the SEM can be used to obtain information about the surfaces of frozen specimens that retain a substantial part of their natural water content. The low-temperature scanning electron microscope (LTSEM) is frequently used as the primary instrumentation for x-ray analytical studies, and many of the preparative techniques discussed here are common to both procedures. The analytical aspects will be considered in in the next chapter. As was the case in the previous chapter, on low-temperature transmission microscopy, it is not proposed to discuss the use of the SEM at ambient temperatures for the examination of specimens prepared by low-temperature methods.

Detailed information about electron optics and image formation in the SEM is available in the now standard texts on the subject (Goldstein *et al.*, 1992; Newbury *et al.*, 1986). In the SEM, a finely focused spot (3–5 nm) of electrons is continuously scanned across the specimen line by line in the form of a raster. Any phenomenon caused by the interaction of this primary beam with the specimen is monitored by suitable detectors and displayed as an intensity variation on a synchronously scanned TV screen. The signal that is generated at a particular point at the surface is determined by the characteristics of the surface, such as its composition, texture, or topography. The image is thus built up point by point in contrast to the TEM where the whole range is formed at the same time. An alternative to continuous beam rastering is discrete rastering using a digital scan generator in which the beam remains for a finite time at each point or pixel on the specimen. This digital method of scan generation may be conveniently linked to digital computers, which are now being used with great effect to manipulate, enhance, and store images and analytical information. As we will show in this and the ensuing chapter on low-temperature microanalysis, digital scanning techniques, which are now available on many modern scanning electron microscopes, provide a useful method for the analysis of images and deconvolution of analytical data.

The versatility of the SEM for the study of solids is derived in large measure from the wide variety of interactions that the primary beam undergoes within the specimen. The inelastic and elastic interactions lead to the generation of back-scattered, secondary, and Auger electrons, x-ray photons, long-wavelength electromagnetic radiation in the visible, UV, and IR regions, phonons, and plasmons. In principle, all these interactions can be used to obtain information about the specimen that relates to its shape, composition, crystal structure, electronic structure, and internal electric or magnetic fields.

In addition to providing a wide range of information about the sample, a finely focused electron probe permits a large depth of field in focus at any one time. This allows easy visualization of the three-dimensional structure of an object without recourse to sections or replicas, and at a spatial resolution better than can be obtained by confocal light microscopy. The large depth of focus means that very rough or highly sculptured objects may be seen in focus across the whole specimen and provides the strong three-dimensional effect and visual appeal of scanning electron micrographs. In this respect, a scanning electron micrograph can more readily be understood by a nonspecialist than a transmission electron micrograph. This is important when it comes to interpreting the rough fracture faces which are frequently examined in LTSEM.

Scanning electron microscopy became commercially available 30 years after the first TEM, yet within five years of their appearance, the first low-temperature images were published (Echlin *et al.*, 1970). There are many reasons for the more rapid advance in low-temperature SEM than in TEM. The large specimen chamber of the SEM made it more easy to construct cold stages, and the late 1960s was at the beginning of the time when people were seriously concerned with using nonchemical methods in sample preparation for electron microscopy.

The singular advantage of LTSEM is that it is the only way to visualize the natural triphasic structure of materials such as lung and leaf tissue, foams and foods, and soils. The conventional preparative techniques remove the liquid phases and distort the solid phases, and even freeze-drying produces a sample in which the gaseous and liquid phases are indistinguishable.

Another attribute is that it is a relatively fast technique for obtaining medium high resolution, i.e., 50–100 nm information about a wide range of samples. With the appropriate equipment, hydrated samples can be frozen, fractured, coated, and imaged within a few minutes, for the first time allowing high-resolution visualization of dynamic processes occurring in the aqueous phase. In addition, the necessary ancillary preparative equipment associated with LTSEM allows a single sample to be sequentially analyzed by a process of repeated fracturing further into or across the sample.

Although it is a relatively young technique, LTSEM is probably more widely used than LTTEM, even though the two techniques are used to study quite different problems. A survey of the literature shows, not surprisingly, that LTSEM has a particularly strong appeal in the plant sciences. Many plant tissues have large and humid internal surface areas, which along with pulmonary tissue is just the type of specimen best studied by these instruments.

LTSEM allows one to study quite large specimens, an important point when the topological and topographical interrelationships of the component parts are an integral part of the investigation. However, LTSEM does not generally have the high spatial resolution afforded by the freeze-fracture replication technique, which is centered on the TEM.

It is claimed that the frozen-hydrated fracture face of a cell or tissue is a true representation of the object in question and that it is the only way we may hope to study the natural properties of the sample. While this may be so, insofar as specimens are prepared without the dubious benefit of chemical intervention, the somewhat larger specimens that are generally examined invariably suffer from problems of ice crystal damage. Low-temperature SEM is not a panacea for all investigators, but like all good techniques should be used in concert with other methods.

There is no comprehensive review of LTSEM that could provide additional background information to what is written in this present chapter. The reviews by Beckett and Read (1986) and Read and Jeffree (1991), while covering many of the fundamental problems of instrumentation and sample preparation, are strongly biased toward biological specimens. In contrast, the review by Sargent (1988), although short on specific techniques, has a wide breadth of application. Bastacky *et al.* (1987b, 1988) have compiled a selective bibliography of LTSEM, while Marshall (1987) contains many of the references to the earlier work on the use of LTSEM for both morphology and analysis.

10.2. LOW-TEMPERATURE MODULES FOR SCANNING ELECTRON MICROSCOPES

10.2.1. Introduction

The specimen stages of SEM are of two general types. The first type is the drawer type, in which the stage runs on rails in the microscope chamber below the final probe-forming lens. In order to load a specimen, the chamber is allowed to come up to atmospheric pressure, the whole stage assembly is retracted from the microscope, and the sample is loaded *al fresco*. The type of stage has certain limitations when used in association with a low-temperature module. The other type of stage is fixed permanently inside the microscope chamber with access via an air lock. Such air locks can, in fact, be used equally well with the drawer-type stage, although in these cases the drawer remains firmly closed. Although some microscope makers, i.e., JEOL, Hitachi, and Zeiss, continue to build their own cold modules, the tendency is to rely on subsidiary suppliers, such as Balzers, VG Microtech Polaron, Gatan, and Oxford Instruments, to provide the low-temperature equipment.

In contrast to TEM stages, SEM stages are of a much sturdier construction and the specimens have a much larger freedom of movement. Typically movement in the x and y might be as much as ± 10 mm from the center point, and the z movement could be as much as 40 mm. In addition the samples can be tilted and rotated, in some cases, eucentrically. All SEM cold stages are in the form of a small module which is fixed on the top of the stage platform, and within certain limitations the specimen retains the same freedom of movement associated with ambient temperature SEM. A number of the cold modules and their associated cooling systems are easily fitted and removed from stage platforms, thus releasing the microscope for other uses.

Low-temperature modules for SEM are of three general types: liquid nitrogen cooled, liquid helium cooled, and Joule–Thompson devices, in which the cooling is achieved as a result of the rapid expansion of a high-pressure gas in an enclosed environment, so-called adiabatic cooling. Figure 10.1 shows the general features of these three types of stage.

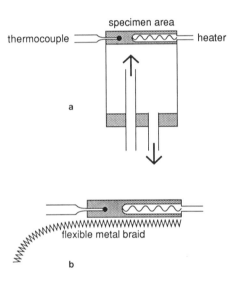

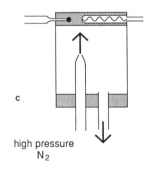

Figure 10.1. The three main types of cold stages used in low-temperature scanning electron microscopy. (a) Cooling by a direct supply of LN_2 or LN_2-cooled nitrogen gas from an external Dewar. (b) Cooling by means of a flexible metal braid connected to a sealed LN_2 reservoir in the specimen chamber. The reservoir is replenished from an externally located Dewar. (c) Cooling by means of a Joule–Thompson refrigerator in which the temperature is lowered by the adiabatic expansion of compressed nitrogen gas. [Modified and redrawn from Robards and Sleytr (1985).]

10.2.2. Liquid-Nitrogen-Cooled Low-Temperature Modules

The majority of SEM cold stages are of this type and they fall into three categories:

1. Liquid nitrogen is fed, either by gravity or low pressure, from a Dewar outside the microscope to a small sealed reservoir close to the low-temperature module sitting on the microscope stage. The reservoir is connected to the cold module by a flexible conductive coupling to allow it to be moved on the stage. Such cold modules, if constructed properly, can give temperatures at the module surface of about 110 K. The final temperature that may be obtained is influenced by a number of factors, of which the nature and form of the flexible coupling is most important:

   a. The flexible coupling should be made of a high thermal conductivity metal such as oxygen-free, high-purity copper, in the form of braid threads or flat leaflets. Robards and Crosby (1979) have shown that strands of fine copper thread are the most effective coupling material.
   b. The size of the flexible coupling should strike a balance between maximal cross-sectional area and maximal flexibility. It should not be stiff at low temperatures.
   c. The flexible copper strands should be very tightly clamped to both the liquid-nitrogen reservoir and the cold module using a minimum of solder, which is much less conductive than pure copper. In the cold stage designed by Saubermann *et al.* (1977), the ends of the copper threads actually lie free within the hollow liquid-nitrogen reservoir.
   d. The liquid-nitrogen reservoir should be placed as close as possible to the cold module while still maintaining sufficient stage movements (Taylor and Burgess, 1977; Marshall *et al.*, 1982a).
   e. If possible, the flexible conductive coupling should be shrouded in some form of insulating material, such as a series of lightweight plastic rings, to diminish heat input from the surrounding parts of the microscope chamber.

The cold module to which the specimen support is attached is usually quite small, and it is customary to fit it with a small electrical heater cartridge and thermocouples. The module is securely offset from the underlying stage platform with insulated supports such as nylon. The temperature control mechanisms are similar to those described in Chapter 8 (Section 8.3.6).

Flexible braid/thread cooled modules are found on many commercial scanning electron microscopes and form the basis of the design of the ancillary modules which may be fitted to existing specimen stages. There is much variation in the freedom of movement of the specimen. The x, y, and to a large extent the z movement is largely unaffected. It is usually possible to tilt the specimen by 30°–45°, but the amount of rotation is frequently limited to between 0°–150°. If the cold module is to be used in connection with x-ray microanalytical studies, it is important to ensure that the sample can be rotated and, if necessary, tilted in the direction of the detector in order to maximize the collection of x rays. Much of the success of these cold modules turns on establishing an effective solid but flexible thermal pathway from the LN_2 reservoir and to ensure that all mechanical connections are minimal in number and highly efficient at contact interfaces.

The flexible braid cold modules are favored by the few microscope manufacturers who supply their own cold stages. In practice, these stages were not very efficient and rarely get much colder than 150 K. The cold modules developed by Echlin and Moreton (1973), Saubermann and Echlin (1975), and Taylor and Burgess (1977) were all modified from commercial equipment. The VG Microtech, Polaron SP200A and the cryosystem described by Fujikawa *et al.* (1988) for use on the JEOL 840A SEM are examples of braid cooled cold stages.

2. A less popular, but nevertheless highly effective cold module, is where liquid nitrogen is circulated directly around the specimen area. Temperatures as low as 80 K can be achieved, but such a system has a number of disadvantages. In order to retain sufficient specimen movement, the liquid nitrogen has to be fed in by flexible lines. Flexible metal hoses are not sufficiently supple, and plastic tubing such as nylon becomes brittle at low temperatures and any sudden movement can break the tubing with spectacular effects on the vacuum of the SEM. Some success has been achieved with coiled plastic tubing which enters and leaves the cold module in wide sweeps. If only limited specimen movement is needed these are clearly the best cold stage, although movement of the cryogen in the system, either as a consequence of flow or bubbling, will cause problems of vibration.

An alternative approach is to use dry nitrogen gas that has been chilled by passage through a metal coil immersed in a Dewar of LN_2. The final temperatures are now as low (ca. 95–100 K) as may be obtained with a liquid-nitrogen-cooled stage but the problem of vibration is diminished. The Oxford Instruments CT 500A and VG Microtech, Polaron E7400 cold modules use this type of gas flow cooling with flexible plastic tubes as the feed lines.

The three types of LN_2-cooled stages can be used in the different reflective modes and with modification in the transmitted mode of operation of the SEM. Echlin and Moreton (1973) redesigned the specimen bucket on the Cambridge series 200 hot–cold stage so that it could be tilted up to more than 90° above the horizontal axis. A specimen holder was constructed with a 5-mm-high platform tongue projecting from its surface. A hole was made in this tongue to accommodate a transmitted electron holder. The specimen stage was simply tilted by 90° to ensure that the specimen was in a horizontal plane over the transmitted electron detector, which was placed below a hole in the bottom of the stage.

Taylor and Burgess (1977) modified the cold stage of a JEOL instrument, and Saubermann *et al.* (1977) have done the same for an AMRAY instrument, so that they could be used in the transmission mode. In these cases it was not necessary to tilt the stage to view the specimen. A similar type of stage has been designed by Ryan *et al.* (1985) for use with the JEOL JSM-35C SEM. The specimen holder is the tip of a TEM (Philips) specimen rod, which fits into a machined recess in the specimen stage. Tensioning springs ensure that there is good contact between the sample holder and the cold stage. This particular modification has the advantage that standard TEM grids can be used as specimen supports. Due attention has been made in all the modifications to ensure that samples can be transferred and examined at low temperatures without fear of melting or contamination.

Some of the new field emission TEM/SEM instruments such as the Hitachi-900 and JEOL JSM-890 microscopes are fitted with side-entry holders. High-resolution single and double tilt-entry cryoholders are available from Gatan (models 613 and 636) and can be made to fit different types of microscopes. A somewhat simpler side-entry cryoholder has been designed for the top stage of the Akashi DS-130 high-resolution SEM by Inoué and Koike (1989). The specimen holder is placed at the end of an insulated rod and is connected to a LN$_2$ tank via a copper braid.

10.2.3. Liquid-Helium-Cooled Modules

A number of liquid-helium cooled modules have been built for use with scanning microscopes. The bath cryostat principle has been employed by Seifert (1983) and Heubener and Seifert (1984) in studies on superconducting materials and crystals. A small tank containing about 30 ml of liquid helium is placed on the microscope stage assembly (Fig. 10.2). The top cover of the tank is in direct contact with liquid helium and is either the specimen itself, e.g., a round disk of sapphire, or it functions as a sample holder. The top cover is clamped into position and sealed with a ring of indium. A standard liquid-helium cryostat outside the microscope is used to replenish the tank inside the microscope, and temperatures between 1.5 and 4.2 K have been obtained. The cryostat and cold module are thoroughly shielded and precooled with liquid nitrogen. Bode *et al.* (1988) used a miniature evaporation helium cryostat mounted on the microscope stage for electron-beam-induced current and cathodoluminescent studies at ca. 5 K, and Bresse *et al.* (1987) used a continuous flow cryostat at 10 K in similar studies on semiconductors.

The Oxford Instruments liquid-helium cold stage unit is a good example of a continuous flow cryostat. Liquid helium passes from a storage vessel through a flexible transfer line to a heat exchanger in direct contact with the sample mounted on the SEM stage. The exhaust gas is passed through a flow controller and pump at room temperature. The sample is surrounded by an efficient radiation shroud cooled

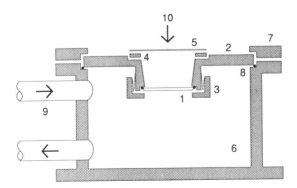

Figure 10.2. Liquid-helium-cooled cold stage for use in a scanning electron microscope. 1, Sample; 2, sample holder; 3, clamping screw to hold sample in position; 4, copper ring for wire heat sink; 5, fenestrated thermal shield; 6, liquid helium tank; 7, clamping ring; 8, indium seal; 9, liquid helium tubes; 10, axis of electron beam. [Redrawn from Heubener and Seifert (1984).]

by the exhaust gas. A base temperature of 20 K is reached in the CF301 module, and 4.5 K in the CF302 module. Depending on the module, movement of the sample is limited to a maximum of 10 mm in the x, y, and z directions with small tilt and rotation. Small electrical heaters and thermocouples are an integral part of all the liquid-helium cooling modules. The present design status of liquid-helium stages for SEM does not allow them to be used in conjunction with transmitted electrons. LTSEM at liquid-helium temperatures appears to be the exclusive preserve of material scientists and is currently being used to study high-T_c superconductors, semiconductors, and beam sensitive materials. The review article by Huebener (1988) provides a useful introduction to SEM at very low temperatures.

10.2.4. Joule–Thompson Refrigerator

This refrigerator allows dry nitrogen gas at a pressure of 1200 psi to expand to atmospheric pressure through a 10-μm nozzle into a vacuum-tight container below the cold module surface. The rapid expansion of the high-pressure gas results in adiabatic cooling, and the cold module can be quickly brought down to its working temperature of 90 K. Pawley and Norton (1978) used an Air Products (Allentown, Pennsylvania) Joule–Thompson refrigerator as the basis of a cold module for the AMRAY scanning electron microscopes. The refrigerator was integrated into a copper cooling module, which also contained a small heater and a thermocouple. Figure 10.3 shows a Joule–Thompson refrigerator attached to the stage of an AMRAY SEM. This type of cold module is of simple design and needs only to be connected to a supply of clean, dry, high-pressure nitrogen gas via a flexible, narrow high-pressure tube and a somewhat larger bore plastic tube to take away the exhaust gas. These two lightweight, highly flexible connections mean that the cold module has a full range of movement, including ±150° rotation. Only the region of the module in contact with the rapidly expanding gas gets cold; the remainder of the module and the supply and exhaust lines stay at ambient temperature. Because the cold module is so small, the temperature can be controlled to a large extent by simply adjusting the flow rate of the high-pressure gas. The module is cooled down to the working temperature using gas at a pressure of about 1500 psi (100 bar). Once the temperature is reached, the gas supply is throttled back to 1000 psi (66 bar) and about two working days may be obtained from a large cylinder of nitrogen at an initial pressure of 3600 psi (240 bar). When not in use, it is advisable to maintain a slight positive pressure of 15 psi (1 bar) in the gas lines to prevent any dust particles lodging in the small orifice in the refrigerator.

It is important to ensure that the inside of the nitrogen gas cylinders are clean and to use clean, dry nitrogen with a very low water content. The gas dryness is more important than its ultimate purity because even small amounts of moisture may condense as ice crystals, which can block the narrow tubes used for the passage of the gas. A number of Joule–Thompson refrigerators have been in use on AMRAY scanning electron microscopes for up to 10 years. The long life and reliability is even more remarkable when one realizes that the Joule–Thompson refrigerator used on the AMRAY instruments was designed for cooling electronic components in a ballistic missile. Under these circumstances a 30-min working lifetime would be quite

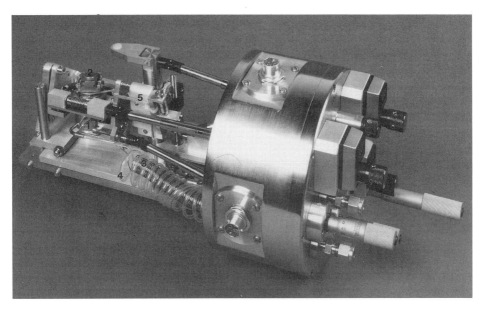

Figure 10.3. Joule–Thompson refrigerator on an AMRAY specimen stage. 1, Base of refrigerator; 2, specimen stub in a groove on the top of the refrigerator; 3, coiled stainless steel high-pressure gas supply to refrigerator; 4, coiled plastic tube for exhaust gas from refrigerator; 5, electrical connections for thermocouple and heater. The refrigerator sits in the stage assembly which retains full x, y and x movement together with rotation and tilt. (Picture courtesy AMRAY Inc.)

satisfactory. Apart from the occasional temporary blocking of the small orifice, these refrigerators are a most effective cold module for LTSEM. Their only disadvantage is that the present design and location of the refrigerator preclude using the SEM in the transmission mode of operation.

10.3. ANTICONTAMINATION DEVICES

In spite of the improved vacuum systems of modern scanning electron microscopes, the large volumes of the specimen chambers of these types of instruments are a potential source of contamination. The copper braid and internal LN_2 reservoir of the flexible braid type of cold module function as effective cold traps because they are invariably colder than the working surface of the cold module. In addition to the usual cold traps at the top of diffusion pumps or the use of oil-free turbomolecular pumps, it is advisable to fit and use LN_2-cooled anticontamination plates in the region around the specimen. Most SEMs can be fitted with such devices, which are usually in the form of a large flat copper plate just below the final lens and offset by a few millimeters with insulated plastic studs. The anticontamination plate should be cooled by its own LN_2 Dewar and can only be used continuously with the pre-pumped type of stage, which remains permanently inside the microscope column.

Wherever possible the cooling sequence should be to first cool down any LN_2 traps in the primary vacuum line of the microscope, then cool the anticontaminator, and finally cool down the cold module. The system will then be in a fit state to accommodate the precooled sample.

10.4. TRANSFER DEVICES

LTSEM of samples that do not contain water is relatively easy. The specimen is placed securely on the cold module on the specimen stage and the microscope pumped down to its working vacuum. Any liquid-nitrogen traps are filled, the specimen chamber anticontaminator cooled to its working temperature, and finally, the sample cooled to the temperature at which the investigations are to be made. It is this final slow cooling of the sample that precludes this approach being used with hydrated samples. Such samples are invariably quench cooled (and fractured or otherwise manipulated) outside the microscope and then transferred as a deep frozen sample onto the precooled cold module. The specimen must be protected during this transfer both from excessive warming and from contamination by condensing vapors.

A number of different types of transfer devices have been built that effectively protect the sample during this critical part of the preparative process. They all work on the same principle. The sample is kept cold by making sure it is in good contact with a previously cooled block of metal with a good thermal capacity. The sample is protected from the environment either by surrounding it with cold metal surfaces and/or dry nitrogen gas, by immersing it under LN_2, or by placing it in a good vacuum environment.

In the initial studies using the Cambridge Instruments (Leica U.K.) cold stage (Echlin et al., 1970), a very simple, but surprisingly effective, transfer method was used. A plastic bag, containing the frozen sample on its specimen stub in a small Dewar of LN_2, was securely taped to the outside of the specimen chamber. The bag was inflated with dry nitrogen by simply placing a small amount of LN_2 in the plastic bag. The microscope chamber was let up to atmospheric pressure using dry nitrogen gas and the drawer-type stage was withdrawn into the inflated plastic bag. The sample was quickly and firmly placed on the precooled stage, which was pushed back into the microscope column. All the manipulations could be done through the thin wall of the plastic bag. The sample transfer was completed within seconds, but it took several minutes for the specimen stage to pump down to its working vacuum.

This type of transfer system has now been discarded, because try as we may, a small amount of ice would sometimes condense on the surface of the sample as the system was not fitted with an effective anticontaminator. However, this simple method is included here because it is a way by which any SEM may be converted to an unsophisticated cryo-SEM. No cold module is needed, although it is necessary to modify both the sample holder and the top of the specimen stage. The sample holder should be constructed of a high heat capacity material (stainless steel is effective) and a thin heat-insulating layer of material (cork, plastic) placed on the specimen stage. The frozen sample is firmly fixed to the precooled modified sample stub and immersed in LN_2. Sample transfer is via the plastic bag and specimens will remain frozen for up to 30 min depending on the size and thermal capacity of the stub. As the sample is uncoated, it should be examined at low voltage in order to reduce surface charging. Although this is far from an ideal experimental process because there is invariably some contamination and surface charging, it is a beginning and will provide an opportunity of examining first frozen hydrated (FH) and then subsequently freeze dried (FD) material in the SEM.

Although the adhesive tape and plastic bag followed the Cambridge tradition of using string and sealing wax as a temporary but effective method of making seals on scientific equipment, it was clear that this would not be a satisfactory long-term solution. Echlin and Moreton (1973) and Marshall (1977a) sealed frozen specimens under metal caps, which fitted over the specimen stubs during transfer to and from the cold stages in an evaporator and an SEM. The caps were only removed when the working vacuum pressure was reached. The method was effective but cumbersome to use. Saubermann and Echlin (1975) made a much simpler transfer device in which the frozen sample, sitting on a large metal block precooled with LN_2, was withdrawn into a tubular metal shroud, which was also cooled with LN_2 (Fig. 10.4). The outer tube is sufficiently long to condense all vapor molecules before they arrived at the surface of the specimen. Saubermann et al. (1981a) made an improved model, which incorporated a vacuum flange and a small plug (Fig. 10.5). The specimens in the capped tubular devices were loaded onto the cold module on the SEM stage via a small port on the side of the microscope. Similar systems were devised by Hutchinson (1974) and Lechner (1974).

As LTSEM studies became more critical, particularly when the instrument was being used in connection with x-ray analytical studies, it was necessary to use an air-lock transfer device which would load specimens directly onto the precooled cold module maintained at high vacuum inside the microscope. A number of portable air-lock devices have been described, each designed for a particular SEM (Fuchs and Lindermann, 1975; Zierold, 1976; Saubermann et al., 1977; Fuchs et al., 1978; Robards and Crosby, 1979; Lichtenegger and Max, 1980; Sitte, 1982; Perlov, 1983). All the devices follow the same general construction, in which the frozen specimen is contained within an evacuated chamber, the walls of which are cooled by LN_2. A gate valve enables the transfer device to be interfaced to the microscope and to the various pieces of preparative equipment (Fig. 10.6). The Robards and Crosby (1979) transfer device provides double protection for the specimen during transfer. The

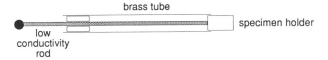

Figure 10.4. A simple device for transferring cold specimens onto the cold stage of an SEM. The lower half of the tube and large thermal mass specimen holder assembly is first cooled with LN_2. The specimen is picked up under LN_2 and retracted into the cold tube, which protects it from environmental contamination. [Redrawn from Saubermann and Echlin (1975).]

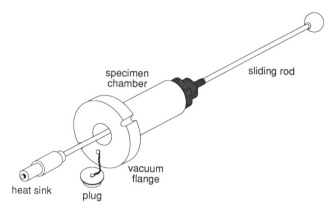

Figure 10.5. Schematic drawing of a sealable transfer system with the heat sink shown in the extended position. The insulating Delrin cylinder with vacuum flange can be sealed with the small attached plug. This type of device is used to transfer frozen sections from the cryomicrotome to the cold stage of the SEM via an airlock. [Redrawn from Saubermann *et al.* (1981).]

specimen stub has its own concentric cold metal shrouds which enclose the specimen while it is inside the evacuated transfer device. The apotheosis of transfer devices is where they are part of more elaborate preparative equipment such as the VG Microtech, Polaron SP200A sputter cryo system, the Oxford Instrument CT1500 Cryotrans, the VG Microtech, Polaron E7400/7450 Cryotrans system, and the Balzers SCU 020 SEM cryo unit, all of which can be made to interface to different SEMs.

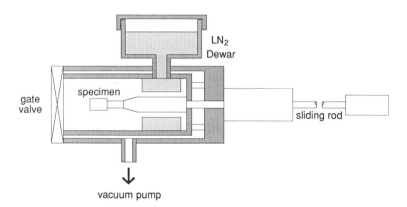

Figure 10.6. A simple vacuum device for transferring cold specimens to the cold stage of an SEM. The gate valve interfaces with the preparative equipment and with an air lock on the side of the microscope. The cold sample is initially picked up under LN_2 and withdrawn into an inner chamber cooled with LN_2, which protects it from contamination. The large thermal mass of the specimen holder keeps the specimen cool. The gate valve is closed and the inner chamber is evacuated to ca. 10^{-3} Torr (133 mPa). Frozen samples may be transferred between different pieces of cryopreparative equipment either directly or via airlocks. [Drawn from Cambridge Scientific Instruments (Leica UK plc).]

The Gatan 626 cryotransfer system, which combines both a cryoholder and a transfer system, is designed for use with side-entry TEM/SEM instruments. A number of the microscope manufacturers also make transfer devices for their particular instruments.

This detailed discussion of transfer devices has been quite deliberate because in many respects transfer is one of the most critical parts of the preparative procedure. Unless due care is taken, a well-prepared specimen can all too easily be ruined during transfer between different components of the preparative instrumentation. The suggestion (Marshall, 1987) that simple transfer systems are as effective as complex systems is misleading. The simple, i.e., non-air-lock systems, may suffice for low-resolution studies, but the vacuum-pumped systems are critical for high-resolution and analytical studies. But even these systems may not be entirely suitable because every time one moves a frozen sample from one piece of equipment to another, there is always a risk of contamination. As we will shortly demonstrate, the preparation involves three distinct phases (quench cooling, fracturing/sectioning, and coating), which would involve three transfers if carried out on separate pieces of equipment. This increased risk of transfer-mediated contamination has prompted the development of preparative devices attached to the microscope column, in which many of the procedures are carried out. The sample remains in a clean, cold environment all the time. The quality of the environment may be conveniently monitored using a small quadrupole mass spectrometer attached to the side of the experiment equipment (Echlin *et al.*, 1975; Saubermann and Echlin, 1975; Ryan *et al.*, 1985). Lawton *et al.* (1989) have modified the crystal oscillator from a quartz thin film monitor as an inexpensive detector for water vapor.

10.5. CRYOPREPARATION DEVICES

The nature and operation of these devices have been discussed at some length in Chapter 5. They are either dedicated and directly attached to the microscope, or nondedicated, where the complete sample preparation is carried out in the cryopreparation unit and the prepared sample transferred to the microscope cold module. There is no dedicated cryopreparation that will handle all aspects of sample preparation, i.e., quench cooling, fracturing/sectioning, coating, and direct transfer into the microscope.

In the dedicated Oxford Instruments cryotrans CT1500, the AMRAY Biochamber, the VG Microtech, Polaron Cryotrans E7400, and the Balzers SCU 020, the frozen sample is transferred to the preparation unit via an air-lock transfer device. The sample is fractured and coated in the preparation unit and transferred directly via a second air lock to the microscope cold module. A similar system is described by Aldrian (1980). Nondedicated systems include the VG Microtech, Polaron 7450 and SP200A and the Oxford Instruments CP2000. In these systems the specimen is first quench cooled and transferred via an air-lock system to the unit where the sample is fractured and coated. A second transfer then moves the sample to the column of the microscope. Most home-made cryopreparation units (see, for example, Pesheck *et al.*, 1981; Marshall, 1987; Marshall, 1980c; Marshall and Carde, 1983;

Oates and Potts, 1985) are based on the two-transfer system and in many respects this is the best way to proceed, provided adequate sample transfer facilities are available. The coating facilities available on the commercially produced cryopreparation units can only be described as adequate. Although they give satisfactory coating layers for low-resolution morphological studies, these thin films are not satisfactory for high-resolution and analytical studies. It would seem far more satisfactory to carry out the fracturing and coating in a dedicated coating unit fitted with low-temperature stages, so that one could take advantage of the progress that has been made in coating technology for nonconductive specimens. Similarly, it is probably better to do the quench cooling under idealized conditions—for details see Chapter 3. These advantages must, however, be set against the disadvantage of faulty transfer between the various pieces of equipment. In situations where frozen sections are to be studied in the SEM, the frozen sections are usually cut *al fresco* and transferred *in vacuo* to the scanning microscope. The examination of frozen sections is most usually associated with analytical studies, and we will return to a more detailed discussion of how such sections should be handled in the following chapter.

10.6. SPECIMEN PREPARATION

10.6.1. Introduction

In comparison of LTTEM, sample preparation for LTSEM is quite complex and involves several discrete but vitally interconnected steps. Some of these steps, viz., quench cooling, fracturing, and sectioning, have already been discussed in considerable detail in earlier chapters. Other steps such as etching and coating will be considered here. Most LTSEM is concerned with looking at bulk samples, either intact or fractured. The size of the specimen, and hence the size of ice crystals that form in the sample, will to a large extent dictate the actual resolution that may be obtained in the SEM. There seems no point in attempting to cool rapidly a large sample that is already fixed to a specimen support, although this must be the approach to adopt with small specimens that are attached later to the specimen stub.

10.6.2. Sample Selection

There are three principal ways samples may be selected for LTSEM:

1. The sample may be fixed to the specimen holder at room temperature and the sample and specimen quench cooled. This invariably results in a large piece of material which can never be cooled fast enough to ensure small ice crystals in the sample. There is little point in trying to cool such samples rapidly and they are best cooled in melting nitrogen at 65 K.
2. Small pieces, i.e., less than 0.5 mm, of samples may be cooled by one of the methods outlined in Chapter 3. The frozen samples are attached to the large specimen holder by some mechanical means such as a vicelike or springlike fitting under liquid nitrogen.

3. Small pieces of tissue are placed inside some form of small container such as thin-walled metal tubes, which are then quench frozen. These small sample holders are loaded into the larger sample holders under LN_2.

10.6.3. Sample Specimen Holders

The sample specimen holders for SEM are usually quite bulky and made of a high conductive metal such as copper, brass, aluminum, or duralium. Although these large pieces of metal (and the associated specimen) cool far too slowly, once they are cold they remain cold and help maintain a sample in the FH state during transfer between cold modules. Another advantage of having large specimen holders is that they may be cut, drilled, threaded, and machined to accommodate much smaller specimen holders containing samples that have been properly quench frozen, and/ or specimens with a particular geometry. The samples must be firmly attached to the specimen holder in order that they remain cold and remain in position during the fracturing process. Firm attachment is achieved either by modifying the flat surface so that the sample is retained in good mechanical contact, or by sticking the sample to the surface with some form of adhesive (see Section 10.6.4).

Grooves may be cut into the sample holder, into which flat specimens can be placed prior to cross fracture. Holes in the sample holder can be used to accommodate drops of fluid samples and can be made to measure for cylindrical samples. Larger holes can be drilled in the sample holder to accommodate a series of smaller specimen holders. Figure 10.7 shows a 12-mm-diameter specimen holder with four

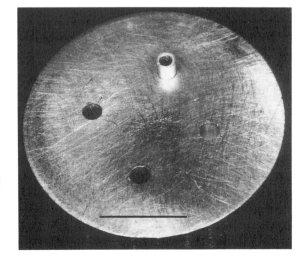

Figure 10.7. Duralium specimen holder precision drilled with tapered holes to take four small specimen tubes. The tubes, which can be made of silver, beryllium, or graphite, are 1×3 mm with an inside diameter of 0.6 mm. The tubes are loaded into the holes under LN_2 and the whole holder transferred under LN_2 and screwed into the pre-cooled cold stage of the microscope. Marker = 5 mm.

holes to accommodate small tubes made of silver, Be, or C. Bastacky *et al.* (1987a, 1990) fitted a movable grooved slide to the surface of the AMRAY specimen holder. Small blocks of tissue, with precisely cut parallel sides, are placed in the well of the sample holder and held in position by the movable grooved slide secured by a small Allen screw. Circular saw blades 0.125 mm thick with embedded synthetic diamond dust were used to carefully trim the samples held under liquid nitrogen. Unlike the much earlier work of Lechene *et al.* (1975), the saw was not used to expose the region of interest; this was done during the ensuing fracturing process. Jeffree *et al.* (1987) has devised a series of different sample holders for flat specimens such as leaves, including a device that gave complementary fractures. A similar hinged device for complementary fractures has also been designed by Beckett and Porter (1988). Wergin and Erbe (1989) have made a series of accessory holders which may be used with the VG Microtech, Polaron SP200A system.

In many cases, the specimen holders remain unaltered during use other than gently abrading the surface in order that the sample remain firmly attached during cooling and fracturing. The specimens may either be placed directly onto the metal surface, or—in the case of very wet specimens such as droplet suspensions—may be first placed on a filter paper or Millipore filter, which can then be attached to the specimen stub after the excess liquid has been drained away.

An important feature of all specimen stubs is that they should be capable of being quickly and easily inserted into the cold module of the SEM. Experience has shown that this is best achieved using grooved holders held in position by spring clips or stubs which can be screwed into a thread inserted on the cold module. The sample must make good thermal (and electrical) contact with the cold module; otherwise the specimen will rapidly dry inside the high vacuum of the microscope. All sample holders should be thoroughly dried before use. A small amount of moisture on a finely threaded screw or in a small hole will form a solid sliver of ice, which will prevent pieces of metal from forming good mechanical contact.

10.6.4. Sample Attachment

Samples can be attached to the specimen stub by mechanical means or by the use of different forms of adhesives. Mechanical attachment requires alteration to the specimen holder; a number of such modifications have been discussed in the previous section. Wherever possible, and this is largely dictated by the nature of the sample, the samples should be placed in small specimen holders, which have a better chance of being more effectively quench cooled. In a study of *Lemna* roots, Echlin *et al.* (1979) inserted 2–3-mm tips of small, ca. 250-μm diameter, roots into short, 1–2-mm lengths of silver tube filled with artificial pond water so that various lengths of root protruded from the end of the tube (Fig. 10.8). These miniature vases of roots were quenched in an SN_2/LN_2 slush and then mounted under LN_2 into the holes of the screw-threaded specimen holder. The holes were carefully tapered to ensure that the specimen ends of the tubes were flush with the top of the sample holder and the base of the tube was held securely in position. In later experiments involving x-ray

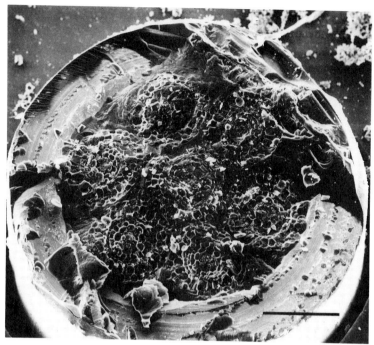

Figure 10.8. A low-temperature SEM image of seven frozen-hydrated cross-fractured root tips of Duckweed (*Lemna minor L.*) previously loaded into the type of small silver tube shown in Fig. 10.7 and quench cooled in LN_2/SN_2 slush. Samples were fractured in vacuo in an AMRAY Biochamber, coated with 8 nm of gold and photographed at 110 K, 15 keV, and 12 pA beam current. Bar = 200 μm.

microanalysis, carbon and beryllium tubes were used (Echlin *et al.*, 1982). Pesheck *et al.* (1981) used the hollowed-out heads of aluminum machine screws as sample holders. They also covered prefrozen samples with very thin copper foils, which acted as deformable shims as the frozen samples were press-fitted into holes in the specimen holders. The thin aluminum foil found in some cigarette packets is ideal for this purpose—cooking foil is too thick and thin gold foil, although ideal, is far too expensive. Carr *et al.* (1986) wrapped trimmed frozen samples in aluminum foil, and then clamped them into position on a copper stub.

The underlying principle that should be adopted in all forms of mechanical attachment is that the sample is held firmly and in good thermal contact with the sample holder. Many biological samples have a natural stickiness due to associated or exuded mucopolysaccharides or proteins. In these circumstances it is only necessary to place the moist sample onto the surface of the sample stub. In our first attempts at LTSEM, we examined tissue culture cells which had been persuaded to grow on the surface of small disks of platinum placed in the cell culture (Echlin *et al.*, 1970). After quench cooling, the platinum disks were attached to the specimen stub with silver dag.

A variety of adhesives have been used to hold specimens on the sample stub. They should have the following properties:

1. They should be conductive.
2. They should not interfere with the sample; this limits the choice of materials either to water-immiscible materials or to buffered salts solutions closely similar to any fluids within the sample.
3. They should solidify at the low temperatures used to examine the sample.

One of the best adhesives for biological material is a mixture of colloidal graphite made up with water. The graphite slurry is reasonably conductive and provides a good support around the sample. It has the additional advantage of being planed smooth after cryofracture to provide a smooth, low background surface suitable for both morphological and analytical studies (Preston et al., 1982). Various modifications have been made to this basic mixture, including incorporating a small amount of polyvinylpyrrolidone (PVP) or hydroxyethyl starch (HES) as an antifreeze agent (Echlin et al., 1982), making up the slurry in a balanced salts solution (Echlin and Taylor, 1986), making up the colloidal carbon in Tissue Tek (Allan-Wojtas and Yang, 1987; Brooks and Small, 1988), and polyvinyl cement (Carr et al., 1980).

A wide range of organic liquids have been used as mounting media including n-heptane (Steinbrecht and Zierold, 1984), n-butylbenzene (Saubermann and Echlin, 1975), toluene (Karp et al., 1982), trichloroethylene (Pearce, 1988), and glycerol, polyethylene glycol, or high-vacuum grease, Apiezon M (Jeffree et al., 1987).

Success has also been achieved using more complex mixtures of materials such as a yeast–sucrose solution (Eveling and McCall, 1983) or commercially available electroconductive silver and carbon paints (Ryan et al., 1988). Liquid metals and metal alloys such as mercury (mp 234 K), gallium (mp 303 K), and Woods metal (343 K) have been used to mount geological samples (Pesheck et al., 1981). Mercury was not satisfactory because of its large exponential coefficient, and gallium attacks some metals. Woods metal was satisfactory although its rather high melting point created problems with small frozen specimens. There are a number of tertiary eutectic alloys of gallium, indium, tin, and zinc which have melting points between 283 and 300 K, and it would be instructive to explore their usefulness as mounting media for frozen specimens.

The aqueous and nonaqueous mounting media are used in separate ways. With aqueous media such as colloidal graphite paste, the samples are either attached to or mounted in the material at room temperature, which is then quench cooled. Samples to be mounted in nonaqueous media are first quench cooled and then mounted in the medium, which is held just above its melting point. This can be conveniently carried out by having a well in the sample holder which is filled with a small amount of the particular mounting medium. The sample holder is cooled to just above the melting point of the mounting media and the prequenched sample, kept in LN_2, is quickly placed in the little well using LN_2-cooled fine forceps. The low temperature of the sample will rapidly cool the mounting medium to below its melting point and the sample will be set in position with some of the material appearing above the solidified surface. The stubs are then immediately returned to the LN$_2$ Dewar.

With biological and hydrated samples it is important to ensure that the specimens are maintained in a fully hydrated state up to the point where they are quench frozen. In many instances the samples can be excised and quench cooled either directly or after mounting in the sample holder within a few seconds. If the sample preparation is likely to be protracted, the specimen mounting area should be enclosed in some form of glove box and the walls lined with absorbent paper soaked in water.

10.6.5. Sample Cooling

These procedures have been considered at great length in Chapter 3 and need not be repeated here. Suffice it to say that in situations where the sample (usually small) has been preloaded onto a specimen holder (usually large), the sample holder should be held in such a way as to ensure that the specimen is the first object to enter the cryogen. SN$_2$/LN$_2$ slush is usually the cryogen of choice, even though it is far from ideal. As already indicated, a large holder and small samples cools slowly, and the use of organic liquid cryogens would only marginally improve the cooling rate. A second, and more important, reason for not using organic liquid cryogens is that they will coat the sample holder and make it difficult to achieve good thermal contact with the precooled cold stage, which is frequently at temperatures below the melting point of the organic cryogen. The thin layer of organic cryogen on the sample holder will solidify on contact with the cold module and result in poor thermal contact between the two metal surfaces. It is more satisfactory to first quench cool small samples, which are then securely mounted under LN$_2$ into the larger sample holders. Bastacky *et al.* (1987a) have built a small cryoprobe which may be used for *in situ* quench cooling of small samples within a larger piece of tissue. The probe is a highly polished copper disk soldered onto a stainless steel cylinder. LN$_2$ is sprayed onto the inner surface of the disk by means of a flexible connection. The probe can be quickly precooled and easily moved to a precise location on the sample. A similar effect has been achieved using a small block of organic cryogen which had been solidified in LN$_2$. The halocarbon or ethane "popsicles" are applied directly to the region of the specimen to be cooled. Before the sample holder and specimen are transferred either to the precooled cold module of the SEM or to the next part of the preparative chain, it should be thoroughly "washed" in clean LN$_2$, and if necessary any surface deposits of ice removed with a fine brush.

10.6.6. Sample Transfer

The actual method of transfer will be determined by the equipment. It is important that all transfer rods, capping devices, tools, etc. are thoroughly dry and the cold module and fracturing device of the next piece of equipment to be used is adequately precooled. The actual transfer should be carried out as speedily and carefully as possible to avoid contamination and/or warming of the sample.

10.6.7. Fracturing

Although some samples are examined intact, many specimens are fractured to reveal their internal contents. The fracturing process has the added advantage of producing a clean surface devoid of any contaminating ice crystals. The process of fracturing has already been described and should only be carried out after the sample has reached thermal equivalence on the cold module of the SEM. The fractures should be made decisively with a micrometer-controlled knife which should not scrape along the sample surface. It may be necessary to make a series of secondary fractures to obtain a relatively planar surface. The most effective fractures are made by allowing the knife to traverse the specimen in a continual upward direction. Alternatively, the frozen surface may be cryoplaned by passing a cold knife over the surface several times. Although this produces a more planar surface, it is invariably severely scoured with knife marks. The pieces of material removed during the fracturing process should remain either on the cold knife or on the underlying cold module. If these small frozen chips are allowed to warm up, they can act as a potential source of contamination of the freshly cleaned surface. Jeffree *et al.* (1987) and Beckett and Porter (1988) describe modifications to the specimen stub used with the VG Microtech, Polaron SP2000 sputter cryo which will allow complementary fractures to be produced. Flat samples (e.g., leaves) may be mounted between two clamping plates onto a hinged holder. The stub is held in the open position during quench cooling and transfer into the preparation unit. Once the sample has been loaded onto the cold stage and has reached thermal equivalence in a clean environment, the hinge is closed to produce two complementary fracture faces. The modifications described by Jeffree *et al.* (1987) allow both complementary transverse and paradermal fractures to be poduced (Fig. 10.9), which have been used to great effect in studies on the way fungal hyphae are attached to host cell walls in infected leaves (Fig. 10.10).

10.6.8. Micromanipulation

In addition to being fractured, frozen samples may also be dissected by micromanipulators. Hayes and Koch (1975) were the first to use this technique and showed that by using cooled micromanipulator tips it was possible to dissect, in real time, single frozen-hydrated cells and break off and collect specific parts of cells, which could be chemically analyzed. More recently, Fujikawa *et al.* (1988, 1990) have shown how the technique can be used to make continuous obervations on the same sample by repeatedly cutting out and replicating specific parts of a frozen sample.

10.6.9. Etching

This is a process where material is sublimed from the surface of the sample, either to remove a layer of contamination or to quite genuinely etch into the sample surface to reveal underlying structural features. Beckett and Read (1986) suggest that samples are etched for the following reasons:

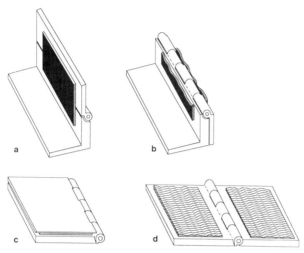

Figure 10.9. Complementary fracture device. The vertical hinge shown in (a) and (b) is used to obtain complementary cross fractures of a leaf. The initial configuration is shown in (a) with the leaf sample mounted prior to quench cooling. In (b) the hinge is closed under LN_2, and shows the two complementary cryofracture surfaces. The hinge assembly shown in (c) and (d) is used to obtain paradermal fractures of flat specimens such as leaves. In (c) the flat specimen is between the two faces of the closed hinge prior to quench cooling. In (d) the hinge has been opened under LN_2 exposing the two complementary fractures. [Modified and redrawn from Jeffree *et al.* (1987).]

1. To remove surface ice, which has been derived either from the natural environment or as a result of contamination.
2. To provide information regarding extracellular spaces.
3. To accentuate the differences between gas filled and fluid filled spaces.
4. To enhance the mass density differential between aqueous and organic regions of fracture faces.
5. To provide information on the amount and distribution of water in and on the sample, based on the segregation zones initiated by ice crystal formation.

The etching (sublimation) occurs when the temperature of the sample is raised to ca. 170 K while maintaining it in a high vacuum, ca. 100 μPa. There are two ways this thermal input may be achieved. The temperature of the whole sample can either be raised from below using the electrical heater embedded in the cold module, or just the surface of the sample can be warmed with a radiant heater situated above the sample (Fig. 10.11). Irrespective of the procedure used, the process of etching is empirical and the extent of the etching is apparent only after the process is completed. There are disadvantages and advantages to both procedures even with the most carefully controlled cold module. Heating the whole sample from below will generate a thermal gradient across the specimen; the top will be colder than the bottom. Only when the critical temperature is reached at the surface of the sample will the ice begin to sublime, even though the temperature inside the sample may be well above the critical sublimation temperature. Once sublimation occurs, it will proceed very

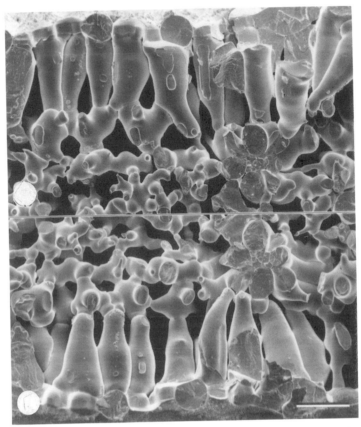

Figure 10.10. Complementary cross fracture of a frozen-hydrated french bean leaf (*Phaseolus vulgaris L.*). The samples were quench cooled by plunging onto nitrogen slush, fractured (but not etched), and sputter coated with gold at low temperature. Samples photographed at low temperature at 5.0 kV. Bar marker = 50 μm. [Picture courtesy Read and Jeffree (1991).]

quickly into the sample because the water vapor from the warmer ice deeper in the specimen now has an easy escape route to the surface. This water vapor either will escape from the surface or may recondense onto the cold outer parts, which may have already freeze-dried. There can be no set etching time, as samples vary in size and water content, and different cold modules have different thermal capacity and controls.

Robards and Sleytr (1985) argue that etching by heating the whole sample is more controllable because the process may be observed in the SEM. Experience by this present author (see Echlin *et al.*, 1979 *et seq.*) shows that this is not the case because once surface sublimation is observed, it is too late to stop the rest of the sample from freeze-drying. Even with the most efficient cold module, the poor thermal conductivity of the sample prevents it from being rapidly recooled to below the sublimation temperature.

Figure 10.11. A diagrammatic representation of the two ways the surface of frozen-hydrated samples may be etched. In (a) the whole specimen is gradually heated from *below* to the etching temperature. In this procedure, the specimen surface is the last region to warm up, which means the whole sample may be freeze-dried. In (b), the surface of the specimen is heated from *above* using a radiant heater while the bulk of the specimen remains cold. In this procedure, only the surface of the specimen is etched, which would allow repeated cryofractures to be made of the underlying frozen-hydrated sample. [Modified and redrawn from Robards and Sleytr (1985).]

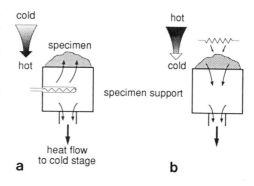

Far more reproducible results have been obtained by subliming away the surface layers of ice using a low-wattage radiant heater placed a few millimeters above the frozen surface. Experiments by Echlin and Burgess (1977), Echlin *et al.* (1979), and Muller *et al.* (1991) showed that the surface etching rate could be controlled either by periodically observing changes in the microscope or by measuring the rise in water vapor in the system. The bulk of the sample remained frozen and it could be refractured to reveal fully frozen-hydrated tissue. Subsequent calculations by Talmon (1980) showed that the rate of sublimation could be calculated as a function of the radiant energy and the vacuum in the system. In practice, the sample is kept as cold as possible and placed under the radiant heater for predetermined period of time.

Irrespective of whether the etching is achieved by heating the bulk or the surface of the sample, it is important that the subliming surface is close to an efficient and cold anticontamination plate to trap the subliming water molecules.

In the early days of LTSEM it was generally considered necessary to etch all fracture faces to be examined in the microscope. In SEM studies, the etching is usually quite deep in comparison to the freeze-fracture procedures used in TEM studies. The principal reason for etching is that it is difficult to see very much in freshly fractured fully hydrated samples. However, with experience these new images of frozen-hydrated material become more easily understood, and we have now reached the point where etching is the exception rather than the rule for morphological studies and, as we will show in the next chapter, is contraindicated for bulk analytical studies. Controlled etching remains an effective way of removing surface contaminants, and in this respect etching by radiant heating is far more easily controlled. Wherever possible, the etching should be carried out in an ancillary chamber to the microscope, although the progress of etching may be periodically checked in the SEM. Once the etching is completed the sample temperature should be quickly reduced to below the recrystallization point.

372

10.6.10. Sample Coating

Frozen-hydrated samples have poor bulk and surface conductivity and generally charge in the electron beam, i.e., they rapidly build up an electrical potential equal to that of the primary electron beam. This is not surprising considering the poor conductivity of the ice at low temperatures (Durand *et al.*, 1967; Hobbs, 1974). This charging results in image distortion, deflection of the primary beam, and a reduction in the energy of the incoming electrons. The amount of charging is dependent on the beam current and voltage and the temperature and state of hydration of the sample, although the experimental evidence is ambiguous at times.

The papers by Marshall (1975, 1980c), Echlin and Moreton (1973), and Echlin (1978a, 1978b, 1978c) contain a detailed discussion of the charging phenomena in frozen samples. The studies by Brombach (1975) and Fuchs *et al.* (1975) have shed some light on what happens when a high-energy electron beam interacts with frozen specimens. These workers placed electrodes in and around model specimens of pure ice in order to measure the specimen current at different positions in the sample. They concluded that charging is not a serious problem, provided that the sample is mounted on a good conductor and that the mean depth of penetration of the electrons is more than half the thickness of the specimen. Thin samples would not charge, a feature borne out by many observations on frozen sections, but bulk samples would charge, even if irradiated at quite low accelerating voltage. Fuchs and colleagues showed that a space charge built up just below the surface of uncoated frozen samples. This space charge had a flat, pancakelike profile unlike the more usual teardrop-shaped electron interactive volume associated with bulk samples (Fig. 10.12). They found that the application of a coating layer to the surface would effectively eliminate surface charging but would not necessarily reduce the internal space charging. The studies of Fuchs and his colleagues, and in particular the work by Brombach (1975), has had a persuasive influence on the way we considered the interaction of a fine probe of electrons with a bulk frozen-hydrated sample. It went some way in explaining the ambiguous charging phenomena occasionally found with frozen samples. Although the electrical conductivity of a frozen-hydrated sample is small, it may be larger than the conductivity of freeze-dried material. Such differences are probably related to the continuity of frozen-hydrated material compared to the spatial discontinuity of the dried material. Marshall and Condron (1985a, b) found that the backscattered electron yield rises rapidly in charging uncoated frozen-hydrated samples, whereas it remains more or less constant in samples that have been

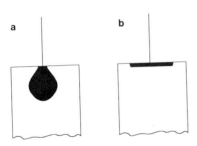

Figure 10.12. Calculated electron diffusion volume is coated and uncoated frozen-hydrated specimens. In (a), the normal, pear-shaped diffusion volume is shown in an uncoated frozen-hydrated sample. In (b), the pancake-shaped diffusion volume is shown in a frozen-hydrated sample which has been lightly coated with a conductive layer. The surface conductive film dissipates the charge and limits the space charge field within the bulk of the sample. [Drawn from data of Fuchs *et al.* (1978).]

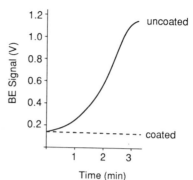

Figure 10.13. Plot of the amplified backscattered electron-signal from coated and uncoated frozen-hydrated gelatine gels at constant beam current. [Redrawn from Marshall and Condron (1985).]

coated (Fig. 10.13). The proposed change in shape of the interactive volume, i.e., pancake shape versus teardrop, remains an enigma. The studies by Brombach *et al.* (1975) were on pure ice and have never been repeated or extended to frozen biological or hydrated organic samples. As we will see in the next section and in the chapter dealing with x-ray microanalysis, the shape and dimensions of the interactive volume have an effect on the quality of the signal(s) that may be obtained from the sample and the ultimate spatial resolution of the microscope system. All the available evidence would suggest that we do need to coat frozen-hydrated samples with a thin conductive layer if we are going to use the SEM to its full advantage. The technology and procedures for coating are commonplace and have been in use for some time; extra care is needed when coating frozen samples.

Space will not permit a detailed discussion of the coating procedures used in SEM, but a good description is given in the papers by Echlin (1978a, b) and standard texts on SEM by Goldstein *et al.* (1992) and Newbury *et al.* (1986). Although a wide range of techniques are available for applying thin conductive films to nonconducting samples examined by conventional SEM, only two techniques are in general use in LTSEM. Evaporative and sputter coating can be used to effectively eliminate the surface charging problems, though the applied coating layer does little to eliminate the space charge which may hold up inside the sample due to electron penetration below the surface.

The evaporative and sputter-coating techniques which will now be described are those that are specifically designed for use with LTSEM that is being carried out at relatively low magnification and hence low spatial resolution. While we should not be complacent about and less careful with such coating procedures, we do not yet have to be generally involved with high-resolution coating methods which involve exotic target materials such as tungsten–rhenium alloys and expensive and complicated equipment such as ion beam sputtering and electron beam evaporation.

10.6.10.1. Evaporative Coating

Many metals, when heated in a vacuum (1.3 mPa or 10^{-5} Torr), begin to evaporate rapidly into a monoatomic state when their temperature has been raised sufficiently for the vapor pressure to reach a value in excess of 1.3 Pa (10^{-2} Torr). The evaporated atoms travel in straight lines and form an effective and continuous coating layer on those parts of the sample that are placed in line of sight to the evaporant source. The noble metals and their alloys are used as coating materials in low-temperature morphological studies, and can be applied either using a conventional vacuum evaporator fitted with an air lock and cold table (see Echlin and Burgess, 1977; or Marshall, 1980d) or using the ancillary evaporation head associated with the AMRAY Biochamber (Pawley and Norton, 1978).

The techniques and processes that are used are the same as for ambient temperature coating. The evaporant, e.g., a predetermined length of gold wire, is wrapped around a refractory conductor, e.g., a piece of tungsten wire. The evaporant source is attached to the electrodes and the vacuum evaporation unit pumped out to its working pressure of 1.3 mPa–130 μPa (10^{-5}–10^{-6} Torr). Sufficient electric current is passed through the evaporant/electrode assembly to heat it to a dull cherry-red color. This will allow the assembly to outgas and burn off any contaminants. The power is turned off and the cold module is cooled to its working temperature. When the temperature of the cold module is low enough, the sample may be introduced into the preparative chamber via an air lock. The frozen sample is allowed to reach thermal equilibrium on the cold module, and if possible a movable shutter is placed between the sample and the evaporant source. The evaporant source is slowly reheated to the point at which the metal melts and forms a shimmering molten bead on the refractory wire support. This process should only be observed through appropriate dark glasses. When this point is reached, the actual evaporation process can take place. If appropriate, the sample should now be fractured and quickly moved into position below the evaporation source. Power is applied to the electrodes and the shutter between the source and the sample is opened. Because the evaporated metal only travels in straight lines, the frozen sample should be rotated and tilted during the coating process to ensure that it receives a complete and even coating. This is one of the disadvantages of evaporative coating and it is frequently difficult to coat complex fracture surfaces adequately. Another disadvantage is that the evaporation source reaches quite high temperatures (ca. 3500 K with carbon) and the radiant energy may be sufficient to cause incipient ice sublimation on exposed surfaces of the frozen sample. For this reason the specimen should not be placed close to the heat source, and the input power to the evaporation source should be kept as low as possible.

The geometric arrangement of the evaporation head on the AMRAY Biochamber overcomes many of these problems. The evaporant source is sufficiently well collimated to ensure that the cone of evaporation just covers the diameter of the specimen stub. The frozen sample sits on a metal shuttle, which may be moved to different positions on the cold module. Immediately after fracturing, the cold shuttle is moved to the top of a large precooled metal cube, which may be rotated and tilted during evaporation. During the initial warm-up and outgassing phase, the specimen remains out of the line of sight between the source and the sample. As the metal evaporation starts, and this may be observed through a port hole on the side of the evaporation head, the cold cube with the shuttle and specimen is moved into position under the evaporant source and immediately rotated and tilted by means of the exterior control mechanism. This ensures that even the most complexly sculptured surface is coated properly, and because the sample is moving rapidly during the coating process the radiant heat load into the sample is minimized. Sheehan (1990) has built a separate coating chamber, which is placed between the cryopreparation chamber and the column of the microscope. This allows continuous, 2-nm layers of chromium to be evaporated onto flat, frozen-hydrated paper surfaces, maintained at high vacuum and low temperatures. There are no facilities to move the sample during coating. A different approach has been adopted by Inoué and Koike (1989), who have fitted a small evaporation coater directly to one of the side ports of the SEM. The sample holder may be temporarily disconnected from its cooling block and tilted at 90° toward the small cooling unit. The specimen is coated with metal from a direction perpendicular to the sample surface. A small fenestrated aluminum plate is placed between the evaporator and sample holder to reduce contamination of the specimen chamber with evaporated metal and to minimize thermal damage to the sample. Presumably the sample can be moved through the tilting arc during coating. The results obtained with a 3–5-mm layer of gold are quite impressive.

Evaporative coating has been used in a relatively unsophisticated manner by a number of workers to coat frozen samples. No attempt is made to measure the coating thickness, but from the appearance of the image, it looks as if most samples are coated with between 10 and 15 nm of gold. The noble metals such as gold, gold-palladium, and platinum are relatively easy to apply; other metals such as aluminum have proved difficult as it invariably forms a poorly conducting black oxide on contact with frozen specimens (Echlin and Moreton, 1973). The noble metals should not be used if x-ray microanalysis is going to be carried out on the specimens. Although we will consider specific coating procedures for low-temperature x-ray microanalysis in the following chapter, the metal chromium is a useful compromise (Marshall, 1977a, 1980d). It is easy to evaporate and provides a good conductive layer with a high yield of the secondary and backscattered electrons which are important in the formation of SEM images. The metal is of sufficiently low atomic number not to give rise to spurious x-ray signals which might interfere with the elements being analyzed. Carbon is not a good material as it requires a very high evaporation temperature (3500 K) and there is a serious risk of heat damage to the frozen sample. This risk is diminished if carbon fiber is used as the evaporative source (Peters, 1984a, b). Although evaporative techniques can provide good conductive thin metal films on frozen specimens without any surface damage, most of the more recent SEM studies on frozen samples have relied on sputter coating to provide the conductive layer. A technique for high-resolution thin-film (1.5 nm) shadowing with a mixture of platinum–iridium–carbon has been recently intoduced by Wepf et al. (1991). The method is designed specifically for biological macromolecular structures which are to be examined by TEM, STM, and SEM at a spatial resolution of ca. 1 nm.

10.6.10.2. Sputter Coating

Sputtering occurs when an energetic ion or neutral atom strikes the surface of a target and imparts momentum to the atoms over a range of a few nanometers.

Some of the target atoms receive enough energy in the collision to break the bonds with their neighbors and they may be dislodged. If the velocity imparted to them is sufficiently high, the target atoms are carried away from the surface by momentum transfer. The bombarding ions or neutral atoms are usually one of the heavier noble gases, such as argon, and for most of the work on LTSEM the target is one of the noble metals or their alloys.

The purple glow discharge normally associated with sputter coating is located some distance from the target and is a result of electrons being ejected from the negatively charged target. Under the influence of an applied voltage, the electrons accelerate and may collide with a bombarding gas molecule, leaving behind an ion and an extra free electron. The positive ions are accelerated toward the negatively biased target where they cause sputtering. The ejected target atoms have a mean free path of ca. 4 mm and suffer multiple collisions with gas molecules in the residual vacuum. As a consequence, the target atoms strike the sample from all directions. At high accelerating voltages, many electrons are ejected per impinging ion and these electrons have sufficient energy to damage delicate targets. This is now no longer a problem as Panayi et al. (1977) showed that electron bombardment of the specimen could be virtually eliminated by placing a permanent magnet at the center of an annular target. Such an arrangement deflects any electrons away from the specimen. Much of the impact energy of the positive ions appears as heat in the target, which makes sputtering a rather inefficient process. Undue temperature rises in the target can also create problems with frozen-hydrated specimens.

A number of factors affect the deposition rate and in turn the likelihood of damage to the frozen specimen. The sputtering yield increases with the energy of the bombarding gas and there is an approximate increase in sputtering with current. Impurities in the bombarding gas such as CO_2 and H_2O decreases the deposition rate. The presence of CO_2, H_2O, and O_2 in the gas plasma can give rise to highly reactive ions which can damage the sample. Finally, the deposition rate is higher the closer the target is to the specimen, but also increases the heat load into the sample.

On the basis of the foregoing discussion, it is possible to list the features that would be desirable in a sputter coater to be used to coat frozen-hydrated samples:

1. The target material should be one of the noble metals or their alloys.
2. The bombarding gas should be pure dry argon.
3. The target geometry should be designed to give high sputtering rates.
4. The sample should be protected from stray electrons by placing a high-flux rare earth permanent magnet behind the target. In addition to sweeping aside any electrons, such magnets localize the energized gas plasma to a small part of the target and ensure high sputtering rates.
5. The sample should be kept cold, i.e., at least 123 K, and protected from any thermal radiation generated in the target. The grain size of sputter-coated films becomes finer at low specimen temperature owing to facilitated coalescence of the metal film (Echlin et al., 1978b; Muller et al., 1991; Walter et al., 1990).

6. In general one should aim at using the lowest possible voltage and current commensurate with obtaining the thinnest satisfactory coating layer within a reasonable period of time in order to minimize the thermal input into the specimen. If sample heating is a problem, it is better to sputter coat intermittently at somewhat higher energy input, rather than coat continuously with a lower energy input. The periodic coating allows the target heat to cool down between sputtering, whereas there is a slow rise in temperature with continuous sputtering.

7. The sputtering should be carried out in a clean, contamination-free environment. Unlike evaporative coating, sputter coating is carried out at a relatively poor vacuum of 13 Pa, ca. 10^{-1} Torr. However, this should be a clean environment, which is easily obtained by repeated flushing with argon.

A number of pieces of equipment are now available which enable FH samples to be sputter coated for LTSEM. The sputtering units available on the VG Microtech, Polaron and Oxford Instruments cryopreparation units appear to be based on the original design concept of Robards and Crosby (1979) and Robards *et al.* (1981). The frozen sample sits on a cold module and is virtually surrounded by a cold shroud, except for a small circular cutout immediately below the retractable gold target. The target is fitted with a quadrupole magnet, which has the effect of increasing the rate of sputtering while at the same time diminishing the input of stray electrons on the sample. The chamber is pumped down to a good vacuum of 130 mPa dry argon. The high-resolution sputter head operates at a low energy input (from 20 mA at 1.0 kV to 25 mA at 0.44 kV) and a pressure of 13 Pa (10^{-1} Torr). During a typical coating run of 3–4 min the sample temperature rises by no more than 10 K. For samples maintained at ca. 110 K, such a small temperature rise would have no effect on the frozen-hydrated surface. No attempt has been made to measure the coating thickness, which is estimated to be about 10 nm thick. The recently introduced Balzers SCU 020 preparation and cooling unit is similar to the other pieces of equipment except that the unit has its own turbo molecular pump to provide a high and clean initial vacuum before backfilling with argon. Sputter coating is carried out using a planar magnetron sputtering head (Muller *et al.*, 1991). These workers found that a 3-nm layer of platinum was sufficient to prevent charging in a frozen-hydrated sample. In dried materials it was found necessary to add an additional 10 nm of carbon to prevent charging. Walter *et al.* (1990) have used this technique to obtain the high-resolution images shown in Fig. 10.14.

Although it is theoretically possible to convert an ambient temperature diode sputter coater to work at low temperatures, this has not in fact been done. The conversion itself would be quite simple; there is, however, the added complication of providing a suitable air lock to enable frozen-hydrated samples to be transferred without fear of contamination and incipient melting. An alternative approach has been adopted by Lindroth *et al.* (1991), who have converted a freeze-drying unit to a low-temperature sputter coater. The system can sputter 1.5-nm layers of tungsten onto samples maintained at 183 K. All the sputter coating of frozen samples is being carried out using coating modules which are part of a more complex cryopreparation

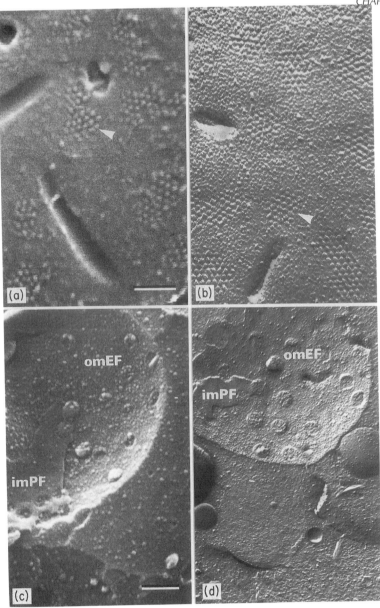

Figure 10.14. Correlative high-resolution low-temperature SEM images, (a) and (c), and ambient temperature transmission electron microscope images of replicas, (b) and (d), of freeze-fractured yeast cells. (a),
(b), protoplasmic fracture faces of plasma membranes showing hexagonally arranged intramembranous
particles. These hexagonal arrays, which have a periodicity of 16.5 nm, are typical of yeast cells in the
stationary growth phase. Bar = 100 nm. (c), (d), cross-fractured yeast cells showing the endoplasmic face
of the outer membrane (omEF) and protoplasmic face of the inner membrane (inPF) of the nuclear
envelope. Note the nuclear pores within the nuclear envelope. Bar = 200 nm. Samples (a) and (c) cryofixed
by plunging into liquid propane and then fractured and coated with 3 nm of platinum in the Balzers
SCU 020 cryopreparation unit. Examined fully frozen-hydrated in a Hitachi S-800 field emission SEM at
10 keV. [Pictures courtesy of Walther *et al.* (1990).]

unit. Sputter coating of frozen-hydrated samples at low temperature is now a more or less routine procedure, and a wide range of results have been obtained. Its ease of use and avoidance of sample damage makes it the coating technique of choice.

10.6.10.3. Uncoated Samples

Some preliminary results from Pawley *et al.* (1991) indicate that good morphological images can be obtained from uncoated materials provided they are examined in one of the newly designed low-voltage SEM. The thin samples are placed on a gold foil, which provides good conductance, and are imaged at between 1.5 and 3.0 keV.

10.7. SPECIMEN EXAMINATION

10.7.1. Electron-Beam–Specimen Interactions

Specimens subjected to electron beam irradiation in an SEM exhibit a series of complex interactions that give rise to a variety of signals which may be detected and imaged in the microscope. In comparison with the TEM, which essentially derives information from scattered electrons as they pass through the thin sample, the SEM derives information from scattered electrons *and* photons as they are reflected from the surface of bulk samples and transmitted through thin samples.

The different types of signal that may be detected in the SEM are shown in Fig. 10.15. All of the signals can be used in association with LTSEM, although the majority of morphological studies have been carried out using the secondary electron signal. We will briefly discuss each of the specimen–beam interactions in turn in

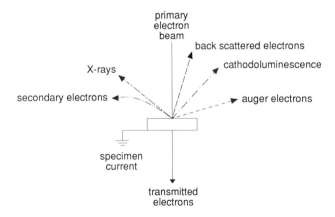

Figure 10.15. An illustration of the different signals that may be generated by the interaction of the primary electron beam with a specimen in the SEM.

relation to low-temperature microscopy, with the exception of x-ray photons, which form the basis of the final chapter in this book.

10.7.2. Signal Generation

In order to understand the different types of specimen–beam interactions it is first necessary to understand the way these signals are generated and detected. The primary beam electrons penetrate the sample to a depth that is directly dependent on their energy and inversely dependent upon the density of the sample. Thus for a given density, a high-energy primary beam (e.g., 30 keV) penetrates deeper into the sample than a low-energy beam (e.g., 3 keV). For a given beam energy, there is deeper penetration into a low-density material, e.g., a freeze-dried organic sample, than into a higher-density material, e.g., a frozen-hydrated organic sample. Figure 10.16 is an idealized diagrammatic representation of the variation of electron scattering in the specimen as a function of the primary beam voltage and density. It is immediately obvious that there are differences in both the size and shape of the excitation or interactive volume, and, depending on the nature of the signal being collected to form the image, there is a considerable difference in the spatial resolution of the acquired signal. Figure 10.17 is a diagrammatic representation of those parts of the interactive volume that give rise to the different signals that may be used in SEM. Thus for a given voltage, the backscattered electron signal is derived from a much larger part of the interactive volume than the secondary electron signal. These differences are reflected in the fact that the spatial resolution that may be obtained with the SE signal is generally better than that which may be obtained using the BSE signal.

10.7.3. The Nature of the Different Signals

Within the volume of primary excitation, a number of different scattering events occur, depending on the relative amount of energy lost by the primary beam electrons during their interaction with the sample. These interactions can generally be divided into two classes:

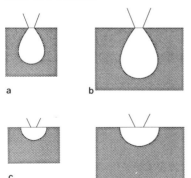

Figure 10.16. Diagrammatic representation of the variations in electron scattering in a specimen as a function of sample density and beam voltage. (a) Low-voltage–low-density; (b) high-voltage–low-density; (c) low-voltage–high-density, and (d) high-voltage–high-density.

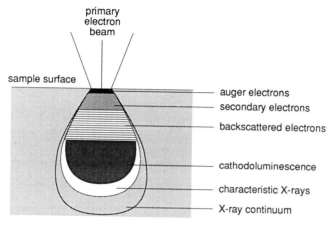

Figure 10.17. Cross section of the microvolume of primary excitation showing the different regions from which signals may be generated. Thus, the secondary electron signal is derived from a much smaller microvolume than the cathodoluminescence and x-ray signal. These variations in the size of the interactive microvolume affect the spatial resolution of the different signals which are used in SEM.

1. Elastic events, which affect the trajectory of the beam electrons within the sample without significantly altering their energy. The electrons that are scattered are referred to as backscattered electrons, and they can be used to provide both morphological and chemical information about the sample.
2. Inelastic events, which result in a transfer of energy to the sample and the generation of a whole range of secondary signals including secondary and Auger electrons, x-ray and longer-wave UV, visible light and IR photons, and other signals, which will not be considered here, e.g., phonons and plasmons.

A large amount of information is available about the generation and detection of these different signals (Goldstein *et al.*, 1992; Newbury *et al.*, 1986). We will not repeat this information here, but will briefly consider the ways these different signals can be used to advantage in LTSEM.

10.7.4. Backscattered Electron Signal

Backscattered electrons (BSE) are those electrons that have undergone single or multiple scattering events in the sample and have escaped through the surface of the specimen. The signal is composed of a wide range of energies, although most are thought to contain at least 80% of their original energy. Because they are so energetic they will produce the same types of signals produced by the primary beam, e.g., secondary electrons on x-rays. The BSE coefficient, i.e., the number of BSE produced per incoming primary beam electron, increases with atomic number and with the tilt angle of the specimen. Because of their high energy, the BSE travel in straight lines and only those electrons directly in line with the detector will be used to form the image. The scintillator-photomultiplier detector used in the SEM will collect both the low-energy secondary electrons (see Section 10.7.5) and the higher-energy backscattered electrons. In order to collect only the BSE, a small negative bias is applied to the front of the detector to repeal the low-energy secondary electrons. The interactive volume from which BSE are emitted is much larger than that from which the SE are emitted. There are several advantages to using the BSE signal to obtain images of frozen samples:

1. True topographic images may be obtained, particularly when a low takeoff angle is used. Because the BSE travel in straight lines, the detector will only image those parts of the sample that are in direct line of sight. This will give rise to useful shadowing effects, particularly on fracture faces. These images are useful in distinguishing those parts of the complexly sculptured fracture face that are going to give a strong x-ray signal (see Chapter 11).
2. Because the BSE have such a high energy, the image they form is less influenced by local charging. It is possible to obtain quite reasonable BSE images from poorly coated or even uncoated samples examined at low kV.
3. Because of the large size of the interactive volume, the spatial resolution is less good, i.e., ca. 100–200 nm, than that which may be obtained with the SE signal.
4. Because the BSE signal is strongly influenced by atomic number, it is possible for these atomic number contrast effects to form images.

In work on calcium oxalate crystals in frozen-hydrated tobacco leaf tissue, it was possible to distinguish the crystals embedded in a matrix composed largely of water (ice) and low atomic number biological material. The study was carried out at low kV on uncoated fractures of frozen-hydrated bulk tissue (unpublished). J. M. Cook (personal communication) has been able to clearly distinguish baryte deposits in frozen-hydrated fractures of wet drilling mud filter cake (Fig. 10.18).

10.7.5. Secondary Electron Signal

Secondary electrons are characterized by having a very low energy, i.e., less than 50 eV, and mostly are emitted from the first 5–10 nm of the specimen surface. Other secondary electrons are formed deeper within the sample as a consequence of BSE-induced ionization. The low-energy SE are collected by applying a small positive bias as the front of the detector. This has the effect of pulling electrons from around corners and is one of the reasons that SEM images have such a marked three-dimensional effect. The SE signal is the principal imaging mode for examining surface structure. Figure 10.19 shows an SE image of frozen-hydrated fracture of meristematic cells from a plant root. Its advantages in LTSEM are as follows:

1. It will show up fine structure detail (3–5 nm) on properly coated samples. A 5-nm layer of gold/palladium is usually sufficient.
2. The weak SE signal is strongly influenced by charging effects although it is possible to obtain images from uncoated samples which are examined at low kV.

Figure 10.18. SEM images of frozen-hydrated oil field mud filtercake. The sample was prepared from a typical water based oil well drilling mud, consisting of water with about 9 vol.% of barium sulfate (a dense mineral) and about 3 vol% of bentonite (a gel-forming clay for control of viscosity and filtration). The mud is filtered through a hard filter paper to form a mud cake which is quench frozen in LN₂, fractured at low temperature under vacuum, lightly etched with an overhead radiant heater and sputter coated with a very thin layer of gold. Photographs taken at 130 K and 15 keV. Figure (a) is a secondary electron image showing a background of ice containing thin sheets of bentonite clay and larger, more equiaxed, particles of barium sulfate. Edge contrast reveals the bentonite. Figure (b) is a back-scattered electron image of the same area. The contrast between the ice and bentonite is low, but the dense barium sulfate shows up with high brightness. Bar marker = 10 μm. (Pictures courtesy John Cook, Schlumberger Ltd, Cambridge.)

3. The SE signal can be used to great effect in obtaining good images from deeply fractured samples.
4. Contaminating ice, i.e., ice deposited on the sample after it has been coated and introduced into the microscope, can be easily observed because of its relatively high emissivity of secondary electrons. The reason for this is as follows. The contaminating ice is usually in poor contact with the underlying conductive surface. The electron beam causes the ice deposit to charge, and it is this bright signal that is observed. Excessive ice contamination can lead to considerable charging and streaking on the recorded image. Recent studies by Pawley *et al.* (1991) and Herter *et al.* (1991), using a field emission SEM equipped with a cold stage, have shown that it is possible to obtain high-resolution information using both the BSE and the SE signal from coated (3 nm Pt) and uncoated FH samples maintained at 123 K. However, the

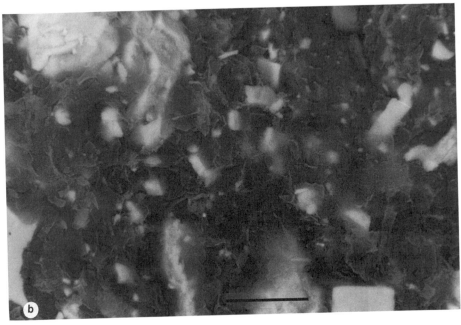

Figure 10.18. Continued.

optimal imaging of cell surfaces at low temperatures was only achieved by a combination of freeze substitution and critical point drying followed by a thin metal coating (Fig. 10.20).

10.7.6. Cathodoluminescence

Certain materials radiate visible, UV, and IR light when irradiated with an electron beam. This arises from the production of electron–hole pairs by the primary beam. When they relax back to an equilibrium state, they emit part of the absorbed energy as photons. The emitted photons may either be collected and displayed or, for more critical work, analyzed spectroscopically. This phenomenon is called cathodoluminescence and is a valuable source of information about dislocations in semiconductors and piezoelectric materials and for materials stained with specific fluorescence stains. As seen from Fig. 10.17, the interaction volume that gives rise to cathodoluminescence is quite large and as a consequence the spatial resolution is only marginally better (i.e., ca. 50 nm) than that which can be obtained with good fluorescent light optics. The intensity of cathodoluminescence is very variable. Plastics, glass, and some metals give a very weak signal, whereas some phosphors and fluorescent dyes can emit up to 10% of the absorbed energy. With biological samples, the signal quality is strongly influenced by preparation and even the thin carbon coating layer used to enhance surface conductivity can reduce the cathodoluminesc-

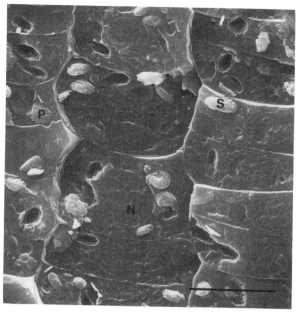

Figure 10.19. Secondary electron image of a frozen-hydrated fracture face of dividing root cortex cells of Maize (*Zea mais. L.*). The sample was very lightly etched at 123 K with an overhead radiant heater and immediately coated with 7 nm of gold. The picture was taken at 110 K, 12 keV, and 15 pA beam current. N, Nucleus; S, starch grains; P, plastids. Bar marker 10 μm.

ence by 50%. In spite of all these negative comments about cathodoluminescence, LTSEM can be used to advantage with this technique.

1. At low temperatures, the spectral peaks of the emitted light, which are generally broad and overlapping at ambient temperatures, are much narrower and more clearly defined. This sharpening of the spectra and the reduction in background facilitates the characterization and identification of the emitted light.
2. Low temperatures offer a degree of protection from thermal damage which can occur with the relatively high beam currents that are necessary to produce a sufficiently strong cathodoluminescent signal.
3. At liquid-helium temperatures, the emission bands are even sharper and more intense (Holt and Saba, 1985).

Papers by Holt and Saba (1985) and Bresse *et al.* (1987) provide a useful introduction to low-temperature cathdoluminescence and demonstrate the type of information that is available for systems using a liquid-helium cryostat.

10.7.7. Transmitted Electron Image

The transmitted electron (TE) detection mode gives an additional dimension to the SEM in that it provides a means of examining the internal structure of the sample.

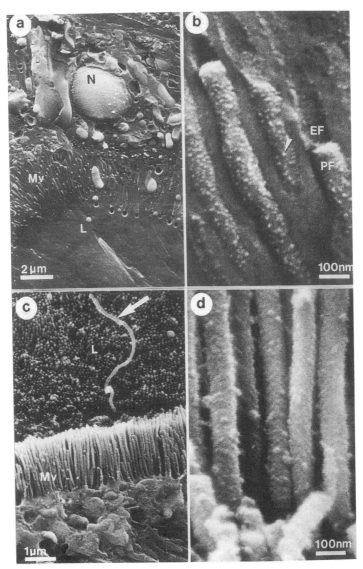

Figure 10.20. (a), (b), Low- and high-resolution secondary electron images of a frozen-hydrated proximal tubule from a kidney epithelial cell. The higher magnification image of the microvillar membrane shows numerous 10-nm particles on the plasmic surface of the membrane. L, lumen; Mv, microvilli; N, nucleus. (c), (d), Low- and high-resolution secondary electron images of freeze substituted and critical-point dried kidney epithelial cells. All samples magnetron sputter coated with 3 nm of platinum. [Pictures courtesy Herter *et al.* (1991).]

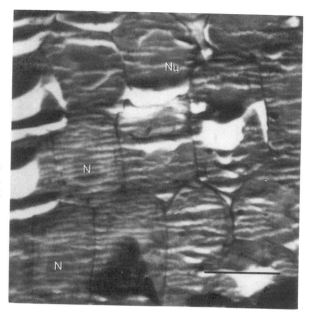

Figure 10.21. Scanning transmission electron image of a 1.0 μm uncoated frozen-hydrated section of dividing pea (*Pisum sativum L.*) root cortex cells. Although there are tears and ripples in the section, it is possible to distinguish the nucleus and cyto-plasm and the intercellular spaces. Photograph taken at 130 K, 30 keV, and ca. 0.45 nA beam current. Bar = 10 μm.

If the sample is thin enough, a substantial proportion of the electrons are transmitted, and in this respect the SEM functions as a TEM. The signal is usually noisier and at lower resolution than the TEM image as the simple detectors collect electrons over a wide range of energies. This mode of operation is, however, particularly useful in the examination and analysis of the somewhat thicker frozen sections (ca.1.0 μm) used in x-ray microanalysis. We will return to these matters in Chapter 11. Figure 10.21 is a transmitted image of a 1.0-μm frozen-hydrated section of pea root tissue taken in the SEM. It is important to distinguish this simple system of an SEM functioning in a transmission mode from a STEM. The STEM is similar to the SEM in that the transmitted electrons are collected and imaged on a point-to-point basis, but it is different in one important aspect: The energy of the electrons is analyzed at each point at which they pass through the thin specimen, and the image can be formed by using electrons whose energy loss lies in any specified range. Because the electron beam scans across the specimen on a point-to-point basis, the electrons in any specified range can be related to a given point on the specimen.

In summary, an SEM operating in the transmission mode gives a relatively low resolution transmitted beam image. A STEM is a high-resolution transmitted beam instrument which gives high-resolution morphological and analytical information.

10.7.8. Other Modes of Imaging by Low-Temperature SEM

Although BSE, SE, and TE, and to a lesser extent cathodoluminescence, are the main imaging modes used in LTSEM, additional information may be obtained from the other signals that are generated. The most important of these additional signals are of course the characteristic and background x-ray photons, which can provide chemical information about the sample. These processes will be discussed in some detail in Chapter 11.

Auger electrons are emitted when an outer shell electron fills an inner shell hole caused by primary electron excitation. These emitted electrons are characteristic of the atom from which they are released and have very low energy. They have an escape depth of 0.5–2.9 nm, and Auger electron spectroscopy is a sensitive method of analyzing surface monolayers. This analytical technique, which is particularly useful for light element analysis, is also very sensitive to contamination and can only be carried out properly at ultrahigh vacuum (13 nPa, 10^{-10} Torr). Although cryopumps and low-temperature anticontamination devices can be used to obtain and maintain a clean environment, there are both advantages and disadvantages to actually cooling the sample. Auger electron spectroscopy is an important tool in surface physics but has only had limited application with organic materials.

Although of peripheral interest to the main focus of this book, mention must be made of other modes of imaging in the SEM which are used at very low temperature with great effect by material scientists, earth scientists, and surface and solid state physicists. These include scanning electron acoustic microscopy or thermal wave imaging; ballistic phonon imaging, and two-dimensional voltage imaging, are all carried out on nonhydrated samples held at liquid-helium temperatures and have important applications in cryoelectronics, thin-film superconductors, and electron beam lithography. The comprehensive review article by Huebener (1988) should be consulted as a useful introduction to these more esoteric forms of LTSEM.

Although there are a wide range of signals that may be used in SEM, their effectiveness in providing genuine information at low temperatures is dependent on the amount of energy it is necessary to put into the sample to obtain a significant signal. We will return to the question of beam damage in a later section, but suffice it to state here that the lowest amount of energy should be used commensurate with obtaining a usable signal. Although no hard and fast rules have been established, the following guidelines may prove helpful in establishing the beam current (measured at the level of the specimen with a Faraday cup) and accelerating voltage that may be safely used in the examination of FH biological and organic samples by LTSEM:

	Current	Voltage
Secondary electrons	1–20 pA	*1–10 keV
Backscattered electrons	10–20 pA	*1–10 keV
Transmitted electrons	10–20 pA	10–30 keV
X-ray photons	1–10 nA	5–15 keV
Cathodoluminescence	10–50 nA	10–30 keV

It is important to remember that with bulk samples the higher the accelerating voltage the greater the size of the interactive volume and the poorer the spatial

---

*These low voltages, i.e., 1 keV and below, are only effective in providing good images in an SEM specifically designed for low-voltage apparatus.

resolution, and that higher beam currents, while giving an improved signal generation, do so at the expense of beam damage. It has been known for some time that
there is an increase in signal current at low keV. The production efficiency of secondary and backscattered electrons is improved and specimen preparation related charging is reduced because the average number of SE emitted from the sample for each
incident primary is greater at low accelerating voltage.

10.7.9. Morphological Appearance of Frozen Images at Low Temperatures

The appearance of frozen specimens in the scanning electron microscope is
dependent on a number of factors. As a general rule, there is an improvement in the
"quality" and "amount" of information as the water is sublimed away from a frozen-
hydrated sample. This is to a large extent an illusion simply because the structural
appearance of the frozen dried sample can be more closely identified with the appearance of samples prepared by more conventional wet chemical means. This interpretation of images is made even more difficult in LTSEM, where the image quality is
vitally dependent on the imaging conditions. Saubermann and Echlin (1975) showed
that contrast formation in the reflected electron and transmitted electron modes of
the SEM is a complex process and the production of "good" or "bad" images is
dependent on signal processing, which may be affected by the subjective impression
of the operator. The quality of the image is adjusted using brightness and contrast
controls, rather in the way the television set is adjusted to give an optimal picture.
The white areas of the picture (minimal electron scattering in the transmission mode,
maximum scattering in the reflective mode) will always appear bright, and the black
areas of the picture (maximum electron scattering in the transmission mode, minimum scattering in the reflective mode) will always appear dark. There is, however,
a wide range of contrast between these two extremes and these contrast levels can
be enhanced, deemphasized, or differentiated as the operator chooses. Some contrast
enhancement can be achieved by signal manipulation: mixing of secondary and
backscattered electron signals, use of dark field imaging, use of dark field imaging
and the addition/subtraction of bright field/dark field images in STEM instruments.
If comparisons are to be made between the contrast in frozen-hydrated and freeze-
dried sections it is important to maximize the topographic and beam induced conductivity contrast of the image in the frozen-hydrated state, allow the specimen to warm
up to the ice-sublimation temperature, and reexamine the specimen without changing
either the contrast or the brightness levels.

In the simplest type of thin specimen, such as a suspension of particulate biological material in an aqueous matrix, the changes in morphological appearance between
the frozen-hydrated and freeze-dried states are usually quite dramatic. There may be
a reversal of contrast in the transmission electron image, and a substantial increase
in topographical contrast in the reflected electron image. Multiple electron scattering
by ice in the hydrated specimen results in a decrease in the contrast of the biological
material. As the ice is removed, electron scattering by the biological material is
increased and the contrast is enhanced. These changes may or may not also be seen
when frozen sections are warmed up. Thus in the early work by Bacaner et al. (1973)

on sections of rabbit muscle, by Moreton *et al.* (1974) on sections of *Calliphora* salivary gland, and by Saubermann and Echlin (1975) on sections of mouse liver, high contrast images with recognizable contents were obtained from fully frozen-hydrated sections. Gupta (1976) was unable to obtain similar high-contrast scanning transmission images with easily recognizable contents of frozen-hydrated sections of *Calliphora* Malpighian tubules. In the articles by Moreton *et al.* (1974) and Gupta (1976) it is stated that contrast *changes* were observed during warmup although no micrographs are presented to demonstrate this point. Saubermann and Echlin (1975) show a pair of micrographs in which it is not possible to observe any significant contrast differences in the 30-kV scanning-transmission images of rat liver between the frozen-hydrated and freeze-dried state, although such changes are clearly apparent in the 50 kV scanning-transmission images of *Calliphora* salivary gland obtained by Berridge and Gupta and shown in the article by Taylor and Burgess (1977). One-micrometer sections of high vacuolate root tissue (Echlin, unpublished data) examined at 20 kV showed a small but significant change in contrast between the frozen-hydrated and freeze-dried states. However, 1.0-μm sections of frozen 10% bovine serum albumen made up in dilute salts solution and examined at 50 kV show only slight changes in contrast as one goes from the frozen-hydrated to the freeze-dried state (Fig. 10.22). This is, however, to be expected because a single protein species in water was being examined where the amount of dehydration would be the same in all areas of the section.

Even in situations where careful comparative mass measurements are made, some variability in the contrast of scanning transmission images and the amount of topographical detail apparent in reflected electron images of sectioned material between the frozen-hydrated and freeze-dried states is to be expected and may be related to one or more of the following factors:

a. *Water content of the tissue.* Sections from tissues with a high water content show a greater transmission electron image contrast change between the frozen-hydrated and freeze-dried states than sections from tissues with a low water content. Differentiation dehydration in adjacent areas of the same section would also give rise to changes in the image contrast.

b. *Accelerating voltage.* Contrast changes in the transmission electron image are more apparent at higher accelerating voltages.

c. *Exposure to the electron beam.* Contrast changes in the transmission electron image are more apparent if the sections are viewed continuously as they pass from the frozen-hydrated to the freeze-dried state than if the sections are allowed to dry without continuous exposure to the electron beam.

d. *Water binding capacity of the tissue.* The water binding capacity is likely to vary in different parts of the specimen. Less dehydration, and hence smaller contrast images, will occur in regions where the water is tightly bound, and more dehydration and greater contrast changes will occur where the water binding capacity of the tissue matrix is diminished or entirely absent.

A careful reexamination of these early frozen-hydrated images in light of more carefully monitored studies—see, for example, Fig. 9 in Saubermann *et al.* (1981b)

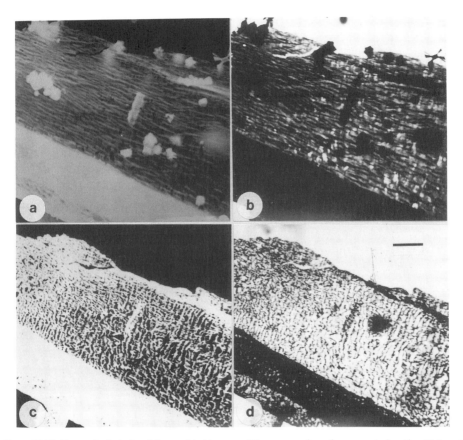

Figure 10.22. Frozen-hydrated and freeze-dried images of the same region of an uncoated section (0.5 μm) cut from a quench-cooled 10% solution of bovine serum albumin, seen in the secondary and transmitted electron mode in the SEM. (a) Secondary image of the frozen-hydrated section; (b) transmitted image of the frozen-hydrated section. Note the virtual reversal in contrast between the two images; (c) secondary image of the freeze-dried section; (d) transmitted image of the freeze-dried section. The frozen-hydrated images were taken at 153 K and the sections allowed to freeze dry in the microscope column before being rephotographed at 300 K. Note how the surface ice crystals disappear as the sections are dried and that there is a significant change in contrast between the freeze-dried and frozen-hydrated sections. All photographs taken at 50 keV and 1.0 nA beam current. Bar marker = 10 μm.

and numerous more recent studies—brings one to the inescapable conclusion that these early frozen-hydrated images were, in fact, already freeze-dried.

The substantive changes in the morphological appearance of images between the frozen-hydrated and freeze-dried states are usually more easily recognized in fractured than in sectioned material. Figure 10.23 shows the difference between the two images in plant cells, while Fig. 10.24 demonstrates the more subtle difference between tissues that are unetched, partially etched, or deeply etched. Such changes are frequently less easily appreciated in intact frozen samples. The phenomenon of etched surfaces and partial dehydration is less of a problem in morphological studies

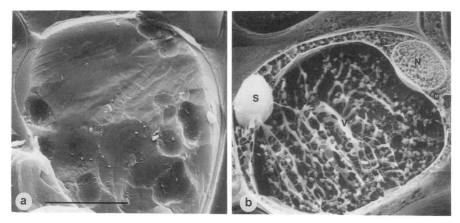

Figure 10.23. Secondary electron images of a frozen-hydrated (a) and a freeze-dried (b) fracture face of outer cortex cells of roots of Duckweed (*Lemna minor L.*). Although the frozen-hydrated fracture face is virtually featureless, it is possible to distinguish the main cellular compartments. The freeze-dried fracture has increased topographic contrast but reveals the large voids once occupied by ice crystals. C, Cytoplasm; N, nucleus; V, vacuole; and S, starch grain. Samples coated with 8 nm of gold and photographed at 110 K, 15 keV, and 15 pA beam current. Bar marker = 10 μm.

than with analytical studies, where an accurate measurement of the local mass of the sample is an important part of the quantitative equation. These matters will be considered in the following chapter, where it will be shown that the hydration state of a sample can be more accurately and more objectively measured and need not be dependent on the vagaries of the appearance of the image.

10.8. BEAM DAMAGE DURING LTSEM

10.8.1. Introduction

In contrast to the large amount of information that is available on beam damage at low temperatures in the TEM, much less information is available about the damage that can occur to samples maintained at low temperature in the SEM. In some respects this is a consequence of the experimental imaging requirements of the SEM. With a few notable exceptions, few SEM images are recorded at magnifications much above 20,000×. As there is a direct relationship between the size of the scanned area on the specimen and the area displayed on the cathode ray tube, much less energy is being put into the sample to obtain the lower-magnification SEM image than is the case with high-resolution TEM investigations.*

---

*The only accurate way to measure the amount of energy going into a sample is to measure the beam current at the level of the specimen. This may be achieved simply either by inserting a Faraday cup at the level of the specimen or by measuring a precalibrated proportion of the current impinging on the final aperture, which has been electrically isolated (Echlin and Taylor, 1986). Full details of the practical aspects of the procedures used to establish this parameter in an SEM are to be found in Goldstein *et al.* (1981).

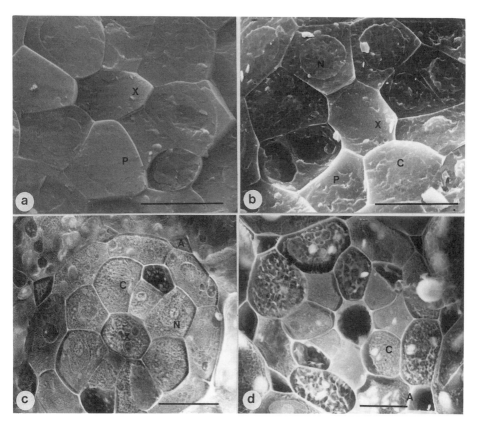

Figure 10.24. Secondary electron images of progressively surface etched fracture faces of frozen-hydrated root tips of duckweed (*Lemna minor L.*). The cross fractures are made approximately 250 μm from the root tip and show a ring of phloem tissue cells surrounding a central xylem cell. (a) Unetched fracture face with little topographic contrast other than cell outlines; (b) lightly etched surface showing details of the major cell compartments; (c) deep-etched sample showing organelles and ice crystal ghosts; and (d) freeze-dried sample. N, Nucleus; C, cytoplasm; X, xylem; P, phloem tissue; CC, companion cell; ST, sieve tube; A, air-space. Samples (b)–(d) etched at 123 K with an overhead radiant heater and all samples coated with 8 nm of gold. Photographs taken at 110 K, 12 keV, and 20 pA beam current. Bar marker = 10 μm.

This is not to suggest that beam damage is not a problem in structural studies carried out in the SEM, for delicate samples can be all too easily destroyed in the microscope. The type of damage that is observed is seen either directly on the viewing screen as *it occurs* or on the recorded micrograph. It is not usually possible to measure the subtle structural changes discussed in the previous chapter. When the SEM is used in connection with analytical studies, beam damage assumes much greater dimensions because there is a thousandfold increase in the amount of energy put into the sample. We will come back to these aspects in the following chapter.

There is a well-rehearsed and dreary catalogue of the different manifestations of beam damage that can occur in ambient temperature structural studies in SEM.

They are mostly due to localized sample heating, and the damage may usually be traced back to inadequate sample preparation and coating. The beam damage takes the form of blisters, cracks, and cavities on the surface and, in some extreme cases, complete vaporization of the sample. Such damage can be distinguished from the transient effects of charging and image characteristics peculiar to SEM, such as the edge effect. Most plastics are notoriously prone to damage, although this may easily be avoided by decreasing the beam current and lowering the temperature.

The problem of beam damage in the SEM is quite different from that in the TEM. In the TEM a large proportion of the beam energy passes right through the sample. For bulk samples examined by SEM, a large amount of the beam energy is dissipated in and absorbed by the sample; the reflected signals used to provide the image represent a very small amount of the energy put into the sample. The energy that is dissipated into the sample has to be drained away, and with conductive samples or samples that have an adequate coating layer this usually does not present a problem, provided the sample is in good thermal/electrical contact with the sample holder. The low conductivity of frozen, organic, and biological material, which may also contain a large amount of ice, will present a problem in this respect and it is necessary to focus our discussion first on beam heating and secondly on damage due to ionizing radiation.

10.8.2. Beam Heating

All the available evidence points to the fact that beam heating in bulk frozen specimens is not a serious problem, provided the specimens are adequately coated and in good thermal contact with the underlying support. Talmon (1982, 1987), on the basis of early studies on samples smaller than the electron range (Talmon, 1977a, b, c, 1978), has calculated that beam heating is also of little consequence in samples that are larger than the electron range. Calculations showed that beam heating that was limited to between 10 and 20 K depended on the following factors: the ratio between the radius and height of the sample, the ratio of total surface area to the scanned area, specimen thickness, specimen conductivity, and the beam accelerating voltage and current. Talmon considers that the main reason for such a small temperature rise is that the samples and the specimen stub offer a large area for heat dissipation. Temperature rises *within* the sample are twice as high as those that develop at the surface. This will be of concern in x-ray microanalysis, which uses high beam current densities. In order to minimize thermal damage to the specimen, the sample should be maintained at the lowest possible temperatures on the cold module, so that even 20 K rise will keep the sample within a safe temperature.

Similar calculations have been made by Pesheck *et al.* (1981) on features standing out on an otherwise flat frozen-hydrated fracture surface in which the beam–specimen interactive volume could be accommodated. These "worst case" calculations were based on frozen mixtures of crude oil and brine in samples of sandstone,

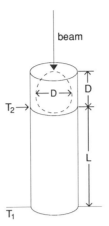

Figure 10.25. Diagram showing the dimensions of a specimen, where d is the diameter of the interactive volume and L is the specimen thickness. The heat input (Q) into the sample is a function of the beam current and the accelerating voltage. Assuming a steady state conduction of this heat, the temperature rise in samples of varying thickness is given by $dT(T_2 - T_1) = QL/AK$, where A is the cross-sectional area of the cylinder and K is the thermal conductivity of the material. If K is known, this formulation can be used to calculate the heat rise in samples of varying thickness. [Data from Pesheck *et al.* (1981).]

which has a better thermal conductivity than pure water, and which had been liberally coated with a 60-nm layer of carbon and gold. Figure 10.25 and Table 10.1 show a summary of their results. Even in 20-μm-thick samples, which at the kV they were using may be assumed to be bulk samples, there was only 4 K temperature rise across a sample maintained at 94 K when irradiated with a 100-pA beam. At these levels of beam currents, which are also used in x-ray microanalysis, the temperature rises are much higher. Provided *samples* are kept cold enough, i.e., 123 K or lower, and

Table 10.1. Steady State Temperature Difference Across a Cylindrical Specimen of the Dimensions Described in Fig. 10.25 at Various Beam Currents and at Varying Depths into the Sample[a]

l (μm)	Beam current				
	10 pA	50 pA	100 pA	1 nA	10 nA
1	0.02	0.1	0.2	2.1	20.9
2	0.04	0.2	0.4	4.2	41.8
5	0.1	0.5	1.0	10.5	105
10	0.2	1.0	4.2	20.9	209
20	0.4	2.1	4.2	41.8	418

[a]The specimen is a frozen mixture of oil and brine in a sandstone plug coated with a thin film of gold and kept at 94 K. The diameter of the interactive volume is 6.5 μm, the accelerating voltage is 20 keV, and the thermal conductivity is 0.29 W m K. [Data from Pesheck *et al.* (1981).]

coated with a thin layer (ca. 5–10 nm) of a noble metal, specimens may be examined in the SEM with beam currents up to 50 pA with little danger of thermal damage to the sample.

10.8.3. Ionizing Radiation Damage

No studies have been carried out on the effects of ionizing radiation on bulk frozen samples examined in the SEM, although some simple calculations will show that at the relatively low magnifications used for LTSEM any effects would be minimal and would pass unnoticed. At a magnification of 10,000×, a 10-μm^2 area on the sample will fill the 100 × 100-mm display cathode ray tube used for recording an SEM image. In order to detect sufficient signal quality and contrast information at say 10% above the background noise, it would be necessary to use a beam current of 10 pA and an exposure of 100 s. At a magnification of 10,000× this would be equivalent to a radiation dose of between 60 and 70 e^- nm^{-2}. This simplistic calculation does not consider the spatial resolution that might be required in the image, a feature that is related to the probe diameter and hence local current density. Neither does it take into account the time taken to examine the specimen before attempting to record in image. However, even this extended time can be balanced by the fact that much lower beam currents, e.g., 1 pA, are frequently used in conventional SEM.

At higher beam currents, i.e., 5–10 nA, it is possible to measure mass loss from frozen samples and it must be assumed that at these levels of radiation considerable damage is taking place. Although there is a marked lack of any definitive information on the subject, it is assumed that decreased sample temperature will ameliorate the effects of radiation damage. Some recent studies (Echlin, unpublished data) on tobacco tissue sheds some light on this problem. Small pieces of dried tobacco, sheet material, approximately 200 μm thick, containing approximately 12% water, were placed onto an aluminum specimen stub covered with thick colloidal carbon slurry. The samples were quench cooled, coated with 10 nm of chromium, and examined in an SEM at a sample temperature of ca. 110 K. A series of experiments were undertaken to establish the beam current necessary to perform digital scanning x-ray microanalysis, during which a static electron probe is slowly stepped across the sample. At a magnification of ×2000, a spot size of 1 μm^2, a beam current of 5 na, and a delay time of 200 m s per pixel point, visible damage was observed in photographs taken of the scanned area (Fig. 10.26). The radiation dosage into each pixel point was calculated to be 6.25 ke^- nm^{-2}. No damage was observed when the delay time per pixel point was reduced to 100 m s, which correspond to a radiation dose of 3.13 ke^- nm^{-2}. Correlative mass loss studies using an energy dispersive x-ray spectrometer revealed a mass loss of between 20% and 30% at the high dose and only about 5% loss at the lower dose. In order to avoid damaging the sample at ambient temperature it was necessary to reduce the beam current to 2 nA.

These experiments would imply a two- to threefold cryoprotection factor, at least as far as visible damage is concerned. This is not to suggest that there is no structural damage at the lower doses; we simply had no way of measuring the effect. Clearly more experiments need to be carried out to establish the amount of energy that can

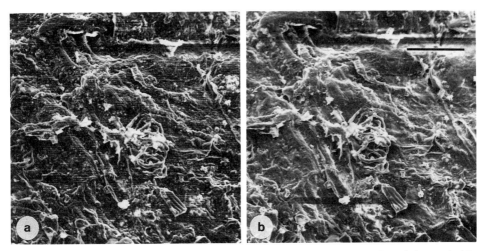

Figure 10.26. 200-μm-thick tobacco sheet material containing ca. 12% water, quench cooled, coated with 10 nm of chromium and maintained at 110 K. (a) Scanning electron micrograph of a surface digitally scanned at a dose of 6.25 ke^- nm^{-2} per pixel. Visible damage may be seen on the surface of the specimen. (b) Same region of the sample scanned with a dose of 3.13 ke^- nm^{-2}. The sample appears undamaged. Bar marker = 50 μm.

be tolerated by a specimen before it is damaged. In the meantime it behooves scanning microscopists to record the beam currents, probe diameter, and frame time of all images taken at high magnifications.

10.9. ARTIFACTS IN LTSEM

LTSEM has its own peculiar set of artifacts, which are quite distinct from artifacts introduced by beam damage or by faulty specimen preparation and transfer. It would be impossible to catalogue all these artifacts, so a few examples are given by way of illustration. Many of these artifacts have been considered elsewhere in this book, and two papers by Read and Jeffree (1988) and Moss *et al.* (1989) list other artifacts associated with the preparation or more specific types of sample.

10.9.1. Frozen-Hydrated and Freeze-Dried Samples

In spite of the obvious differences between frozen-hydrated and freeze-dried and/or etched samples of biological and hydrated organic materials, mistaken interpretations persist. At its most obvious, frozen-hydrated material contains less structural detail in both the reflective (SE and BSE) and transmitted (TE) imaging modes. The difficulties arise when the natural water content is low, and it is then necessary to carefully examine the sample for the presence of ice-crystal ghosts where solutes are pushed to the edge of the growing ice crystals. The sublimation of this ice is the sure and only way of providing structural evidence that incipient dehydration has taken place.

These structural changes are more difficult to see in tissue fractures where the cells remain intact, for example, fractured leaf samples, where the cells appear turgid in both the frozen-hydrated and freeze-dried state. In these circumstances it is necessary to examine the coating layer. In a fully frozen-hydrated state it should be continuous. Any cracks or discontinuities in the surface are usually indicative of the small amount of shrinkage that accompanies freeze-drying. Bastacky *et al.* (1987a) have shown that air–liquid interfaces appear smooth in the frozen-hydrated state, and are more irregular in freeze-dried specimens. Pearce (1988) suggests that it is useful to follow the process of freeze-drying by periodic examination in the SEM as it can provide information on cell collapse and the way the ice crystals have formed during the initial cooling.

10.9.2. Extracellular Ice Crystals

Extracellular deposits have been seen on the surface of freeze-fractured, frozen-hydrated leaves (Jeffree *et al.*, 1987; Pearce and Beckett, 1985; Beckett and Read, 1986). In nearly all cases, these extracellular deposits disappeared when the samples were freeze-dried. It was first thought that these were ice deposits formed by condensation of the water vapor present in the extensive intercellular spaces of leaves, particularly as no droplets were found in severely drought-stressed leaves (Pearce and Beckett, 1985, 1987). A careful examination of the phenomenon by the workers concerned (*loc. cit.*) revealed that in nearly all the cases the extracellular ice was an artifact of the preparative technique. Jeffree *et al.* (1987) found that up to 4% of the leaf water content could be in the form of extracellular droplets and that the amount increased as the cooling rate was reduced. Jeffree and his colleagues, on the basis of the size and form of the ice crystallites, proposed that the extracellular ice deposits arose in four main ways:

1. As the volume of intracellular water increased during freezing, sufficient pressure was exerted to push water out of the cells.
2. As the cooling front passed through the tissue, water evaporating from warmer cells condensed onto colder cells.
3. The growth of extracellular ice withdrew water from inside the cell.
4. Water vapor that existed naturally in the intercellular spaces condensed.

This phenomenon has been seen in a number of different leaves, and one example is reported in aerenchymatous root tissue (Webb and Jackson, 1986). An examination of SEM images of frozen hydrated lung tissue (Bastacky *et al.*, 1987a) suggests that the same effect may also be occurring in this material. It is not known whether the same phenomenon occurs in tissues without extracellular spaces. Jefferies (1988) has made a cryo-SEM study of the water droplets on leaf surfaces that have been sprayed under the conditions used to apply herbicides, fungicides, insecticides, etc. It was found that there were high losses of the larger droplets under conditions of fast cooling. Slow cooling, which results in poor internal structural preservation, gave a marked improvement in droplet retention on unwettable surfaces. Improvements in the specimen preparation methods allows cryo-SEM to be used as an accurate means of monitoring the size and distribution of foliar sprays.

10.9.3. Ice Crystal Deposits

Nearly all frozen-hydrated specimens have surface ice crystals, and some thought should be given regarding their origin:

1. They may be of natural origin, i.e., condensation from a humid air environment or solidification from an aqueous environment.
2. They may be condensation from residual water vapor in the preparative equipment and the microscope. It is useful to remember that at a pressure of 1.3 mPa (10^{-5} Torr), approximately 70% of the residual air in the microscope column is water vapor.
3. They may be a consequence of the redistribution of water that occurred during cryofixation (see Section 10.9.2).
4. They may be derived from the water vapor that condenses in cryogens and onto the various handling devices used in cryofixation.

These artificial ice deposits need either to be prevented (which is difficult) or recognized as such (even more difficult) or selectively removed (a possibility) before ice crystals can be ascribed to a natural phenomena. Some success has been achieved in removing extraneous ice crystals by local irradiation with the electron beam. The spurious ice crystals are usually in poor contact with the underlying substrate so they rapidly charge up and are repelled from the specimen surface.

10.9.4. Specimen Shrinkage

This is usually only seen in freeze-dried samples and is, to a large extent, a consequence of the expansion of water during freezing. In cells, the shrinkage is manifest as a space at the cell periphery, and in whole tissues the local variation in shrinkage can be related to the connections that exist between cells. It is important to distinguish the shrinkage caused by freeze-drying from shrinkage that has arisen as a consequence of deliberately introduced dehydration. This is particularly important in the examination of frozen-hydrated and freeze-dried emulsions, soils, muds, and latex suspensions, which can have a variable water content.

10.10. APPLICATIONS

Many studies have been made using LTSEM, and the papers by Bastacky *et al.* (1987b, 1988), Marshall (1987), Sargent (1988), and Read and Jeffree (1991) provide a useful introduction to the more recent literature. We will only consider the broader applications, which fall into two main classes:

1. LTSEM used solely as a convenient means of rapidly examining the structure of a large number of similar samples which could equally well be studied by other means. A few recent examples will be given of this type of application.
2. LTSEM used to provide new information which could not be studied by other means. These types of application fall into three broad groups: (a) samples in which the juxtaposition of the gas, liquid, and solid phases are important, i.e., biological tissues; (b) samples that are so labile that this is the only way the information may be obtained, i.e., polymers and plastics; (c) where a particular phenomenon can only be observed at low temperature, i.e., superconductivity and cryoelectronics.

10.10.1. LTSEM as a Convenient Imaging System

The relative speed of preparation makes LTSEM a convenient way of examining a large number of specimens at low spatial resolution. Brooks and Small (1988) have used the technique in a study on the taxonomic significance of the microstructure of several agricultural forage crops. There are distinct advantages in using fresh material rather than herbarium material (Fig. 10.27). Blackmore and Barnes (1984) have taken a similar approach in their studies on pollen morphology. Guggenheim *et al.* (1991) have used low temperature for detection of fungicide activity. Gardner *et al.* (1981) have used LTSEM as a rapid technique for examining surgically removed animal cartilage tissue, and Rubinsky *et al.* (1990) have used cryo-SEM to study the mechanisms of cryosurgery. Campbell and Porter (1982) have studied the distribution of microorganisms in soil samples. The common factor in these and related studies is speed and convenience.

10.10.2. LTSEM as an Innovative Imaging System

The real advantage of LTSEM is revealed in the situations when it is the only way to obtain information. Examples may be found in both the physical and biological sciences.

10.10.2.1. Higher Plant Tissues

Examination of frozen-hydrated specimens is the only way mucigel secretion by plant roots may be localized (Sargent, 1986) (Fig. 10.28); the surface cementing matrix in mycorrhiza may be seen (Alexander *et al.*, 1987), and the initiation and development of aerenchyma (cortical gas-filled spaces) may be followed (Webb and Jackson, 1986). Mention has alrady been made of the way LTSEM is used in studies on the form and development of intercellular spaces in leaves. It has also been used to study cell shape in drought-stressed tissues (Pearce and Beckett, 1987), the nature of the connections that form between plant cells in roots, stems, and leaves (Barnett and Weatherhead, 1988; Carr *et al.*, 1980), and the variation in stomatal aperture in Avena leaves (Gardingen *et al.*, 1989). Cryo-SEM has been used with great effect to study the freezing responses in plants and frost stress tissues (Robson *et al.*, 1988;

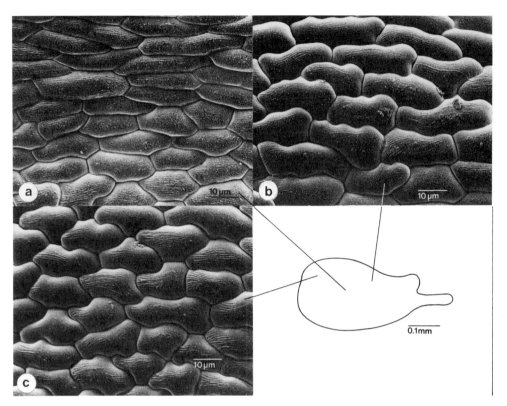

Figure 10.27. Low-temperature scanning electron micrographs of different regions of the frozen-hydrated surface of a flower petal of Alfalfa (*Medicago hypogaea*). Samples were coated with a thin layer of gold and photographed at 113 K and 5 keV. [Pictures courtesy Brooks and Small (1988).]

Ashworth *et al.*, 1988; Pearce, 1988) (Fig. 10.29). LTSEM appears to be the only way of examining the epicuticular waxes which form an important interface between the surface of a leaf and its immediate exterior environment (Eveling and McCall, 1983; Sargent, 1983; Williams *et al.*, 1987). The technique has also been used to study the controlled desiccation of lichens (Brown *et al.*, 1987) and the effect of ozone on developing birch leaves (Scheidegger *et al.*, 1991). Conventional, wet chemical procedures would have involved rehydrating the samples and would have defeated the purpose of the experiment.

10.10.2.2. Animal Tissue

Pulmonary tissue can only be studied properly if LTSEM is included in the investigation procedures. Two papers summarize the pioneering work that has been done on this tissue by the research group in Berkeley (Bastacky *et al.*, 1987a; Finch *et al.*, 1987). The low-temperature techniques have proved particularly useful in

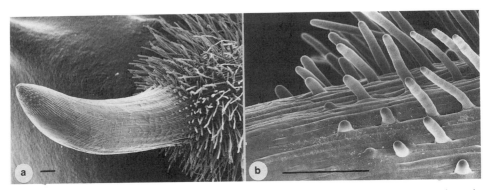

Figure 10.28. Low-temperature scanning electron micrographs of frozen-hydrated seminal roots of *Lepid-ium sativum*. Samples coated with a thin layer of gold and examined at 110 K. (a) Distal part of the root showing the ablative root cap, zone of elongation, and the zone of root-hair development. (b) Details of the zone of root-hair development showing the progressive extension of root hairs. The surface details of the root hairs and the thin layer of mucigel-covered epidermis are preserved in the fully hydrated condition. Bar marker = 100 μm. [Pictures courtesy Sargent (1986).]

studying pulmonary oedema (Hook *et al.*, 1987) and the association of environmental pollutants with pulmonary tissue. Intestinal tissues have been studied by Carr *et al.* (1986), cardiac tissue by Bullock *et al.* (1980), and skeletal tissue by Gardner *et al.* (1981). Inoué and Koike (1989) have found high-resolution LTSEM an effective way to study the three-dimensional configuration of lipid droplets in a wide range of gastrointestinal tissues (Fig. 10.30).

10.10.2.3. Fungal Tissues

Fungal mycelium and many of the reproductive structures of the microfungi are very susceptible to dehydration. LTSEM provides a means of examining these deli-cate structures either in the fully hydrated state or at low temperature in the FD state. The papers by Jones *et al.* (1989), Read *et al.* (1983), and Beckett and Read (1984, 1986) provide a good introduction to what may be achieved with LTSEM. The same techniques have been used to study the host–parasite relationships that fungi develop with higher plants (Beckett and Porter, 1988; Beckett and Woods, 1986; Williamson and Duncan, 1989).

10.10.2.4. Freezing Injury

The processes of freeze damage and injury associated with equilibrium cooling, which until recently have only been studied by cryo-light microscopy, are now being studied by LTSEM. Bischof *et al.* (1990) used a direct solidification stage to study the mechanisms of ice propagation in kidney tissue as a function of cooling rate and glycerol concentration. Selected samples, with a known thermal history, were quench

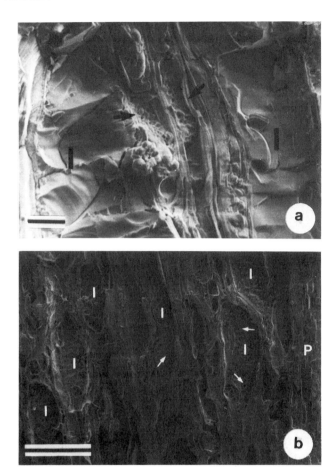

Figure 10.29. Low-temperature scanning electron micrographs of apple bark tissue. Samples were quench cooled in LN$_2$, fractured under vacuum, etched or unetched, and coated with a thin layer of gold. Photographs were taken at 140 K and at varying low voltages. (a) Large extracellular ice crystals are visible in the cortical tissue (C) of bark in plants exposed to temperatures of 268 K. The cortical cells are collapsed and distorted in response to freezing (arrows). Sample unetched and viewed uncoated at 2 keV. Marker = 10 μm. (b) Bark tissue stressed at 263 K showing large ice crystals (I) in the cortical cells. Adjacent cortical cells are distorted and collapsed. Large ice crystals are absent from the smaller periderm cells (P). Specimen etching reveals eutectic margins on extracellular ice crystals (arrows). Sample surface etched, coated with 12 nm of gold and examined at 5.5 keV. Bar marker = 100 μm. [Pictures courtesy Ashworth *et al.* (1988).]

cooled and studied in detail by LTSEM. Fujikawa (1990) has investigated the relationship between cellular deformation and intramembrane particle aggregation in the plasma membrane in slowly cooled fungal hyphae. Quench-cooled samples were studied both in the frozen-hydrated state and as replicas. Murase *et al.* (1991) used

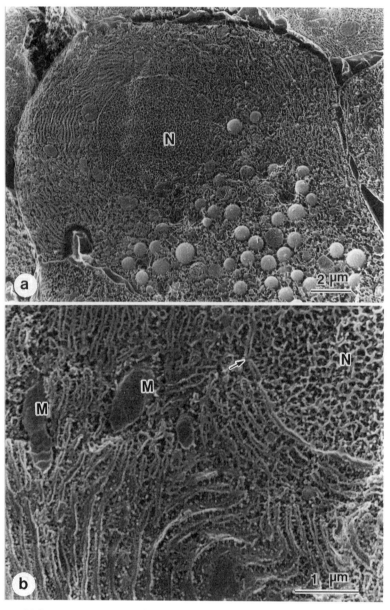

Figure 10.30. (a) Low-temperature scanning electron micrographs of the fractured surface of a quench-frozen mouse pancreatic acinar cell. The nucleus (N), endoplasmic reticulum, and zymogen granules are clearly visible. (b) A higher magnification of the intracellular structures in a pancreatic acinar cell. The cisternae of the endoplasmic reticulum are in parallel arrays with ribosomes on the membrane faces. The outer and inner membranes of the mitochondria (M) are visible. A nuclear pore may be seen on the nuclear envelope. The sample has been fractured, deeply etched, and coated with 3–5 nm of gold. Small ice-crystal ghosts may be seen throughout the sample. Photographs taken at ca. 120 K and 15 keV. [Pictures courtesy Inoue and Koike (1989).]

LTSEM in combination with differential scanning calorimetry to study slowly freeze-concentrated and freeze-dried phosphate solutions.

10.10.2.5. Food Materials

As all food materials are derived from plant and animal materials, it is not surprising that LTSEM has found broad application in the food industry. Kalab (1981) and Kalab and Modler (1986) describe a number of applications with milk

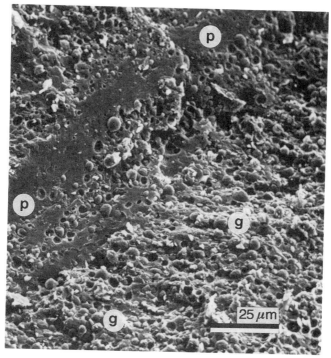

Figure 10.31. Low-temperature scanning electron micrograph of frozen-hydrated Yugoslav cream cheese called Kajmak. The sample was fractured under LN_2, coated with 20 nm of gold, and photographed at 173 K, 10 keV, and 40 pA beam current. Individual fat globules (g) and fat globule clusters surrounded by protein (P) can be seen in the hydrated matrix. (Picture courtesy Miloslav Kalab, Ottawa.)

products (Fig. 10.31), and Saito (1981) has carried out cryomicroscopy of soya products such as tofu. Other studies have been made on meat, cheese, mayonnaise, chocolate, and of course ice cream. The journals *Studies on Food Microstructure* and *Journal of Food Science* are good sources of up-to-date information.

10.10.2.6. Organic Materials

Pulp and paper have been investigated using LTSEM by Howard and Sheffield (1987) and more recently by Moss *et al.* (1989), although difficulties remain with the interpretation of some of the images. Sawyer and Grubb (1987) show how low-temperature SEM may be used in studies on automobile fuels (Fig. 10.32) and plastics and polymers, and Rowe and McMahon (1989) give a number of examples of the use of cryo-SEM in the pharmaceutical industry. It would appear that water is the principal constituent of most of the medical and cosmetic creams, lotions, and gels. Wilson (1989) provides a useful introduction to the way LTSEM can be used to study foams, including the head on a pint of beer (Fig. 10.33).

10.10.2.7. Inorganic Specimens

LTSEM has been used to study the distribution of brine and oil in rock (Pesheck *et al.*, 1981), of oil in tar sands (Mikula and Munoz, 1988), and of the poor geometry in a variety of soils and sedimentary rocks (Sutanto, 1988) (Fig. 10.34). Cryo-SEM has shown the grain boundaries in Antarctic ice (Mulvaney 1988), the porous structure of hardening concrete (Aldrian, 1980), the distribution of the components in oil well mud cake (Cook and Murphey, 1988), and the orientation and cluster aggregation of particulate coatings for paper products (Sheehan and Scriven, 1988).

10.10.2.8. Materials Science and Surface Physics Applications

These applications are outside the general remit of this book, but are very briefly included here to illustrate that very low-temperature SEM, i.e., 2–20 K, has important uses in studying the properties of materials. The applications are primarily directed towards the study of semiconducting and superconducting materials where different

Figure 10.32. Low-temperature scanning electron micrograph of wax deposits filtered from diesel fuel. Samples are not fractured or etched but are coated at low temperatures with a thin layer of gold before being photographed at 120 K. The wax deposits can only be imaged at low temperatures as they contain occluded volatile solvents which are beam sensitive at room temperature. Bar marker = 20 μm. (Picture courtesy John Sargent, Oxford.)

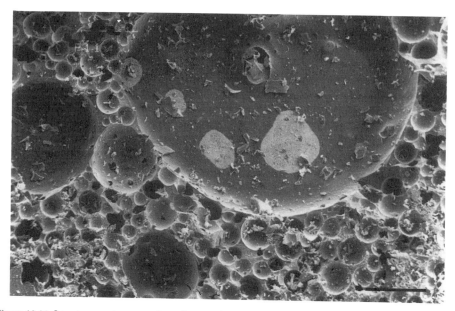

Figure 10.33. Low-temperature scanning micrograph of frozen-hydrated beer foam. The foam was quench cooled, fractured, and sputter coated with a thin layer of gold before being examined and photographed at 113 K. The cryopreservation retains the thin walls of the foam bubbles, which contain 99.9% water. When the foam is fractured, tiny fragments of the shattered bubble wall fall onto the fracture surface. Bar marker = 20 μm. (Picture courtesy Ashley Wilson, York).

modes of imaging in the SEM are used at low temperatures to obtain information. Thus electron beam induced conductivity and cathodoluminescence are powerful tools to study the electronic structure of single defects in materials. Sample temperature is one of the parameters that is changed during the measurements. The various imaging techniques that are used with enhanced effect at low temperatures rely on the localized heating effect caused by the electron beam during the scanning which generates the important response signal providing the structural information about the sample. Jenkins *et al.* (1990) describe a simple device that enables electron beam testing to be carried out at LN_2 temperatures. The recent review by Huebener (1988) is a good introduction to this type of work.

10.11. ADVANTAGES AND DISADVANTAGES OF LTSEM

In concluding this chapter it is useful to consider the advantages and disadvantages of low-temperature SEM (as distinct from operation at ambient temperature) as a means of assessing the value and importance of the technique.

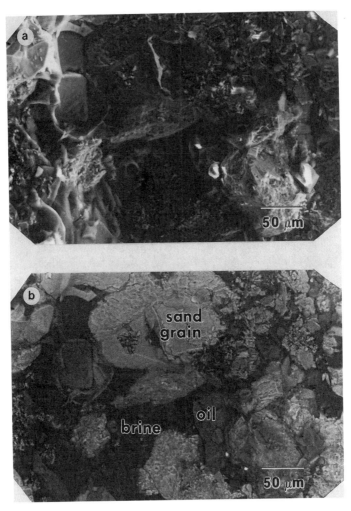

Figure 10.34. Low-temperature scanning electron micrograph of a frozen-hydrated liquid-bearing rock fracture face. Samples were quench cooled in liquid ethane, fractured under vacuum in an Oxford Instruments CT 1000A cryo-system, and coated with 4–7 nm of chromium. (a) Secondary electron image showing brine, crude oil, sand grains, and core filling clay. (b) Backscattered electron image of the same area in which the brighter areas correspond to the mineral components and the darker areas correspond to the brine and oil. [Pictures courtesy Sutanto (1988).]

10.11.1. The Advantages

(a) Speed: This is largely dependent on the amount of water present in the specimen, but generally information would be available within 1–2 h as distinct from 1–2 days in specimens prepared by more conventional wet chemical procedures. Using one of the dedicated cryopreparation units described earlier, it would only

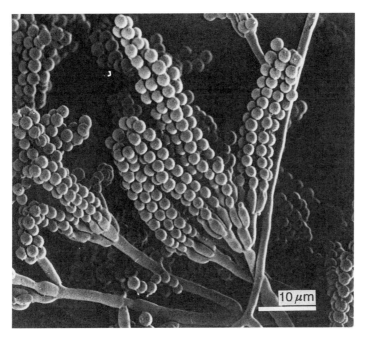

Figure 10.35. Frozen-hydrated, unetched metal-coated sample of fungus *Penicillium nalgiovense* showing the delicate conidiophores and the very labile spherical asexual conidiospores. The specimen was fractured and sputter coated with a thin layer of gold *in vacuo* at 113 K in a Hexland CT 1000A Cold Stage System (Oxford Instruments) and examined at 5 keV at 113 K in a AMRAY 1000A scanning electron microscope. (Picture from unpublished results of Paula Allan-Wojtas and Ann Fook Yang, Ottawa.)

take about 10 min to quench cool, fracture, (etch), and coat hydrated biological or organic material. It would be necessary to add another 90 min for instrument pump down, and to cool the anticontamination devices and the cold modules. However, once the equipment is cooled down, the turn around time between samples would be about 15 min. Unhydrated samples such as polymers and plastics may be examined within a few minutes of the cold modules reaching their working temperature.

(b) Convenience: This arises from the speed of operation, and allows dynamic processes to be followed by taking repeated samples from the same material. Beckett and Read (1986) have used this approach to study fungal infection processes and fungal growth.

(c) Avoidance of deleterious preparative procedures: This is the "raison d'être" of low-temperature preparation procedures and enables one to examine the changing role water plays in the size and shape of hydrated specimens. Such procedures are particularly important in studies on biological material, which are largely dependent on water as a structural component.

(d) Avoidance of sample extraction: Many wet chemical procedures give good structural preservation but at the expense of solubilizing and removing different components of the specimen. Thus cuticular waxes are removed from leaves and insects, and oils and tars disappear from oil-bearing rock strata. Low temperature avoids these problems.

(e) Low-temperature SEM, in common with other low-temperature microscopical techniques, reduces the thermal damage to samples. The situation regarding ionizing radiation damage is less clear, but there is every reason to believe that the same parameters operate in scanning beam as in fixed beam instruments.

(f) The electron beam can act as a highly localized heat source which generates important response signals in samples maintained at very low temperature. A variety of different signals can be used to provide structural information about semiconducting and superconducting material. These diagnostic and analytical signals are only obtained at low temperatures and are quite distinct from the signals that are more usually associated with ambient SEM.

(g) LTSEM is unique in providing a simple and inexpensive way of imaging semithick (1–2 μm) frozen sections—albeit at reduced spatial resolution. This imaging mode has important applications in analytical studies.

(h) Low melting point materials may be examined by decreasing their temperature. This stabilizes the sample surface and lowers the effective vapor pressure of otherwise volatile materials.

10.11.2. The Disadvantages

(a) Hydrated samples have to be small in order to avoid substantial ice crystal damage and dimensional changes brought about by freezing.

(b) In line with studies carried out in the TEM, the presence of ice increases the chance of damage by ionizing radiation.

(c) LTSEM cannot yet be considered a high-resolution imaging system, although the preliminary evidence from scanning electron microscopes fitted with field emission guns and comprehensive cold modules suggests that this disadvantage will quickly disappear (Herter *et al.*, 1991).

10.12. SUMMARY

The scanning electron microscope, with its wide variety of specimen–primary beam interactions, is a particularly versatile electron beam instrument for studying the structure of frozen bulk specimens. The finely focused electron probe provides a large depth of field in focus at any one time and allows easy and direct visualization of the three-dimensional structure of an object without recurse to sections or replicas at a spatial resolution approaching that which may be obtained in the TEM. A wide variety of cold stages and dedicated and nondedicated cryopreparation chambers are available for scanning electron microscopes and allow frozen samples to be fractured and/or microdissected, etched, and finally coated with a thin layer of evaporated or sputter-coated metal. The initial sample preparation is straightforward and centers on quench cooling well-secured samples in such a way as to ensure that the regions of interest are adequately preserved. Samples may be examined in the fully frozen-hydrated or freeze-dried state and information obtained about their elemental molecular and structural constitution depending on the nature of the interaction of the primary beam with the specimen.

The interperetation of images obtained from frozen samples of biological and hydrated organic material is strongly influenced by their state of hydration. There is an apparent increase in both the quality and quantity of morphological information as a fully frozen-hydrated bulk specimen is dehydrated by sublimation of the ice matrix. In sectioned material, dehydration is frequently accompanied by reversals in contrast. Beam damage remains a problem with low-temperature SEM, with mass loss being the principal concern, particularly with high-resolution studies. Low-temperature SEM has a very wide range of applications, both from the ability of the technique to provide new information and from its capacity to rapidly provide images of hydrated samples.

11

Low-Temperature Microanalysis

11.1. INTRODUCTION

The term *microanalysis* is used here to describe procedures that can provide *in situ* qualitative or quantitative chemical information about a specific part of the sample as distinct from more general chemical information about the whole of the sample. The essential feature is that, to varying degrees of spatial resolution, the procedures allow analysis of one or more chemical species—atoms, molecules, macromolecules— in relation to an identifiable feature of the sample ranging from subcellular compartments, inclusions or polymers, cells, or tissues.

A wide range of *in situ* analytical procedures are available for the study of materials (Echlin, 1984) although significantly fewer can benefit from the added dimension of being used at low temperatures. As discussed earlier in the book, the effectiveness of many of the analytical procedures is enhanced by using low-temperature preparative techniques, which serve to retain the chemical(s) being analyzed at their natural location in the sample.

This chapter will consider only those analytical methods in which the sample is kept at sub-zero temperatures during analysis. Such methods center on microprobe instrumentation in which a finely focused beam of high-energy radiation is used simultaneously to localize the site of analysis and, by interacting with the sample, to generate secondary signals, which can be processed to provide the chemical information. The most significant and widely used of these methods is electron probe x-ray microanalysis (EPXM), in which a beam of electrons is used to excite x-rays locally in the specimen. Electron probe x-ray microanalysis is particularly useful in studies on biological samples, as it allows identification and quantitation of diffusible, physiologically active elements in whole cells and subcellular compartments and thus provides the basis for explaining structural responses to changing functional states. The application of low-temperature procedures during the preparation and actual analysis of the samples helps to ensure that the specimens are maintained in a physiologically defined state. The review by Le Furgey *et al.* (1988) provides ample evidence that this is probably the most important application of low-temperature EPXM. Much of the content of this final chapter will be directed toward the special procedures that have been developed to enable this method to be used at low temperatures as well as considering the advantages and limitations of this approach.

In particular, the discussion will center on the algorithms that have been devised for quantitative elemental analysis and for measuring the water content of samples. In addition, we will consider problems of image formation, spatial resolution, sensitivity, and beam damage. Considerable emphasis will be placed on hydrated samples and freeze-dried samples which are analyzed at ca. 100 K. Much less emphasis will be placed on considering samples that have been prepared using low-temperature techniques but which are subsequently analyzed at ambient temperatures. It is not intended to discuss the immunochemical and cytochemical procedures that are now providing so much important information about the molecular constitution of samples. This is not to belittle such procedures, but they are excluded simply because such methods, while benefiting from low-temperature sample preparation, have to be carried out at temperatures at which water is in a liquid state.

11.2 ELECTRON PROBE X-RAY MICROANALYSIS

11.2.1 Introduction

Electron probe x-ray microanalysis is a method of nondestructive chemical analysis in which it is possible to relate the elemental composition of a small region of cells or tissue with their morphological appearance. One usually achieves this analysis by focusing a fine beam of electrons onto a preselected area of a specimen and collecting, measuring, and analyzing the x-rays that are emitted from a specified microvolume of the sample. These x-rays are characteristic of the atoms that make up the sample. Alternatively, the beam of electrons can be made either to simultaneously illuminate all parts of the sample, or progressively scan the sample in a series of discrete rastering points. In these two cases the emergent x-rays may be used to give an elemental map or image of the specimen. The main advantages of electron probe x-ray microanalysis are that the analysis is carried out *in situ*, and a photographic record can be made of the precise region that is analyzed. Although electron probe x-ray microanalysis has been in existence for more than thirty years, it was not until the late 1960s that it was first applied to biological and organic samples (Hall, 1968) and five years later to frozen specimens (Bacaner *et al.*, 1973). A brief explanation will be given of the physics and technology of EPXM and an outline of the instrumentation that is needed. Reference should be made to the books by Goldstein *et al.* (1992), Newbury *et al.* (1986), and Lyman *et al.* (1990) for a detailed explanation of the physical principles and practical procedures of x-ray microanalysis. The small book by Morgan (1985) provides an excellent introduction to biological electron probe microanalysis.

11.2.2. The Physical Principles of X-Ray Microanalysis

X-rays are emitted from *any* specimen when it is irradiated with high-energy radiation such as a flux of electrons or x-rays. Nearly all x-ray microanalysis is carried out by means of a high-energy beam of electrons, which can be focused easily onto the area of interest on the sample. The same beam of electrons can also be used to provide topographical information about the specimen. In order for the x-rays to be produced, the energy of the incident or primary electron beam has to be greater than a minimum value known as the critical excitation potential. This increases with atomic number, but because most of the elements of biological interest are of low atomic number (i.e., $Z = 11–20$) the primary-beam energy need be no more than 10–15 keV. The interactions of electrons with matter can generally be divided into two classes.:

1) Inelastic scattering, where the incident electron transfers energy to the sample with only a small change in its trajectory.
2) Elastic scattering, where there is a significant change in trajectory of the incoming electron without any substantial change in energy.

During inelastic scattering of the electron beam, x-rays can be formed by two distinctly different processes.

(1) Characteristic x-rays. These are produced when the incident electron beam displaces or removes an inner-shell orbital electron and leaves the atom in an excited state with a vacancy in the electron shell. The vacancy is filled by an electron from one of the series of shells of orbiting electrons that surround the nucleus. These shells are designated *K*, *L*, *M*, etc., decreasing in energy as the distance from the nucleus increases. Each shell has a set number of permitted energy states, which may be occupied by orbital electrons. The energy difference resulting from these electron transitions is emitted as an x-ray photon characteristic of the electron orbital and the atom from which it was derived. The energy of the x-ray photon is the potential

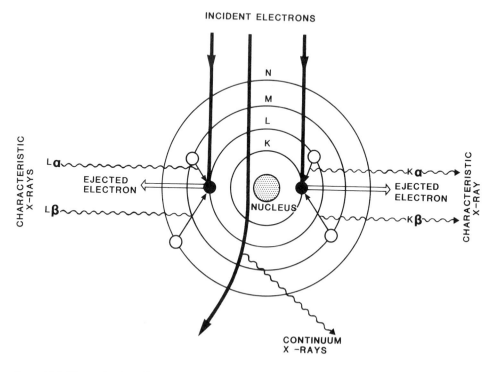

Figure 11.1. The production of x-rays from a specimen irradiated with a high-energy beam of electrons.

energy difference between the two orbitals involved in the transition and has an important bearing on whether it can be detected. Very low-energy x-ray photons (e.g., from lithium) would be largely adsorbed by the specimen; a few x-ray photons would escape from light elements up to $Z = 10$ and many of them would be absorbed before they reached the detecting system. It is only when we get to elements such as sodium and above that there is a sufficient flux of emitted x-rays to make x-ray microanalysis a practical proposition.

Figure 11.1 shows that a number of different x-rays can be derived from a given element and they can be conveniently named. If an electron was ejected from the inner K shell and the vacancy filled from the L shell, the resulting x-ray is referred to as a K_α x-ray photon; a transition from the M shell to the K shell is a K_β x-ray photon. Vacancies in L shells give L x-rays; in M shells, M x-rays; etc.

Thus each element has a distinct x-ray fingerprint depending on the number of electron orbitals, a consequence of its atomic number. For example, nitrogen ($Z = 7$) only has K and L electron orbitals and so can only yield K_α x-rays; phosphorus ($Z = 15$) has K, L, and M orbitals, and so yields K_α and K_β x-rays; and calcium ($Z = 20$) has four orbitals and so yields K and L x-rays.

The x-ray fingerprints for all the elements have been characterized and are available as tables or slide rules, or are embedded in the software of the computer-driven analytical programs. For light elements it is not possible to detect L and M x-rays, although K and L x-rays are a characteristic feature of most metals, and heavy elements such as lead or gold have a bewildering array of K, L, and M x-ray lines. The characteristic x-ray fingerprint for each element is the basis of the x-ray analytical technique, as there is a direct relationship between the intensity of characteristic x-rays from an element in the specimen and the number of atoms of that element within the irradiated microvolume.

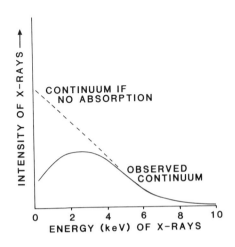

Figure 11.2. The production of x-rays from a specimen irradiated with a high-energy beam of electrons.

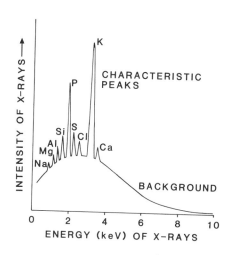

Figure 11.3. An idealised spectrum from biological material showing characteristic peaks and background (continuum) radiation.

(2) *Continuum Radiation or Background.* As seen in Fig. 11.1, the primary beam electrons may pass close to the nucleus of the sample atom, and experience a large change in direction but suffer only a small energy loss. In the majority of collisions the energy loss is less than 1 keV, although the energy distribution extends all the way to the energy of the incoming beam. Most of these low-energy x-rays are either absorbed by the specimen or remain undetected by the x-ray analyzer (Fig. 11.2). The continuum intensity is proportional to the total mass of the irradiated microvolume of the sample.

It will be seen that x-ray production from the irradiated microvolume is primarily a consequence of the characteristic and continuum radiation. Figure 11.3 shows the visible manifestation of these interactions as a spectrum of the background radiation overlaid with a the characteristic peaks of the elements that are sufficiently high on the Periodic Table to be detected.

11.2.2.1. The Intensity of Characteristic X-Rays

Although the intensity of x-ray emission is linearly proportional to the irradiating beam current, several other factors also play a part and their effects vary according to whether the sample is in the form of a bulk sample (i.e., does not transmit the incoming beam) or a thin section (transmits the incoming electron beam with negligible loss of energy).

(1) *Bulk Specimens.* The intensity of x-rays is proportional to (a) the number of atoms encountered by the primary electron beam; (b) the fraction of electron interactions that result in the emission of characteristic x-rays (the fluorescent yield); and (c) the ionization cross section (the probability that ionization of the inner-shell electron will occur), which increases rapidly with energy from zero to the critical excitation potential and reaches a plateau potential for a given element.

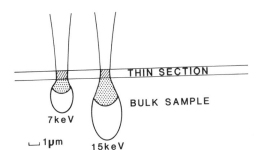

Figure 11.4. The relative sizes of the x-ray interactive volume in thin sections and bulk samples
as a function of the accelerating voltage of the
incident electron beam.

Below this figure the x-ray yield drops away rapidly. Above this figure there is
no improvement in x-ray emission and the primary beam penetrates deeper into the
sample. This penetration has two important consequences for x-ray microanalysis:

1. The further the beam penetrates into the sample the larger the size of the
   volume from which x-rays are emitted, which in turn decreases the spatial
   resolution.
2. The deeper the electrons penetrate into the sample, the longer the path length
   for the x-rays, many of which are absorbed by the specimen.

Absorption is a particular problem for light elements and for this reason the voltage
of the primary beam for bulk samples should be no more than 15 keV. Figure 11.4
shows these effects in bulk samples and thin sections.

(2) Thin Sections. Most of the electrons pass through the section without loss
of energy and absorption is negligible. The characteristic x-ray yield is simply a
function of fluorescent yield, ionization cross section, and the number and atomic
weight of the atoms in the sample.

11.2.2.2. The Production of X-Rays

The first golden rule of x-ray microanalysis is that the x-rays generated within
the specimen are not necessarily the x-rays emitted from the specimen. There are
cogent reasons for this state of affairs that depend on three interrelated factors: the
atomic number effect on the generation of x-rays; the absorption of primary x-rays;
and the fluorescence effect resulting in the secondary generation of additional x-rays.

1. The Atomic Number Effect (Z). This is a measure of the backscattering of
the primary beam from the surface of a bulk specimen and the amount of stopping
power of the specimen matrix. The latter is important in determining how many
characteristic x-rays are likely to be generated before the energy of the incident
electron beam decays to zero. In x-ray microanalysis of biological and organic
material, the atomic number effect has only become a problem with bulk specimens
of high atomic weight. The atomic number effect is of little consequence in thin
sections.

2. The Absorption Effect (A). The primary generated x-rays passing through the samples are absorbed exponentially with distance. This can be a problem with bulk samples, where low-energy x-rays may be generated deep in the sample. In thin sections absorption is negligible.

3. The Fluorescence Effect (B). This effect occurs when the generated x-rays have sufficient energy to interact with other inner-shell orbital electrons of the specimen and produce secondary x-rays of a low energy. This is only a serious problem in bulk specimens in which a small amount of high atomic weight material is embedded in a low atomic weight matrix. The fluorescence effect is of negligible proportions in thin sections.

11.2.2.3. Spatial Resolution

The spatial resolution in bulk samples depends on the volume through which electrons diffuse and spread in the specimen. Figure 11.4 shows that this diffusion is a function of accelerating voltage and specimen density, and to a much lesser extent of the beam diameter. Thus, for a given accelerating voltage, the spatial resolution is much lower in a bulk organic sample (between 1 and 3 μm). In most biological thin sections the mean atomic number and density is so low that beam spread is negligible and the spatial resolution depends primarily on beam diameter. In these circumstances the resolution is typically 20–100 nm.

11.2.2.4. Specimen Surface Topography

X-ray photons travel in straight lines and care must be taken to ensure that there is a direct line of sight between the region of the specimen being analyzed and the x-ray detecting system. Most metallic and geological specimens can be prepared with a smooth surface, but that is rarely the case with bulk biological and organic specimen, and surface irregularities can easily obscure and absorb emitted x-rays. The x-ray collection efficiency may be substantially improved if the specimen is tilted toward the detector. The surface topography effects rarely present a problem with thin section, which are assumed, sometimes erroneously, to be flat.

In terms of the production, collection, and (as we shall see later) quantification of x-rays, the use of thin sections is preferable to the use of bulk samples. There are minimal problems of absorption, the spatial resolution is improved, and the quantitation is more straightforward than with bulk samples. However, bulk samples are much easier to prepare, and when this preparation is centered on one of the low-temperature methods, there is a much greater degree of certainty of keeping the sample cold and retaining the elements of biological interest than in a thin section.

11.2.3. Instrumentation for X-Ray Microanalysis

Two pieces of equipment are needed to carry out x-ray microanalysis: a source of high-energy electrons and a means of detecting and measuring the x-ray photons. The source of high-energy electrons is conveniently provided by one of the various types of electron beam instruments such as transmission and scanning electron microscopes and their derivatives. The only requirement for these instruments is that they produce a narrow beam of electrons that may be brought to focus at various points on the specimen. Two features of x-rays, their energy and wavelength, are the principal determinants behind the design of instruments to detect, measure, and quantify the x-ray photons.

11.2.3.1. Wavelength-Dispersive Analysis

X-rays, in common with all electromagnetic radiation, have a range of wavelengths and radiate in all directions from the specimen. The instrumentation is designed so that a small proportion of these emitted x-rays impinge on a diffracting crystal, which reflects and focuses a given wavelength of x-ray photons into a proportional counter. The counter converts the x-ray photons into electrical signals of varying intensity that may be processed, counted, and quantified. The range of wavelengths and thus the number of elements that can be detected can be varied by use of a series of crystals of varying lattice spacings and by changes in the angle of the crystal relative to the specimen and the detector,

Because the analysis is restricted to one element at a time per spectrometer, most instruments have more than one spectrometer, each with several different crystals. Although the analysis is a time-consuming process, wavelength spectrometers have a useful, if limited, part to play in the study of biological and organic samples.

11.2.3.2. Energy-Dispersive Analysis

These instruments take advantage of the fact that x-ray photons also have discrete energies. The detector is a lithium drifted, *p*-type silicon Si(Li) crystal that converts the x-ray photons into a series of variable electrical charges, which are amplified, measured, counted, and quantified. The detector, which is kept under vacuum and cooled with liquid nitrogen to reduce electronic noise, is isolated from the vacuum of the microscope by a thin, 8.0-μm beryllium window. The presence of this window prevents the cold detector from becoming contaminated and absorbs the backscattered electrons. The window unfortunately also absorbs some of the light-element x-ray photons, which limits the usefulness of these detectors for the analysis of elements below sodium. However, most energy-dispersive detectors also operate in a so-called windowless and thin-window mode, which obviates some of these problems. As we will see later in this chapter, windowless detectors are of particular value in the analysis of frozen-hydrated samples, primarily for the determination of water content by measuring the local concentration of oxygen.

Energy-dispersive analysis, which is more frequently used than wavelength analysis, has a number of advantages:

1. All elements may be analyzed simultaneously.
2. The analysis time is short, 1–3 min.

3. The overall efficiency of collection is improved because the detector can be moved to within a few millimeters of the sample and thus intercepts a much larger solid angle of x-ray photons. A consequence of this improved efficiency is reduced beam currents and smaller probe sizes.

However, energy-dispersive analyzers also have their limitations. The x-ray resolution is poor (ca. 150 eV) and decreases with increasing energy of x-ray photons. Although this dependence creates problems with identification of all the peaks, spectral overlaps, poorer peak-to-backgrodund ratios, and hence poorer minimum detection limits, these effects may be partially alleviated by the use of computers to massage the resultant spectrum. Low-energy x-ray photons are absorbed by the beryllium window, and at high rates of x-ray production the counting statistics and the subsequent processing of the signal may give rise to spectral artifacts.

Ideally, biological and organic material microanalysis should be carried out on an electron-beam instrument fitted with both types of spectrometers. The high spatial resolution and improved limits of detection (particularly with light elements) of the wavelength analyzer complements the multielement analytical capabilities of the energy-dispersive systems.

11.2.4. Cold Stages for X-Ray Microanalysis

The cold stages and specimen transfer devices used for x-ray microanalysis are generally the same as the stages used for morphological studies with low-temperature TEM and SEM.

The stages must be mechanically stable and maintain the sample at low temperatures (ca. 100 K) and low contamination rates ($1-2$ nm/h^{-1}). Stage design and sample transfer have already been discussed at great length in Chapters 9 and 10 and need not be repeated here. We should, however, consider two aspects of cold stage design in equipment that is to be used for x-ray microanalysis.

It is important that the microscope producing the electron beam and the x-ray detectors do not contribute any extraneous x-ray signals. Even with the best electron beam instrument there is going to be some contribution from extraneous sources as the primary electron beam is scattered within the column, the poles pieces, the stage assembly, and the sample holder. Williams (1987) has shown how this x-ray fluorescence in the microscope can be minimized by using special condenser apertures and non-beam-defining hard x-ray apertures. Every attempt should be made to eliminate the extraneous instrument-derived x-ray signals so that the background is no more than 1% of the background generated by the specimen (Bentley *et al.*, 1979). It is equally important that the cold stage has a low x-ray background, and in most instruments this is achieved quite simply by providing a low-background insert made of beryllium or carbon in the immediate region of the sample. The bulk of the cold stage is usually made of high-purity copper to ensure good thermal conductivity. Most TEM manufacturers provide their own low-background holders for use with analytical studies. Similar holders are also available from Gatan Inc. and Oxford

Instruments. This is generally not the case with scanning microscopes, and it is usually necessary to fabricate low-background holders from such materials as Duralium (an alloy of aluminum) (Gupta *et al* 1977), graphite (Echlin *et al*., 1982), or beryllium (Saubermann *et al*. 1981a). Low-background holders are not an absolute requirement for x-ray microanalysis carried out in the SEM, as shown by the extensive studies carried out by Marshall and his colleagues. The breadth of these investigations, which were carried out using specimens either held in a metal vice or stuck to a carbon planchette, are summarized in two papers by Marshall (1987, 1988).

The final source of extraneous x-rays is the spectrum holder or support—which for sectioned material is usually a transmission electron microscope grid. Opinion is divided as to how to handle the extraneous x-rays that might be engendered by the specimen holder, which by necessity must be in close contact with the sample. For studies on sections in either the TEM or the STEM many workers use large mesh or slot copper, nickel, or titanium grids covered by a thin carbon or plastic film, which are inserted into low-background holders. Care has to be taken to avoid analyzing sections close to the grid bars. Good examples of these grid assemblies may be seen in the study by Andrews and Leapman (1989), who use a copper grid in a beryllium insert, and the earlier investigations by Rick *et al* (1982), who used nickel grids in a carbon insert. The proponents of this approach argue that it is relatively easy to make spectral corrections for the extraneous x-rays derived from the specimen support, which appear as recognizable characteristic peaks and associated background. The alternative approach is to use low-background grids and holders such as aluminum (Nicholson, 1982) or beryllium (Saubermann *et al*., 1981). In the case of beryllium this avoids the introduction of extraneous characteristic peaks and one is only left with a very low background, which may be accounted for in the appropriate analytical algorithms. Hall *et al*. (1972) pioneered the use of plastic support films coated with a thin layer of aluminum and stretched across a 3-mm annulus in a Duralium support. The presence of the characteristic peak of aluminum interferes with sodium detection by EDS and the aluminum contributes more extraneous background than carbon.

Such specimen holders were widely used in the early low-temperature x-ray studies by Hall and his colleagues (see, for example, Gupta *et al*., 1978) but have now been superseded by the metal grid approach, which provides more certain thermal stability at low temperatures. All three approaches appear to work, although none of them are ideal in terms of low extraneous x-rays and thermal stability. The use of copper and nickel grids, while giving a recognized x-ray signal, which may be removed by spectral processing, can give problems of peak overlaps with light elements. This is a particular problem with frozen-hydrated sections, which have to be close to the grid bar if they are to remain hydrated during analysis. The grid bar will make a considerable contribution to the total x-ray spectrum. It would seem prudent to avoid such complications in the first place and use low-background grids such as beryllium or pyrolytic graphite. Frozen sections can be placed closer to the grid bar for good thermal contact without having to contend with the characteristic x-ray peaks from the material. But an additional problem arises in that it is difficult to assess accurately the small contribution the low-background material may make to the background of the total x-ray spectrum. The thinly aluminized plastic film support, while contributing a calculable addition to the total spectrum, would also engender peak overlaps with light elements analyzed by EDS. Personal experience with this method has shown that frozen samples have to be close to the edge of the holder if they are to stand any chance of remaining fully hydrated throughout examination and analysis at low temperature.

It is also helpful to diminish the extraneous radiation in x-ray analytical studies on bulk samples examined in the scanning electron microscope. Marshall (1987) glues his samples to the specimen support using 10% methyl cellulose, while Zs. Nagy *et al.* (1977) and Preston *et al.* (1982a) use graphite slurry cements. Echlin *et al.* (1982) and Echlin and Taylor (1986) have mounted samples either into carbon or beryllium tubes, or into holes in a graphite support using a buffered graphite slurry. Fuchs and Fuchs (1980) surrounded frozen specimens with soft indium metal, which could be gently squeezed to form a good electrical and thermal contact. Carr *et al.* (1986) have used aluminum foil to pack frozen specimens into the sample holder. The other modification it may be necessary to make to a TEM or SEM cold stage is to ensure that the x-ray detector–sample geometry is optimized for maximum signal collection. This may be achieved by placing a large-area (30-mm^2) Si(Li) detector as close as possible, i.e., a few millimeters, to the sample in order to subtend a large solid angle of x-ray collection. The usual configuration is to locate the x-ray detector more or less at right angles to the electron beam. The specimen is then tilted toward the detector at between $10°$ and $40°$ which, while maximizing signal collection, may cause parallax effects with transmitted images. Care must be taken that the housing of the x-ray detector does not contribute to the extraneous x-rays in the region around the sample. Much attention has been taken by instrument manufacturers to ensure that the designs of the cold stage and of the detector are fully compatible. This is a particular problem with the TEM, where there is much less space to maneuvre in the region of the specimen. In some instances it may be necessary first to load the sample and tilt it toward the detector, and then wind in the detector so that it is close to the specimen.

11.2.5. The Process of X-Ray Microanalysis

To the novice, the process of x-ray microanalysis can appear to be fraught with difficulty, although these difficulties are readily overcome once they are recognized and appreciated. The difficulties center around the second golden rule of microanalysis, which states that x-rays emitted by the specimen are not necessarily solely those collected and analyzed by the detector. This rule implies that the total x-ray signal is not representative of the sample, in that certain signals are deleted, and/or contains spurious signals that are not derived from the specimen.

The principal reasons for the loss of x-rays are faulty specimen preparation, absorption either within the specimen or as a result of poor specimen detector geometry, mass loss due to high beam currents during analysis, and charging due to an excess overvoltage being applied to a poorly conducting sample. These problems may be alleviated by adequate sample preparation and by minimizing the energy input into the sample sufficient to give an adequate signal.

The spurious additional signals are derived from changes made to the sample during preparation (i.e., fixatives, buffers, stains, and thin films of coating material) and scattering from the regions of the microscope immediately surrounding the specimen (i.e., support grid and film, microscope stage and column, and even parts of the x-ray detector). At best, all these spurious signals can be eliminated; at worst, they can be considerably alleviated.

The instrumentation and procedures by which the collected x-ray photons are displayed as a usable signal for the analyst need not concern us here, although the appropriate details may be found in one or more of the references cited in Section 11.2.1. Suffice it to say that the x-ray signal can be suitably manipulated and displayed either visually as a spectrum on a CRT (cf. Fig 11.3) or numerically as a printout. The subsequent handling of this signal will depend on whether qualitative or quantitative information is required.

1. Qualitative Analysis. X-ray microanalysis can be carried out at varying degrees of sophistication. In many instances it is sufficient to perform a simple qualitative analysis in which the presence (or absence) of an element is confirmed and directly related to a morphological feature. One can obtain the information by mapping the distribution of the element in the whole sample, running transect scan lines across various regions of the sample and expressing the differences in elemental concentration as varying peak heights, or analyzing discrete parts of the sample by means of a point or reduced raster. This approach can tell us, at best, that some regions of a sample contain more of an element than another; at worst, that an element is either present or absent.

2. Quantitative Analysis. Many analyses require a more sophisticated approach, in which detailed information is needed on either the relative or absolute concentration of elements in the sample. This process is called quantitative analysis and requires that considerable attention be paid to specimen preparation, the process of x-ray acquisition, and the way of ensuring data are corrected and quantified.

A wide range of quantitative procedures are available for converting the collected x-rays to relative or absolute concentrations. The various algorithms that have been devised to produce quantitative analytical information will be discussed later in the chapter. Particular emphasis will be placed on the low-temperature analysis of frozen-hydrated and freeze-dried samples.

11.3. TYPES OF SPECIMENS

11.3.1. Introduction

The ideal specimen for x-ray microanalysis would be perfectly flat, have a clear and recognizable morphology, and contain local concentrations of elements within a well-defined matrix. Such samples are rarely found among biological and organic materials studied by low-temperature x-ray microanalysis. One of the main problems with biological x-ray microanalysis is that the elements of interest are invariably of low atomic number ($Z = 11–20$). This problem is exacerbated by the fact that these elements are frequently at low concentration and in a low atomic number hydrated organic matrix of C, H, O, and N. One way around this problem is to use higher atomic number marker ions as tracers for their low atomic number counterparts. Calcium ($Z = 20$) may be tracked by strontium (38); potassium (19) by rubidium (37), caesium (55), or thallium (81); chlorine (17) by bromine (35). These marker elements, some of which are quite toxic, are applied at low concentrations during experiments prior to the start of any low-temperature protocols. The papers by Wroblewski *et al.* (1989) and Dorge and Rick (1990) provide an outline of these, not entirely satisfactory, procedures.

There are several types of specimens, and they can exist either in a frozen hydrated (ice present) or freeze-dried (ice absent) state. The two major specimen types are bulk specimens, which also includes microdroplets and whole cells, which do not transmit any portion of the electron beam, and sections, which transmit varying proportions of the beam. The sections, which may be cut from freeze substitution or freeze-dried resin-embedded material, or from fresh frozen material, may be subdivided into thin (less than 100 nm) and thick (more than 100 nm) specimens. We need to consider briefly the advantages and disadvantages of each specimen type because their form and constitution have an important bearing on the way they should be analyzed and the nature of the information that will be available. Such information will be of value in designing experimental protocols for x-ray microanalysis.

11.3.2. Bulk Specimens

These are most usually small pieces of biological or hydrated organic material which have been excised from a larger specimen and then quench cooled. The ideal frozen samples are usually fractured or microplaned at low temperatures to expose a fresh surface, which may be examined and analyzed. The same procedure may be used for plastics and polymers.

The advantages of bulk specimens are as follows:

1. They are prepared easily and quickly.
2. Because of their relatively large thermal capacity, they remain fully hydrated at low temperatures.
3. They are less readily damaged by the electron beam (see Section 11.10).
4. They have a readily recognizable morphology at low spatial resolution which may be further enhanced by surface etching.
5. Large areas of the sample may be examined and analyzed.
6. They can be used for the analysis of fluid (frozen) filled spaces.
7. The peak/background ratios for most elements are good and are enhanced with freeze-drying.

The disadvantages of bulk specimens are the following:

1. The fractured surfaces are not smooth, which may create problems with quantitative x-ray microanalysis due to x-ray absorption. This is much less of a problem with cold stages that have rotation and tilt facilities. A given fracture face may be maneuvred into an optimum position for analysis.
2. The x-ray spatial resolution is between 1 and 2 μm. In the biological sciences this would limit analysis to the major compartments of the cell.
3. The analytical programs for quantitation are less accurate than those devised for sectioned material.
4. Light element analysis is difficult owing to absorption of the generated x rays within the sample.

The analytical procedures for bulk samples are continuing to evolve but have already shown themselves to be capable of producing valid physiological data, e.g., Marshall and Wright (1972), Marshall (1983), Marshall *et al.* (1985), Echlin *et al.* (1982), Echlin and Taylor (1986), Echlin (1990), and Lustyik and Zs. Nagy (1985). The reviews by Marshall (1987) and Zs. Nagy (1988, 1989) provide further applications of the low-temperature analysis of bulk material.

11.3.3. Microdroplets

Nanoliter volume samples are collected from various tissue compartments using micropipettes. These are specialized techniques in which all handling procedures are carried out under paraffin oil to avoid drying and changes in the concentration of electrolytes in the droplet. Small volumes (10–100 pl) of the liquid sample and appropriate standards are placed on a polished beryllium support and freeze-dried. The miniscule piles of dried salts are analyzed by electron probe x-ray microanalysis using well-tested algorithms. Bearing in mind the caveats at the beginning of this chapter, this is not a low-temperature analytical procedure as the actual analysis is carried out at ambient temperature. The method, which was originally devised by Ingram and Hogben (1967), Morel and Roinel (1969), and Lechene (1970), has been reviewed by Quamme (1988).

11.3.4. Whole Cells and Particles

These are a particular type of bulk sample which exist as discrete entities, i.e., microorganisms, isolated or cultured cells, or hydrated polymer complexes, and which do not need any preparation other than quench cooling. Kirk *et al.* (1978) have analyzed intact spray frozen red blood cells, but found that it was necessary to use cryosections to be able to identify the cell types. Colonna and Oliphant (1986) have studied intact sperm cells and Zierold (1991) has analyzed whole amebas. The large size and paucity of structural detail have not made this a popular approach to low-temperature x-ray microanalysis. More studies need to be carried out and in particular to see if the same analytical procedures that have been developed for particle analysis in the material sciences can be applied to these types of specimen. The advantage of these specimens is that they are very easy to prepare and maintain at low temperature. The disadvantages center on the fact that one is analyzing the natural surface (and below) of the sample, which may be contaminated. The spatial resolution is poor and there are uncertainties about the quantitative procedure. The use of x-ray microanalysis of whole cells and particles has been reviewed by Wroblewski and Roomans (1984) and Sigee (1988).

11.3.5. Sectioned Material

As mentioned earlier, sectioned material comes in a range of thicknesses, and 100 nm can usually be imagined in a conventional 100-kV transmission electron microscope; thicker sections, up to 2.0 μm, can only be properly visualized in STEM instruments or in an SEM working in the transmission mode.

11.3.5.1. Thin Sections

These can be cut from a variety of specimen types, the only limitation being that the use of interventionalist chemical procedures invariably causes losses of diffusible elements. It is unfortunate that such methods (freeze substitution and low-temperature embedding) provide samples that are easy to cut and provide the highest spatial resolution. Resin-embedded, freeze-dried material remains an uncertain compromise and for the moment fresh frozen sections (hydrated or dried) are a better approach to low-temperature x-ray microanalysis.

The advantages of thin sections are as follows:

1. They can provide high spatial resolution, ca. 10–25 nm.
2. In the dried state they provide good, recognizable images.
3. The procedures for quantitative x-ray microanalysis are well established and accurate.
4. In the dried state they are relatively resistant to radiation damage.
5. In the dried state they provide reasonably good peak-to-background ratios.
6 They provide flat surfaces suitable for analysis.
7. Morphological information may be obtained using transmitted electrons as well as reflected electrons, i.e., secondaries and backscattered primaries.

The disadvantages of thin sections are the following:

1. It is sometimes difficult to prepare thin frozen-hydrated sections of fresh material that has not been subject to any chemical treatment. This is a particular problem with samples containing large vacuoles and aqueous cell free spaces.
2. Image contrast in the frozen-hydrated state is poor and makes it very difficult to recognize details within the specimen.
3. Frozen-hydrated material is very susceptible to radiation damage.

4. Considerable care has to be taken to prevent frozen-hydrated material drying out during examination and analysis.
5. The x-ray yield is very low and requires a sensitive detection system and careful elimination of extraneous radiation.

11.3.5.2. Thick Sections

These too can be cut from a variety of samples, but for analytical purposes are most usually cut at a thickness of 0.5–1.0 μm from fresh frozen material. The advantages of thick sections are the same as those given for thin sections. The spatial resolution is not quite so good, i.e., 50–100 nm rather than 10–25 nm. The peak-to-background ratios are better in both the hydrated and the dried states because there is more material to analyze in the thicker sections. They are a little easier to cut than thin sections, and in the hydrated state they are probably the best type of material for the quantitative analysis of large aqueous spaces. Thick sections are subject to the same disadvantages as thin sections, although the radiation sensitivity in the frozen-hydrated state may not be quite so high and they are a little easier to maintain in a fully hydrated state.

11.3.6. Frozen-Hydrated versus Freeze-Dried Samples

Before leaving this section on the types of specimens that may be used for low-temperature x-ray microanalysis, we should briefly consider the advantages and disadvantages of using frozen-hydrated and/or freeze-dried material. It will be convenient to direct our discussion to what one would hope to achieve with low-temperature microanalysis rather than consider it in relation to the different specimen types.

11.3.6.1. Spatial Resolution

With sectioned material there is very little difference in the spatial resolution between the dried and hydrated state. There is only a small change in the size of the x-ray interaction volume between the two hydration states of sections. This is not the case with bulk material, where the removal of water by freeze-drying leads to voids throughout the specimen and a decrease in the average density. This decrease in density can result in up to a tenfold decrease in spatial resolution in a bulk sample.

11.3.6.2. Image Visualization

The image quality, i.e., contrast and recognition, is lower in frozen-hydrated material than in the dried material. This is a particular problem in sectioned material, where the extensive electron scattering in fully frozen-hydrated samples makes it difficult to see anything within the sections. All is revealed, however, by simply drying the sample. Image visualization is less of a problem with bulk material, where the uneven freeze-fracture surfaces provide a measure of topographic contrast. Sublimation of the surface ice (etching) improves image quality, but care must be taken to restrict the etching just to a surface layer. We will return to the vexing question of partial drying and surface etching at the end of this section.

11.3.6.3. Beam Damage

Frozen-hydrated material is much more susceptible to ionizing radiation damage than the dried counterpart. This presents a major problem in the high-resolution analysis of cell free spaces and is a matter we will return to again later in this chapter.

11.3.6.4. Analytical Integrity

The *raison d'être* of low-temperature microanalysis is that the cryopreparation method retains the sample in its natural—albeit frozen—state. It is the intregrity of the frozen-hydrated state we wish to study, for all the elements to be measured by x-ray microanalysis will, we hope, be more or less *in situ*. It is doubly unfortunate that it is difficult to see anything in such material and that it is also very susceptible to beam damage. Although freeze-drying overcomes these two problems, it does, in turn, create an additional problem. Elemental relocation occurs during the sublimation of ice, and it is assumed that most ions will either react with counter ions or bind to the closest organic molecule. In cells with a 10–15% dry weight content of organic material, the movement will probably be less than the average dimension of the original ice crystals, 20–50 nm. In large water-filled compartments, such as the vacuoles of mature plant cells and extracellular spaces and tissue lumens, the distances will be much greater. It is doubtful whether quantitative analysis can be carried out in such compartments after freeze-drying (Zierold and Steinbrecht, 1987), as nothing remains to trap the translocated elements. The accepted wisdom is that during freeze-drying of well-frozen material, the elements are moved to the boundary of the compartment, i.e., the tonoplast of the plant vacuole, the surrounding cell boundaries of the extracellular space or tissue lumen. This may well be what occurs in bulk material, but in sectioned material the element may equally well be deposited *in situ* on the underlying plastic or carbon film used to support the section. Analysis of supported and unsupported frozen sections before and after freeze-drying would be a useful exercise. It is assumed, of course, that there has been minimal elemental relocation during the initial quench cooling and subsequent handling procedures.

11.3.6.5. Quantitative Analysis

There is no doubt that quantitative elemental analysis is more easily achieved in freeze-dried samples than in frozen-hydrated material provided a way is found to accurately measure the local water content of the compartment being analyzed. Such procedures are considered in Section 11.8.

11.3.6.6. Structural Integrity

Fragile specimens appear more readily to retain their structural integrity when in a frozen-hydrated rather than a freeze-dried state. It is not certain whether this is due to the collapse of macromolecules as water is removed, or whether it is due to the inevitable shrinkage and distortion that occurs during freeze-drying.

11.3.7. Partially Dried Samples

It has been suggested (Hall and Gupta, 1983) that one way around the problem of poor image contrast and low peak-to-background ratios in frozen-hydrated sections would be to partially dry the section. This approach is fraught with difficulties. It assumes that water is distributed equally in all parts of the sample, which is patently not the case in biological material. It assumes a constant and equal drying rate, which will only occur in homogeneous samples. There is no evidence that partial drying can be accurately controlled or measured. Partially dried samples will also create severe problems with the quantitative analytical procedures. In spite of the difficulties, it would be more expedient to either have fully frozen-hydrated or completely freeze-dried material. The various criteria that can be used to establish whether material is in a frozen-hydrated or freeze-dried state will be considered in Section 11.8.3.

The situation is probably not quite so exacting with bulk material, where surface etching of a 100-nm layer of ice from a sample that is going to be analyzed to a depth of 2–3 μm will have little effect on the accuracy of the quantitative analysis. Careful etching can be used in some circumstances to remove contaminating layers of ice. Although surface sublimation improves the image quality of bulk frozen-hydrated material, such etching procedures should only be used as a last resort.

It should be clear that there is no perfect sample type for the low-temperature x-ray microanalysis of diffusible elements. The frozen-hydrated, fractured surface of bulk material is best suited for the low spatial resolution of the major cell compartments and for different cell types within a tissue. The thick frozen-hydrated section would be most likely to give valid information about the water content and elemental constitution of the aqueous compartments. The thin freeze-dried section appears to be the only way to achieve high spatial resolution quantitative analysis of subcellular components.

11.4. INTERACTIONS WITH THE ELECTRON BEAM

11.4.1. Introduction

An electron beam will interact with solid material in a variety of ways. Some of these interactions, i.e., elastic and inelastic scattering, which have been discussed earlier, form the basis of the signals we use in microscopy and analysis. The depth the electrons penetrate into the sample, and the interactions of the beam with nonconducting specimens, will influence the quality and spatial resolution of the analytical information. We need to consider these phenomena in relation to the analytical spatial resolution in both bulk and sectioned material in the frozen-hydrated state.

Somlyo and Shuman (1982) define analytical spatial resolution as the minimum distance over which a difference in composition can be detected. It is a function of probe size, accelerating voltage, and beam current. Ideally the diameter of the area sampled by the electron beam should be equal to the beam diameter. This is rarely, if ever, the case because the electron beam is scattered in all directions as it interacts with the atoms and electrons of the sample. It is possible to directly visualize and/or model the interaction of the electron beam with the specimen.

An electron beam penetrating a bulk sample forms a pear-shaped or teardrop-shaped microvolume. Close to the surface the microvolume is cylindrical but rapidly becomes cone shaped as the electron beam penetrates and becomes scattered in the depth of the sample. Because sections correspond to the regions near the surface, the interactive microvolume is much smaller than in bulk samples.

The probability of a scattering event, the so-called cross section, and the average distance an electron travels between scattering events, the so-called mean free path, can be calculated. The basis of these calculations need not concern us here; full details may be found in Goldstein *et al.* (1981), but the following facts emerge:

1. Scattering is more probable in high atomic number materials.
2. There is more scattering at low incident beam energy.
3. Elastic scattering causes the beam electrons to deviate from their original direction of travel, causing them to diffuse into the solid.
4. Inelastic scattering reduces the energy of the electron beam and thus limits its range and travel.
5. For light elements, the elastic scattering cross section decreases relative to the inelastic cross section so that inelastic scattering dominates. In high atomic number specimens the reverse is true.

The volume within which the electron beam interacts with the solid, depositing its energy and giving rise to the different forms of secondary radiation we use in microscopy and analysis, is called the interactive volume. An understanding of the size and shape of the interactive volume as a function of the specimen and the beam parameters is thus vital for a full appreciation of x-ray microanalysis of both sections and bulk materials. Although most of the original calculations were made on high atomic number materials with good conductivity, i.e., metals, the same principles may be applied to the low atomic number specimens which are the main preserve of low-temperature analysis.

11.4.2. Dimensions of the X-Ray Analytical Microvolume and Spatial Resolution in Sections

Most sections studied by low-temperature microanalysis are analyzed at relatively high electron energy, i.e., 20–30 kV in the SEM working in the transmission

Table 11.1. *Parameters Determining Elastic Electron Scattering of Substances of Biological Interest*[a]

Material	Z	A (g)	D (kg m^{-3})	E (10^{-23} m^2)	MFP (nm)
Carbon	6.00	12.01	2000	7.89	126
Dextran	4.14	7.72	1500	4.56	187
Protein	3.83	7.10	1300	4.29	211
Lipid	3.21	5.75	1000	3.60	265
Nucleic acid	5.26	10.24	1600	6.65	160
Ice	3.33	6.01	920	3.22	337

[a]Z, Mean atomic number; A, mean atomic weight; D, density; E, elastic electron scattering cross section; MFP, mean free path for elastic electron scattering. [From Carlemalm *et al.* (1985).]

mode and up to 100 kV in STEM and TEM instruments. Table 11.1 compiled from data from Carlemalm *et al.* (1985), provides some useful information on this question and may be used in conjunction with other data (Echlin *et al.*, 1982; Echlin, 1988; Goldstein *et al.*, 1981) to estimate the size and shape of the interaction volume in frozen sections. Let us assume that a frozen section is 80% ice, 18% organic matter, and 2% inorganic minerals. To a first approximation, the mean mass density of this material is 1000 kg m^{-3} which would give an elastic electron scattering cross section of ca 3.6 × 10^{-23} m^2 and elastic electron mean free path of 265 nm. In simple terms, a very large proportion of the electron in an 80–100-keV electron beam would pass right through a thin (ca. 100 nm) frozen section without any scattering. Most of the scattering that would occur would be in a forward direction, rather than a lateral direction, so that to a first approximation the interactive volume of the electron beam in a thin section would be in the form of a cylinder whose diameter in a very thin section would be only slightly wider than the diameter of the electron beam. The situation would only show a marginal improvement in spatial resolution if the frozen-hydrated section is freeze-dried. For although there might well be an 80% reduction in section thickness as the section is dried (Zierold, 1988), there would be a concomitant increase in the mass density of the remaining organic material and hence an increase in the electron scattering cross section. It is important to note that thin sections of uncoated frozen sections do not charge,* provided they are in good contact with the underlying conductive support. An analytical spatial resolution of 8.7 nm has been demonstrated in two-dimensional Fourier transforms of molybdenum x-ray maps of catalase crystals stained with ammonium molybdate and using a probe size of between 3.5–5.0 nm (Somlyo and Shuman, 1982). An instrumental analytical resolution of better than 10 nm is quite exceptional and has only been achieved in a very thin crystalline object using a fine electron probe only available

---

*An electron charge builds up rapidly in nonconducting samples when they are scanned with a beam of high-energy electrons. The charging can lead to image distortion and considerable deflection and deceleration of the incident electron beam, which would make it impossible to carry out even qualitative x-ray microanalysis.

with a field emission gun. In nonperiodic objects a spatial resolution of ca. 30 nm has been obtained in ultrathin freeze-dried sections of muscle fiber (Somlyo and Shuman, 19823.

In sections 100–200 nm thick, a spatial resolution of approximately 100 nm can usually be obtained using a 50-nm probe. Although values have been calculated for the spatial resolution one might expect to obtain in soft biological and organic material, these have generally only been obtained from sections of known thickness and ignoring electron scatter within the specimen (Hall 1968, 1971; Shuman *et al.*, 1976). Hall and Gupta (1983) have extended these studies and set out the relationship between the spatial resolution and section thickness and provide estimates of the optimal instrumental conditions that give the best analytical spatial resolution and the best limits for the measurement of elemental concentrations. On the assumption of a single scattering event, Hall and Gupta (1983) provide a useful model of the volume of x-ray generation within a thin section (Fig. 11.5). This model, when applied to section thickness up to 1.0 μm and accelerating voltages above 30 kV, indicates that to a first approximation the expected spatial resolution would be 1–2 times the incident probe diameter. Above 1.0 μm thickness, and with probe sizes generally used on such material (ca. 100 nm), the spatial resolution becomes progressively poorer and the shape of the interactive microvolume becomes increasingly more conelike. These figures assume that the probe current is sufficiently high to give a statistically significant count rate from a given element and that the sample is homogeneous,. With frozen material we know that this is not the case as the formation of ice crystals segregates the nonaqueous compartments of the sample into spaces between the ice crystals that vary in size between 50 and 1000 nm.

Whereas a static probe may be used for the high-resolution analysis of very thin freeze-dried section, i.e., less than 100 nm, such a probe would give erroneous information from thicker sections. In the thicker sections (500–1000 nm) most commonly used in the x-ray microanalysis of frozen material it is more appropriate to use a reduced raster (ca. 0.5–1.0 μm²), which should be larger than the biggest ice crystals. The increase in size of the incident probe would have a concomitant effect on the size and shape of the analytical interaction volume and spatial resolution. A further complication to this simple relationship between probe diameter and spatial resolution comes from additional studies by Hall and Gupta (1983). Table 11.2,

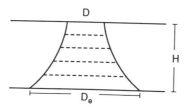

Figure 11.5. Volume of x-ray generation within a thin specimen. *D*, diameter of the electron probe in the plane of incidence; D_e, diameter in plane of emergence; *H*, specimen thickness. [Redrawn from Hall and Gupta (1983).]

Table 11.2. Relationship Between Instrumental Parameters for Optimal Resolution and the Local
Concentration of Potassium in a Section Analyzed with a 50-keV Electron Beam

C_{K+} (mmol kg DW)	R (μm)	D (nm)	H (μm)	I (nA)	N_{K+}	N_B
20	0.129	116	0.99	2.42	469	1733
30	0.105	94	0.86	1.37	346	852
40	0.090	81	0.78	0.93	285	526
50	0.081	73	0.73	0.70	248	366
60	0.074	67	0.69	0.55	223	275
80	0.065	58	0.63	0.39	192	178
100	0.059	53	0.59	0.30	174	128
120	0.055	49	0.56	0.25	161	99

C_{K+}, Local concentration of potassium; R, resolution; D, probe diameter; H, section thickness; I, beam current; N_{K+}, characteristic counts for potassium; N_B, background counts. [From Hall and Gupta (1983).]

which is taken from their paper, shows that better analytical spatial resolution may be obtained when the local concentration of the element is higher.

In summary, it has been shown that the analytical spatial resolution decreases with increasing probe size and section thickness. As the beam penetrates deeper into the sample there is an increase in the amount of scattering, particularly in the lateral direction and the shape of the interactive microvolume progressively changes from cylinder to a truncated cone with an ever increasing diameter. However, the spatial resolution that may be obtained in practice is also closely related to the local concentration of the element being analyzed, the voltage and the beam current of the incident electron probe, and ultimately to the size of the ice crystals in the frozen sample.

11.4.3. Dimensions of the X-Ray Analytical Microvolume and Spatial Resolution in Bulk Samples

Although it is known that the interactive microvolume is much larger in bulk material, the exact dimensions and shape in frozen samples remains unclear. The uncertainty arises from the extent to which these essentially nonconducting samples charge up in the electron beam. Early studies by Brombach (1975) and Fuchs and Lindemann (1975) showed that frozen-hydrated bulk samples rapidly acquire a charge when examined at low temperatures in an SEM. These studies, which were carried out at low voltage (5 kV) on uncoated samples of pure ice, showed that the space charge, which initially developed as a pear-shaped microvolume within the sample, quickly assumed a pancake-shaped electron diffusion volume immediately below the sample surface (Fig. 11.6). This phenomenon appears to be quite unique for ice and has not been reported in other insulators when irradiated with an electron beam. If the sample is coated with a thin layer of evaporated carbon, the charging

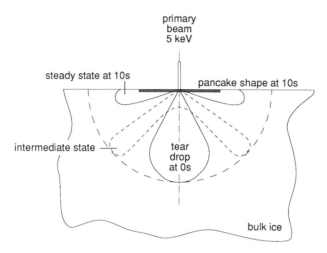

Figure 11.6 A model for the steady-state interaction volume of electrons in ice at 123 K. [Redrawn from Brombach (1975).]

is diminished (although not eliminated) and the internal charge stage within the sample is negligible.

Brombach (1975) and Fuchs and Fuchs (1980) provide a detailed explanation of what is considered to occur during this interesting interaction of electrons with ice. If the phenomenon in pure ice is truly representative of what happens in frozen-hydrated samples of biological and organic material, there are important implications for the size and shape of the x-ray interactive volume. It would suggest that the spatial resolution in frozen-hydrated specimens is better than that suggested by the generally accepted pear-shaped microvolume. Thus, at an accelerating voltage of 5 kV, the pancake is only 100 nm deep and 400 nm wide. It is unfortunate that the studies of Brombach (1975) and Fuchs and Lindemann (1975) were never extended or repeated. Investigations by Marshall (1982b) and Marshall and Condron (1985b) suggest that the internal space charge is rapidly conducted away via a surface coating layer and that no gross distortion of the electron diffusion volume occurs. They show that the cutoff in continuum energy has the same value as that in a spectrum from a high conductive sample. If the Brombach model is correct, the electron backscatter yield would be expected to increase as the electron interaction volume is pushed toward the surface by the developing space charge. Experimental evidence from Marshall and Condron (1985) has shown this only to occur in uncoated samples, which are not normally used for x-ray microanalysis because the surface charging leads to image distortion and analytical inaccuracies.

The size and shape of the interactive x-ray volume can be calculated in two ways. The more pessimistic results are derived from electron range equations, which can be used to calculate the depth and width of electron penentration into the sample.

More optimistic results are obtained by calculating the size of the microvolume from which 90% of the x-rays are derived.

11.4.3.1. Spatial Resolution as a Function of Electron Penetration

Following the suggestions by Boekestein *et al.* (1980), the analytical spatial resolution can be divided into two components. The *lateral resolution* is the maximum width of electron penetration into the sample assuming that the interactive volume is in the form of a half sphere. The *depth resolution* is the maximum depth of electron penetration assuming that this is in the form of an elongated tear drop. Because the electron beam diffuses randomly in the specimen, there is still considerable discussion concerning the various formulations of the range equation, and the actual shape of the interactive volume. The limits of electron penetration are also approximately equivalent to the volume of the sample in which x-rays may be *generated.* This is not the same as the volume from which x-rays are *emitted,* because the x-rays of all the elements analyzed will be absorbed to a varying extent in the frozen matrix. This would have the effect of making the size of the interaction volume an overestimate. However, the calculations in this chapter have assumed the worst possible case for both the lateral and depth resolution. A measure for the lateral resolution may be obtained from the equation derived by Reed (1975):

$$R^{\text{lateral}} = 0.02331 \frac{E_0^{1.5} - E_c^{1.5}}{\Delta} \tag{11.1}$$

and a measure for depth resolution from the equation derived by Anderson and Hasler (1966):

$$R^{\text{depth}} = 0.64 \frac{E_0^{1.68} - E_c^{1.68}}{\Delta} \tag{11.2}$$

where E_0 and E_c are, respectively, the energies of the incident electron beam and the critical absorption edge for the x-ray radiation of interest and Δ is the density of the sample in g/cm^{-3}.

Using these two equations, the variations in lateral and depth resolution for sodium (the lowest atomic number element usually analyzed in studies of biological and hydrated organic material) with changes in accelerating voltage are plotted in Fig. 11.7. The x-ray spatial resolution for the other elements will be marginally better than the figures for sodium. The density for a fully hydrated sample at 100–110 K is taken to be 1.0 g/cm^{-3} (1000 kg m^{-3}). In the absence of any direct measurements on the density of cells, the figure of 1000 is arrived at after making the following assumptions. Kamiya and Kuroda (1957) measured the density of *Nitella* cell sap and obtained a figure of 1014 kg m^{-3}. The density of water at 110 K is 940 kg m^{-3} and the *average* dry weight of plant cell cytoplasm, for which these calculations were made, is ca. 15%. Thus a combination of a decrease in density of the water phase at low temperatures and a small increase in the density due to the solid phase gives a figure of 1000 kg m^{-3}

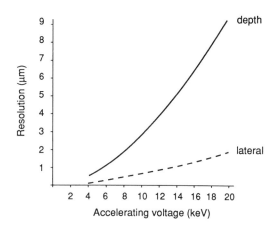

Figure 11.7. Calculated relationship between accelerating voltage and generated x-ray spatial resolution for sodium in frozen hydrated meristematic plant material.

At accelerating voltages of between 12 and 15 kV on frozen-hydrated material, the maximum depth resolution of sodium x-rays is 6.0 μm and the lateral resolution is 1.1 μm. If the analysis is carried out using a reduced raster (1.0 μm^2) the lateral resolution should properly be increased to 2.5 μm. These depth resolution calculations are much worse in freeze-dried samples, and some studies by Echlin and Taylor (1986) on vacuolate plant cells demonstrate the extent of this problem. The vacuole, a large water and dilute solutes filled bag, may easily occupy 90% of the volume of a mature leaf cell. The water is an important structural component of these cells, contributing to cell turgor and hence to tissue integrity. Water has a density of 1000 kg mm^{-3}, and is slightly lighter (940 kg m^{-3}) in the solid form. Add to this frozen phase the density of the dissolved components; the density of the vacuole is approximately 1000 kg m^{-3} when in the fully frozen-hydrated state. It is thus possible to produce reasonably accurate calculations of electron penetration. A graphic example of such calculations is shown in Fig. 11.8 and demonstates that at 15 keV and in a 8000-μm^3 cell, one can probably analyze separately the vacuole, cytoplasm, and cell wall with only a small change of an overlap of interactive volumes. The precision is improved by tilting the specimen toward the detector and working at a lower acceleration voltage. This precision is, however, only achieved if a static beam is used for analysis. As discussed earlier, a reduced raster is more frequently used in order to sample both the ice crystals and the non-ice boundary phases. Simple calculations have shown that provided the raster is not placed at the edge of the cell, the excited x-ray microvolume remains within the cell concerned even when the flat fracture face is tilted at 45° to normal.

The situation changes quite dramatically if these large vacuolate cells are freeze-dried. It is difficult to calculate the precise variations in density, but we estimate that the cytoplasm density decreases by 50%, the vacuole by 90%, while the cell wall may show a small increase (1300 kg m^{-3}) as the cellulose microfibrils collapse on each other. Figure 11.8 gives an estimate of the relative sizes of the interactive volumes and demonstrates the impracticality of attempting to analyze either the vacuolar

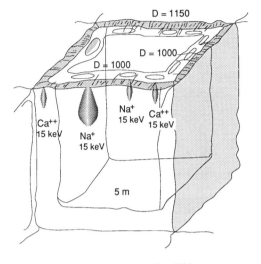

a

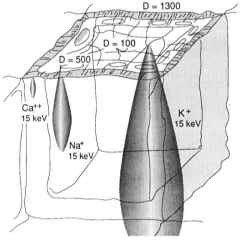

b

Figure 11.8. Diagrammatic representation of the relative range of electron penetration into a mature, vacuolate, frozen hydrated (a) and in the same freeze-dried (b) plant cell. The dark tear drop shapes represent the varying sizes of x-ray interactive volume in relation to both the element concerned and the beam accelerating voltage. The varying values for density (D) are estimations of the density of the different cell compartments. The larger of the two sodium interactive volumes in (a) is that to be expected from using a 6-μm^2 scanning raster as distinct from a stationary probe illustrated by the smaller of the two volumes. [From Echlin and Taylor (1986).]

contents or cytoplasm of a freeze-dried cell. In freeze-dried samples the electrons pass right through the vacuum which fills the lumen of ice-crystal ghosts remaining after the sublimation of the ice. The electrons are only scattered when they come into contact with the solid phase consisting of the biological material and the salts that have been swept aside by the initial growth of the ice crystals. In the "solid" parts of the plant cell, i.e., wall, cytoplasm, and organelles, the ice crystals can be quite small, i.e., ca. 100 nm, provided the cooling is fast enough. But there is nothing we can do to preserve the structural integrity of the contents of a freeze-dried vacuole. The water, which makes up ca. 90% of the contents of a mature plant cell, has been removed and the remaining solid phases such as salts, organic molecules, and macromolecules fall to the lower side of the vacuole as viewed from the image of the surface.

It is virtually impossible to obtain any significant spatially related x-ray data from the jumble of cell detritus lying at the bottom of a plant vacuole, or the extracellular and intercellular spaces that are a common feature of most biological material. Only fully hydrated samples can provide valid x-ray data.

These resolution figures are pessimistic and possibly misleading for the following reasons:

1. The electron range equations are probably an overestimate since most of the primary beam energy is scattered in a forward direction into the sample with much less lateral scattering.
2. Studies by Anderson (1967) have shown that there is a decreasing number of x-rays emitted from the sample the greater the depth of penetration of the electron beam. Thus, 50% of the emitted x-rays came from the top 25% of the sample/beam interaction volume.
3. Electron range equations take no account of x-ray absorption.
4. The different range equations give widely different results. Marshall and Condron (1985a, b), using three different range equations, found that the maximum range for 12-kV electrons exciting aluminum x-rays in a frozen hydrated gelatine gel varied from 1.7 to 3.7 μm

11.4.3.2. Spatial Resolution as a Function of X-Ray Emission

The depth resolution analysis can also be calculated from the x-ray depth distribution. Parobek and Brown (1978) derived an empirical formula for x-rays emitted at low incident electron energies, and Marshall (1982a) has used this to calculate the x-ray depth resolution. The ionization function $\phi(\rho z)$, where ρ is the density of the sample and z the depth in the sample, describes the depth distribution of x-rays generated in a target by electron interaction. Marshall (1982a) has calculated $\phi(\rho z)$ curves over 2000 depth intervals in an analytical program that handles up to ten elements. A figure for depth resolution is given by integrating 90% of the $\phi(\rho z)$ curve for the emitted fraction of x-rays and calculating the depth limit of the integral. The calculated curve is corrected for the effects of absorption.

Figure 11.9 shows the calculated depth resolution for light elements in a protein–water system as a function of accelerating voltage. As expected, the spatial resolution decreases with an increase in accelerating voltage. The results of all these calculations show that the spatial resolution for the x-ray microanalysis for elements Na to K at 15 kV is 2 μm and appears to show little variation with changes in the protein concentration (5–20%) of the matrix. This spatial resolution is considerably smaller than the spatial resolution that may be calculated from electron range equations. There is experimental evidence that supports the notion that the microvolume in the sample from which 90% of the x-rays are generated (although not all necessarily emitted) is smaller than the microvolume described from electron range calculations. Marshall and Condon (1985a, b), have used photographic film as a model. The film,

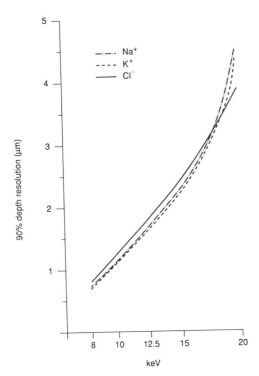

Figure 11.9. Calculated depth resolution for light elements, Na⁺, K⁺, and Cl⁻ in a frozen protein–water matrix as a function of accelerating voltage. [Redrawn from Marshall (1982).]

as seen in cross section, is made up of a 1.5-μm gelatine layer, a 10-μm-thick emulsion layer containing Br and Ag, a thick polyester base, and finally a thick gelatin layer. Small pieces of film were swollen in water and quench frozen. X-ray intensities were obtained from the flat surface of the film over a range of voltages. At 10 kV, no Br or Ag could be detected, at 12.5 kV a small signal was obtained from the emulsion layer, and at 15 kV a clear signal could be obtained. The signal at 12.5 kV corresponded to the $\phi(\rho z)$ predicted depth resolution for a 1.5 μm thick, 30% gelatin layer. For measurements of lateral resolution, pieces of swollen, frozen-hydrated film were microtomed to a smooth surface in cross section. X-ray intensities were taken at different distances from the widely dispersed silver grains in the emulsion. No Br or Ag signal could be detected at sites more than 2 μm from a grain, but were present in spectra taken within 1 μm from a grain. Similar results were obtained from frozen-hydrated sections of gelatin gels containing electrolytes and from some earlier studies using fine wedges of a gelatine–salts mixture (Marshall, 1981).

Oates and Potts (1985) have confirmed the work of Marshall and his colleagues. These workers deposited thin layers of ice of known thickness onto surfaces coated with Sb_2S_3 and were able to detect sulfur signals under a 0.75-μm ice layer at 10 kV, under a 3-μm layer at 15 kV, and under a 4-μm layer at 20 kV. These figures correspond very closely to the $\phi(\rho z)$ predictions. Some rather ambiguous results have been obtained by Zs Nagy *et al.* (1977), who placed Araldite sections of varying thickness over copper and aluminum surfaces. From these and other results they conclude that at an accelerating voltage of 10 kV sodium will be excited to a depth of 4–5μm into the sample. However, if the sample is tilted at 45°, most (90%?) of the x-ray comes from a depth of 2–3 μm. It is difficult to assess the quality of this work as there are variations in the thickness of the Araldite sections and the x-ray microanalysis was performed under such low dose conditions as to question the statistical significance of the results. It is useful to compare the depth and lateral resolution measurements obtained by the two methods, even though they actually measure two different parameter (electron penetration versus x-ray emission). Such a comparison is made in Table 11.3 and shows that the depth profile calculated from the $\phi(\rho z)$ curves is three times smaller than that from electron range equations. The figures for lateral resolution are marginally smaller at each voltage for the $\phi(\rho z)$ data but show a considerable difference for the electron range calculation. This is to be expected, for it is known that there is an increase in the forward direction electron scattering at higher beam voltages. The data in Table 11.3 have been calculated for sodium in a frozen-hydrated organic (biological) matrix containing 75% water. This would represent the worst possible case, for at higher z values the interactive volumes become progressively smaller. If these data are to be put to any practical use in analyzing bulk frozen material it would seem prudent to take some middle value for calculating the size of the interactive volume. In that way one can be reasonably certain that the analysis is being truly made from the particular compartment in question. Most analyses of bulk frozen organic and biological material are made at between 12 and 15 kV. This is a good compromise voltage as it ensures that the overvoltage is sufficiently above the K shell absorption edges to excite x-rays from the elements being analyzed, i.e., $Z = 11$–20 and allows a sufficiently large analytical microvolume to be sampled from frozen material to ensure statistically significant results. Because of the effects of x-ray absorption, all the light elements ($Z = 11$–20) have approximately the same depth resolution at 12–15 kV in frozen-hydrated material (Marshall, 1982a, b). This may be fortuitous from the point of measuring the spatial resolution of analysis, but it is necessary to correct for the absorption of the lighter elements in any quantitative analysis of bulk samples. These are matters we will return to later in this chapter. For really critical studies, where spatial resolution is important, it would be appropriate to use lower voltages, i.e., from 5–7 kV

Table 11.3. Depth (D) and Lateral (L) Resolution for Sodium Calculated from Electron Range Equations and $\phi(\rho z)$ Curves as a Function of Accelerating Voltage[a]

		Primary beam accelerating voltage						
		4 kV	6 kV	8 kV	10 kV	12 kV	15 kV	20 kV
Electron	Depth	0.5	1.0	2.0	3.0	4.5	6.0	10.5
penetration	Lateral	0.2	0.3	0.4	0.6	0.8	1.1	1.8
(μm)								
90% X-ray	Depth	0.6	0.7	0.8	1.3	1.7	2.4	3.5
emission (μm)	Lateral	0.5	0.6	0.7	1.1	1.4	2.0	2.9

[a]Data from Marshall (1982) and Echlin et al. (1982).

for sodium to 8–10 kV for calcium. The choice of the operating kV for studying bulk frozen material is critical. Higher accelerating voltages favor increased x-ray intensity but at the expense of a large x-ray interactive volume. Lower accelerating voltages diminish the size of the interactive volume but cause a decrease in x-ray intensity. Marshall (1987) has calculated that for elements of biological interest ($Z = 11$–20) a 1 kV reduction results in a 11–15% drop in x-ray intensity.

11.5. IMAGING PROCEDURES USED IN CONJUNCTION WITH LOW-TEMPERATURE X-RAY MICROANALYSIS

11.5.1. Introduction

One of the essential features of x-ray microanalysis is that the analysis can be performed in relation to a recognizable structure. It is equally important to see the specimen as to measure its chemical composition. In early microanalytical studies the imaging procedures were based on reflected or transmitted light, which imposed severe limitations of morphological spatial resolution. Modern x-ray microanalyzers make use of reflected and transmitted electrons to image the sample, which permits a considerable improvement in spatial resolution. Some of these imaging procedures, i.e., transmitted electrons, and reflected secondary and backscattered primary electrons, have been discussed in the two previous chapters. In this section we will again need to consider briefly these imaging mechanisms in relation to the special problems that arise in trying to analyze frozen-hydrated and freeze-dried bulk and sectioned material. We will also need to briefly consider additional imaging procedures such as electron energy loss spectroscopy and scanning transmission microscopy, which offer distinct advantages for viewing frozen specimens, which invariably have a low topographic and atomic number contrast. All these imaging procedures can provide very high morphological spatial resolution of the order of a few nanometers. The previous section in this chapter considered the analytical spatial resolution that may be achieved in sections and bulk samples, and provided figures that were considerably greater than a few nanometers. It is appropriate to consider what morphological spatial resolution is necessary in analytical studies. As one might expect, there is no single answer, but a good rule-of-thumb that should be adopted is that the morphological spatial resolution should be ten times better than the analytical spatial resolution that theory predicts from the instrumental and experimental parameters. There is thus little need to try to achieve 1–2-nm spatial resolution in a system in which the analytical information is derived from a 10-μm^3 microvolume. But it is important to choose an imaging procedure that can allow an unambiguous identification of the region being analyzed.

11.5.2. Transmitted Electron Images

This is the only imaging mode available in conventional transmission electron microscopes and we have already discussed the imaging mechanisms at some length in Chapter 9. Contrast is intrinsically poor in frozen-hydrated sections but improves dramatically when such material is freeze-dried. Contrast within images of ultrathin frozen hydrated sections is now generally considered to be too poor for it to be useful for morphological studies. In the few instances where it has been possible to obtain recognizable images from ultrathin frozen-hydrated section, the sections have been cut at ca. 110 K from vitrified material (McDowall et al., 1983). The production of such material invariably involves cryoprotectants whose use is eschewed in micro-analytical studies. The fact that ultrathin frozen-hydrated sections are very sensitive to beam damage has virtually precluded their use in connection with x-ray micro-analytical studies. Images of freeze-dried sections, with their increased contrast and improved radiation resistance, remain the principal model of visualization for high spatial resolution x-ray analytical studies.

11.5.3. Secondary Electron Images

This is the principal imaging mode employed in the SEM and can provide a wealth of morphological information from both bulk samples and sections. Cryo-micrometry rarely produces smooth sections and the small discontinuities in the surface of frozen-hydrated sections reflect changes in the underlying structure. With experience these discontinuities can be related to specific features within the section. The suggestion by Hall and Gupta (1983) that it is not possible to see subcellular details even as large as nuclei in secondary electron images of fully hydrated bulk samples is mistaken. Extensive studies by Echlin and Marshall and their numerous collaborators over the past ten years have shown that it is relatively easy to locate and identify the major compartments within cells and tissues. The images are a result of atomic number contrast and an edge effect coupled with some topographic contrast from the backscattered electrons which invariably form part of the signal. As with sectioned material, the image quality improves with surface etching and freeze-drying. Secondary electron images have the advantage that they may be taken at low beam currents, i.e., 5 pA and hence very low total electron dose per picture point, ca. 60 e^- nm^{-2}. Secondary electron images are one of the principal ways morphological information may be obtained from frozen-hydrated and freeze-dried bulk samples.

11.5.4. Backscattered Electron Images

The signals that give rise to these images are another important consequence of the electron beam interacting with the sample surface. Unlike secondary electrons, the backscattered electrons travel in straight lines from the sample to the detector, and this feature has been used to great advantage in obtaining images of bulk specimens. Marshall (1982b) fitted a solid state backscattered detector onto the energy dispersive detector used to collect the x-ray signal. This gave a backscattered image with high topographic contrast. Since both the backscattered electrons and the x-ray photons travel in straight lines, the BSE image may be used to determine that an uninterrupted line of sight exists between the sample and the x-ray detector. This is of paramount importance for the x-ray microanalysis of rough surfaces where the low-energy x-rays of light elements may be absorbed by surface discontinuities. Marshall (1982a, b) has combined the signal from such a system with the signal from an unbiased secondary electron detector placed at the opposite side of the specimen. The combined signals from the two detectors gives a quasi-three-dimensional image with high topographic contrast, which are invaluable for assessing the surface topography of fracture surfaces of frozen-hydrated bulk material. Marshall (1982b, 1984) has devised procedures in which a quantitative comparison may be made of the x-ray and BSE signals to show whether any changes in the x-ray signal are due to changes in the local concentration of the element concerned or to changes in the surface topography. For the first time we have a quantitative measure of surface roughness.

The slope of the sample toward or away from the BSE detector can either be judged from visual differences in the grey level or by directly measuring changes in the signal output from the detector. As a result of these extensive studies, Marshall and Condron (1985a) have shown that considerable angular deviation from the horizontal can be tolerated in the analysis of element P to Ca, but less for Na-Mg. Backscattered electron imaging is proving to be very useful in ensuring that fully quantitative analysis may be carried out on the rough surfaces of frozen bulk samples.

11.5.5. Scanning Transmitted Electron Images (STEM)

These images are the scanning electron microscope equivalent of the transmitted electron microscope image and are the most usual way to obtain information from thick sections, i.e., 0.5–1.0 μm. The transmitted electron signal from thick sections is a multiscattering event and may be processed in two ways. The entire signal may be collected using a scintillator–photomultiplier system or a solid state diode. Alternatively, the constituent parts of the transmitted signal may be collected separately. The unscattered and inelastically scattered electrons are collected on a central bright field collector and the elastically scattered electrons are collected on an annular collector.

In most cases, the whole transmitted signal is used to form images of thick sections. Frozen-hydrated sections of most hydrated biological material have virtually no contrast owing to the fact that the electron scattering in ice is very close to that in biological material. If there are local concentrations of organic material or heavy elements it is claimed that these components can be identified in hydrated sections (Lubbock *et al.*, 1981; Gupta and Hall, 1983). Hall and Gupta (1982) list a number of instrumental changes that may be tried in order to improve the problem of imaging frozen-hydrated thick sections. In all cases there is a dramatic improvement in image contrast when sections are freeze-dried.

There is a marginal improvement in image contrast if the constituent parts of the transmitted electron beam are treated separately. Zierold (1982, 1985, 1986a, 1988) has demonstrated that image quality in thin, ca. 100 nm frozen-hydrated sections, may be improved by using different combinations of the bright field and dark field images. Contrast is very poor in bright field alone but increases as the dark field component is included in the imaging. This may be achieved by using either the pure dark field, ratio imaging of the bright field/dark field or by reversing the dark field

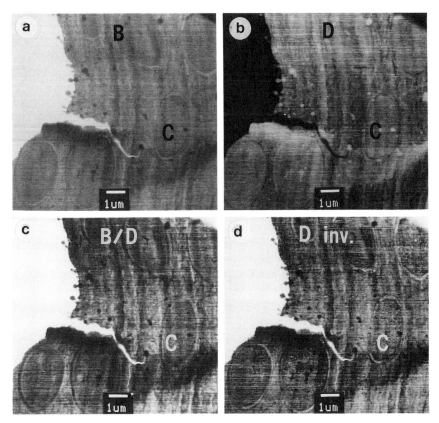

Figure 11.10. The same frozen-hydrated cryosection from a yeast suspension imaged by different conditions in a STEM. (a) Bright field; (b) dark field; (c) ratio of bright field to dark field; (d) inverse dark field. [Picture courtesy Zierold (1985).]

contrast. Figure 11.10 shows that these procedures do result in a significant change in image contrast and provide a useful way of improving the image quality of frozen-hydrated sections. There is some debate (see discussion in Zierold, 1986a) regarding the maximum thickness of sections that may be imaged by these procedures and it would appear that 0.25 μm would be the upper limit. It is doubtful whether this procedure could be used for imaging the somewhat thicker sections most frequently used in x-ray microanalysis. As we will discuss later, these dark field/bright field images can also be used as a basis to determine the wet weight concentration of elements in thin frozen sections.

The STEM can also be used to analyze material as well as provide images. This powerful analytical technique is nowhere better illustrated than in the recent study by Leapman and Andrews (1991), who used a field emission, high-resolution STEM in combination with an ED x-ray system to analyze thin frozen-hydrated and freeze-dried sections kept at 100 K. They were able to quantitate the calcium content of

400-nm^2 regions within freeze-dried sections with an uncertainty of ± 2 mmol/kg dry wt, equivalent to ± 12 atoms.

11.5.6. Electron-Energy-Loss Imaging (EELS)

Although not a strict imaging procedure in that it only uses a small fraction of the total signal available from the interaction of the electron beam with the sample, electron-energy-loss imaging appears capable of producing very high-resolution analytical images (Somlyo and Shuman, 1982). The energy-loss electrons are those electrons that have suffered a loss of energy during inner-shell ionization which result in the production of characteristic x rays. They may be collected by an electron-energy spectrometer after they have passed through the section. Thus it would be possible to collect all the electrons that have undergone a characteristic energy loss in the production of sodium x-ray photons. An elemental map can be produced in which the analytical spatial resolution may in some circumstances be less than 1.0 nm (Adamson-Sharpe and Ottensmeyer, 1981). This mode of imaging, which is very sensitive to low atomic number elements ($Z = 6$–9) is only applicable to ultrathin sections, c. 30–50 nm at 100 keV, although thicker sections, c. 100–200 nm, could be studied at 1 meV (Egerton, 1982b). A paper by Probst et al. (1989) indicates that images may be obtained from thicker sections using electron spectroscopic imaging at 80 kV. They show images from 150-nm-thick frozen-hydrated sections using zero-loss electrons with an energy width of 10 eV. They also show electron spectroscopic and bright field/dark field STEM images of conventionally prepared material. No details are given of the temperature of the frozen-hydrated sections or of any checks that were made to ensure it was in a fully hydrated state. The high contrast of the image suggests that it may well be freeze-dried rather than frozen hydrated. A recent paper by Michel et al. (1991) shows that images of frozen-hydrated sections recorded without energy filtering have much lower contrast than those taken in the low-loss filtering mode. Comprehensive up-to-date reviews of the procedure may be found in the recent papers by Johnson et al. (1988), Leapman and Ornberg (1988), and Leapman and Swyt (1989). The whole of a recent issue of *Ultramicroscopy* [Vol. 32(1) 1990] is devoted to papers on energy filtering imaging and analytical techniques.

In spite of a number of different imaging procedures that may be used in conjunction with the x-ray microanalysis of frozen material, the central problem of poor image contrast in frozen-hydrated materials is not entirely solved. The simple and easily achieved procedure of improving image contrast by artificially introducing high atomic number material as in fixation and staining is precluded for obvious reasons. The only alternative is the skill and experience of the operator. It is a common experience with nearly all the people working in the field that with time there is an improved familiarity with the poor image quality of frozen-hydrated material. This fact, coupled with the important correlative microscopical studies that must accompany these investigations, allow us to find our way around these poor but nevertheless pristine images.

11.6 SAMPLE PREPARATION FOR LOW-TEMPERATURE X-RAY MICROANALYSIS

11.6.1. Introduction

The success or failure of quantitative x-ray microanalysis turns on achieving proper preservation of the sample. The earlier chapters in this book focus on different aspects of these techniques, quench cooling, cryosectioning and fracturing freeze substitution, and freeze-drying, and we do not need to repeat this information here. While the debate continues as to whether freeze substitution can produce specimens in which quantitative analysis of diffusible elements can be achieved, most preparative techniques for x-ray microanalysis are centered on nonchemical intervention procedures. The earlier chapters on TEM and SEM provide information on how specimens should be transferred into such intruments and subsequently maintained. All the procedures have been discussed except for sample coating, and this section will briefly discuss the need for such a preparative step and, more specifically, how it may be applied to the types of frozen specimen we use in x-ray microanalysis.

11.6.2. Charging

Nearly all nonconductive specimens examined in an electron beam instrument will rapidly acquire an electric charge. The charging phenomenon is complicated, particularly in frozen-hydrated material. It is made up of two components: the surface charging, which can be eliminated by applying a thin layer of a suitable conductor and bulk charging which is difficult to assess and to eliminate. Charging is primarily a problem with bulk samples and its effects range from minor distortion of image quality as shown by shifts in the imaging raster to a full blown flare up of the sample in which it assumes the same potential as the incident electron beam and acts as an electron mirror. Charging will reduce the energy of the incident electrons and the emission of signals from the specimen and makes it difficult to compare the analysis from one part of the specimen to another. These effects influence both the size and shape of the interactive microvolume and the nature of the x rays generated (but not necessarily detected) in the sample. The amount of charging is related to sample temperature (the electrical conductivity of ice decreases with temperature), sample conductivity, beam current, beam voltage, and the presence of dissolved ions and electrolytes (Echlin, 1978c; Echlin et al., 1982). While it might be instructive to better understand the charging process it would be more pragmatic to discuss ways of reducing the effect. Although some of the problems associated with charging may be alleviated by working at lower voltages, it is impracticable to work at accelerating voltages much below 5.0 kV as the x-ray yield from light elements within a bulk matrix is reduced to the point where the counting statistics are impaired. Any attempt at increasing the bulk conductivity of the sample by impregnation with conducting materials such as salts of osmium and silver will immediately negate any analysis of light elements within a frozen matrix. One is left with a single choice, and a thin surface layer of a suitable conductive material appears to do the trick.

11.6.3. Coating Techniques for Low-Temperature X-Ray Microanalysis

The coating procedures that are used are no different from those generally employed with nonconductive samples. A good review of these techniques may be found in the books by Goldstein *et al.* (1992), Newbury *et al.* (1986), and Lyman *et al.* (1990). The discussion that follows is directed primarily towards applying a thin 10–15-nm coating layer to the surface of a bulk frozen sample. The same methods could be used to coat the thin plastic films which are frequently used to support thin sections of frozen material.

The usual coating materials of noble metals and their alloys were impracticable for most biological samples for three reasons:

1. The *L* and *M* x-ray lines of these heavy metals Au, Pd, Pt, etc. may interefere with the *K* lines of the light elements.
2. A film of sufficient thickness to overcome the problems of charging (and thermal damage) will possibly attenuate the incoming electron beam and absorb the emitted x-ray photons.
3. X-ray photons from the coating material may make a contribution to the general background associated with the sample. This could compromise the accuracy of analytical procedures which rely on a background measurement as part of the quantitative algorithm.

There are only four elements that may be used as coating materials in connection with quantitative analytical studies. These materials, together with the properties that are relevant to the x-ray microanalysis of light elements, are shown in Tables 11.4 and 11.5. Each material has both advantages and disadvantages.

11.6.3.1. Beryllium

This is probably the best material to use. It has medium thermal and electrical properties, negligible x-ray interference and contribution to background, low evaporation temperature, and is easy to apply. Unfortunately it is very toxic when in the

Table 11.4. A Comparison of Some of the Physical Properties of Four Elements Commonly Used to Coat Biological and Organic Materials Prior to X-Ray Microanalysis

	Beryllium	Carbon	Aluminum	Chromium
Density (kg m^3)	1800	2300	2700	7200
Thermal conductivity (W cm)^a	2.00	1.29	2.40	0.93
Resistance $(\Omega \text{ cm})^a$	4.57	3500	2.80	13.0
Evaporation temperature (K)	1520	3000	1275	1450
X-ray lines (keV)	K_α 0.109	K_α 0.227	K_α 1.487	
	K_α 5.415	K_β 5.946	L_α 0.598	

^{a}Figures given for 300 K. Values improve at lower temperatures.

Table 11.5. Percentage X-Ray Absorption by a 20-nm Layer of Different Coating Materials, Assuming an X-Ray Emergence Angle of 40° to the Plane of the Coating[a]

Emitter		Absorber			
Element	Ka (keV)	Beryllim	Carbon	Aluminum	Chromium
O	0.515	2.23	7.45	5.49	6.48
Na	1.041	0.30	1.22	0.89	13.97
Mg	1.254	0.17	0.73	0.54	9.07
P	2.014	0.04	0.15	1.93	2.89
S	2.308	0.03	0.10	1.33	2.00
Cl	2.622	0.02	0.08	0.51	1.34
K	3.314	0.01	0.04	0.49	0.75
Ca	3.692	0.01	0.03	0.37	0.52

[a]After Marshall (1987).

finely particulate and thus potentially ingestible form. Marshall and his colleagues have made extensive use of beryllium as a coating material in connection with their analytical work on insect tissue. Any potential users of beryllium should consult their papers (Marshall and Carde, 1984; Marshall et al., 1985) for details of the procedures and precuations.

11.6.3.2. Carbon

This is the most popular coating material. It is easy to apply and has little effect on the x-ray yield from the sample. The electrical properties are rather poor. However, carbon suffers from one grave disadvantage: the high temperature of evaporation, which can easily cause thermal damage to the specimen. Marshall (1977a) measured a 40-K temperature rise in samples held at 140 K which had been continuously coated with carbon for 30 s, and Vesley (quoted in Echlin and Taylor, 1986) has measured a 473-K temperature rise in samples during carbon coating. Echlin et al. (1980a, 1981) found evidence of surface etching to a depth of 150 nm in samples held at 110 K and heavily coated with carbon for 15 s while the sample was being continually rotated and tilted. Echlin et al. (1980a) placed a small aperture between the evaporation source and the sample in order to reduce the thermal load into the frozen sample. Although a thin layer of carbon appears to diminish charging in bulk samples, it is not the best coating material for such specimens. Carbon coating has been used extensively to coat freeze-dried sections and to provide a thin conducting layer to support such specimens.

11.6.3.3. Aluminum

In spite of its good electrical properties and low evaporation temperature, aluminum has not been extensively used to coat bulk frozen materials. Echlin and Moreton (1973) and Saubermann and Echlin (1975) found that, while it was possible to evaporate metallic aluminum onto metal and plastic surfaces maintained at low temperatures, the evaporated material was deposited as a nonshiny material on frozen-hydrated samples. There is no clear explanation as to why this should happen because the coated samples displayed no charging artifacts. It is thought either that ice deposits on the sample surface layer could have induced the formation of an oxide, or that the specular appearance of the element changed when it was deposited at low temperature. These effects were not noted in a later study by Oates and Potts (1985), who evaporated 25 nm of aluminum onto layers of ice maintained at 91 K. Subsequent studies by Oates (personal communication) have shown that temperature and pressure are of critical importance during the evaporation of aluminum onto flat frozen specimens. The evaporation temperature should be between 1200 and 1300 K and the pressure should be no greater than 780 μPa (6×10^{-6} Torr). The degree of shininess of the deposited aluminum layer is a function of surface roughness. Oates has also had success using magnesium as a coating material, although the sample temperature rose by 20 K during coating. The real disadvantage to using aluminum as a coating material in connection with x-ray microanalysis is that its x-ray emission at 1.487 kV would create difficulties with the analysis of some of the light elements using ED spectrometers because of the peak overlaps and the contribution to the general background. Aluminum has only a minimum effect on the absorption of x rays from light elements. In spite of these problems, aluminum has been used extensively to coat the large area plastic films that are used to support thick frozen sections; cf. the many studies by Hall and Saubermann and their colleagues.

11.6.3.4. Chromium

This material is in many respects a compromise. Its use was first suggested by Marshall (1977a) and extended in the more recent work by Echlin and Taylor (1986). The thermal and electrical properties are sufficient to prevent specimen charging and diminish beam-induced thermal damage. The low evaporation temperature does not cause ice sublimation from the surface of frozen samples, and the characteristic x-ray lines are sufficiently far away from the characteristic lines of the elements of biological interest to avoid causing any problems with their identification.

However, the higher density of chromium could well present problems of beam attenuation and absorption of soft-rays emitted from frozen-hydrated samples. Echlin and Taylor (1986) have investigated this and related problems using standards of known provenance and have proposed that a small correction should be made for the absorption of x rays from low atomic number elements. This involves measuring the concentration of chromium at each area of the specimen and making a small correction to the measured concentrations of each of the other elements. This correction can be used to cover attenuation of the incoming beam, the absorption of emitted x rays, and the contribution the chromium coating makes to the general background of the spectrum. The corrections procedures may be embedded easily into the analytical software. The coating procedures for all four elements are relatively easy to carry out, provided due care is taken with beryllium. All four materials are applied using evaporative techniques at high vacuum with the sample maintained at low temperatures and, wherever possible, rotated and/or tilted during the actual coating process. Such movements of the sample will ensure an even coating on rough surfaces. The relatively low sputtering yield of these materials, primarily due to the formation of an oxide layer on the target, precludes using this method to apply the thin films. The long sputtering times, i.e., minutes at the low vacuum, that are used would cause the target material to warm up, and thus encourage sample contamination and surface etching.

11.7. Quantitative Procedures for Low-Temperature X-Ray Microanalysis

11.7.1. Introduction

X-ray microanalysis can provide fully quantitative elemental analysis at local sites within a wide range of inorganic, organic, and biological materials. The peak intensities and background radiation observed on x-ray spectra can be transformed into concentrations that may be expressed either on a wt/wt or a wt/vol basis. The analytical procedures that have been developed over the past 30 years to bring about these transformations can be applied more or less equally well to all materials. Good reviews of the processes that are involved and the basis of the calculations may be found in the books by Chandler (1977), Goldstein *et al.* (1992), Morgan (1985), and Newbury *et al.* (1986).

Within these general approaches more precise methods have been developed for the analysis of biological and organic materials. These procedures have been further refined to handle the special case of frozen material. It is not proposed to follow the development of these various analytical algorithms but broadly to consider the computational routines, which now permit fully quantitative analysis to be performed on frozen sections and frozen bulk samples. We will briefly discuss the salient features of these different techniques together with their advantages and disadvantages. One of the unique features of low-temperature x-ray microanalysis is that it allows measurements to be made on the local concentration of water in a sample. Although these measurements are made in concert with the elemental analysis, it is proposed to consider them separately in the following section of this chapter as they are complex (and contentious) issues.

11.7.2. Quantitative Analysis of Frozen Sections

Quantitative procedures for thin sections are reasonably straightforward because in nearly all biological thin sections the measured x-ray intensity is proportional to the amount of element in the analyzed microvolume. The very thinness of the sample precludes any absorption of generated x-rays; and since most biological and organic samples are of low atomic number, the atomic number of fluorescence effects are negligible. The spatial resolution is good because there is little beam spread within

a thin sample. The quantitative procedure makes use of two well-established relationships of x-ray optics:

1. The intensity of the continuum or background radiation is proportional to
   the total mass thickness of the analyzed microvolume (Kramers, 1923).
2. The peak intensity of the characteristic x-ray radiation of an element is
   proportional to the concentration of that element in the analyzed microvolume (Castaing, 1951).

The Kramers relationship may be utilized to give a measure of the total size of the
microvolume, and the Castaing equation provides a measure of the concentration
of element(s) within the same microvolume. The quantitative analysis of diffusible
elements in frozen sections is based on one or both of these two relationships.

11.7.2.1. Elemental Ratio Method

This involves calculating elemental ratios in which the measured x-ray counts
of any two elements per unit area are proportional to their relative elemental concentrations. These ratios can be determined without standards, and while this method
does not give absolute concentrations, it does provide a measure of the relative
concentration of one element to another or the relative concentration of the same
element in different parts of the specimen. The method was devised by Hall (1968)
and Hall and Hohling (1969) and the proportionality is based on the simple
formulation

$$I = KM \qquad\qquad (11.3)$$

where I is the number of characteristic x rays counted from an analyzed microarea,
M is the local mass of the element in the analyzed microareas, and K is a constant
related to beam current, probability of x-ray excitation, time, and instrumental detection effficiency.

11.7.2.2. Peripheral Standard Method

This is a logical extension of the elemental mass ratio method which compares
the characteristic x-ray counts obtained from a region of the section with the characteristic x-ray counts from a standard. The procedure was first applied to frozen
sections by Dorge et al. (1975, 1978), and elaborated by Rick et al. (1979, 1982).
Immediately prior to freezing the specimens are covered with a thin layer of a 20%
albumin solution containing known concentrations of electrolytes. This peripheral
standard is quench frozen along with the sample and then sectioned. The albumin
layer forms a layer around the section of the sample. Elemental concentrations are
calculated by dividing the characteristic x-ray counts from a microarea in the cell
with those obtained from a similar microarea in the peripheral standard and multiplying this ratio with the known wet weight concentration of the particular element
in the standard. It is not necessary to obtain data from the peripheral standard for
every element. The concentration of some elements can be determined from their

characteristic x-ray counts in the specimen on the basis of the relative elemental sensitivities calculated using other types of standards. This procedure can be used on both frozen-hydrated and freeze-dried sections using the following formulation:

$$\left(\frac{M}{I}\right)_{sp} = \left(\frac{M}{I}\right)_{st} \tag{11.4}$$

Where M is the element mass per unit area at the time of analysis in the specimen (sp) and the standard (st) and I is the number of characteristic x-rays from the standard and the specimen, after background correction and peak overlap removal.

If one assumes that the specimen thickness is the same for the specimen and the peripheral standard, the elemental mass per unit area can be simply converted to concentration of the element mass per unit volume:

$$C_{sp} = C_{st}\frac{I_{sp}}{I_{st}} \tag{11.5}$$

Where C is the concentration of a given element per unit volume in standard (st) and specimen (sp) and I is the corrected intensity* of x-rays

The success of this particular method depends on several important assumptions:

1. That the specimen and standard have the same thickness.
2. That the section and standard show the same amount of shrinkage as they go from a frozen-hydrated to a freeze-dried state.
3. That the section and standard are analyzed under the same instrumental conditions.
4. That the standard is not altered by the specimen.
5. That the specimen is not altered by the standard.

The first three assumptions are usually correct, although there may be some differential shrinkage between regions of high water content and low water content, i.e., extracellular spaces and cell cytoplasm.

In a series of careful and detailed studies, Saubermann et al. (1986a, b) showed that the last two assumptions are not always met and as a consequence the method is unreliable.

11.7.2.3. Continuum-Normalization Method

This is probably the most widely used analytical method for frozen sections. It was originally introduced by Hall and Werba (1969), Hall (1971, 1979), and Hall et al. (1973) and has been progressively refined by Hall and his colleagues and other

---

*The corrected intensity is a measure of x-ray peak integral that has been corrected for any additional signals, including local background peak overlaps, extraneous background, variations in collection time and beam current, and changes in instrumentation. Every attempt should be made to collect the x-rays under standard conditions.

workers during the past 20 years. There is an extensive literature on the theoretical derivation of this method, and papers by Shuman *et al.* (1976), Hall and Gupta (1979, 1982, 1983), and Saubermann (1988a, and b) provide a useful introduction to this topic.

The continuum normalization method is based on the premise that in sections up to $1.0 \, \mu m$ thick, the continuum x-ray intensity provides a measure of the local mass of the analyzed microvolume and the characteristic x-ray count provides a measure of the element mass fraction. For sections thinner than $1 \, \mu m$ the effects of x-ray absorption can be neglected. The ratio that is obtained can be standardized either with independent or peripheral standards. In practice, it is necessary to make several measurements:

1. The extraneous continuum from the instrument, support grid, and film.
2. The total continuum generated by the spectrum together with extraneous continuum.
3. The total x-ray generation in the region of the peak of interest.

It is thus possible to calculate the actual continuum contribution from the region of the specimen being analyzed and the local mass fraction of the element being analyzed. The concentration of a given element, i.e., mass of the element per unit mass of the specimen, can be calculated using the following formulation:

$$C_{sp} = C_{st} \frac{(I/W)_{sp}(Z^2/A)_{sp}}{(I/W)_{st}(Z^2/A)_{st}} \tag{11.6}$$

Here C is the concentration of a given element in the specimen (sp) and standard (st), I is the corrected intensity of characteristic x rays from the element in the specimen and standard, W is the corrected intensity of continuum x rays, and Z^2/A is the average weighted measure of all constituents in the specimen (sp) and standard (st), where Z is atomic number and A atomic weight. The continuum is measured in a region of the spectrum away from the characteristic peaks, and because the elemental peak intensity and the continuum intensity are measured at each point analyzed, it is not necessary to have uniformly thick sections and standards. The continuum method can be used to measure elemental concentrations in both the frozen-hydrated and freeze-dried states, which may be expressed either as millimoles of an element per kilogram of wet tissue or millimoles per kilogram of dry tissue. There are a number of potential sources of error in the continuum method:

1. Accurate measurement of the mass of the specimen may present difficulties, particularly in frozen-hydrated sections where the presence of ice makes them very sensitive to beam damage and mass loss.

2. There must be an absolute assurance that frozen-hydrated sections have not lost any water.

3. There may be inaccurate corrections for the contribution to the specimen continuum measurements from the extraneous counts generate from outside the specimen. A recent paper by Roomans (1988) provides some useful background information on the corrections that must be applied for extraneous background. He

found, for example, that geometric factors connected with the position of the analysis relative to the grid bars were of prime importance.

4. The average weighted measure of all constituents in the specimen (Z^2/A) is usually not known and is difficult to measure. In biological and hydrated organic specimens, this material is primarily made up of C, H, O, and N. However, Hall (1979) has shown that the matrix of most biological soft tissue does not vary very much.

5. Although the continuum method can be used for ultrathin sections, i.e., ca. 100 nm, as the section thickness is decreased below 0.5 μm, the extraneous continuum becomes a larger part of the total signal, and the total x-ray yield (characteristic and continuum) decreases. This may necessitate using higher beam currents, long counting times, and careful correction procedures.

6. The interpretation of the analytical results in terms of net changes in intercellular element composition may be complicated by changes in the cellular volume either as a result of the preparative technique—e.g., freeze-drying causes 10–20% shrinkage—or a particular experimental treatment—e.g., the effect of a drug. Bostrom *et al.* (1991), in a study of the effect of the drug cisplatin on electrolytes in the kidney tubule, showed that the analytical results were a combination of changes in element level and drug-induced shrinkage of the tubule cells. The shrinkage changes could be accounted for by morphometric measurements

7. In the analysis of thin sections of freeze-dried and embedded material, the measured quantity is millimoles of element per kilogram of embedded specimen. For any one analyzed region there is a wide range in the amount of embedding material that might be present. This variation could seriously affect any comparisons that might be made of data collected from different parts of the section. Hall (1991) has recognized this ambiguity and suggests that an analyzable label absent from the specimen, such as the dibromoacetophenome used in an epoxy resin by Ingram and Ingram (1983), should be incorporated, homogeneously, into the embedding medium. This would permit the x-ray data to be expressed unambiguously as millimoles of element per kilogram of dehydrated tissue. The results would be unaffected by shrinkage during freeze-drying or by incomplete replacement of the water by the embedding medium.

The continuum method continues to be refined and modified in various ways and has now been sufficiently widely used to show that it is a reliable method of quantitation, particularly for the analysis of matrix free extracellular spaces in frozen-hydrated sections. The different algorithms for the analysis of frozen sections are well characterized and are available as software programs either for real time or subsequent computer analysis. The programs are usually menu driven and allow the operator to deconvolute the x-ray spectra, make correction to extraneous signals, and introduce data from previously calculated standards for absolute quantitation.

X-ray continuum intensity is not the only way to measure the local mass per unit area that is needed for normalization of the characteristic x-ray signal. In the conventional TEM and in the STEM, it is possible to obtain a measure of the mass per unit area from the proportion of incident electrons that are transmitted through the specimen. In this beam attenuation method the transmitted intensity I_T of an electron beam passing through an object of mass thickness ρt is related to the initial intensity I_0 by the following expression:

$$I_T = I_0 \exp(-S_T \rho t) \qquad (11.7)$$

Where S_T is the total electron scattering cross section, which is constant over a limited range of mass thickness. Provided one can obtain an independent means of measuring S_T, a measure of the incident and transmitted beam intensities will allow the mass thickness to a be calculated from

$$\rho t = \frac{ln(I_0/I_T)}{S_T} \qquad (11.8)$$

These intensities can be readily measured using the exposure meter/photometer fitted to most transmission microscopes and/or the transmission detectors found on scanning electron microscopes. A paper by Halloran and Kirk (1979) describes how this quantitative measure of mass thickness may be carried out. This method of measuring section thickness complements the x-ray continuum method, for whereas the latter is useful for sections up to 3 μm thick (depending on beam kilovoltage) and is not as good for very thin sections, the beam attenuation method works well for the ultrathin sections. The total mass thickness of thin specimens can also be determined with high efficiency using backscattered electrons (Niedrig, 1978). Most solid state detectors used for recording the BSE signal subtend a large solid angle up to 2 sr. This means that a much more effective use is made of the potential signal, and is up to 60 greater than the small acceptance angle (0.1 sr) of most ED spectrometers. The local mass may also be determined using electron energy loss spectroscopy (Egerton, 1982b; Leapman et al., 1984).

Linders et al (1982) have made a comparison of the different methods that have been designed to measure specimen mass thickness and conclude that methods based on transmitted electron intensity are more accurate than the continuum x-ray method. However, as far as low-temperature x-ray microanalysis of thin frozen sections is concerned, the continuum method has the advantage that it can be done at the same time as the measurements are being made for the elements of interest.

Hall (1975) considers the x-ray yields are linearly proportional up to a thickness of about 4 μm for biological material. Above this thickness the linearity of emission no longer holds and it may be necessary either to apply the ZAF corrections mentioned earlier or to use some of the correction procedures described by Warner and Coleman (1975). The use of thick sections (2–10 μm) on very thin supports is no longer a popular way of carrying out microanalysis because of the limited spatial resolution and problems of quantitation.

In all three quantitative procedures for frozen thin sections which are described here, it is assumed that absorption of x-rays is negligible because the mean atomic number and x-ray absorption cross section of frozen biological and hydrated organic material are low. Shuman et al. (1976) found that the low-energy continuum radiation is absorbed in sections of freeze-dried quantitative procedure. This finding, coupled with the fact that freeze-dried sections are only about one third the thickness of their frozen-hydrated counterparts, would put an upper limit of about $4\,\mu$m for negligible absorption in frozen-hydrated sections and about $1.5\,\mu$m for freeze-dried and resin-embedded material. Hall and Gupta (1979) find it necessary to multiply all the recorded sodium counts from a fully hydrated $1\,\mu$m section by a factor of 1.3 to correct for absorption in the section. They outline a simple procedure for calculating absorption corrections and estimating the mass per unit area using a thin film of known composition and thickness. The standard that is used is a 2-μm film of a polycarbonate plastic Makrofol supplied by Siemens, coated with a measured thickness of aluminum. Separate counts are made on the specimen and the standard with constant probe current and counting times. The mass per unit area in the specimen M can be measured using the following equation:

$$M = \frac{C_{sp}}{C_{st}}(M)\,\frac{(Z^2A)_{sp}}{(Z^2A)_{st}} \tag{11.9}$$

where C is the corrected continuum radiation in the specimen (sp) and standard (st), (M) is the mass per unit area of the standard, and Z^2/A are defined in equation (11.6) as correction factors which account for differences in the continuum in the standard and specimen.

11.7.3. Quantitative Analysis of Frozen Bulk Samples

The quantitative analysis of frozen bulk samples is less straightforward than the analysis of frozen sections. The difficulties center around the fact that as the electron beam penetrates the sample it generates x-rays, some of which are absorbed by the sample. Not all the x-rays that are generated are collected, and it is difficult to obtain accurate measurement of the dimensions of the microvolume in which the x-rays are being generated. These problems have meant that a number of corrections have to be made to the classical equation of Castaing to ensure quantitative analysis, and a number of different approaches have been made to ameliorate these problems. Marshall (1980c) has used peak intensities with matching standards and corrected peak intensities (Marshall, 1982a, 1984, 1987). Boekestein et al. (1980, 1983, 1984) have used peak-to-background ratios with matching standards; Zs-Nagy et al. (1977) have used peak to defined background; Potts and Oates (1983) peak to total background; and Boekestein et al. (1983) have used conventional ZAF procedures, while Echlin and Taylor (1986) have applied such procedures to peak to local background ratios. Not all these methods are successful and it is now proposed to briefly discuss why some methods have limited applications whereas other methods can be used for the quantitative analysis of frozen bulk samples.

11.7.3.1. The ZAF Corrected Peak Intensities Method

This method, in one or other of its derivative forms, is one of the standard analytical procedures used by material scientists, who invariably have the privilege of analyzing flat, highly polished samples. The details and derivation of the method need not concern us here. The formation is as follows:

$$C_{sp} = (ZAF)_{sp} \frac{I_{sp}}{I_{st}}$$ (11.10)

Where C_{sp} is the local mass fraction of the element being analyzed; I_{sp} and I_{st} are the corrected x-ray counts from the specimen (sp) and standard (st), and ZAF are correction factors. Z is a correction factor related to differences in the mean atomic number between specimen and standard, A takes account of internal absorption of x-rays generated in the specimen, and F is a correction for x-rays generated in the specimens by other x-rays.

These three correction factors have been well characterized in materials science applications, but they are much more difficult to measure accurately in biological and organic samples. Some of these problems could be overcome if the standards could be made up in an identical matrix to that of the specimen, because the correction factors Z, A, and F could be cancelled out. Another difficulty arises from the fact that it is very difficult to obtain flat, highly polished samples of frozen biological and hydrated organic material. In spite of these difficulties, a number of attempts (Boekestein *et al.*, 1980, 1983; Roomans, 1981) have been made to see if the ZAF method, or modifications of the method, could be applied to the analysis of light elements in a low-density matrix. There is general agreement that the classical approach of applying ZAF matrix corrections is not a satisfactory procedure for the quantitative analysis of biological and organic material. The reasons are as follows:

1. The ZAF algorithms that have been developed are primarily aimed at heavy-element analysis in a well-defined matrix. These procedures do not work satisfactorily for light-element analysis in an ill-defined light-element matrix.
2. It is difficult to measure accurately the composition of a multielement (CHON) matrix.
3. It is difficult to measure the background induced fluorescence in a light-element matrix.
4. There are difficulties in obtaining reliable estimates of the absorption effects of light-element matrices.
5. Frozen fracture faces present an analytically "rough" surface and such variations of local tilt angle give rise to inaccurate quantitation.'

For these reasons, the ZAF corrected peak intensities method of quantitation has found little favor in the analysis of biological and organic material generally and frozen samples in particular.

11.7.3.2. Peak-to-Local-Background Ratio Method

This is a fairly straightforward method, which reduces the problem of variations in the surface topography of the specimen and minimizes the corrections that have to be made with respect to absorption, atomic number, and secondary fluorescence.

Millner and Cobet (1972) found that if an elemental characteristic peak minus background is ratioed to its local background, the correction factors Z, A, and F are minimized. This concept of ratioing the peak to its local background has been developed and extended by Statham and Pawley (1978) and Small *et al.* (1978) first to the analysis of particles and latterly to bulk materials. Echlin *et al.* (1982), Echlin and Taylor (1986), and Echlin (1989) have used this procedure for the analysis of frozen-hydrated fracture faces of bulk samples.

The rationale of the procedure is based on the finding that since the characteristic and background x-rays of the same energy are generated within nearly the same depth distribution, they are subject to the same local composition-related absorption, secondary fluorescence, and atomic number effects. The percentage of characteristic radiation absorbed by the sample is the same as the percentage of background that is absorbed, so the absorption factor (A) cancels out. In frozen-hydrated samples the atomic number effect (Z) is low, and in any event affects the peak and background in the same way. Frozen-hydrated samples have a low atomic number, so the secondary fluorescence effect (F) is low. These assumptions are particularly pertinent for homogeneous samples as the characteristic and background x-ray signals vary with changes in the average atomic number of the sample. This particular method is most appropriate for frozen-hydrated bulk biological material, where water forms by far the most abundant constituent of the tissue, so that we have what may be considered a homogeneous sample. The ratio of peak area to background immediately under the peak is more or less independent of sample geometry, detector efficiency, live time correction inaccuracies, and small variations in the beam current. Provided the peaks are properly characterized and the background below the peak is measured correctly, the accuracy of the method approaches that which may be obtained by use of the continuum method. The formulation for this procedure is as follows:

$$\frac{C_{sp}}{C_{st}} = K \frac{[(P - B)/B]_{sp}}{[(P - B)]_{st}} \tag{11.11}$$

Where C_{sp} and C_{st} are the local mass fracture of the element being analyzed in the specimen (sp) and the standard (st), $P - B$ is the x-ray peak intensity minus local background, B the same local background correlated for differences in matrix composition, and K, a correlation factor, can be obtained from measurements on standards of known composition. The two critical steps in this procedure are the accurate measurement of the peak to local background ratios from the x-ray spectrum and an accurate assessment of the matrix composition, i.e., those elements not normally analyzed, C, H, O, and N.

The peak to local background ratio algorithms are available on a number of software packages. The following procedure is adopted in the Link System ZAFPB programming to deconvolute the raw x-ray spectrum Escape peaks are removed and the spectrum subjected to a digital filter to produce a series of profiles that are independent of background. These profiles are modified to match the current spectrometer gain and resolution and then subjected to the same digital filter. Peak areas are determined from a least-squares fit of the filtered profiles to the filtered spectrum and a fit index calculated for the filtered region only and not for the whole spectrum. The complete profiles of peak plus background are used in this process. The background subjected profiles are then scaled to match the unknown spectrum and stripped away from it and the background under the peak determined.

The continuum is calculated from the first estimate of concentration and the true background fitted to it. The ratio Z^2/A, absorption and fluorescence corrections are calculated in an iterative process using a series of standard factors previously calculated and stored on disk. Such information includes details of the sample/spectrometer geometry, pure element peak profiles and, if available, the matrix composition* of the sample.

Once convergence has been achieved, the peak areas, peak-to-background ratios, and corrected local concentration [mmol/kg fresh weight (FW)] are printed out (see, for example, Boekestein *et al.*, 1984; Marshall, 1987; Hall, 1988). There is general agreement that the peak to local background ratio method is a reliable means of analyzing bulk frozen samples.

11.7.3.3. Peak-to-Continuum Ratio Method

Zs-Nagy and his colleagues (Zs-Nagy *et al.*, 1981; Zs-Nagy, 1983, 1988) have used a variation of the peak to background ratio method in which they used the background from a well-defined energy region away from the characteristic peaks and not from immediately below the characteristic peak of interest. Samples are prepared by their freeze-fracture, freeze-dry, technique and elemental concentrations are calculated using the continuum method of Hall *et al.* (1973), which was designed for thin sections where the effects of absorption, atomic number, and fluorescence could be neglected. It is claimed that these procedures work provided proper criteria are respected. These include using noncharging bulk organic crystals of known composition as standards without *ZAF* correction and the criterion that the elements of interest are contained within a low atomic number matrix, i.e., C, H, O, and N. It is very difficult to evaluate the effectiveness of the approach of Zs-Nagy and co-workers primarily because their analysis is carried out either on frozen dried tissue in which the size of the interactive volume is much enlarged, or on surface-etched material in which "a sort of equilibrium can be expected at between 163 and 188 K." Yet in spite of these misgivings, Zs-Nagy and Casoli (1990) claim that there is a close similarity in the concentration of Na and K in rat brain and liver tissue analyzed using the peak to local background ratio and the peak to continuum ratio methods.

11.7.3.4. The $\phi(\rho z)$ Corrected Peak Intensities Method

In this method, the x-ray distribution $[\phi(\rho z)]$ of Parobek and Brown (1978) has been used by Marshall and his colleagues (Marshall, 1982a, 1984; Marshall and

---

*In our work on tobacco leaves (Echlin and Taylor, 1986; Echlin, 1988) we carried out an independent organic chemical analysis of the dried sample to obtain the relative amounts of C, H, O, and N and coupled this with fresh and dry weight measurement to obtain the wet weight values.

Condron, 1987) to make corrections with respect to absorption and atomic number in bulk frozen-hydrated samples. The corrections are generated from a set of empirical equations for both sample and standard in which an assumed matrix is used for the sample. They find that the variation in matrix composition of cells and tissue only has an effect on sodium and magnesium. An algorithm is used to derive corrections factors for a calculated matrix composition for these light elements.

The ionization function $\phi(\rho z)$ describes the x-ray distribution in a solid target. The x-ray intensity for a given element in a specimen (I_{sp}) is given by

$$I_{sp} = \int_0^\infty \phi(\rho z)d(\rho z) \tag{11.12}$$

where ρ is the density and z is the depth of x-ray production in the sample.

The path taken to the detector by x-rays generated at a depth z in the sample is $z \operatorname{cosec} \theta$, where θ is the x-ray takeoff angle. Any absorption of x-rays may be calculated from

$$\exp(-\mu\rho z \operatorname{cosec} \theta) \tag{11.13}$$

where μ is the mass absorption coefficient for the element concerned. Marshall (1982a) provides a series of values of mass absorption coefficients which may be used with the $\phi(\rho z)$ program. The total x-ray intensity reaching the detector is given by

$$I_{sp}^* = \int_0^\infty \phi(\rho z)\exp(-\mu^s\rho z \operatorname{cosec} \theta)d\rho z \tag{11.14}$$

These corrected x-ray intensities can be substituted in the following formula to give the concentration of the element in the sample:

$$C_{sp} = \frac{I_{sp}I_{sp}^*}{I_{st}I_{st}^*}C_{st} \tag{11.15}$$

Where C_{sp} and C_{st} are the concentrations of the element concerned in the specimen (sp) and standard (st), I_{sp} and I_{st} are the measured x-ray counts from the specimen and standard, and I_{sp}^* and I_{st}^* are the measured x-ray counts for the specimen and standard corrected for absorption (A) and atomic number (z) effects. The fluorescence effect (F) is not included in these calculations because it will be very small in frozen-hydrated bulk samples. This method is sufficiently accurate to permit quantitative analysis to be carried out on frozen-hydrated bulk samples. It has the added advantage that nonideal standards may be used where there is a marked difference in the composition of the matrix between standard and sample.

11.7.4. Standards

The success of any quantitative analysis depends on there being suitable standards whose characteristic x-ray counts may be compared with those from the specimen. Although quantitation can be carried out using standards that do not closely resemble the composition of the specimen, it is usually desirable in biological applications to use standards of a nature similar to the specimen. As far as possible the standards should be of similar surface texture and the same thickness and internal homogeneity as the specimen. In addition one should aim at having a standard with approximately the same organic matrix and hence dry weight fraction as the specimen. Similarly, standards for use with frozen-hydrated samples should contain the same amount of water (ice) as the specimens. As might be expected, a very wide range of materials and formulations have been used for standards, and good reviews on the general preparation of standards may be found in the articles by Chandler (1977), Spurr (1975), de Bruijn (1981), Roomans (1981), and Warley (1990). Standards for frozen samples present their own set of problems, and three general approaches have been adopted:

1. The peripheral standard where the specimen is encased in standard elemental concentrations made up in a solution of artificial organic material, which, together with the specimen, is quench cooled and either fractured or sectioned. We have already discussed the problems associated with this type of standard (Section 11.7.2.).
2. Independent matrix-matching standards in which standard elemental concentrations are prepared in a hydrated organic matrix which closely resembles the matrix of the frozen specimen, i.e., electrolytes incorporated into a gelatin solution. This avoids making corrections for differences in the matrix composition between specimen and standard.
3. Independent non-matrix-matching standards in which standard elemental concentrations are prepared in a matrix that is different from that of the frozen specimen, i.e., electrolytes incorporated into an artificial organic matrix such as an epoxy resin. It may be necessary to either make corrections for differences in the matrix or enter the matrix composition into the quantitative algorithm.

The first two types of standards may be used either frozen hydrated or freeze-dried, whereas the third type of standard is independent of the state of hydration. Two types of standards are used in connection with the quantitative analysis of frozen sections. Solutions of electrolytes made up in glycerol, gelatin, albumin, dextran, or polyvinylpyrrolidone, which are then quench cooled and cryosectioned. The sections may be used either in the frozen-hydrated or (more usually) in the freeze-dried state. Care must be taken in using frozen-hydrated standards to quantify compartments with different degrees of hydration. The other type of standard is where electrolytes are incorporated into resin mixtures such as the macrocyclic polyethers or epoxy resin. There is a risk of inhomogeneity with both types of section, phase separation during cooling of aqueous standards, and incomplete mixing with the resin standards (Condron and Marshall, 1987).

The same types of standards used for sections are also used in connection with the quantitative analysis of bulk specimens, i.e., blocks of gelatin solutions and blocks of doped epoxy resins thinly coated with a layer of conducting material. There are, however, some important differences which take into account the poor conductivity of frozen salt solutions and the difficulty of quench-cooling bulk samples without the formation of large ice crystals. The conductivity problem has been cleverly circumvented by Whitecross *et al.* (1982), who incorporated a small amount of finely powdered, high-purity graphite into the electrolyte solution. The same procedure has been used by Echlin *et al.* (1983), where, in addition to graphite, an organic cryoprotectant, hydroxyethyl starch, was added to ensure microcrystalline ice formation during the quench cooling. Small droplets of the graphite-slurry standards are quench cooled and fractured. The fracture faces are analyzed at low temperatures in both the frozen-hydrated and freeze-dried state. Marshall (1977b) has spray-dried isoatomic droplets, which are then coated with a thin layer of carbon, aluminum, or chromium.

The usual procedure with standards is to make up a range of concentrations and analyze them under the appropriate hydration states using the same instrumental parameters that would be used to analyze the specimens. After making any necessary corrections to the x-ray data, i.e., matrix differences, etc., standard calibration curves are made for each element, which in modern x-ray analyzers may be permanently embedded in the analytical software program.

11.7.5. Detection Limits

In any analytical procedure one is interested to know what is the smallest amount of a substance that can be detected. In the context of low-temperature x-ray microanalysis, the detection limit must be linked to some measure of precision to ensure reproducibility of results. Thus, it would be inappropriate to assume that the presence of a small characteristic elemental peak on an otherwise noisy and high background would indicate that the element in question had been detected. It is necessary to establish some criterion, e.g., the peak is 3–5 times the background, before accepting that a particular element has been detected. It will be necessary to establish different criteria for sections and bulk materials and for frozen-hydrated and freeze-dried samples.

The question of detection limits in cryosections has been addressed by Hall and Gupta (1983), who considers a precision of five times the probable error as an appropriate criterion to use in connection with detection limits. This could be considered a rather extreme figure, as Liebhafsky *et al.* (1960) suggest that an element may be considered to be present when the number of peak counts exceeds the background N_B by $3(N_B)^{1/2}$. The detection limits will vary for each element and it is appropriate to consider the limits for sodium. This is in many ways the worst possible case because the concentration of sodium is low in most biological tissues and the background is relatively high in the region of the characteristic peak for sodium. Hall and Gupta (1983) consider that with both an energy-dispersive and a wavelength-dispersive spectrometer it should be possible to analyze 20 mmol kg^{-1} FW sodium in a hydrated section at a spatial resolution of 100 nm. This can only be achieved with a quite high probe current and any improvement in electron limits would necessitate putting an unacceptably high electron dose into the sample. The damage induced in the sample by such high electron doses is what ultimately limits the sensitivity of the method. The dose needed to record information must not exceed the damage dose (Isaacson, 1989). At lower resolution, and at a precision of 3 times the possible error, the figure for sodium in a frozen-hydrated section is closer to 10 mmol kg^{-1} FW. In freeze-dried samples it should be possible to detect sodium at a concentration of 5 mmol kg^{-1} DW.

The detection limits with bulk samples are much less certain because they have not been subjected to such systematic experimentation. Marshall and Condron (1987) used a windowless detector and were able to detect 10 mmol kg^{-1} FW sodium with a precision of three times the possible error, but were only able to detect the element at a concentration of 25 mmol. If the precision is increased to five times the possible error the detection limit is in excess of 25 mmol. Zierold (personal communication) has provided a useful comparison of the relative and absolute detection limits that may be expected in the different types of specimen used in low-temperature x-ray microanalysis (Table 11.6). The values for higher elements will be considerably improved, particularly if small-diameter probes are used with ultrathin freeze-dried specimens. Thus, Bond (1984) has been able to detect 300 μmol/kg DW of calcium in freeze-dried muscle cells, and for the heavier elements of biological interest, Fiori et al. (1988) have estimated that K, Ca, and Fe could be detected at between 0.1 to 0.5 mmol/kg wet weight. These detection limits are strongly influenced by beam damage and whether or not water is present in the sample.

Table 11.6. Comparison of the Detection Limits in the Types of Frozen Specimens Used in Low-Temperature X-Ray Microanalysis[a]

	Frozen-hydrated bulk specimens	Freeze-dried bulk specimens	Frozen-hydrated 1-μm-thick sections	Freeze-dried 100-nm-thick sections	Freeze-dried ultrathin (50-nm) sections
Lateral analytical resolution	~5 μm	~2 μm	~500 nm	≤50 nm	20 nm
Relative detection limit	~10 nmol/liter cell water	~10 mmol/kg dry weight (3 mmol/1 liter cell water)	~10 mmol/liter cell water	~5 mmol/kg dry weight (3 mmol/liter cell water)	~0.5 mmol/kg dry weight (~0.1 mmol/liter cell water)
Absolute detection limit	1.5×10^8 atoms in $1 \times 5 \times 5$ μm^3	5×10^7 atoms in $2 \times 2 \times 2$ μm^3	1.5×10^6 atoms in $1 \times 0.5 \times 0.5 \mu$m^3	500 atoms in $100 \times 50 \times 50$ nm^3	40 atoms in $50 \times 20 \times 20$ nm^3
Ion distribution	Preserved (?) if vitrified	Redistribution possible	Preserved (?) if vitrified	Redistribution possible	Preserved (?) if vitrified
Imaging	With some difficulty	Good-depends on ice crystal damage	Difficult	Good, depends on ice crystal damage	Good, depends on quality of cryofixation
Problems	Possible radiation damage (mass loss)	Recrystallization, shrinkage, radiation damage (mass loss)	Radiation damage (mass loss)	Recrystallization, shrinkage, some radiation damage (mass loss)	Recrystallization, shrinkage, some radiation damage (mass loss)

[a]Based on data supplied by Karl Zierold, Max Planck Institute, Dortmund.

11.8. THE WATER CONTENT OF FROZEN-HYDRATED SAMPLES

It has already been intimated that one of the unique advantages of low-temperature x-ray microanalysis is that it may be used to measure the local concentration of water in specimens. We need to consider some of the different analytical procedures that have been used. As a corollary, we also need to consider the various criteria that may be used to establish that a given sample is either in a fully hydrated state or freeze-dried. Some of these criteria, which are based on the morphological appearance of the sample, have already been considered in Chapters 9 and 10. We will consider here some of the more objective analytical measurements, in particular those that may be made directly on the specimen.

Before entering into these discussions it is appropriate to ask why it is necessary to measure the local concentration of water. After all, the low-temperature technology and the analytical algorithms allow us to preserve the natural hydrated state, and all elemental concentrations can be simply expressed on a wet weight basis. Although it is possible to make wet weight elemental measurements, in practice it is more accurate and easier to carry out the elemental analyses on freeze-dried samples maintained at low temperatures. The wet weight mass fractions may be calculated from the measurements made on the local concentration of water.

The advantages of analyzing freeze-dried sections are as follows:

1. They are much less susceptible to radiation damage.
2. They have enhanced peak-to-background ratios for particular elements.
3. They have sufficient contrast to permit unambiguous identification of features within the section.
4. It is much easier to keep a section fully freeze-dried than fully frozen hydrated.

11.8.1. The Water Content of Frozen-Hydrated Sections

There are four broadly different methods that may be used to measure the local water fractions in frozen specimens:

1. Mass thickness may be estimated first in frozen-hydrated then in freeze-dried sections. -
2. Mass thickness changes can be estimated using sections of external water content standards.
3. Local water fractions can be calculated from element concentrations per mass in the frozen-hydrated and dried state.
4. Changes in the oxygen content in frozen-hydrated and freeze-dried specimens can be measured directly.

Within these four broad categories there are a number of specific procedures that have been devised. Papers by Hall and Gupta (1982), Zierold (1986b), Saubermann (1988c, 1988b), and Von Zglinicki (1991) serve as a useful introduction to the problems associated with therse different types of measurements.

11.8.1.1. The Peripheral Standard and Continuum Normalization Method

This method was introduced by Gupta *et al.* (1977, 1978) and relies on making mass fraction measurements on the specimen and peripheral standard whilst they are in the partially dehydrated state, i.e., 10–25% water loss. This dehydration produces sufficient change in image contrast to permit morphological identification of different compartments and is particularly useful for preserving the extracellular spaces in a partially hydrated state.

Characteristic and continuum counts are made from the same selected cellular compartments and from the peripheral standard in both the partially hydrated and fully freeze-dried state. The water content is calculated from the changes in the spectra and from independent standardization using peripheral standards of known water content. The success of the method depends on the following assumptions:

1. That the standard and specimen dry at a standard rate. It is very difficult to provide controls to check this assumption.
2. That the standards and specimen do not change their water composition during sample preparation. The detailed studies by Saubermann *et al.* (1986a, 1986b, 1986c) have shown that exchanges can occur between the standard and the sample.
3. That uniform shrinkage occurs between the two hydration states. This is usually quite small so may be ignored. If shrinkage is a problem it is more accurate to obtain the local dry mass fraction by normalization against the characteristic x-ray signal from some convenient element (Gupta *et al.*, 1977).
4. There is no need to have uniform thickness of the sample and standard because the continuum normalization procedure determines mass fraction, not volume fractions.

11.8.1.2. Continuum-Normalization of Frozen-Hydrated and Freeze-Dried Sections

Dow *et al.* (1984) have modified the previous method and analyzed the same fully frozen hydrated and fully dried areas of the specimen. The water content may be calculated either by comparing the continuum generation rate in the wet and dry states or by comparing changes in the elemental mass fraction between the wet and dry values. The method relies on being able to precisely identify regions within the sample while in the fully hydrated state. It is generally accepted that this is usually difficult to achieve and then only at a low level of spatial resolution. It is equally important to be able to identify the same area in the freeze-dried state. The problematic aspect of this method is that there is no assurance that the section is in a fully hydrated state as even a slight increase in contrast indicates loss of water from the specimen.

11.8.1.3. Comparison of Continuum Generation in Frozen-Hydrated and Freeze-Dried Sections

This method was introduced by Saubermann *et al.* (1981a, 1981b) and Bulger *et al.* (1981) and simply compares the continuum generation rate between hydrated and dried samples. Continuum counts, as a measure of specimen mass, are first carried out in the hydrated state over large areas using very low beam currents. By dividing these large areas into smaller analytical areas, it can be shown that the large area is of uniform mass thickness. The section is then freeze-dried and reanalyzed in the dried state. The water content can be calculated by comparing the "wet" and "dry" continuum generation rates. Characteristic x-rays were collected on dried samples and wet weight concentration calculated from the local measurements of water. The advantages of this method are that it is not necessary to positively identify the areas being analyzed in the hydrated state. In addition, morphological identification of cellular compartments is easier in the dried section, and there is an improvement in the peak/background ratio of the elements being analyzed. Another advantage that has become apparent since this work was carried out is that freeze-dried sections are much less beam sensitive than their frozen-hydrated counterparts. The disadvantage is that it does not permit the analysis of extracellular spaces whose contents are in completely aqueous solution and seriously dislocated on drying. Hall and Gupta (1983) consider that this method could suffer from potentially large errors if the local dry mass fraction is not measured accurately. They suggest that two measurements of the continuum in the same area, i.e., before and after drying, would cause beam damage to the sample. Saubermann and his colleagues use very low beam current for these measurements, which are made at low temperatures.

11.8.1.4. Water Content from Freeze-Dried Cryosections

In contrast to the three previous methods, the technique devised by Dorge *et al.* (1978) and Rick *et al.* (1979) involves calculating the water content from the dry mass of freeze-dried sections. This makes use of peripheral standards of known water content, and local differences in continuum x-rays will reflect differences in water content. The water content may be determined by reference to the known standard. The validity of this technique depends on there being equivalent shrinkage during drying, equivalent mass loss during analysis, and that the standard and sample have an equivalent mean atomic number. It also assumes that there are no changes in water concentration of the sample and standard during tissue preparation. This we have shown is not always the case. Warner (1986) has made a careful analysis of the validity of the Dorge–Rick method and has applied corrections for differences in the dry weight content in the tissue specimen.

11.8.1.5. Combined Interferometry and Dry Mass Measurements

Von Zglinicki (1988), von Zglinicki *et al.* (1987) and von Zglinicki and Bimmler (1987) have devised a rather complicated way to estimate the organelle water fractions from freeze-dried samples. It involves measuring the dry mass concentration of mitochondria with an interference microscope, together with stereological measurements of the volume densities of the organelles. These data are combined with local dry mass measurements of intracellular compartments using x-ray microanalysis. Accuracy of the method depends on constant section thickness, no net movement of water between compartments, and no differential shrinkage between compartments during freeze-drying. This method has been tested using isolated mitochondria in a concentrated albumin solution.

11.8.1.6. Calibrated Measurement of Intracellular Water

Ingram and Ingram (1983) have devised a technique for measuring intracellular water that avoids having to use frozen-hydrated samples. Freeze-dried samples are infiltrated with a resin containing a bromine compound as a marker element. It assumes that all the water is removed by freeze-drying and that the doped embedding medium faithfully replaces this water. The Ingrams were able to show that there is a linear relationship between water content and the Br La x-ray signal and that the uncertainty of water measurement is about 5%.

11.8.1.7. Water Content by STEM Dark Field Imaging

Zierold (1986b, 1988) has shown how it is possible to measure the water content in intracellular compartments by using the dark field signal of the dry mass after freeze-drying. Calibration is achieved by measuring the dark field intensity of cryosections from varying concentrations of dextran–water solutions. As mentioned earlier in this chapter, data obtained from attenuation of the transmitted electron signal are more accurate than those derived by x-ray microanalysis.

11.8.1.8. Water Content by Electron-Energy-Loss Spectroscopy

Leapman et al. (1984), Leapman and Ornberg (1987), Probst et al. (1989), and Strain et al. (1989) have shown how electron-energy-loss spectroscopy can be used as a means of measuring mass thickness in frozen-hydrated and freeze-dried sections. The EELS inelastic scattering method has a number of advantages over the dark field STEM technique and x-ray continuum method. It can be applied to sections that are normally too thick to utilize dark field STEM imaging and the beam current ca. 2 pA is much lower than that normally used to make the x-ray continuum measurements. The calibration studies that have been made on sections of polyvinyl-pyrrolidone and gelatine standards show considerable variation. This would indicate that more studies need to be done both to establish the thickness at which nonlinear measurements begin and to account for the wide and differential shrinkage between compartments. It is interesting to note that both Leapman and Ornberg (1987) and Strain et al. (1989) obtain higher dry weight fractions than the original preparations, which could mean that the estimates of relative inelastic cross sections may need improving.

11.8.1.9. Water Content Measured by Changes in the Oxygen Concentration

Andrews and Leapman (1989), using thin window detectors, have demonstrated a high oxygen peak in hydrated sections. This peak almost disappeared when the section was freeze-dried. It would be interesting to see if these changes, which occur in a region of high background, could form the basis of a quantitative procedure for measuring the local water concentration in frozen sections. As we will see, this is one of the methods used to measure the local concentration of water in frozen bulk samples.

In spite of the large amount of work that has been carried out on measuring the local concentration of water, there does not seem to be an ideal answer to the problem. The methods involving the use of x-rays suffer the disadvantage of high radiation doses and the techniques involving STEM and EELS imaging have not been sufficiently well tested on a wide range of samples to know whether the method has a broad application.

The Saubermann method is probably the best we have at the moment, if only for the fact that it is the only method that has been rigorously tested against other similar techniques (see Saubermann *et al.*, 1986a, b, for details). It is the only technique that provides an assurance that the samples remain unaltered, fully hydrated, and subjected to low beam current during the initial analysis. Its only disadvantage is that it is difficult to use with large aqueous extracellular spaces, and in these cases the method described by Dow *et al.* (1984) has certain advantages.

11.8.2. The Water Content of Frozen-Hydrated Bulk Samples

Fewer studies have been carried out to devise procedures for measuring the local concentration of water in frozen bulk samples. This may either reflect the greater assurance one has of being able to maintain a bulk sample in the fully hydrated state, or that the analytical algorithms for frozen bulk material are less accurate than those for frozen sections. Two methods have emerged, one based on the peak-to-background method developed for use with frozen sections, and the other based on measuring the local concentration of oxygen.

11.8.2.1. Peak-to-Background Method

This method was originally developed by Zs-Nagy *et al.* (1977), and the following references give the current status of their work (Zs-Nagy, 1988; Lustyik and Zs-Nagy, 1985, 1988). The mass fraction of water (W) can be calculated from the change in concentration of some evenly distributed element such as potassium or phosphorus during drying as follows:

$$W = 1 - \frac{C_w}{C_d} \qquad (11.16)$$

Where C_w and C_d are the wet and dry weight concentrations of a particular element, e.g., K^+. C_w and C_d are derived from the peak-to-background ratios of the element using the continuum method of Hall, where the background is not the local background but that measured in a region away from the characteristic peaks (4–6 kV). It will be remembered that, when using this method, it is necessary to include corrections for any changes in the matrix composition between the wet and dry state.

The wet and dry concentrations of the target element(s) in the wet and dry state may be calculated as follows:

$$C = F(R \cdot Z^2 A) \tag{11.17}$$

where F is a constant, R is the peak-to-background ratio of the element, and $Z^2 A$ is the average weighted measure of all constituents in the specimen. By definition $Z^2 A_w = (1 - W) Z^2 A_d$ where appropriate values for $Z^2 A_w$ and $Z^2 A_d$ are 3.666 and 3.28, respectively (Hall et al., 1973).

The mass fraction of water (W) in a bulk sample is obtained without using standards as follows:

$$W = \frac{R_d - R_w}{R_d + 0.1179 R_w} \tag{11.18}$$

The constant 0.1179 is derived from

$$[(Z^2 A)_w - (Z^2 A)_d]/(Z^2 A)_d$$

The actual experimental procedures used by Zs-Nagy and colleagues are rather complicated and involve taking continuous wet mass fraction measurements from uncoated samples while the sample is slowly warmed (0.7 K/min) from 160 to 185 K. During this time, the peak-to-background ratios for the target elements gradually increase until they reach a plateau at 175–185 K. It is quite clear from the work of Lustyik and Zs-Nagy (1985) that the wet weight mass fractions are made from surface etched (dried) material—or the "superficial exsiccation layer" to use their terminology. They claim that this layer is only 100 nm deep, but no experimental evidence is presented to enable one to check this figure. The dry mass fraction measurements are taken from the 30–40-μm superficial layer samples which have been maintained in the high vaccum (500 μPa) of the microscope for 2–3 h at 210 K. The analyses are carried out at between 10–20 kV and very low beam current (1–2 pA), which gave count rates of 400–500 cps. The method has been tested on a number of different mammalian tissues and it is claimed to be able to reveal differences in water content as small as 1%–2%.

The uncertainties associated with this method center on the charging artifacts that must occur in uncoated samples, on the comparatively large size of the interactive volume at the higher accelerating voltages, and the fact that dry mass fraction measurements are made from deep-etched samples and not from fully freeze-dried material. Additional problems arise from the very large difference in the size of the interactive volume in frozen-hydrated and freeze-dried material. Equation (11.18) is only valid if the target element being used for water quantitation is distributed uniformly in the two different interactive microvolumes. This would not be the case in tissues with aqueous extracellular spaces or in cells with large aqueous vacuoles.

11.8.2.2. Measurement of Local Oxygen Concentration

Marshall (1982a) makes the assumption that cell cytoplasm can be considered to be a protein solution and that a linear relationship would exist between oxygen concentration and the protein concentration. From this he argues that local measurements of the changes in oxygen concentration between the wet and dry state would be representative of the mass fraction of water.

The oxygen intensities are measured using a windowless detector (Marshall, 1984). Valid oxygen peak intensities can be obtained after corrections have been made in relation to background intensities and peak overlaps from carbon and nitrogen. If ice is used as an oxygen standard no atomic number corrections are necessary, and the necessary absorption corrections may be obtained from the $\phi(\rho z)$ calculations made on protein solutions. Absorption of oxygen x-ray photons decreases as the protein concentration decreases, and is also very dependent on local takeoff angle. This problem is tackled by initially making approximations for the O_2 concentration value by regression of protein concentration on oxygen intensity from frozen standards. By an iterative process the initial oxygen estimates are used to calculate an absorption correction which is used to calculate the second oxygen estimates. The final oxygen concentration is used to give the local dry weight mass fraction and the water mass fraction.

The effects of topography on oxygen x-ray absorption are reduced by normalizing the O intensities by determining their ratio to the backscattered electron intensity (Marshall and Condron, 1985a). The concentration of oxygen (C_{sp}) may be obtained as follows:

$$C_{sp} = \frac{I_{sp}F_{st}}{I_{st}F_{sp}} C_{st} \tag{11.19}$$

where C is the oxygen concentration in the specimen (sp) and standard (st), I the measured peak intensities for oxygen in the specimen and standard, and F the absorption factor. Marshall and his colleagues have tested this method on gelatin solution models and have obtained a reasonable level of accuracy.

11.8.3. Criteria to be Used to Confirm the Fully Frozen-Hydrated State

In any quantitative procedures it is important to first establish that the component being analyzed is present in an unaltered state. This is nowhere more important than in studies on frozen-hydrated samples, where the presence and absence of water can have profound effects on local concentration of elements.

Although it is possible to obtain measurements of local concentrations of water, it would also be reassuring to have other independent criteria that may be used to check that samples really are in the fully hydrated state. A number of such criteria have already been established (Echlin, 1978c) and it is useful to review the effectiveness of these and other criteria in confirming the existence of the fully frozen-hydrated state.

11.8.3.1. Morphological Changes

These have already been considered at some length in the previous chapter (Section 10.7.9). In essence the image "contrast" and "quality" is poor on frozen-hydrated material and better in freeze-dried material. The reasons for this have been well documented and need not be repeated here.

11.8.3.2. Changes in X-Ray Spectra

There are four changes that may be observed to varying extents in both frozen sections and frozen bulk samples:

(a) The spectrum from a frozen-hydrated sample has a high background, particularly at the low-energy end of the spectrum. This high background region virtually disappears as the material is dehydrated. This high background is enhanced if a windowless detector is used, and for obvious reasons is more easily observed in bulk samples than in frozen sections. Table 11.7 and Fig. 11.11 show this effect in frozen tobacco leaves.

(b) There is a dramatic increase in the peak-to-background ratios of the elements as the water is removed from a frozen-hydrated sample. Table 11.7 shows some comparable data from frozen-hydrated and freeze-dried plant mesophyll cells.

(c) If a windowless detector is used, there is a decrease in the oxygen peak as the specimens are dehydrated.

(d) Frozen-hydrated samples will undergo a measurable mass loss as they are freeze-dried.

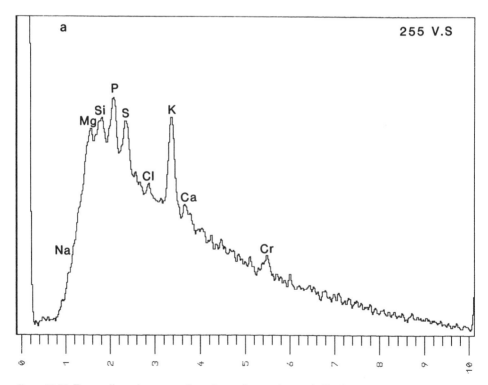

Figure 11.11. Energy dispersive spectra from frozen fractured mesophyll cells of *Nicotiana tabacum* leaves in a frozen-hydrated (a) and freeze-dried (b) state. Note how the low-energy hump in the frozen-hydrated spectrum disappears as the specimen is dried. The verticle axis (VS) is the number of x-ray counts; the horizontal axis is the energy (keV) of the x-ray photons.

Table 11.7. Relative Peak-to-Local-Background Ratios of a Number of Elements in the Same Nicotiana tabacum Mesophyll Cell in the Frozen-Hydrated (FH) and Freeze-Dried (FD) State

	Na	Mg	P	S	K	Ca
Frozen-hydrated	0.06	0.04	0.29	0.26	0.52	0.09
Freeze-dried	0.13	0.28	0.37	0.47	0.72	0.56

11.8.3.3. Changes in the Dimensions of the Sample

Nearly all frozen-hydrated samples shrink as they are freeze-dried. The amount of shrinkage will vary depending on the natural water content of the sample. It is often easier to measure the changes in the dimensions of a frozen section than in a bulk sample.

11.8.3.4. Changes in Surface Conductivity

In spite of some earlier evidence to the contrary (Echlin, 1978c), it is now generally accepted that uncoated, frozen-hydrated samples are more likely to show charging artifacts than their freeze-dried counterparts. This is not a very good cri-

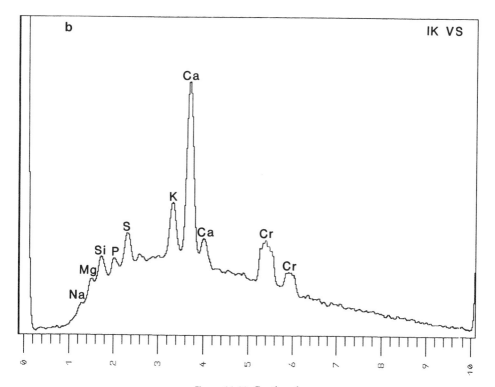

Figure 11.11. Continued.

terion for the amount of charging will vary depending on the nature of the sample, how well it is grounded, and the instrumental conditions, i.e., voltage, beam current, and contamination.

In the early studies on the analysis of frozen-hydrated samples, insufficient emphasis was placed on ensuring that the specimens were in a fully hydrated state. A reexamination of some of these data in light of what we now know about frozen-hydrated samples shows that a number of these samples were far from fully hydrated. This misinterpretation is probably of little consequence because the instrumentation and analytical algorithms were not sufficiently advanced to give fully quantitative data. But now we have the full capability for such studies and it is important that we make as many checks as possible to ensure that the samples are fully hydrated if accurate measurements are to be made on the local concentrtation of water.

11.9. LOW-TEMPERATURE DIGITAL ELEMENTAL MAPPING

11.9.1. Introduction

So far in our discussions on low-temperature x-ray microanalysis we have been concerned solely with procedures that use a static probe (TEM and STEM) or reduced raster (SEM) to localize the sites of analytical interest. In such procedures the microscopist preselects the area to be analyzed and places the probe or reduced raster over this region of interest. There are obvious advantages and disadvantages to this procedure. An alternative procedure with scanning beam instruments is to use x-ray mapping, in which the electron beam is moved over the specimen area in a predetermined pattern and the x-ray photons collected at each picture point or pixel.

X-ray mapping, or two-dimensional scanning, takes the output of a single-channel analyzer and uses it to modulate the brightness of the cathode ray tube during normal secondary electron raster scanning. The analyzer can be set to collect a particular x-ray energy or wavelength and each x-ray photon detected appears as a dot on the cathode ray tube. Regions of high elemental concentration are characterized by high dot density. This technique has been applied to sections and bulk samples, although data from rough samples must be treated with caution, as those parts of the sample that are in line of sight with the x-ray detector will appear to have a higher x-ray signal.

The appearance of the x-ray map differs from a secondary electron image since the latter uses an analog rather than a digital signal to modulate the intensity on the cathode ray tube. The x-ray intensity is usually several orders of magnitude less than the secondary electron signal and it is unusual to have more than one event recorded per picture element. This can be overcome by repeatedly scanning the same area and/or using higher beam currents. This leads to a loss of precision and accuracy and at best conventional x-ray mapping can only be used for qualitative analysis.

The advent of computer-controlled digital scanning devices (Gorlen *et al.*, 1984) and the commercial availability of digital scanning electron microscopes has opened up a new procedure for quantitative analysis, which has been applied with great effect to the quantitative analysis of frozen sections and flat bulk specimens.

11.9.2. Digital Elemental Analysis of Frozen Sections

An alternative to repeated beam rastering is discrete beam rastering, which may be carried out using a digital scan generator. The x and y velocity components of the electron beam are zero for a finite period of time, i.e., the probe stops at a given geometrically assigned point on the specimen for a given period of time. Each point is referred to as a pixel, and the beam is moved from pixel point to pixel point across the specimen. During the time the beam is at each pixel point, signals are generated which may be used for both structural and analytical studies. The delay time at each pixel varies and can be as short as 100 ms and as long as several seconds depending on the required accuracy and precision of the analysis. For x-ray imaging purposes, the raster is generally made up of a low number of pixels, i.e., 64–512, along each line and the same number of lines. Thus a digital x-ray map would be from as low as 64 × 64 pixels to as high as 512 × 512 pixels. This means that the time to acquire one analytical image can be quite long, i.e., a 128 × 128 raster with a delay time of 1 s per pixel takes approximately 4.5 h. Samples have to be kept stable, cold, and uncontaminated for long periods of time. This is a major disadvantage of this method because even with the most stable microscope instrumental drift can be a problem. Statham (1987) has described a computer correction procedure which compensates for some shifts. Another approach is to analyze more pixels over a shorter period of time (Gorlen *et al.*, 1984). For high-resolution morphological imaging it might be necessary to go up to a larger number of pixels, i.e., 1024 × 1024. However, it is doubtful whether for morphological imaging alone the digital image could compete with the analog image with its much larger number of grey levels. For details of computer-aided imaging, one should refer to the excellent chapter written by Fiori in the book by Newbury *et al.* (1986).

The real advantage of this approach is that the data accumulated with discrete rastering can be stored on a digital computer and mathematically manipulated. Such manipulations can be used to process and deconvolute spectra, remove artifacts of background and peak overlaps and finally express the data as quantitative x-ray intensity maps where the concentration of elements (and water) are expressed as a series of numerically coded gray levels or colors. Foster and Saubermann (1991) have devised a system based on a personal computer which provides a relatively inexpensive way of quantitative elemental analysis of cryosections. Details of these procedures, and some of the pitfalls that may occur, are found in the paper by Heyman and Saubermann (1987), Saubermann and Heymann (1987), Ingram *et al.* (1987), Fiori *et al.* (1988), Johnson *et al.* (1988), Statham (1988), Saubermann (1988c), Ingram *et al.* (1989), and Foster and Saubermann (1991). Two examples will be given to show how these procedures have been applied to frozen sections.

A. In the procedure devised by Saubermann and used in his laboratory (see Saubermann, 1988; LoPachin *et al.*, 1988, 1989), x-ray intensity maps have been made for a number of elements in frozen-hydrated sections of leech ganglia and rat sciatic nerve. Flat sections are analyzed first in the fully frozen-hydrated state at low temperature and low dose (ca. 2 ke$^-$ nm^2) in an analog mode in the SEM to obtain the continuum generation counts, which are used for measuring the water content [see Section 11.8.1, Method 3]. The sections are then freeze-dried *in situ* and recooled to 90 K and a 64 × 64 pixel x-ray intensity map made for both continuum and characteristic elements. The x-ray mapping is carried out at a beam current of 1.4.nA and a delay time of 4 s live time per pixel. The maps are completed in ca. 4.5 h and if a 150-nm probe is used the electron beam dose is ca. 2.0 Me$^-$ nm^2. The 64 × 64 pixel image is used as it provides reasonably good morphology and sufficient x-ray signal for the statistically significant x-ray analysis of different regions of the same sample.

An example of the type of x-ray map that can be produced by this technique is shown in Fig. 11.12. Similar data from rat peripheral axons are presented as color-coded digital images in the paper by LoPachin *et al.* (1988).

B. Ingram *et al.* (1989), LeFurgey *et al.* (1989), and Davilla *et al.* (1989) use a scanning transmission electron microscope to produce x-ray intensity maps from kidney, nasal epithelium, avian heart, and skeletal muscle sections. Thin (100-nm) freeze-dried sections have been imaged, at room temperature, with a 64 × 64 pixel raster using a 40-nm^2 probe, a beam current of ca.1.nA, and a delay time of 1 s per pixel point. Each image took approximately 74 min to produce at an electron dose of ca. 2.5 Me$^-$ nm^2. The procedures are of interest because the authors have used a general purpose multiprocess minicomputer for the real time display of graphic summaries of the results. Quantitative x-ray intensity mapping has also been carried out on thin cryosections in STEM instruments by Johnson *et al.* (1988) on rat parotid gland and by Fiori *et al.* (1988) on mouse cerebellar cortex.

11.9.3. Digital Elemental Analysis of Frozen Bulk Samples

There is no reason in principle why the same procedures should not be used to analyze the surfaces of frozen samples, although in practice there are two severe drawbacks. As we have already shown, the x-ray spatial resolution is less good in bulk materials than in sections and the analytical algorithms are not so precise or accurate. An additional hazard is that very small variations in the surface topography will seriously influence the collection of x rays, either by signal enhancement or x-ray absorption.

Echlin (unpublished data) has used digital x-ray intensity mapping at low temperatures in a qualitative study on the co-precipitation of mixed salt solutions dried under controlled conditions on a filter paper matrix. A 128 × 128 pixel raster was used at 90 K with a 0.2-μm^2 probe, a beam current of 2 nA, and a delay time of 200 ms per pixel point. The images took ca. 55 min to produce at an electron dose

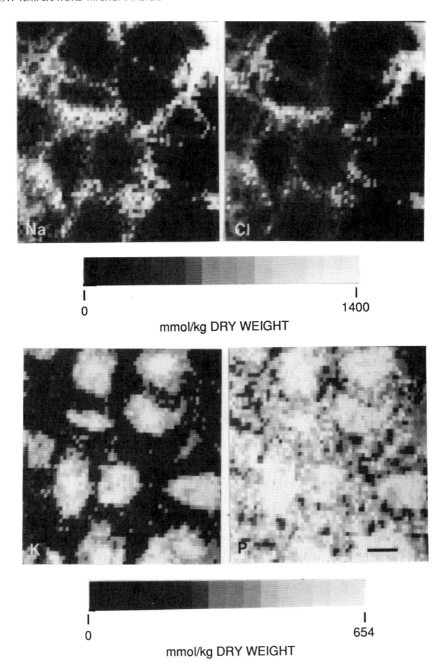

Figure 11.12. Quantitative digital x-ray images of Na, Cl, K, and P obtained from a 0.5-μm-thick leech ganglia cryosection (frozen hydrated then freeze-dried) displayed as mmol kg DW using a pseudo-gray scale. Each pixel value can be retrieved and expressed as actual concentration. Image is 64 × 64 pixels. Bar marker = 25 μm. [Picture courtesy of Saubermann and Heyman (1987).]

of ca. 50 ke$^-$ nm^2. The analysis was carried out in an AMRAY scanning electron microscope using a Link System 860 Series II x-ray microanalyzer. The computer software associated with this spectrometer allows the spectrum to be processed and deconvoluted, and the x-ray intensity maps that could be produced were analyzed by the peak to local background method, which includes corrections for background stripping and peak overlaps. It was found that the effects of variations in local topography could be ameliorated to some extent by tilting the flat sample to 25° towards the x-ray detector and stripping the entire background from the spectrum. An example of the type of information that can be found is shown in Fig. 11.13.

Digital x-ray intensity mapping procedures have a number of advantages over static probe analysis—advantages that are not just confined to the analysis of frozen samples:

1. The method provides an enormous amount of quantitative analytical information about the sample in a relatively short period of time. Thus, in a period of 4–5 h, 6–8 elements may be analyzed at anything from 4×10^3 to over 10^6 points on the sample depending on how many pixels are used to form the image. This data can be stored in a digital form for quick recall, further processing, and comparison.

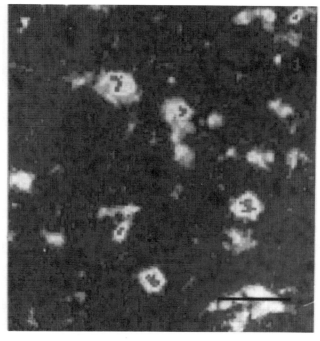

Figure 11.13. X-ray intensity Digimap showing the distribution of calcium precipitated from mixed calcium phosphate and calcium chloride solutions dried on a piece of filter paper. The background has been stripped away at each pixel point and the relative concentration of calcium expressed as gray scale intensities with white at the highest concentration through gray to black at the lowest concentration. Image is 128×128 pixels. Bar marker $= 100 \, \mu$m.

2. The x-ray intensity maps provide easily interpretable images in which areas of high and low concentration may be quickly recognized. The remarkable power of the human brain to recognize patterns circumvents the need for statistical processing.

3. The interrelation between structural and analytical data becomes more easily recognizable.

4. It helps to remove any subjective bias regarding exactly where the analysis should be carried out. The whole area of the sample may be mapped.

11.10 BEAM DAMAGE DURING LOW-TEMPERATURE X-RAY MICROANALYSIS

11.10.1. Introduction

Beam damage during x-ray microanalysis is a problem of the same dimensions as that met with during high-resolution electron microscopy. Unless great care is taken, the results of analytical studies on organic samples will be completely negated by the conditions under which the analysis is carried out. A modern electron probe instrument with a field emission source is capable of producing a 1-nA beam current in a 0.5-nm probe. Isaacson (1989) has calculated that with a typical 100-s counting time, analysis of a sample under such conditions would impart a dose of $3 \times 10^{11} \, e^- \, nm^{-2}$ to the specimen. Isaacson provides a table of the electron beam doses which cause different types of damage to a range of organic and inorganic materials. It is quite clear that a dose of 300 g e^- nm^2 far exceeds the minimal damage dose for all organic and biological materials.

As with high-resolution structural studies, low-temperatures are both a help and a hindrance in ameliorating these effects. The presence of ice in frozen-hydrated samples maintained at temperatures well below their recrystallization temperature provides a ready source of free radicals, although the diffusion of these free radicals is diminished. Conversely, in freeze-dried samples, beam damage is considerably reduced if the analysis is carried out at low temperatures. High-resolution structural damage, as measured by changes in the diffraction patterns of ordered molecules, are generally not a problem in x-ray microanalytical studies even though the total electron dose used for electron probe analysis is several orders of magnitude greater than that required for low dose electron microscopy. With a few notable exceptions, most analytical studies are carried out at low spatial resolution, and although these structural changes occur they are not observed. However, the long times needed for signal acquisition (50–200 s) and the increased beam current (nA) above that needed to acquire structural information (pA) bring with them their own peculiar set of problems, which center on etching and a diminution of the general mass of the sample, dislocation, and/or loss of specific elements from the samples and contamination.

At very high beam currents and long counting times visible changes may be observed in specimens. Sections either appear more electron transparent ("brightening or thinning") because of mass loss or become more electron dense because of contamination (mass gain). In extreme conditions the sections may melt, fray, or disappear entirely. Holes will appear in bulk samples and/or the surface may develop fine cracks. There is nothing subtle about the appearance of these effects; they are the gross manifestations of beam damage and one would not even begin to attempt any analysis on such samples.

It is the nonobservable but measurable beam damage that represents the real problem. For unless care is taken to constantly monitor any changes that may be occurring in the sample, the analysis may be carried out under varying and unaccountable conditions.

In many qualitative analytical studies these losses are of little consequence, as it is assumed they occur to an equal extent in both the specimen and the accompanying standard. This assumption is, of course, only valid if the analysis of sample and standard is carried out under identical conditions and if the matrix of the two sets of samples is the same. Mass loss, element loss, and contamination assume a much greater importance in quantitative studies, and we will briefly consider the extent to which these three processes influence quantitative low-temperature x-ray microanalysis.

11.10.2. Mass Loss of Organic Material

The extent to which heavier elements are removed from a specimen depends on the nature of the sample, although where this loss occurs it is at a much slower rate than observed in organic samples. Mass loss is primarily a problem of the loss of light elements from the organic and/or hydrated matrix. Mass loss from organic samples was first systematically measured by Bahr, Johnson, and Zeitler (1965), who weighed uncooled thin organic films before and after a total electron exposure of $650 \, e^- \, nm^2$ in an electron microscope. The mass loss was strongly influenced by the chemical composition of the organic film, but most samples, including gelatin films, showed a 20%–30% mass loss before leveling off. Hall and Gupta (1974) extended these studies to freeze-dried tissue sections, ca. 1.5 μm thick, maintained at low temperatures inside a scanning electron microscope. The local mass loss was measured as a function of the change in x-ray continuum intensity. Hall and Gupta found that at a total dose of ca. $2500 \, e^- \, nm^2$, a large proportion of the mass was removed from a section of freeze-dried material. These changes are characterized by an initial rapid loss, which occurred within ca. 30 s, with a stable residual mass once the dose of $2500 \, e^- \, nm^{-2}$ has been reached.

This mass loss is substantially reduced in samples cooled to 90 K. (The exact relationship could not be precisely given because of the difficulty of accurately measuring the sample temperature.) In practical terms, this would mean that an uncooled 1–2-μm-thick tissue section would lose 30%–40% of its mass during a typical 100-s counting time, whereas this loss would be only 5% if the sample was analyzed at 90 K. It should be noted that Hall and Gupta (1974) utilized electron exposures (ca. $7 \, ke^- \, nm^{-2}$) well below those most frequently used during a 100-s counting time for conventional biological x-ray microanalysis. Mass loss may be calculated more accurately by measuring the change in dark field intensity as a function of electron dose (Carlemalm *et al.*, 1985; Egerton, 1982a; Reichelt and Engel, 1984, 1985).

Figure 11.14 shows the mass loss from a frozen-hydrated 100-nm-thick section of yeast (Zierold, 1988). In spite of substantially higher electron doses than those used by Hall and Gupta (1974), the rate of mass loss appears slower. Zierold found that a thin layer of carbon reduced the mass loss, although Hall and Gupta (1974) could find no such effect in samples coated with either carbon or aluminum. This may well be related to differences in the total section thickness and total electron dose in the two sets of experiments and the fact that in the Hall and Gupta experiments the minimum coating thickness only represented 5% of the total sample thickness; in the Zierold experiment the coating layer represented ca. 17% of the total section thickness. Zierold considers that the carbon coating layer prevented the loss of the volatile component of the reaction products of electron beam sample interaction.

More recent data by Joy (personal communication) show that a substantial proportion of the mass loss from a sample occurs very early on in the process of irradiation. Joy has investigated the mass loss in fatty acids irradiated with precisely measured electron doses. After a very low dose of only ca. 350 e nm^2, there was an approximately 30% mass loss, and as Fig. 11.15 shows this mass loss continued at an alarming rate as the electron dose increased. Although these samples were not cooled, and although higher molecular weight fatty acids may be particularly radiation sensitive, losses of this magnitude must be of grave concern in the x-ray microanalysis of organic material.

Joy *et al.* (1989) consider that any measurements of mass loss should consider the amount of energy absorbed by the sample rather than the irradiated (incident) energy, which is the more usual way beam doses are quoted in the literature. It is not how many electrons hit the sample that is important, but how many interact

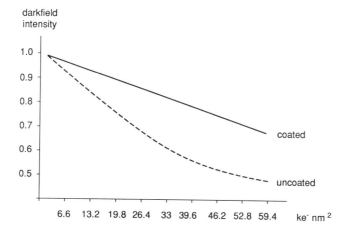

Figure 11.14. Dark field intensity as a measure of the mass thickness of a frozen-hydrated ultrathin cryosection of yeast as a function of irradiated electron dose. [Redrawn from Zierold (1988).]

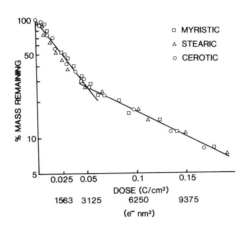

Figure 11.15. Mass loss from fatty acids as a function of accumulated electron dose. Sample at 300 K (uncooled). (Redrawn from unpublished data supplied by David Joy, University of Tennessee.)

with the sample. An electron that does not interact with the sample will not damage it and it is necessary to calculate the fraction of electrons at a given energy that pass through a section of given thickness without losing energy by inelastic scattering events. Joy has made the appropriate calculations using the formula proposed by Egerton (1987):

$$P = \left(\frac{t}{\lambda_i}\right)^n \exp\left(-\frac{t}{\lambda_i}\right) \qquad (11.20)$$

Where P is the probability of a given electron undergoing no inelastic events, t is the section thickness, and λ_i is the inelastic mean free path for the materials and energies of interest. Joy has calculated values for λ_i for both freeze-dried and frozen-hydrated biological material, which are given in Table 11.8. These data suggest that thin sections (e.g., 100 nm) should not suffer too much mass loss when examined at high keV. Joy has provided a series of graphs (Fig. 11.16) in which the percentage

Table 11.8. Inelastic Mean Free Path (nm), of Electrons of Varying Energy in Frozen Hydrated (70% Water) (FH) and Freeze-Dried (FD) Tissue Sections[a]

Energy (keV)	Mean free path (FH) (nm)	Mean free path (FD) (nm)
30	38	73
50	69	109
70	95	130
100	128	160
120	218	330

[a]Unpublished data supplied by David Joy, University of Tennessee.

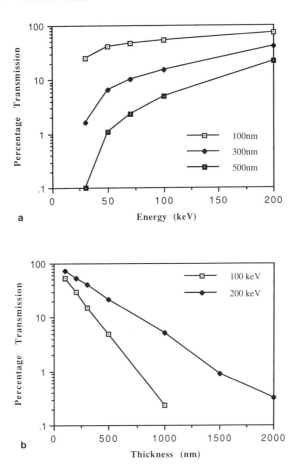

Figure 11.16. Percentage transmission of inelastically scattered electrons through frozen-hydrated (70% water) and freeze-dried tissue sections as a function of section thickness and electron energy. (a) Transmission through tissue sections of varying thickness. (b) Transmission through tissue sections. (c) Transmission through frozen-hydrated sections. (d) Transmission through frozen-hydrated sections of varying thickness.

transmission of inelastically scattered electrons is plotted as a function of beam energy and section thickness.

There is a wide variation in the percentage transmission, but at 100 keV about half the electrons tranverse a thin (100-nm) section without any interaction with the sample. As might be expected, the percentage transmission falls steeply at lower accelerating voltages and with thicker sections, although, interestingly, doubling the section thickness does not double the energy absorbed. The actual amount of energy absorbed depends on the original section thickness.

These new data supplied by Joy may go some way to explain why it is possible to obtain valid x-ray microanalytical data from sectioned material in situations where inordinately high electron doses are employed during analysis. In thick sections and

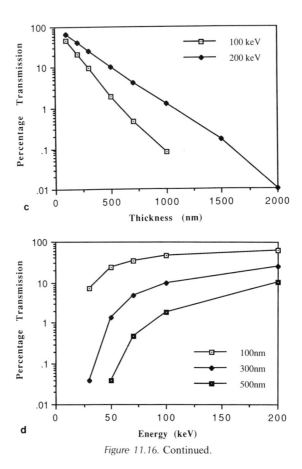

Figure 11.16. Continued.

bulk material, although inelastic scattering events and volume diffusion may still play a part, the data by Dubochet *et al.* (1982) suggest that the mass loss, particularly in frozen-hydrated samples, is due to surface etching and that as such it would be independent of sample thickness.

Somlyo and Shuman (1982) consider that there is a clear distinction between electron beam radiation damage and etching and consider the latter to be due either to the presence of ions in the microscope column, or, when there is a vacuum leak, to ice deposited on an otherwise dry sample. Lamvik and Davilla (1988) found that mass loss from collodion films maintained at 90 K could be either *enhanced* by the presence of water, i.e., specimen etching, or *masked* by the active absorption of water to the surface, i.e., surface contamination. It would thus appear that the quality of the vacuum inside the microscope is an important factor in any measurements on mass loss from irradiated samples. Lamvik and Davilla (1988) also conclude that the configuration and temperature of the cold stage and anticontaminators have an important influence on the layers of water that may become adsorbed on the specimen. Heide (1984) and Hall (1988) consider that the mass loss by erosion from a

100-nm frozen-hydrated sample during the time normally needed to obtain sufficient characteristic counts for x-ray microanalysis, far exceeds the total thickness of the section. The problem is less exacting with thicker sections where the calculations show that approximately 10% of the sample thickness is removed by etching processes. Hall (1986a) quite correctly concludes that it is impracticable to attempt high-resolution quantitative analysis on thin (less than 100 nm) frozen-hydrated sections. In a later study, Hall (1988) calculates that even the initial act of localizing a static probe of 100-nm diameter onto a selected area of a 200-nm section deposits a damaging dose even before any analysis has commenced.

Although it is universally accepted that some degree of mass loss occurs during x-ray microanalysis of frozen samples (dried or hydrated), the extent of this damage, the mechanism(s) by which it occurs, and the cryoprotection effect offered by low temperatures remain uncertain. There are just too many variables for a cogent analysis to be carried out. It is well known that certain molecules are more sensitive to mass loss, whereas others are more resistant. Hydrated material is more sensitive than dried material. Local variations of density within a sample will influence the electron scattering coefficient. In spite of efficiently constructed cold stages, there are

Table 11.9. Range of Incident Electron Doses Used at Varying Temperatures in Low-Temperature X-Ray Microanalysis of Thin (Less than 100 nm), Thick (0.1–2.0 µm), and Bulk Samples in Both the Frozen-Hydrated (FH) and the Freeze-Dried (FD) State

Temperature	Electron dose	Section/bulk	Tissue	Reference
140 K	80 k e^- nm^{-2}	Thin section FH	Rat liver	Zierold (1988)
170 K	50 k e^- nm^2	Thin sections FD	Mouse brain	Fiori et al. (1988)
300 K	2.5 M e^- nm^{-2}	Thin section FD	Rat kidney	Le Furgey et al. (1989)
300 K	5 M e^- nm^{-2}	Thin section FD	Euglena	Cameron et al. (1986)
100 K	550 e^- nm^{-2}	Thick section FH	Gelatine	Marshall and Condon (1985)
100 K	20 k e^- nm^{-2}	Thick section FH	Rat liver	Saubermann et al. (1981)
90 K	2 k e^- nm^{-2}	Thick section FH	Leech ganglion	Saubermann (1988)
100 K	125 k e^- nm^{-2}	Thick section FH	Bioorganic	Hall (1988)
90 K	150 k e^- nm^{-2}	Thick layers FH	Pure ice	Oates and Potts (1985)
100 K	3.5 M e^- nm^{-2}	Thick sections FH	Rat nerve	Lopachin et al. (1989)
90 K	3.6 M e^- nm^{-2}	Thick section FH	Rat liver	Saubermann and Echlin (1975)
300 K	1.125 k e^- nm^{-2}	Thick section FD	Plastic (PVC)	Hall (1985)
300 K	70 k e^- nm^{-2}	Thick section FD	Epithelial	Rick et al. (1982)
100 K	675 k e^- nm^{-2}	Thick section FD	Plastic (PVC)	Hall (1985)
90 K	2.0 M e^- nm^{-2}	Thick section FD	Leech ganglion	Saubermann et al. (1988)
100 K	275 e^- nm^{-2}	Bulk FH	Gelatine	Marshall and Condon (1985)
90 K	50 k e^- nm^{-2}	Bulk FH	Filter paper	Echlin (unpublished)
90 K	62 k e^- nm^{-2}	Bulk FH	Pure ice	Fuchs and Fuchs (1980)
100 K	100 k e^- nm^{-2}	Bulk FH	Bioorganic	Marshall (1984)
90 K	125 k e^- nm^{-2}	Bulk FM	Plant leaves	Echlin/Taylor (1986)
90 K	190 k e^- nm^{-2}	Bulk FM	Plant roots	Echlin et al. (1980)
90 K	250 k e^- nm^{-2}	Bulk FH	Insect	Marshall (1987)
90 K	625 k e^- nm^{-2}	Bulk FH	Plant roots	Echlin et al. (1982)
90 K	1.3 M e^- nm^{-2}	Bulk FH	Mouse gut	Carr et al. (1986)
133 K	0.63–63 k e^- nm^{-2}	Bulk FD	Salts	Roinel (1982)
100 K	3.13 M e^- nm^{-2}	Bulk FD	Plant roots	Harvey et al. (1985)

going to be wide variations in the temperature of the sample. This variability in both cause and effect is exemplified in the wide range of incident electron doses that have been used for low-temperature x-ray microanalysis. Some representative data are shown in Table 11.9 and in nearly all cases the investigators concerned made no mention of mass loss or beam damage and all assumed that valid x-ray data were obtained from their samples, in spite of electron doses ranging over four orders of magnitude.

11.10.3. Loss of Specific Elements from the Sample

Whereas mass loss is primarily concerned with the loss of light elements (C, H, O, and N) from samples because of irradiation with the electron beam, there is also a related problem of the loss of specific elements. Such losses have long been recognized as a potential problem in x-ray microanalysis generally, although there have been few systematic studies on the effects of low temperatures on this problem. Harvey *et al.* (1985) reported losses of potassium during the low-temperature analysis of roots of *Plantago coronopus*. This was not an entirely surprising finding as the electron doses used were in excess of $10^6 \, e^-$ nm^2. Hall and Gupta (1982) consider that a discrepancy in results for chlorine might be due either to loss of chlorine during analysis of frozen-hydrated section or contamination by chlorine during freeze-drying of frozen sections in the electron beam column. Hall (1986) reported severe losses of chlorine from uncooled polyvinylchloride films whereas no losses were recorded at 100 K. Rick *et al.* (1982) report no changes in the concentration of potassium, chloride, and sodium, in spite of 1000-fold increases in the dose of the applied current, whereas sulfur showed a significant decrease.

These various pieces of data only serve to show the wide variability in the extent of specific element losses during low-temperature x-ray microanalysis. The very uncertainty of the problem should put the analyst on his or her guard, and it is probably best to assume the worst will happen. It is perhaps useful to commend the procedure where, once the optimal analytical procedures have been established, the characteristic x-ray count rate for individual elements is measured at different parts of the sample for the same period of time used for analysis. A constant count rate would indicate that specific elemental losses are probably minimal.

11.10.4. Contamination

Local buildup of contamination material in the region of the electron beam during analysis will seriously compromise any quantitative algorithms that rely on mass normalization procedures. The contamination can arise from a number of different sources, and care should be taken to minimize this effect. Specimen cooling will increase the rate at which hydrocarbons and water vapor will condense onto the specimen, and it is important to ensure that the cold specimen is surrounded by even colder anticontaminating surfaces. Contamination affects frozen samples in different ways. A thin layer of beam-induced polymerized hydrocarbons may provide a convenient conducting surface to an otherwise uncoated bulk sample. The same thickness of material deposited on a 100-nm frozen section could easily be equivalent to the original mass per unit area of the section itself. How convenient it would be if the mass loss from the specimen could be assuaged by the contamination deposit on the sample! In addition to adding to the general mass thickness of the sample, contamination may also add specific elements, i.e., S and Cl to the sample.

11.10.5. Cryoprotection Factors

All the studies that have been carried out show that analysis at low temperatures confers a beam radiation cryoprotection factor on the sample, i.e., a sample will exhibit a significantly smaller loss for a given element dose at 100 K than at 300 K. The cryoprotection factor for low-temperature x-ray microanalysis has not been as extensively studied as the cryoprotection factor for structural studies, but the generally accepted figure is between 5 and 8. The data of Hall and Gupta (1974) suggest a 6× cryoprotection factor between 300 and 100 K for mass loss from frozen section, whereas data by Hall (1986) suggest a ×600 cryoprotection factor between 300 and 100 K for the mass loss from thin films of polyvinylchloride. Zierold (1988) claims a further improvement of 2.5× if the temperature is lowered from 140 to 10 K, but there is insufficient evidence to show whether this is a general effect of working near liquid-helium temperatures.

11.10.6 General and Future Approach to Low-Temperature X-Ray Microanalysis

There is now a sufficient body of evidence to suggest that high-resolution quantitative x-ray microanalysis of thin frozen hydrated sections of biological and hydrated organic material maintained at ca. 100 K are unlikely to yield significant results. The experimental data and theoretical calculations show that beam damage conspires against one obtaining meaningful data. The electron dose that is needed to obtain a statistically significant characteristic signal from a thin section, which only contains a small amount of material, inevitably causes extensive mass losses and elemental displacement from the specimen. This is not, however, the case with freeze-dried thin sections, which can withstand a much higher electron dose, presumably due to the absence of water. X-ray microanalysis of frozen-hydrated material can only be accomplished on thick sections (1.0 μm) and bulk samples, and in both cases the use of digital beam analysis may help to reduce the damage to the sample. Bulk samples may be less sensitive to beam damage than thick sections. The reasons for this are not entirely clear, but they probably center on the fact that it is much easier to keep a bulk sample cold than a thick section suspended over a cold metal grid. Full advantage may thus be taken of the cryoprotection factor offered by the low temperature, and higher electron doses may be used during analysis. The presence of a thin conductive layer over the surface of the sample will help to diminish any heating effects and leak away any excess charge. These factors, together with the fact that for a given accelerating voltage a much larger microvolume is sampled in a bulk sample than in a thin section, means that a greater flux of x-rays are emitted from the sample, albeit at diminished spatial resolution.

In the light of the recent work directed toward optimizing conditions for low-temperature x-ray microanalysis it would be appropriate to list the approach that should be adopted.

1. Wherever possible the sample should be maintained at ca. 100 K and should be surrounded by efficient anticontaminators maintained at ca. 80 K.
2. The analysis should be carried out in a clean, water-free, high-vacuum environment.
3. The electron dose should be optimized for maximal signal output with minimal sample damage.
4. Attempts should be made to maximize the amount of information that can be derived from the sample for each interaction event, i.e., use multiple signal collection.
5. One should seek alternative analytic procedures such as electron spectroscopy, which can be achieved using low electron doses, and make use of mass normalization procedures that are not based on measuring the x-ray continuum. Alternate procedures include attenuation of the zero-loss signal in an electron-energy-loss spectroscopy spectrum and using backscattered electron intensity. Both these methods allows mass measurements to be made at much lower electron doses (Hall, 1988).

11.11 APPLICATIONS OF LOW-TEMPERATURE X-RAY MICROANALYSIS

To conclude this chapter it is appropriate to discuss briefly the applications of low-temperature x-ray microanalysis to hydrated samples. In nearly all the several hundred published papers, the emphasis has centered on two main types of investigation. The analysis of low atomic number diffusible elements (Na^+, Mg^{2+}, S, Cl, P, K^+, and Ca^{2+}) and the analysis of cell free spaces and aqueous lumens. Both types of investigation have had an important impact on our understanding of the transport of ions and water across cells and tissues. Another closely related application is concerned with the accurate measurement of local concentrations of water in cells and tissues. Most of the published studies have been carried out in animal cells, and the recent book edited by Zierold and Hagler (1989) and the papers by LeFurgey et al. (1988), Ziergold and Steinbrecht (1987), Marshall (1988), Saubermann (1988c), and Zglinicki (1988) provide useful starting points to review the current range of interests and specific physiological applications. There have been far fewer studies on plant material and nearly all the investigations have been carried out on fractured bulk samples and have centered on either relating changes in ion compartmentation as a function of growth or physiological state (Wiencke et al., 1983; Leigh et al., 1986; Echlin, 1990), storage (Ockenden and Lott, 1990) or the effects of salinity on ion distribution in different plant organs (Harvey *et al.*, 1985; Koyro and Stelzer, 1988; Storey and Walker, 1987).

Low-temperature x-ray microanalysis is also being used on nonbiological samples, primarily to take advantage of the cryoprotection afforded by working at low temperatures. In a few instances, frozen-hydrated samples have been analyzed, i.e., the detection of high concentrations of sulfur at the grain boundaries of Antarctic ice (Wolff *et al.*, 1988).

Low-temperature x-ray microanalysis is now an established analytical technique which is providing a wide range of new and valid information. The acceptance and utilization of the technique has fortunately been accompanied by a critical evaluation of its limitations. It has already been shown that high spatial resolution analytical data can only be obtained from dried specimens (Hall, 1988). Echlin (1989, 1991) has outlined other limitations due to inadequate sample preparation and beam damage. Although most rapid cooling procedures will arrest physiological process in small samples, they are too slow to prevent some diffusion prior to solidification. Only the outermost layers of a small sample will contain ice crystals smaller than 100 nm. The rapidly diminished cooling rates within the sample become manifest as large ice crystals in the cryosections and cryofractures cut from such material. In biological samples it is assumed (hoped?) that structural components of the cytoskeleton will entrap elements swept aside by the growing ice crystals. Such entrapment does not occur in cell free spaces and aqueous lumens and the presence of ice crystals may limit x-ray spatial resolution to 1–3 μm. Freeze drying such samples only exacerbates the problem. The electron doses used in most low-temperature x-ray microanalysis is between 10^4 and 10^5 e^- nm^2 although much lower doses may be used in assocation with digital scanning procedures. Mass loss from frozen samples continues to be a major problem even at low temperatures, particularly where the organic material is in intimate contact with ice. Thin sections cut from resin-embedded freeze-dried material may, after all, be the most productive type of sample for low-temperature quantitative analysis. There is now a much better awareness of what constitutes a frozen-hydrated and freeze-dried samples, and the mistaken analyses are now a thing of the past.

Low-temperature x-ray microanalysis provides one of the very few ways one may hope to obtain *in situ* analytical information about the distribution and local concentration of diffusible ions in biological and hydrated organic material. Electron probe x-ray microanalysis is a reasonably noninvasive analytic tool, although the inadequacies of sample preparation and the ever-present problems of beam damage will always make this a less than perfect method for high spatial resolution studies. The real advantage lies in the fact that it is one of the few techniques that have been developed to the point where a few tens of atoms of an element may be detected and, in turn, directly related to structural features of the sample. Low temperatures and their associated techniques add enormous emphasis to such analytical studies. At liquid-nitrogen temperatures and below, the effects of radiation damage are lessened. The solidification of water is an effective way of virtually stopping diffusion of electrolytes, and arresting physiological processes, dynamic activities, and most chemical reactions.

11.12. SUMMARY

Techniques and instrumentation are available to analyze frozen-hydrated and freeze-dried bulk samples, single cells, thick and thin sections and suspended particles. The physical processes of x-ray microanalysis are understood and well characterized, and the two types of instrumentation, energy dispersive and wavelength dispersive spectrometers, can provide precise and accurate analytic data. X-ray microanalysis has benefitted from extensive and close association with computers, which have made it so much easier to process the x-ray spectra. A wide range of analytical algorithms are available, which when properly applied can swiftly and effectively convert crude, uncorrected x-ray spectra into local elemental concentrations. Such local concentration can be expressed on a numerical wt/wt, or wt/vol basis, or when combined with digital imaging processes, can be presented as maps of elemental concentration with a direct, real time association with the structural components of the sample. An added advantage of low-temperature microanalysis is that samples can be examined and analyzed in both wet and dried state and measurements made of the local water concentration within and outside multicompartmented specimens. In spite of seemingly insurmountable problems of radiation-induced mass loss from organic and hydrated samples, and ice crystallization induced redistribution of soluble material, low-temperature microanalysis continues to offer the best prospect of accurately localizing the position of chemical element in a wide range of organic and inorganic samples.

Current Status of Low-Temperature Microscopy and Analysis

The main part of this book was completed in the spring of 1991. Science, like time, does not stand still, and in the past six months new data and information has appeared which warrants consideration and inclusion in this book. Although this additional information does not change the *causa causans* or the basic premises on which the book is based, it does provide new insights into the theory and practice of low-temperature microscopy and analysis. It is intended only to discuss these new insights and not to consider the now burgeoning literature which describe the use of many of the established low-temperature methods and techniques described in the earlier chapters. For the reader's convenience, this updated material will be discussed in the order it appears in the book.

Our views on the structure of the liquid and solid phases of water remain largely unchanged although there has been a perceived change in the way in which the food, pharmaceutical and polymer industry view the state of the low concentrations of water associated with many carbohydrates and proteins. Most of the interest centers on the relationship between glass transitions and the stability of low-moisture products. There are matters far removed from the much higher concentrations of water usually associated with the samples studied by low-temperature microscopy and analysis. But it is of interest to note that at low concentrations water may be unfreezable, because the presence of solutes, such as sugar, increases the viscosity at room temperature and further increases on cooling so ice-crystal growth cannot occur. The extent of crystallization of water will depend on the rate of crystallization during cooling before the glass is formed. This unfrozen water may act as a softener or plasticizer when in association with hydrophilic polymers and would have important implications for the way we manipulate samples before they are examined and analyzed. An introduction to this revised thinking about the state of water may be found in the paper by Noel *et al.* (1990).

There is continued interest in biological ice-nucleating agents, although we are some way from being able to use them in preparing samples for low-temperature microscopy and analysis. Luquet *et al.* (1991) have obtained cell-free active nucleating agents (mol. wt. 300 kDa) from naturally lysed cells of *Pseudomonas syringae* at the stationary growth phase.

There have been a few improvements to the procedures used to rapidly cool samples. Chang and Baust (1991) have devised a thermophysical model to analyze the influence of precooling biological specimens in the cold gas layer associated with quench- and spray-cooling techniques. They show that precooling is one of the major limiting steps in obtaining an ultrarapid cooling rate. The specimen size strongly affects the critical depth of the gaseous layer, and ice crystallization in the cold gas layer is reduced by increasing the entry speed. Propane jet cooling has been given new impetus from recent studies by Ding *et al.* (1991) using the RMC MF7200 propane jet freezer. They are able to obtain good structural preservation to a depth of 80–100 μm in samples in which buffered 200-mM sucrose solution was used as a cryoprotectant, compared to a depth of 15 μm using conventional quench cooling. Hanya *et al.* (1992) have improved the impact cooling method by irradiating specimens with microwaves at 2.45 Ghz for 50 ms while the sample is pressed onto the surface of a copper block cooled with LN_2. Hanyu and colleagues consider that brief exposure to microwave irradiation during rapid cooling causes a selective increase in the entropy of water so that the molecules do not relax to form microclusters. More and more scientists are using high-pressure cooling as a means of obtaining excellent structural preservation in a wide range of biological and hydrated samples. The technology remains unchanged but it is now a routine rather than an experimental procedure. High-pressure cooling is also being used in connection with the x-ray microanalysis of diffusible elements and immunocytochemical procedures (Kandasamy *et al.*, 1991).

There have been very few careful studies on changes in molecular orientation as a function of differences in the cryofixation procedure. One such investigation has been carried out by Leforestier and Livolant (1991) who compared the freeze-fracture images of the liquid-crystalline phases of DNA rapidly cooled either on a copper block cooled to 4 K, or by immersion into Freon 22 at 113 K or propane at 93 K, with and without glycerol as a cryoprotectant. The cholesteric stratifications are only seen in noncryoprotected impact-cooled samples.

The vexed question of where in the sample we should measure the cooling rate has been investigated in some detail by Hartmann *et al.* (1991). In a computer simulation study, directed more toward cryopreservation than cryofixation, they found that the geometric center of the sample, commonly used as the location for determining cooling rates is in fact the least representative for the whole volume of the sample. Assuming surface cooling, they found that a region one third away from the center and two thirds away from the cooling surface would be representative of 80% of the entire surface volume. These findings have important implications for the way we measure cooling rates in the wide variety of samples we use in low-temperature microscopy and analysis.

The debate continues regarding the process of cryosectioning. Kirk *et al.* (1991) have made platinum carbon replicas of the surfaces of both cryosections and the complementary surface of the frozen block face, cut under different conditions. Samples sectioned at 152 K with large (1-μm) knife advances, produced chips consisting of a series of fracture faces. The same features were seen in thin sections cut with a blunt knife. True sections, i.e., free of fracture faces, were only obtained with sharp knives and small (0.1-μm) knife advances. Kirk and colleagues revive the idea that micromelting of a superficial layer as a result of pressure exerted by the knife edge may be the mechanism by which true cutting occurs. They also found that variations in ice-crystal size did not influence their results, which is contrary to the dictum that good thin cryosections can only be cut from well-cooled samples.

Woods *et al.* (1991) describe a simple freeze-drier for preparing grid-mounted expanded chromosomes and nuclei. The advantage of the process is that the plunge cooling, and drying carried out either by sublimation or substitution, occurs sequentially in the same piece of equipment and thus avoids sample compression during cooling, and sample shrinkage during dehydration.

Although there have been no significant additions to the chemicals used in freeze substitution and low-temperature embedding, a number of papers have appeared in the past few months which provide new information about how these two procedures may be used to advantage in low-temperature microscopy and analysis. A review by Hobot (1990) emphasizes the importance of the progressive lowering of temperature (PLT) technique in combination with embedding in Lowicryl and the partial dehydration procedures used in combination with the acrylic resin LR white. The latter resin can tolerate up to 12% by weight of water. Hobot (1990) and Hobot and Newman (1991) emphasize the importance of retaining a degree of hydration in samples in order to improve both the preservation of ultrastructure and immunoreactivity. Hobot (1990) provides detailed instructions on how these procedures may be applied to bacteria.

A broader review of the PLT and Lowicryl embedding procedures is given by Carlemalm (1990), and a good comparison of the different procedures of freeze substitution and low-temperature embedding is provided in a recent paper by Martinez and Wick (1991). They confirm the earlier studies of Humbel *et al.* (1983) that methanol is a better substitution fluid than acetone and that diethylether gives generally poor results. Their study shows that rapid substitution at 193 K is imperative. In a comparison between the acrylic resin, LR gold (polymerized at 253 K) and the epoxy resin, Quetol 651 (polymerized at 358 K), the LR gold gave far superior results as judged by sectioning properties, beam stability, and contrast.

The low mass density and consequential low inherent contrast in thin sections from material fixed solely in aldehydes remain a problem, although Roth *et al.* (1990) have successfully modified the uranyl acetate/methyl cellulose staining procedure originally applied to thawed cryosections. The advantage of the modified method is that it does not obscure the colloidal gold particles used in imunocytochemistry.

Weibull *et al.* (1990) have made a comparative study of the ultra-low-temperature (188 K) embedding procedure using Lowicryl HM23. High-resolution images of unfixed chloroplasts substituted in pure acetone at 186 K and embedded in Lowicryl HM23 at 188 K were excellent and not significantly different from material which had been chemically fixed and/or embedded at high temperatures. These results have important implications for immunocytochemical studies.

Nearly all the applications of freeze substitution have been directed toward TEM studies. Wharton (1992) has developed a simple method of substituting nematodes in pure methanol, followed by critical-point drying as a means of preserving samples for SEM. The method is particularly useful for specimens which have sufficiently strong walls to withstand the drying process. The process of freeze substitution may be more rigorously controlled using the new apparatus described by Humbel *et al.* (1991). A versatile control system based on an IBM compatible PC, controls all steps of the freeze substitution protocol and also manipulates the whole preparation from initial quench cooling to final UV polymerization.

The kinetics of intracellular ice formation under different cooling conditions have been studied by Toner *et al.* (1991) using a new cryostage with improved optical performance and an isothermal temperature field. The new cryostage which is based on the original design of Cosman *et al.* (1989) has a thinner (2.54 mm), optically flat sapphire window which has improved optical and thermal properties. The maximum thermal gradient across the window is less than 0.1 °C/10 mm at sample temperature near 273 K.

An increasing number of papers are appearing in which cryotransmission electron microscopy is being applied to solve particular structural problems. This near-routine use of the cryo-TEM is also being accompanied by improvements to sample preparation, instrumentation and analysis of images. Sakata *et al.* (1991) describe an elegant and simple technique to produce 1-nm-thick carbon support films. A 20-nm-thick carbon film is deposited onto the grid bars only of micro-grids which have very small grid openings. The coated microgrid is then covered with a 20-nm colloidion film which in turn is covered by a 1-nm-thick carbon film. The carbon–colloidion–carbon sandwich is annealed at 700 K and a pressure of 10^{-5} Torr which removes the colloidion film leaving the nanofilm covering the small apertures on the microgrid. The resultant films are thin enough to be able to resolve (2–3)-nm carbon particles on their surface.

It is difficult to avoid evaporation of water and concentration of solutes when an unsupported, thin, vitrified film of an aqueous suspension is prepared for cryoelectron microscopy. Cyrklaff *et al.* (1990) have investigated this problem by measuring the contrast of polystyrene spheres in a metrizamide solution (an iodinated carbohydrate solution). Drying effects which are minimal on hydrophilic films are exacerbated on hydrophobic films. A system has been constructed which surrounds the sample with a laminar flow of humid air. A double-sided blotting device is pressed against the grid for a preselected time after which the plunger is released while the blotting paper is withdrawn. Tests show that the increase in solute concentration is limited to less than 20%.

A different approach to the problem of sample drying has been taken by Bailey *et al.* (1991) who have constructed a simple environmental chamber coupled to a plunger-driven rapid cooling device. The temperature control is ±0.1 K and relative humidities of >90% can be obtained over the range 268 to 333 K. Control of sample temperature and vapor pressure allows one to fix the chemical potential of the sample and hence its thermodynamic equilibrium in a particular state.

Rapid cooling allows dynamic events to be fixed with a millisecond time resolution. Knoll *et al.* (1991) have developed a quench flow device for rapid mixing and quench cooling of cells and reactants in the millisecond range with a dead time of ~30 ms. The equipment has been used to follow the lag time for the extrusion of trichocysts in *Paramecium* following stimulation with aminoethyldextran. Time resolved cryo-electron microscopy has also been used by Mandelkow *et al.* (1991) in a study of the dynamic instability of growing, shrinking, and oscillating microtubules.

The kinetics were checked by light or x-ray scattering and 5-μl aliquots withdrawn at various time points, and thin suspensions quench cooled into liquid ethane. The samples could be blotted and quench cooled within 1–2 s; vitrification occurred within 100 μs. Because microtubules are thermolabile, it was necessary to have accurate temperature control during sample preparation. The grids were prepared in a warm room at 310 K, and the problem of evaporative cooling of the thin suspension of microtubules could be avoided simply by keeping the humidity in the room at 95–100% by boiling water.

One of the problems in obtaining high structural resolution with cryo-electron microscopy of frozen-hydrated specimens is the low contrast of micrographs taken close to the electron optical focus. In Chapter 9, it was shown that electron-energy-loss spectroscopy was one way around this problem. A recent paper by Schröder et al. (1990) suggests that zero-loss energy filtering can be used to reduce the background, and the consequent gain in contrast can be utilized to reduce defocusing while recording the image (at low dose 700 e$^-$ nm^{-2}) to a few Scherzer focus values.

Another factor which affects image quality is beam-induced specimen motion. A recent review article by Downing (1991) discusses the ways in which spot-scan imaging in the TEM can overcome this problem. Cattermole and Henderson (1991) give details of a simple circuit which provides a noise-free linear ramp of variable current which can be used for manual drift compensation. Image drift frequently occurs in cryo-electron microscopy and the device reduces the drift to no more than 0.1 to 0.2 Å/s within 30 s.

Cryo-electron microscopy of aperiodic structures has not been able to achieve a spatial resolution of much better than 2 nm whereas cryocrystallography of periodic structures gives a spatial resolution of ca. 0.6 nm in projection for a 3-D frozen-hydrated crystal. In the past few months a number of papers have appeared which show the advantage of this latter technique. Yase et al. (1990) have used a superconducting cryo-electron microscope (JEOL JEM-2000 SCM) to directly observe thin films of cadmium arachidate with two different molecular orientations. They also found a 11.5-fold improvement in the cryoprotection factor for beam damage by working at 4 K compared to 300 K. Lattice fringes corresponding to the (110) spacing at 0.4 nm were confirmed. Misra et al. (1990) have studied unstained, vitrified, frozen-hydrated, two-dimensional crystalline sheets of Na, K-ATPase by cryo-electron microscopy and image processing. The projection maps reveal an asymmetry between the pair of protometers which make up the dimer molecule. Erk et al. (1991) have used a combination of cryo-x-ray diffraction and cryo-electron microscopy in studies of the structure of muscle and muscle components.

Considerable effort continues to be directed toward showing that high-resolution cryo-SEM can achieve the same spatial resolution that can be achieved with a platinum carbon replica of the free-fracture surface. In spite of the application of both field-emission cryo-SEM and low-voltage cryo-SEM, the replica still has the edge—but only just. It is only a matter of time before it will not be possible to distinguish between the two types of images. The improvements in instrumentation are being matched by advances in sample preparation. Kusamichi et al. (1991) have investigated the influence of changes in the surrounding media on the preservation of all ultrastructure during sample preparation for low-temperature SEM. Yeast cells, which are generally considered to be quite robust, were treated with various media; solutions of different pH and at different salt concentrations. Not surprisingly, the best preservation was found with solutions that had an affinity with the molecular structure of the surface features. It was surprising to find the narrow band over which these changes occurred, and it provided a salutary warning to the effects of incipient drying prior to rapid cooling.

Van Doorn and colleagues (1991) have developed a cryo-ultramilling procedure to produce smooth fracture surfaces on bulk frozen samples. The quench-cooled samples were maintained at 100 K and milled using a Reichert Jung Polycut E microtome. During milling, cold nitrogen gas at 120 K was blown onto the milling head and the sample to prevent contamination. The procedure preserved ultrastructural features which were absent from samples prepared by conventional fixation and critical-point drying.

Hermann and Muller (1991) have made a careful examination of the low-temperature metal-coating techniques used in high-resolution biological SEM. They compared double-axis rotary shadowing with electron-beam evaporated chromium, chromium planar magnetron sputter coating, and shadowing with electron-beam evaporated platinum and carbon. Metal deposition was carried out on freeze-dried samples maintained at low temperatures during all three procedures. Optimal visual definition of structural detail was achieved with double-axis rotary shadowing at 188 and 123 K using chromium at a thickness (5–6 nm) greater than the minimum required to produce a continuous layer (4 nm). The visual definition was reduced at 23 K. This is in agreement with the reduced thickness at which metal films coalesce and with their higher nucleation density at lower temperatures. The unidirectional platinum carbon shadowing gave acceptable results, but the sputter-coating procedures gave less acceptable results with small filamentous structures. Extremely fine grained coating layers were found with ion beam sputtered chromium, but the fine structure of the specimens could not be resolved due, it is thought, to the high energy of the ion-sputtered metal atoms. Their findings confirmed earlier results which have shown that the most important parameters affecting the quality of low-temperature coating are film thickness and deposition temperature.

In spite of the dramatic emergence of noninvasive techniques based on vital fluorescent dyes with known target specifity, confocal microscopy, and ratio imaging (see *Journal of Microscopy*, April 1992), low-temperature microanalysis continues to provide highly specific analytical information. Further evidence has appeared in the last few months which shows how this technique can be improved and the applications broadened. Malone *et al.* (1991) have shown how the microdroplet technique can be applied to the quantitative analysis of inorganic ions within the vacuoles of individual cells of intact wheat leaves. The approach utilizes a modified pressure capillary probe to extract the vacuolar sap from individual epidermal cells. Following osmotic pressure measurements based on melting-point depression, precise picolitre-size microdroplets of the extracted sap together with microdroplets of standards are placed on ultrathin Pioloform films and freeze-dried and elemental concentrations measured by x-ray microanalysis using the now standard methods described in Chapter 11. This is the first time the microdroplet technique has been used on plant tissue and it promises to provide greater insight into the factors controlling solute composition in individual cells of a multicellular tissue. Hopkins *et al.* (1991) gives details of procedures for the standardization of frozen-hydrated bulk specimens using a gelatin-based standard. This careful study also provides details of the corrections it is necessary to apply when aluminum is used as a coating material. Variations in uncorrelated standard curves are found to be due to changes in aluminum coating thickness. There is an inverse relationship between coating thickness and elemental x-ray counts. The authors suggest that any errors in quantitation due to inconsistent aluminum coating can be overcome by mounting and analyzing specimen and standard together. Provided the same coating thickness is deposited on the frozen-hydrated sample and standard, this correction procedure seems easier to apply than that used by Echlin and Taylor (1986).

There have been no substantial improvements to either the analytical algorithms or quantitative procedures used in low-temperature analysis of inorganic ions and sample water content. A recent paper by Roomans (1990) provides a concise review of the methods to measure the local water concentration.

Further information is given on real-time quantitative elemental analysis and mapping by x-ray microanalysis in a paper by LeFurgey *et al.* (1992) and the characterization of biological macromolecules by combined mass mapping and electron-energy-loss spectroscopy (Leapman and Andrews, 1992). Both papers are an extension of work described earlier in this book and show how freeze-dried thin-film suspensions (Leapman and Andrews) and freeze-dried thin sections (LeFurgey *et al.*) can be used to provide high-resolution chemical and structural information. Although the low-temperature techniques used in both papers are routine, they form an integral part of a sophisticated analytical package. LeFurgey *et al.* put a great deal of emphasis on the use of a graphics-based microcomputer to acquire spectra and display real-time images, and correct post-acquisition image maps. The software allows static probe acquisition from selected areas and provides a new dimension to microchemical imaging. Leapman and Andrews show how a high-resolution STEM instrument (VG Microscopes HB501) can be used to determine the mass of macromolecular assemblies by elastic scattering and provide sensitive elemental analysis using inelastic scattering.

Finally, some reports have appeared which provide both hope and despair for the procedures used in the x-ray microanalysis of frozen-hydrated bulk samples. Lazof and Lauchli (1991a, b) have devised what they consider to be novel procedures for the analysis of frozen-hydrated and freeze-dried tissue. While providing precise information on the minimal detectable limits of a number of elements in frozen-hydrated standards prepared in a novel way to correct for changes in the composition of the matrix, the authors are either painfully unaware or sadly chose to ignore a large amount of published data relating to the analysis of frozen-hydrated samples.

The authors suggest that scant attention has been paid to electron-probe microanalysis of calcium in plant tissues and that little consideration has been given to the problems of peak overlap between this element and potassium. While the authors have developed a new analytical program to make the necessary corrections to peak overlaps, this procedure does not seem to have been evaluated against methods already described in the literature. The frozen-hydrated samples were analyzed after being lightly etched and coated with chromium. No details are given of the correction procedures which were used either to check the hydration state of the samples or the errors due to inconsistencies of the coating layer. Both matters have received considerable attention in the literature. Finally, the suggestion that arbitrary methods of peak determination are unsatisfactory ignores the wealth of subjective information which accrues as a consequence of experience. Optimized and calibrated instrumentation most certainly provides a nonarbitrary and entirely objective means of analysis but it would be naive to take the data that is generated on face value alone. Unless all variants in instrumentation and the vagaries of the specimen are fully quantified, it is probably more accurate, and certainly quicker, to subjectively analyze the derived data using the enormous piece of software which sits on our shoulders.

In conclusion, and on a more positive note, a recent paper by Van Steveninck *et al.* (1990) provides reassurance that x-ray microanalysis may be carried out on frozen-hydrated bulk material provided the limitations of the method are recognized. Van Steveninck and his colleagues have analyzed zinc containing globules and calcium oxalate crystals in the floating roots of duckweed. Their studies with bulk frozen-hydrated material only yielded qualitative results and it was necessary to use thin sections of freeze-substituted material for quantitative results.

November 1991

13

References

Abermann, R., Salpeter, M. M., and Bachmann, L. (1972). In *Principles and Techniques of Electron Microscopy*. M. A. Hayat, ed. New York: Van Nostrand Reinhold, pp. 197–221.

Abrahamson, D. R. (1986). *J. Histochem. Cytochem.* **34**:847–853.

Acetarin, J. D. (1981). Nouvelles recherches sur les resines d'inclusion pour la microscopie electronique. Thesis, University Louis Pasteur. Stasbourg, France.

Acetarin, J. D., Villiger, W., and Carlemalm, E. (1986). *J. Electron. Microsc. Tech.* **4**:2257–2264.

Acetarin, J. D., Carlemalm, E., Kellenberger, E., and Villiger, W. (1987). *J. Electron. Microsc. Tech.* **6**:63–79.

Adachi, E., Nakatani, T., and Hashimoto, P. H. (1987). *J. Microsc.* **147**:205–208.

Adamson-Sharpe, K. M., and Ottensmeyer, F. P. (1981). *J. Microsc.* **122**:309–314.

Adrian, M., Dubochet, J., Lepault, J., and McDowall, A. W. (1982). *Nature* **308**:32–36.

Akahori, H., Okamura, S., Nishiura, M., and Uehira, K. (1972). *Proc. Electron Microsc. Soc. Am.* **30**:294–295.

Akahori, H., Ishii, H., Nonaka, I., and Yoshida, H. (1989). *J. Electron Microsc.* **38**:158–162.

Aldrian, A. F. (1980). *Micron* **11**:261–265.

Alexander, C., Jones, D., and McHardy, W. J. (1987). *New Phytol.* **105**:613–617.

Allakhverdov, B. L., and Kuzminykh, S. B. (1981). *Acta Histochem. Suppl.* **23**:75–82.

Allan-Wojtas, P., and Yang, A. F. (1987). *J. Electron. Microsc. Tech.* **6**:325–333.

Allen, E. D., and Weatherbee, L. (1979). *J. Microsc.* **117**:381–394.

Allen, E. D., and Weatherbee, L. (1980). *Cryobiology* **17**:448–457.

Allen, E. D., Weatherbee, L., and Permoad, P. A. (1978). *Cryobiology* **15**:375–381.

Allison, D. P., Daw, C. S., and Rorvik, M. C. (1987). *J. Microsc.* **147**:103–108.

Altmann, R. (1890). *Die Elementarorganismen und ihre Beziehungen zu den Zellen.* Leipzig: Veit.

Anderson, C. A. (1967). *Methods Biochem. Anal.* **15**:147–163.

Anderson, C. A., and Hasler, M. F. (1966). In Proc. 4th Intl. Congr. X-ray Optics and Microanalysis. R. Castaing, P. Deschampes, and J. Philibert, eds. Paris: Herman, p. 310.

Andrews, S. B., and Leapman, R. D. (1989). In *Microbeam Analysis 1989*. P. E. Russell, ed. San Francisco: San Francisco Press, pp. 4–5.

Angell, C. A. (1982). *Recherche* **133**:584–593.

Angell, C. A. (1982). In *Water: A Comprehensive Treatise*, F. Franks, ed.Vol. 7. New York: Plenum Press, pp. 1–81.

Angell, C. A. (1983). *Ann. Rev. Phys. Chem.* **34**:593–630.

Angell, C. A. (1988). *Nature* **331**:206–207.

Angell, C. A., and Choi, Y. (1986). *J. Microsc.* **141**:251–261.

Aoki, Y., Kihara, H., Harada, Y., and Fujiyoshi, Y. (1988). In Proc. 11th Intl. Cong. Electron Microsc. Kyoto. pp. 1827–1832.

Appleton, T. C. (1973). In *Electron Microscopy and Cytochemistry*. E. Wisse, W. Th. Daems, I. Mollenaar, and P. van Duijn, eds. Amsterdam: North-Holland, pp. 229–242.

Appleton, T. C. (1974). *J. Microsc.* **126**:317–332.

Appleton, T. C. (1978). In *Electron Probe Analysis in Biology*. D. A. Erasmus, ed. London: Chapman Hall, p. 148.

Appleyard, S. T., Witkowski, J. A., Ripley, B. D., Shotton, D. M., and Dubowitz, V. (1985). *J. Cell Sci.* **74**:105–117.

Armbruster, B. L., Carlemalm, E., Chiovetti, R., Garavito, R. M., Hobot, J. A., Kellenberger, E., and Villiger, W. (1982). *J. Microsc.* **126**:77–85.

Armbruster, B. L., Kellenberger, E., Carlemalm, E., Villiger, W., Garavito, R. M., Hobot, J. A., Chiovetti, R., Acetarin, J. D. (1984). In *Science of Biological Specimen Preparation.* J. P. Revel, T. Barnard, and G. H. Haggis, eds. Chicago: SEM Inc., pp. 77–81.

Armitage, W. J., and Rich, S. J. (1990). *Cryobiology* **27**:483–491.

Arnold, W., and Von Mayersback, H. (1975). *Arzneim. Forsch.* **25**:453–454.

Arsenault, A. L., Ottensmeyer, F. P., and Heath, I. B. (1988). *J. Ultrastruct. Mol. Biol.* **98**:32–47.

Artymink, P. J., Blake, C. C. F., Grace, D. E. P., Oatley, S. J., Phillips, D. C., and Sternberg, M. J. E. (1979). *Nature* **280**:563–568.

Asahina, E. (1961). *Nature* **191**:1263–1265.

Ashford, A. E., Allaway, W. G., Gubler, F., Lennon, A., and Sleegers, A. (1986). *J. Microsc.* **144**:107–126.

Ashworth, E. N., and Ables, F. B. (1984). *Plant Physiol.* **76**:201–204.

Ashworth, E. N., Echlin, P., Pearce, R. S., and Hayes, T. L. (1988). *Plant, Cell Environ.* **11**:703–710.

Asquith, M. H., and Reid, D. S. (1980). *Cryoletters* **1**:352–359.

Athey, B., Langmore, J., Williams, S., Smith, M., Chang, C. F., Grant, R., and Chiu, W. (1987). In *Proc. 45th. Ann. Meet. EMSA.* G. W. Bailey, ed. San Francisco: San Francisco Press, pp. 648–649.

Aurich, F., and Foster, T. (1984). *Cryoletters* **5**:231.

Baba, M., and Osumi, M. (1987). *J. Electron. Microsc. Tech.* **5**:249–261.

Bacaner, M., Broadhurst, J., Hutchinson, T., and Lilley, J. (1973). *Proc. Natl. Acad. Sci. USA* **70**:3423–3430.

Bachhuber, K., Bohme, H., Westphal, C., and Frosch, D. (1987). *J. Microsc.* **147**:323–328.

Bachmann, L. (1987). In *Cryotechniques in Biological Electron Microscopy.* R. A. Steinbrecht and K. Zierold, eds. Berlin: Springer, Chap. 9.

Bachmann, L., and Schmitt-Fumian, W. W. (1973). In *Freeze Etching, Techniques and Applications.* E. L. Benedetti and P. Favard, eds. Paris: Société Française de Microscopie Electronique, pp. 73–79.

Bachmann, L., and Talmon, Y. (1984). *Ultramicroscopy* **14**:211–218.

Bachmann, L., and Mayer, E. (1987). In *Cryotechniques in Biological Electron Microscopy.* R. A. Steinbrecht and K. Zierold, eds. Chap. 1. Berlin: Springer-Verlag.

Bachmann, L., Becker, R., Leupold, G., Barth, M., Guckerberger, R., and Baumeister, W. (1985). *Ultramicroscopy* **16**:305–320.

Bahr, G. F., Johnson, F. B., and Zeitler, E. (1965). *Lab. Invest.* **14**:115–121.

Bailey, J. K., and Macartney, M. L. (1988). In *Proc. 46th Ann. Meet. EMSA.* G. W. Bailey, ed. San Francisco: San Francisco Press, pp. 110–111.

Bailey, S. M., Chiruvolu, S., Longo, M. L., and Zasadzinski, J. A. N. (1991). *J. Electron. Microsc. Tech.* **19**:118–126.

Baker, I. (1986). *J. Electron. Microsc. Tech.* **3**:357–358.

Bald, W. B. (1983). *J. Microsc.* **131**:11–23.

Bald, W. B. (1984). *J. Microsc.* **134**:261–270.

Bald, W. B. (1985). *J. Microsc.* **140**:17–40.

Bald, W. B. (1986). *J. Microsc.* **143**:89–102.

Bald, W. B. (1987). *Quantitative Cryofixation.* Bristol: Adam Hilger.

Bank, H., and Mazur, P. (1973). *J. Cell Biol.* **57**:729–742.

Barbi, N. C. (1978). *Scanning Electron Microsc.* **2**:193–200.

Barlow, D. I., and Sleigh, M. A. (1979). *J. Microsc.* **115**:81–89.

Barnard, T. (1980). *J. Microsc.* **120**:93–103.

Barnard, T. (1982). *J. Microsc.* **126**:317–332.

Barnard, T. (1987). *Scanning Microsc.* **1**:1217–1224.

Barnard, T., Gupta, B. F., and Hall, T. H. (1984). *Cryobiology* **21**:559–569.

Barnett, J. R., and Weatherhead, I. (1988). *Ann. Botany* **61**:581–587.

Bartl, P. (1962). In *Electron Microscopy*, Vol. II 4, S. S. Breese, ed. London: Academic Press.

Bastacky, J., Hayes, T. L., and Gelinas, R. (1985). *Scanning* **7**:134–140.

Bastacky, J., Hook, G. R., Finch, G. L. Goerke, J., and Hayes, T. L. (1987). *Scanning* **9**:57–70.

Bastacky, J., Wodley, C., LaBrie, R., and Backus, C. (1987). *Scanning* **9**:219–225.

Bastacky, J., Wodley, C., LaBrie, R., and Backus, C. (1988). *Scanning* **10**:37–38.

Bastacky, J., Goodman, C., and Hayes, T. L. (1990). *J. Electron. Microsc. Tech.* **14**:83–84.

Baumeister, W., Gross, H., and Zeitler, E. (eds.) (1985). *Ultramicroscopy* **16**:85–258.

Beall, P. T. (1983). *Cryobiology* **20**:324–334.

Bearer, E. L., and Orci, L. (1986). *J. Electron. Microsc. Tech.* **3**:233–241.

Beckett, A., and Porter, R. (1988). *Can. J. Bot.* **66**:645–652.

Beckett, A., and Reiad, N. D. (1986). In *Ultrastructure Techniques for Microorganisms*. H. C. Aldrich, and W. J. Todd, eds. New York: Plenum Press, pp. 45–86.

Beckett, A., and Woods, A. M. (1986). *Can. J. Bot.* **65**:1998–2006.

Beckett, A., Read, N. J., and Porter, R. (1984). *J. Microsc.* **136**:87–95.

Beckmann, J., Korber, Ch., Rau, G., Hubel, A., and Cravalho, E. G. (1990). *Cryobiology* **27**:279–287.

Beeuwkes, R., Saubermann, A. J., Echlin, P., and Churchill, S. (1982). In Proc. 40th Ann. Meeting E.M.S.A. Washington, D.C., G. W. Bailey, ed. San Francisco, San Francisco Press, pp. 754–755.

Behrman, E. J., (1984). In *The Science of Biological Specimen Preparation*. J. P. Revel, T. Barnard, and G. H. Haggis, eds. Chicago: SEM Inc., pp. 1–5.

Bellare, J. R., Davis, H. T., Scriven, L. E., and Talmon, Y. (1988). *J. Electron. Microsc. Tech.* **10**:87–111.

Bellare, J. R., Bailey, J. K., and Mecartney, M. (1989). In Proc. 47th Ann. Meet. EMSA, G. R. Bailey, ed. San Francisco: San Francisco Press, pp. 356–357.

Belous, M. V., and Wayman, C. M. (1967). *J. Appl. Phys.* **38**:5119–5124.

Belton, P. S., Jackson, R. R., and Packer, K. J. (1972). *Biochim. Biophys. Acta.* **721**:16–27.

Bendayan, H., and Shore, G. G. (1982). *J. Histochem. Cytochem.* **30**:149–147.

Benichou, J. C., Frehel, C., and Ryter, A. (1990). *J. Electron. Microsc. Tech.* **14**:289–297.

Bentley, J., Zaluzec, N. J., Kenik, E. A., and Carpenter, R. W. (1979). *Scanning Electron Microsc.* **2**:581–584.

Bernal, J. D., and Fowler, R. M. (1933). *J. Chem. Phys.* **1**:515–548.

Bernhard, W. (1965). *Ann. Biol.* **4**:5–18.

Bernhard, W. (1971). *J. Cell Biol.* **49**:731–742.

Berthe, R., Hartmann, U., and Heiden, C. (1988). *J. Microsc.* **152**:831–839.

Biddlecombe, W. H., Jenkinson, D. M., McWilliams, S. A., Nicholson, W. A. P., Elder, H. Y., and Demster, D. W. (1982). *J. Microsc.* **126**:63–75.

Binnig, G., Rohrer, H., Gerber, C., and Weibel, E. (1982). *Phys. Rev. Lett.* **49**:57–61.

Bischof, J., Hunt, C. J., Rubinsky, B., Burgess, A., and Pegg, D. E. (1990). *Cryobiology* **27**:301–310.

Blackmore, S., and Barnes, S. H. (1984). *Grana* **23**:157–162.

Blackmore, S., Barnes, S. H., and Claugher, D. (1984). *J. Ultra. Res.* **86**:215–219.

Bode, M., Jakubowicz, A., and Habermeier, H.-U. (1988). *Scanning* **10**:169–176.

Boehler, S. (1975). *Artifacts and Specimen Preparation Faults in Freeze-Etch Technology*. Liechtenstein: Balzers AG.

Boekestein, A., Stols, A. L. H., and Stadhouders, A. M. (1980). *Scanning Electron Microsc.* **2**:321–334.

Boekestein, A., Stadhouders, A. M., Stols, A. L. H., and Roomans, G. M. (1983). *Scanning Electron Microsc.* **2**:725–736.

Boekestein, A., Thiel, F., Stols, A. L. H., Bouw, E., and Stadhouders, A. M. (1984). *J. Microsc.* **134**:327–334.

Boller, T., and Wiemken, A. (1986). *Ann. Rev. Plant Physiol.* **37**:137–164.

Bond, M. (1984). *J. Physiol.* **355**:677–686.

Boonstra, J., van Belzen, N., van Maurik, P., Hage, W. J., Blok, F. J., Weigent, F. A. C., and Verkleij, A. J. (1985). *J. Microsc.* **140**:119–129.

Boonstra, J., Maurik, P. V., and Verkleij, A. J. (1987). In *Cryotechniques in Biological Electron Microscopy*, R. A. Steinbrecht and K. Zierold, eds. Berlin: Springer. Chap. 14.

Boonstra, J., van Belzen, N., van Bergen en Henegouwen, P. M. P., Hage, W. J., van Maurik, P., Weigent, F. A. C., and Verkleij, A. J. (1989). In *Immunogold Labeling in Cell Biology*. A. J. Verkleij, and J. L. M. Leunissen, eds. Boca Raton, Florida: CRC Press, pp. 259–276.

Boonstra, J., van Bergen en Henegouwen, P. M. P., van Belzen, N., Rijken, P., and Verkleij, A. J. (1991). *J. Microsc.* **161**:135–148.

Bostrom, T. E., Field, M. J., Gyory, A. Z., Dyne, M., and Cockayne, D. J. H. (1991). *J. Microsc.* **163**:319–333.

Bou-Gharios, G., Adams, G., Moss, J., Shore, I., and Olsen, I. N. (1988). *J. Microsc.* **150**:161–163.

Bourgeois, C. A., Zanchi, G., Khin, Y., Lacaze, J. C., Zalta, J., and Bouteille, M. (1980). In *Proc. 7th Eur. Cong. E.M.* P. Brederoo and W. de Priester, eds. Leiden, Vol. 2, pp. 118–119.

Boutron, P. (1984). *Cryobiology* **21**:183–191.

Boyde, A. (1974). *Scanning Electron Microsc.* **1**:1063.

Boyde, A. (1975). In *Publication P1*, (*Edwards High Vacuum*), Crawley, Sussex.

Boyde, A. (1976). *Scanning Electron Microsc.* **1**:683–691.

Boyde, A. (1978). *Scanning Electron Microsc.* **2**:303–310.

Boyde, A. (1980). In . *Proc. 7th Eur. Reg. Conf. EM*, P. Brederoo and W. de Priester, eds. Leiden, Vol. 2, pp. 768–772.

Boyde, A., and Echlin, P. (1973). *Scanning Electron Microsc.* **1**:759.

Boyde, A., and Franc, F. (1981). *J. Microsc.* **122**:75–83.

Boyde, A., and Machonnachie, E. (1979). *Scanning* **2**:149–156.

Boyde, A., and Wood, C. (1969). *J. Microsc.* **90**:221–232.

Boyne, A. F. (1979). *J. Neurosci. Methods* **1**:353–364.

Branton, D. (1966). *Proc. Natl. Acad. Sci.* (USA) **55**:1048–1054.

Branton, D., Bullivant, S., Karnovsky, M. J., Moor, H., Packer, L., Steere, R. L., Muhlethaler, K., Northcote, D. H., Satir, B., Satir, P., and Staehlin, L. A. (1975). *Science* **190**:54–56.

Bresse, J., Papadopoulo, A. C., and Henoc, P. (1987). *Scanning Microsc.* **1**:205–209.

Bridgeman, P. C., and Reese, T. S. (1984). *J. Cell Biol.* **99**:1655–1668.

Brink, J., and Chiu, W. (1991). *J. Microsc.* **161**:279–296.

Brink, J., Booy, F. P., and Van Bruggen, E. F. J. (1989). *Ultramicroscopy* **27**:91–100.

Brombach, J. D. (1975). *J. Microsc. Biol. Cell.* **22**:233–238.

Brookes, B., and Small, E. (1988). *Scanning Microsc.* **2**:247–256.

Brower, W. E., Freund, M. J., Baudino, M. D., and Ringwald, C. (1981). *Cryobiology* **18**:277–291.

Brown, M. S. (1980). *Cryobiology* **17**:184–186.

Brown, D. H., Rapsch, S., Beckett, A., and Ascaso, C. (1987). *New Phytol.* **105**:295–299.

Bruggeller, P., and Mayer, E. (1980). *Nature* **288**:569–571.

Buchheim, W., and Welsch, U. (1978). *J. Microsc.* **111**:339–348.

Bulger, R. E., Beeuwkes, R., and Saubermann, A. J. (1981). *J. Cell Biol.* **88**:274–280.

Bullivant, S. (1970). In *Some Biological Techniques for Electron Microscopy*. D. F. Parsons, ed. New York: Academic Press.

Bullivant, S. (1977). *J. Microsc.* **111**:101–116.

Bullivant, S. (1984). In *The Science of Biological Specimen Preparation*. J. P. Revel, T. Barnard, and G. H. Haggis, eds. Chicago: SEM Inc., pp. 175–180.

Bullivant, S., and Ames, A. (1966). *J. Cell Biol.* **29**:435–442.

Bullock, G. R., Wilson, A. J., Harris, A. J., and Robards, A. W. (1980). *Proc. R. Microsc. Soc.* **15**:18–19.

Bullough, P., and Henderson, R. (1987). *Ultramicroscopy* **21**:223–230.

Burstein, N. C., and Maurice, D. M. (1979). *Micron* **9**:191–198.

Callow, R. A., and McGrath, J. J. (1982). In *Proc. 10th Ann. N.E. Bioengineering Conf.* E. W. Hanson, ed. IEEE 82CH1747-5, pp. 269–275.

Cameron, I. L., Hardman, W. E., Hunter, K. E., Haskin, C., Smith, N. R. K., and Fullerton, G. D. (1990). *Scanning Microsc.* **4**:89–102.

Campbell, R., and Porter, R. (1982). *Soil Biol. Biochem.* **14**:241–245.

Carlemalm, E. (1980). In *Low Denaturation Embedding for Electron Microscopy of Thin Sections*. E. Kellenberger, E. Carlemalm, W. Villiger, J. Roth, and R. M. Garavito, eds. Waldkraiberg: Chemische Werke Lowi. G.M.bH.

Carlemalm, E. (1990). *J. Struct. Biol.* **104**:189–191.

Carlemalm, E., Garavito, R. M., and Villiger, W. (1980). In *Proc. 7th Europ. Cong. E.M.*, P. Brederoo and W. de Priester, eds., Leiden, Vol. 2, p. 656.

Carlemalm, E., Garavito, R. M., and Villiger, W. (1982). *J. Microsc.* **126**:123–143.

Carlemalm, E., and Kellenberger, E. (1982). *EMBO J.* **1**:63–70.

Carlemalm, E., Villiger, W., Hobot, J. A., Acetarin, J. D., and Kellenberger, E. (1985). *J. Microsc.* **140**:55–63.

Carlemalm, E., Colliex, C., and Kellenberger, E. (1985). *Adv. Electr.* **63**:269–334.

Carlemalm, E., Villiger, W., Acetarin, J.-D., and Kellenberger, E. (1986). In *The Science of Biological Specimen Preparation.* M. Muller, R. P. Becker, A. Boyde, and J. J. Wolosewick, eds. Chicago: SEM Inc., pp. 147–154.

Carmen, P. C. (1948). *Trans. Faraday Soc.* **44**:529–536.

Carr, D. J., Oates, K., and Carr, S. G. M. (1980). *Ann. Bot.* **45**:403–413.

Carr, K. E., Hayes, T. L., Watt, A., Bastacky, J., Klein, S., and Fife, M. (1986). *J. Electron. Microsc. Tech.* **4**:371–379.

Castaing, R. (1951). Ph.D. thesis University of Paris.

Cattermole, D., and Henderson, H., (1991). *Ultramicrosc.* **35**:55–57.

Ceska, T. A., and Henderson, R. (1990). *J. Mol. Biol.* **213**:539–560.

Chandler, D. E. (1986). *J. Electron. Microsc. Tech.* **3**:305–335.

Chandler, J. A. (1977). In *Practical Methods in Electron Microscopy.* A. M. Glauert, ed. Amsterdam: North-Holland.

Chandler, J. A. (1985). *Scanning Electron Microsc.* **2**:731–744.

Chandra, S., Morrison, G. H., and Wolcott, C. C. (1986). *J. Microsc.* **144**:15–37.

Chang, C. F., Ohno, T., and Glaeser, R. M. (1985). *J. Electron. Microsc. Tech.* **2**:59–65.

Chang, J. J., Walter, C. A., and Dubochet, J. (1981). In *Proc. 39th Ann. Meet. EMSA.* W. C. Bailey, ed. San Francisco: San Francisco Press, pp. 642–643.

Chang, J. J., McDowall, A. W., Lepault, J., Freeman, R., Walter, C. A. and Dubochet, J. (1983). *J. Microsc.* **132**:109–123.

Chang, Z.-H., and Baust, J. G. (1991). *J. Microsc.* **161**:435–444.

Chestnut, M. H., Siegel, D. P., Burns, J. L., and Talmon, Y. (1991). *J. Electron. Microsc. Tech.* (in press).

Chiovetti, R., Little, S. A., Brass-Dale, J., and McGuffee, L. J. (1986). In *Science of Biological Specimen Preparation.* M. Muller, R. P. Becker, A. Boyde, and J. L. Wolosewick, eds. Chicago: SEM Inc. pp. 138–154.

Chiovetti, R., McGuffee, L. J., Little, S. A., Clark, E. W., and Dale, J. B. (1987). *J. Electron. Microsc. Tech.* **5**:1–15.

Chiu, W. (1986). *Ann. Rev. Biophys. Chem.* **15**:237–257.

Chiu, W., and Jeng, T. W. (1982). *Ultramicroscopy* **10**:63–70.

Chiu, W., Downing, K. H., Hobbs, L. W., Shuman, H., and Talmon, Y. (1988). *EMSA Bull.* **18**:16–25.

Christensen, A. K. (1971). *J. Cell Biol.* **51**:772–804.

Christensen, A. K., and Komorowski, T. E. (1985). *J. Electron. Microsc. Tech.* **2**:497–507.

Clarke, J., Echlin, P., Moreton, R., Saubermann, A. J., and Taylor, P. (1976). *Scanning Electron Microsc.* **1**:83–90.

Clarke, A. (1987). In *The Effects of Low Temperatures on Biological Systems*, pp. 315–348. B. W. W. Grout, and G. J. Morris, eds. London: Edward Arnold.

Clay, C. S., and Peace, G.W. (1981). *J. Microsc.* **123**:25–34.

Clegg, J. S. (1982). In *Biophysics of Water*, F. Franks and S. Mathias, eds. pp. 365–383. Chichester: John Wiley.

Cohn, F. (1871). *Die entwicklungensgesichte der Gattung Volvox.* Festschrift dem Prof. Dr. Goppert, University Breslau.

Cole, R., Matuszek, G., See, C., and Rieder, C. L. (1990). *J. Electron. Microsc. Tech.* **16**:167–173.

Coleman, R. A., and Wade, J. B. (1989). *J. Electron. Microsc. Tech.* **13**:216–227.

Colonna, K., and Oliphant, G. (1986). In *Microbeam Analysis–1986.* eds: A. Romig, and W. F. Chambers, eds. San Francisco: San Francisco Press. pp. 214–216.

Colquhoun, W. (1984). *J. Ultrastruct. Res.* **87**:97–105.

Colquhoun, W., and Cassimeris, L. (1985). *J. Ultrastruct. Res.* **91**:138–148.

Colquhoun, W., and Sokol, R. (1986). *J. Electron. Microsc. Tech.* **3**:169–176.

Colquhoun, W., Sokol, R., Davison, E., and Cassimeris, L. (1985). *J. Electron. Microsc. Tech.* **2**:353–370.

Condron, R. J., and Marshall, A. T. (1986). *J. Microsc.* **143**:249–255.

Condron, R. J., and Marshall, A. T. (1990). *Scanning Microsc.* **4**:439–447.

Cook, J. M., and Murphy, H. (1988). *Inst. Phys. Conf. Ser.* 93 **2**:563–564.

Cooke, R., and Kuntz, I. D. (1974). *Annu. Rev. Biophys. Bioeng.* **3**:95–126.

Cope, G. H. (1967). *J. R. Microsc. Soc.* **8**:235.

Cosman, M. D., Toner, M., Kandel, J., and Cravalho, E. G. (1989). *Cryoletters* **10**:17–38.

Cosslett, V. E. (1978). *J. Microsc.* **113**:113–129.

Costello, M. J. (1980). *Scanning Electron Microsc.* **2**:361–370.

Costello, M. J., and Corless, J. M. (1978). *J. Microsc.* **112**:17–37.

Costello, M. J., and Fetter, R. D. (1986). *Methods Enzymol.* **127**:704–718.

Costello, M. J., Fetter, R., and Hoechli, M. (1982). *J. Microsc.* **125**:125–136.

Costello, M. J., Fetter, R., and Corless, J. M. (1984). In *Science of Biological Specimen Preparation*. J. P. Revel, T. Barnard, and G. H. Haggis, eds. Chicago: SEM Inc., pp. 105–115.

Costello, M. J., Fetter, R. D., and Frey, T. G. (1986). In *The Science of Biological Specimen Preparation*. M. Muller, R. P. Becker, A. Boyde, and J. L. Wolosewick, eds. Chicago: SEM Inc. pp. 95–101.

Coulter, H. D. (1986). *J. Electron. Microsc. Tech.* **4**:315–328.

Coulter, H. D., and Terracio, L. (1977). *Anat. Rec.* **187**:477–494.

Craig, S., and Staehelin, L. A. (1988). *Eur. J. Cell Biol.* **46**:80–93.

Crewe, A. V., Langmore, J. P., and Isaacson, M. S. (1975). In *Physical Aspects of Electron Microscopy and Microbeam Analysis*. B. M. Siegel and D. R. Beaman, eds. New York: Wiley, pp. 47–62.

Cutler, A. J., Saleem, M., Kendall, E., Gusta, L. V., Georges, F., and Fletcher, G. L. (1989). *J. Plant. Physiol.* **135**:351–354.

Cyrklaff, M., Adrian, M., and Dubochet (1990). *J. Electron. Microsc. Tech.* **16**:351–355.

Dahl, R., and Staehelin, L. A. (1989). *J. Electron. Microsc. Tech.* **13**:165–174.

Dankberg, F., and Persidsky, M. D. (1976). *Cryobiology* **13**:430–432.

Davilla, S. D. (1989). *Scanning*.

Davilla, S. D., Ingram, P., LeFurgey, A., and Lamvik, M. K. (1987). *J. Microsc.* **149**:153–157.

de Briujn, W. C. (1981). *Scanning Electron Microsc.* **2**:357–367.

Degn, L. T. (1987). In *Proc. 45th Ann. Meet. EMSA.* G. W. Bailey, ed. San Francisco: San Francisco Press, pp. 446–647.

Dempsey, G. P., and Bullivant, S. (1976). *J. Microsc.* **106**:261–272.

Derbyshire, W. (1982). In *Water: A Comprehensive Treatise*, Vol. 7, F. Franks, ed. pp. 339–430. New York: Plenum Press.

Dietrich, I., Fox, F., Knapek, E., Lefranc, G., Nachtrieb, K., Weyl, R., and Zerbst, H. (1977). *Ultramicroscopy* **2**:241–249.

Dietz, T. E., Davis, L. S., Diller, K. R., and Aggarwal, J. K. (1982). *Cryobiology* **19**:539–549.

Dietz, T. E., Diller, K. R., and Aggarwal, J. K. (1984). *Cryobiology* **21**:200–208.

Diller, K. R. (1975). *Cryobiology* **12**:480–485.

Diller, K. R. (1979). *Cryobiology* **16**:125–131.

Diller, K. R. (1982). *J. Microsc.* **126**:9–28.

Diller, K. R. (1985). *Cryoletters* **5**:339–342.

Diller, K. R. (1990). *Cryoletters* **11**:75–88.

Diller, K. R., and Aggarwal, S. J. (1987). *J. Microsc.* **146**:209–219.

Diller, K. R., and Cravalho, E. G. (1971). *Cryobiology* **7**:4–6.

Diller, K. R., and Knox, J. M. (1983). *Cryoletters* **4**:77–92.

Diller, K. R., Cravalho, E. G., and Huggins, C. E. (1976). *Med. Biol. Eng.* **14**:321–333.

Ding, B., Turgeon, R., and Pathasarthy, M. V. (1991). *J. Electron. Microsc. Tech.* **19**:107–117.

Dore, J. C. (1985). In F. Franks, ed., *Water Science Reviews. I.* Cambridge: Cambridge University Press, pp. 3–92.

Dorge, A., and Rick, R. (1990). *Scanning Microsc.* **4**:449–455.

Dorge, A., Rick, R., Gehring, K., Mason, J., and Thurau, K. (1975). *J. Microsc. Biol. Cell* **22**:205–212.

Dorge, A., Rick, R., Gehring, K., and Thurau, K. (1978). *Pflügers Arch.* **373**:85–97.

Dorset, D. L., and Zemlin, F. (1985). *Ultramicroscopy* **17**:229–236.

Doty, S. B., Lee, C. W., and Banfield, W. G. (1974). *Histochem. J.* **6**:383–393.

Douzou, P. (1973). *Mol. Cell Biochem.* **1**:15–27.

Douzou, P. (1977). *Cryobiochemistry.* New York: Academic Press.

Douzou, P., Balny, C., and Franks, F. (1978). *Biochimie* **60**:151–158.

Dow, J. A. T., Gupta, B. F., Hall, T. A., and Harvey, W. R. (1984). *J. Membr. Biol.* **77**:223–241.

Dowell, L. G., and Rinfert, A. P. (1960). *Nature* **188**:1144–1148.

Downing, K. H. (1983). *Ultramicroscopy* **11**:229–238.

Downing, K. H., (1991). *Science* **251**:53–59.

Downing, K. H., and Chiu, W. (1990). *Electron Microsc. Rev.* **1**:123–143.

Downing, K. H., and Glaeser, R. M. (1986). *Ultramicroscopy* **20**:269–278.

Draenert, Y., and Draenert, K. (1982). *Scanning Electron Microsc.* **4**:1799–1804.

Dubochet, J. (1975). *J. Ultrastruct. Res.* **52**:276–288.

Dubochet, J. (1984). *Proc. 8th Eur. Cong E.M.* **2**:1379.

Dubochet, J., and McDowall, A. W. (1981). *J. Microsc.* **124**:RP3–RP4.

Dubochet, J., and McDowall, A. W. (1984). In *The Science of Biological Specimen Preparation.* J. P. Revel, T. Barnard, and G. H. Haggis, eds. Chicago: SEM Inc., pp. 147–152.

Dubochet, J., Knapek, E., and Dietrich, I. (1981). *Ultramicroscopy* **6**:77–80.

Dubochet, J., Chang, J. J., Freeman, R., Lepault, J., and McDowall, A. W. (1982). *Ultramicroscopy* **10**:55–62.

Dubochet, J., Lepault, J., Freeman, R., Berriman, J. A., and Homo, J. C. (1982). *J. Microsc.* **128**:219–237.

Dubochet, J., Groom, M., Muller, M., and Neuteboom, S. (1982). In *Advances in Optical and Electron Microscopy*, Vol. 8. V. E. Cosslett, and R. Barer, eds., pp. 107–135, London: Academic Press.

Dubochet, J., McDowall, A. W., Menge, B., Schmid, E. N., and Lickfield, K. G. (1983). *J. Bacteriol.* **155**:381–390.

Dubochet, J., Adrian, M., Lepault, J., and McDowall, A. W. (1985). *Trends Biochem. Sci.* **10**:143–146.

Dubochet, J., Adrian, M., Chang, J. J., Lepault, J., and McDowall, A. W. (1987). In *Cryotechniques in Biological Electron Microscopy.* R. A. Steinbrecht, and K. Zierold, eds. Berlin: Springer-Verlag, Chap. 5.

Dubochet, J., Adrian, M., Chang, J. J., Homo, J. C., Lepault, J., McDowall, A. W., and Schultz, P. (1988). *Q. Rev. Biophys.* **21**:129–228.

Dufour, L., and Defay, R. (1963). *Thermodynamics of Clouds.* New York: Academic Press.

Dunlap, J. R., Luo, S., Bunn, R. D., and Joy, D. C. (1989). In *Microbeam Analysis—1989*, P. E. Russell, ed. San Francisco: San Francisco Press, pp. 127–128.

Durand, M., Deleplanque, M., and Kahane, A. (1967). *Solid State Commun.* **5**:750–759.

Durrenberger, M., Bjornsti, M. A., Uetz, T., Hobot, J. A., and Kellenberger, E. (1988). *J. Bacteriol.* **170**:4757–4768.

Echlin, P. (1971). *Scanning Electron Microsc.* **1**:225–232.

Echlin, P. (1978). In *Advanced Techniques in Biological Electron Microscopy II.* J. K. Koehler, ed. Berlin: Springer-Verlag, pp. 89–122.

Echlin, P. (1978). *J. Microsc.* **112**:225–232.

Echlin, P. (1978). *Scanning Electron Microsc.* **1**:109–132.

Echlin, P. (1984). *Analysis of Biological and Organic Surfaces.* P. Echlin, ed. New York: Wiley.

Echlin, P. (1987). In *Proc. 45th Ann. Meet. EMSA.* G. W. Bailey, ed. San Francisco: San Francisco Press, p. 658.

Echlin, P. (1988). In *Microbeam Analysis 1988.* D. E. Newbury, ed. San Francisco: San Francisco Press, p. 451.

Echlin, P. (1988). *Inst. Phys. Conf. Ser.* **93**(3):589–590.

Echlin, P. (1989). *Beitr. Tabakforsch.* **14**:297–312.

Echlin, P. (1990). *Proc. 12th Int. Cong. EM. Seattle*, Vol. **3**, pp. 324–325. San Francisco: San Francisco Press.

Echlin, P. (1991). *J. Microsc.* **161**:159–170.

Echlin, P., and Moreton, R. (1973). *Scanning Electron Microsc.* **1**:325–331.

Echlin, P., and Moreton, R. (1974). In *Microprobe Analysis as Applied to Cells and Tissues*. T. A. Hall, P. Echlin, and R. Kaufmann, eds. New York: Academic Press, pp. 159–174.

Echlin, P., and Saubermann, A. J. (1977). *Scanning Electron Microsc.* **1**:621–637.

Echlin, P., and Taylor, S. E. (1986). *J. Microsc.* **141**:329–348.

Echlin, P., and Burgess, A. (1977). *Scanning Electron Microsc.* **1**:491–500.

Echlin, P., Paden, R., Dronzek, B., and Wayte, R. (1970). *Scanning Electron Microsc.* **1**:51–56.

Echlin, P., Saubermann, A. J., and Taylor, P. J. (1975). *Scanning Electron Microsc.* **1**:679–687.

Echlin, P., Skaer, H. LeB., Gardiner, B. O. C., Franks, F., and Asquith, M. H. (1977). *J. Microsc.* **110**:239–255.

Echlin, P., Pawley, J. N., and Hayes, T. L. (1979). *Scanning Electron Microsc.* **3**:69–76.

Echlin, P., Lai, C. E., Hayes, T. L., and Saubermann, A. J. (1980). *Cryoletters* **1**:289–298.

Echlin, P., Lai, C. E., Hayes, T. L., and Hook, G. (1980). *Scanning Electron Microsc.* **2**:383–394.

Echlin, P., Lai, C. E., and Hayes, T. L. (1981). *Scanning Electron Microsc.* **2**:489–498.

Echlin, P., Lai, C. E., and Hayes, T. L. (1982). *J. Microsc.* **126**:285–306.

Echlin, P., Hayes, T. L., and McKoon, M. (1983). In *Proc. 10th Int. Cong. X-Ray Optics and Microanalysis*. Toulouse, p. 121.

Edelmann, L. (1978). *J. Microsc.* **112**:243–248.

Edelmann, L. (1979). *Mikroskopie* **35**:31–36.

Edelmann, L. (1981). *Fresenius Z. Anal. Chem.* **308**:218–220.

Edelmann, L. (1984). *Scanning Electron Microsc.* **2**:875–888.

Edelmann, L. (1986). *Scanning Electron Microsc.* **4**:1337–1356.

Edelmann, L. (1989). *Scanning Microsc.* **3**:241–252.

Edelmann, L. (1991). *J. Microsc.* **161**:217–228.

Edwards, H. H., Mueller, T. J., and Morrison, M. (1986). *J. Electron. Microsc. Tech.* **3**:439–451.

Egerton, R. F. (1980). *Ultramicroscopy* **5**:521–523.

Egerton, R. F. (1982). *J. Microsc.* **126**:95–100.

Egerton, R. F. (1982). *Ultramicroscopy* **10**:297–300.

Egerton, R. F. (1987). *Electron Spectroscopy in the Electron Microscope*. New York: Plenum Press.

Egerton, R. F., and Cheng, S. C. (1987). *Ultramicroscopy* **21**:231–244.

Eisenberg, D., and Kauzmann, W. (1969). *The Structure and Properties of Water*. New York: Oxford University Press.

Elder, H. Y. (1989). In *Techniques in Immunocytochemistry*, Vol. 4, pp. 1–28, G. Bullock, ed. London: Academic Press.

Elder, H. Y., Gray, C. C., Jardine, A. G., Chapman, J. N., and Biddlecombe, W. H. (1982). *J. Microsc.* **126**:45–61.

Elgsaeter, A., Espevik, T., Kopstad, G., Mikkelsen, A., and Stokke, B. T. (1984). In *Science of Biological Specimen Preparation*. J. P. Revel, T. Barnard, and G. H. Haggis, eds. Chicago: SEM Inc., pp. 195–201.

Engelhardt, H., Gaub, H., and Sackmann, E. (1984). *Nature* **307**:378–380.

Englich, S., Korber, Ch., Schwindke, P., and Rau, G. (1986). *Cryoletters* **7**:13–22.

Eranko, O. (1954). *Acta Anat.* **22**:331–336.

Erickson, H. P., and Klug, A. (1971). *Phil. Trans. R. Soc. London B* **261**:105–118.

Erk, I., Delacroix, H., Nicolas, G., Ranck, J.-I., and Lepault, J. (1991). *J. Electron. Microsc. Tech.* **18**:406–410.

Escaig, J. (1984). In *Science of Biological Specimen Preparation*. J. P. Revel, T. Barnard, and G. H. Haggis, eds. Chicago: SEM Inc. pp. 117–122.

Escaig, J., and Nicholas, G. (1976). *C.R. Acad. Sci., Paris* **283**:1245–1252.

Escaig, J., Geraud, G., and Nicolas, G. (1977). *C.R. Acad. Sci. Paris* **85**:689–695.

Escaig, J., Gauche, D., and Nicholas, G. (1980). In *Proc. 7th Eur. Cong. E.M.*, P. Brederoo and W. de Priester, eds. Leiden, Vol. **2**, p. 654.

Espevik, T., and Elasaeter, A. (1984). *J. Microsc.* **134**:203–212.

Eusemann, R., Rose, H., and Dubochet, J. (1982). *J. Microsc.* **128**:239–249.

Evans, C. D., and Diller, K. R. (1982). *Microvascular Res.* **24**:314–325.

Eveling, D. W., and McCall, R. D. (1983). *J. Microsc.* **129**:113–122.

Fahy, G. M., Levy, D. I., and Ali, S. E. (1987). *Cryobiology* **24**:196–213.

Falls, A. H., Davis, H. T., Scriven, L. E., and Talmon, Y. (1982). *Biochim. Biophys. Acta.* **693**:364–378.

Falls, A. H., Wellinghoff, S. T., Talmon, Y., and Thomas, E. L. (1983). *Mater. Sci.* **18**:2752–2764.

Farrant, J., Walter, C. A., Lee, H., Morris, G. J., and Clarke, K. J. (1977). *J. Microsc.* **111**:17–34.

Fassel, T. A., and Hui, S. W. (1988). *J. Microsc.* **149**:37–50.

Favard, P., Lechaire, J., Maillard, M., Favard, N., Djabourov, M., and Leblond, J. (1989). *Biol. Cell* **67**:201–207.

Feder, N., and Sidman, R. L. (1958). *J. Biophys. Biochem. Cytol.* **4**:593–599.

Fernandez-Moran, H. (1957). In *Metabolism of the Nervous System.* D. Richter, ed. Oxford: Pergamon.

Fernandez-Moran, H. (1959). *J. Appl. Phys.* **30**:2038–2043.

Fernandez-Moran, H. (1960). *Ann. N.Y. Acad. Sci.* **85**:689–696.

Fernandez-Moran, H. (1960). *J. Appl. Phys.* **31**:1840–1846.

Fernandez-Moran, H. (1964). *J. R. Microsc. Soc.* **83**:183–195.

Fernandez-Moran, H. (1965). *Proc. Natl. Acad. Sci. USA* **53**:445–451.

Fernandez-Moran, H. (1966). *Proc. Natl. Acad. Sci. USA* **56**:801–808.

Fernandez-Moran, H. (1982). In *Proc. 10th Int. Cong. E.M. Berlin.* Vol. **1**, pp. 751–761.

Fernandez-Moran, H. (1986). *Adv. Electron. Electron Phys. Suppl.* **16**:167–223.

Fetter, R. D., and Costello, M. J. (1986). *J. Microsc.* **141**:277–290.

Finch, G. L., Bastacky, J., Hayes, T. L., and Fisher, G. L. (1987). *J. Microsc.* **147**:193–203.

Fiori, C. E., Leapman, R. D., Swyt, C. R., and Andrews, S. B. (1988). *Ultramicroscopy* **24**:237–250.

Fisher, K. A. (1974). *J. Cell Biol.* **63**:100–108.

Fisher, K. A. (1975). *Science* **190**:983–985.

Fisher, K. A. (1976). *Proc. Natl. Acad. Sci. USA* **73**:173–177.

Fisher, K. A. (1978). In *Proc. 9th Int. Cong. Electron Microsc. Toronto* **3**:521–532.

Fisher, K. A. (1989). *J. Electron. Microsc. Tech.* **13**:355–371.

Fisher, K. A., and Branton, D. (1976). *J. Cell Biol.* **70**:453–461.

Fisher, K. A., Yanagimoto, K. C., Whitefield, S. L., Thompson, R. E., Gustafsson, M. G. L., and Clarke, J. (1990). *Ultramicroscopy* **33**:117–126.

Fisher, K. A., Whitefield, S. L., Thompson, R. E., Yanagimoto, K. C., Gustafsson, M. G. L., and Clarke, J. (1990). *Biochem. Biophys. Acta* **1023**:325–334.

Flenniken, R., Kopf, D., LeFurgey, A., and Ingram, P. (1990). *Proc. 12th Int. Cong. EM Seattle.* Vol. **3**, pp. 320–321.

Fletcher, N. H. (1970). *The Chemical Physics of Ice.* Cambridge: Cambridge University Press.

Fletcher, N. H. (1971). *Rep. Prog. Phys.* **34**:913–994.

Flood, P. R. (1980). *Scanning Electron Microsc.* **1**:183–200.

Andersson Forsman, C., and Pinto da Silva, P. (1988). *J. Cell Sci.* **90**:531–541.

Foster, M. C., and Saubermann, A. J. (1991). *J. Microsc.* **161**:367–374.

Fotino, M., and Giddings, T. H. (1985). *J. Ultrastruct. Res.* **91**:112–126.

Fotino, M. (1986). *J. Microsc.* **143**:283–298.

Frank, H. S., and Wen, W. Y. (1957). *Discuss. Faraday Soc.* **24**:133–140.

Franklin, R. M., and Martin, M. T. (1980). *Stain Technol.* **55**:313–316.

Franks, F. (1977). *Phil. Trans. R. Soc. London* B **278**:89–96.

Franks, F. (1979). In *Characterization of Protein Conformation and Functions*, F. Franks, ed. London: Symposium Press.

Franks, F. (1972–82). *Water: A Comprehensive Treatise.* Vols. 1–7. New York: Plenum Press.

Franks, F. (1980). *Scanning Electron Microsc.* **2**:349–360.

Franks, F. (1982). In *Water: A Comprehensive Treatise.* Vol. 7, pp. 215–338. F. Franks, ed. New York: Plenum Press.

Franks, F. (1983). *Cryobiology* **20**:335–345.

Franks, F. (1983). *Water.* London: The Royal Society of Chemistry.

Franks, F. (1985). *Biophysics and Biochemistry at Low Temperatures.* Cambridge: Cambridge University Press.

Franks, F. (1986). *J. Microsc.* **141**:243–249.

Franks, F. (1990). *Cryoletters* **11**:93–110.

Franks, F. and Mathias, S. (1982). *Biophysics of Water.* Chichester: John Wiley.

Franks, F., and Skaer, H. LeB. (1976). *Nature* **262**:323–325.

Franks, F., Asquith, M. H., Hammond, C. C., Skaer, H. LeB., and Echlin, P. (1977). *J. Microsc.* **110**:223–238.

Frauenfelder, H., Petsko, G. A., and Tsernoglu, D. (1979). *Nature* **280**:558–560.

Frederik, P. M. (1982). *Scanning Electron Microsc.* 1982 **2**:709–721.

Frederik, P. M., and Busing, W. M. (1980). *Proc. 7th Cong. E.M.* P. Brederoo and W. de Priester, eds. Leiden, Vol. 2, pp. 712–713.

Frederik, P. M., and Busing, W. M. (1981). *J. Microsc.* **121**:191–199.

Frederik, P. M., and Busing, W. M. (1981). *J. Microsc.* **122**:217–220.

Frederik, P. M., and Busing, W. M. (1986). *J. Microsc.* **144**:215–221.

Frederik, P. M., Busing, W. M., and Persson, A. (1982). *J. Microsc.* **125**:167–175.

Frederik, P. M., Busing, W. M., and Persson, A. (1984). *Scanning Electron Microsc.* **1**:433–443.

Frederik, P. M., Stuart, M. C. A., Bomans, P. H. M., and Busing, W. M. (1988). *J. Microsc.* **153**:81–92.

Frederik, P. M., Stuart, M. C. A., and Verkleij, A. J. (1989). *Biochim. Biophys. Acta.* **979**:275–278.

Frederik, P. M., Stuart, M. C. A., Bomans, P. H. H., Busing, W. M., Burger, K. N. J., and Verkleij, A. J. (1991). *J. Microsc.* **161**:253–262.

Freeman, R., Leonard, K. R., and Dubochet, J. (1980). In *Proc. 7th Eur. Cong. E.M.* P. Brederoo and W. de Priester, eds. Leiden, Vol. 2, pp. 640–641.

P. Brederloo, and G. Boom, eds. pp. 640–641. Leiden.

Fritz, E. (1989). *Scanning Microsc.* **3**:517–526.

Fritz, E. (1980). *Ber. Dtsch. Bot. Ges.* **93**:109–121.

Fryer, J. R., and Holland, F. (1983). *Ultramicroscopy* **11**:67–70.

Fuchs, W., and Fuchs, H. (1980). *Scanning Electron Microsc.* **2**:371–382.

Fuchs, W., Lindemann, B. (1975). *J. Microsc. Biol. Cell.* **22**:227–238.

Fuchs, W., Brombach, J. D., and Trosch, W. (1978). *J. Microsc.* **112**:63–74.

Fujikawa, S. (1990). *J. Electron. Microsc. Tech.* **39**:80–85.

Fujikawa, S., Suzuki, T., Ishikawa, T., Sakurai, S., and Hasegawa, Y. (1988). *J. Electron Microsc.* **37**:315–322.

Fujikawa, S., Suzuki, T., and Sakura, S. (1990). *Scanning* **12**:99–106.

Fukami, A., and Adachi, K. (1965). *J. Electron Microsc. (Jpn)* **14**:112–118.

Fukami, A., Adachi, K., and Katoh, M. (1972). *J. Electron Microsc. (Jpn)* **21**:99–108.

Gallup, A. L., Greenwood, B., and Wilson, J. F. (1975). *J. Microsc.* **103**:285–287.

Gardner, D. L., O'Conner, P., Oates, K., Bereton, J., and Middleton, J. (1981). *J. Pathol.* **134**:306–312.

George, M. F., Becwal, M. R., and Burke, M. J. (1982). *Cryobiology* **19**:628–639.

Gersh, I. (1932). *Anat. Rec.* **53**:309–337.

Giambattista, B., Slough, C. G., and Bell, L. D. (1987). *Proc. Natl. Acad. Sci.* **84**:4671–4677.

Gibson, J. M., and McDonald, M. L. (1984). *Ultramicroscopy* **12**:219–222.

Gilkey, J. C., and Staehelin, A. (1986). *J. Electron Microsc. Tech.* **3**:177–210.

Giordano, P., Bottiroli, G., Prenna, S., Doglio, S., and Baudini, G. (1978). *J. Microsc.* **112**:274–280.

Glaeser, R. M. (1971). *J. Ultrastruct. Res.* **36**:466–482.

Glaeser, R. M. (1975). In *Physical Aspects of Electron Microscopy and Microbeam Analysis.* B. M. Siegel, and D. R. Bearman, eds. p. 205. New York: Wiley.

Glaeser, R. M. (1979). In *Introduction to Analytical Electron Microscopy.* J. J. Hren, J. I. Goldstein, and D. C. Joy, eds. New York: Plenum, pp. 423–436.

Glaeser, R. M., and Taylor, K. A. (1978). *J. Microsc.* **112**:127–138.

Glaeser, R. M., Zilker, A., Radmacher, M., Gaub, H. E., Hartmann, T., and Baumeister, W. (1991). *J. Microsc.* **161**:21–46.

Glauert, A. M. (1974). In *Practical Methods in Electron Microscopy*, Vol. III. A. M. Glauert, ed. Amsterdam: North-Holland, Part I.

Glauert, A. M., and Young, R. D. (1989). *J. Microsc.* **154**:101–113.

Gold, L. W. (1958). *Can. J. Phys.* **36**:1265–1275.

Goldblith, S. A., Rey, L., and Rothmayer, W. W. (1975). *Freeze Drying and Advanced Food Technology*. London: Academic Press.

Goldstein, J. I., Newbury, D. E., Echlin, P., Joy, D. C., Lyman, C. E., Romig, A. D., Fiori, C., and Lifshin, E. (1992). *Scanning Electron Microscopy and X-Ray Microanalysis: A Text for Biologists, Material Scientists, and Geologists*. 2nd Ed. New York: Plenum Press.

Gorlen, K., Barden, L., DelPriore, J., Fiori, C., Gibson, C., and Leapman, R. (1984). *Rev. Sci. Instrum.* **55**:912–921.

Grant, R. A., Schmid, M. F., Chui, W., Deatheridge, J.F., and Hosoda, J. (1986). *J. Biophys. Soc.* **49**:251–258.

Green, D. M., and Diller, K. R. (1981). In *Frontiers in Engineering in Health Care*. B. A. Cohen, ed. New York: IEEE, p. 105.

Griffiths, G. (1984). In *Science of Biological Specimen Preparation*. J. P. Revel, T. Barnaed, and G. H. Haggis, eds. Chicago: SEM Inc., pp. 153–159.

Griffiths, G., Simons, K., Warren, G., and Tokuyasu, K. T. (1983). *Adv. Enzymol.* **96**:466–485.

Griffiths, G., McDowall, A., Back, R., and Dubochet, J. (1984). *J. Ultrastruct. Res.* **89**:65–78.

Gross, H. (1987). In *Cryotechniques in Biological Electron Microscopy*. R. A. Steinbrecht, and K. Zierold, eds. Berlin: Springer. Chap. 10.

Gross, H., Bas, E., and Moor, H. (1978). *J. Cell Biol.* **76**:712–718.

Gross, H., Muller, T., Wildhaber, I., Winkler, H., and Moor, H. (1984). In *Proc. 42nd Ann. Meet. EMSA*. G. Bailey, ed. San Francisco, San Francisco Press. p. 12–15.

Grout, B. W. W. (1987). In *The Effects of Temperatures on Biological Systems*. B. W. W. Grout, and G. J. Morris, eds. London: Arnold.

Gruijters, W. T. M., and Bullivant, S. (1986). *J. Microsc.* **141**:291–301.

Guggenhein, R., Duggelin, M., Mathys, D., and Grabski, C. (1991). *J. Microsc.* **161**:337–342.

Gullasch, J., and Kaufmann, R. (1974). In *Microprobe Analysis as Applied to Cells and Tissue*. T. A. Hall, P. Echlin, and R. Kaufmann, eds. New York: Academic Press, pp. 175–190.

Gupta, B. F., and Hall, T. A. (1983). *Am. J. Physiol.* **244**:R176–R186.

Gupta, B. F. (1976). In *Perspectives in Experimental Biology*. P. Spencer-Davies, ed. Oxford: Pergamon Press, pp. 25–42.

Gupta, B. F. (1979). In *Microbeam Analysis in Biology*. C. P. Lechene, and R. R. Warner, eds. New York: Academic Press, p. 375.

Gupta, B. F., Hall, T. A., and Moreton, R. B. (1977). In *Transport of Ions and Water in Animals*. B. F. Gupta, R. B. Moreton, J. L. Oschman, and B. J. Wall, eds. London: Academic Press, pp. 83–143.

Gupta, B. F., Berridge, M. J., Hall, T. A., and Moreton, R. B. (1978). *J. Exp. Biol.* **72**:261–264.

Haggis, G. H. (1982). *Scanning Electron Microsc.* **2**:751–763.

Haggis, G. H. (1985). *J. Microsc.* **139**:49–55.

Haggis, G. H. (1986). *J. Microsc.* **143**:275–282.

Haggis, G. H., and Bond, E. F. (1979). *J. Microsc.* **115**:225–234.

Haggis, G. H., and Pawley, J. B. (1988). *J. Microsc.* **150**:211–218.

Haggis, G. H., Schweitzer, I., Hall, R., and Bladon, T. (1983). *J. Microsc.* **132**:185–194.

Hagler, H. K., and Buja, L. M. (1984). In *The Science of Biological Specimen Preparation for Microscopy and Analysis*. J. P. Revel, T. Barnard, and G. H. Haggis, eds. Chicago: SEM Inc., pp. 161–166.

Hall, T. A. (1968). In *Quantitative Electron Probe Analysis*. K. F. J. Heinrich, ed. National Bureau of Standards. USA Special Technical Publication. 298, pp. 269–299.

Hall, T. A. (1971). In *Physical Techniques in Biological Research*. 2nd Ed. Vol. 1A. G. Oster, ed. New York: Academic Press, pp. 157–275.

Hall, T. A. (1975). *J. Microsc. Biol. Cell.* **22**:271–281.

Hall, T. A. (1979). In *Microbeam Analysis in Biology*. C. P. Lechene, and R. R. Warner, eds. New York: Academic Press, pp. 183–203.

Hall, T. A. (1986). *Micron Microsc. Acta* **17**:91–100.

Hall, T. A. (1986). *J. Microsc.* **141**:319–328.

Hall, T. A. (1988). *Ultramicroscopy* **24**:181–184.

Hall, T. A. (1991). *J. Microsc.* **164**:67–80.

Hall, T. A., and Gupta, B. F. (1974). *J. Microsc.* **100**:177–188.

Hall, T. A., and Gupta, B. F. (1979). In *Introduction to Analytical Electron Microscopy.* J. J. Hren, J. I. Goldstein, and D. C. Joy, eds. New York: Plenum Press, pp. 169–177.

Hall, T.A., and Gupta, B. F. (1982). *J. Microsc.* **126**:333–345.

Hall, T. A., and Gupta, B. F. (1983). *Q. Rev. Biophys.* **16**:279–339.

Hall, T. A., and Hohling, H. J. (1969). In *X-Ray Optics and Analysis.* G. Mollenstadt, and H. H. Gaukler, eds. Heidelberg: Springer, p. 582.

Hall, T. A., and Werba, P. (1969). In *Proc. 5th Int. Cong. X-Ray Optics and Microanalysis. Tubingen.* G. Mollenstadt, and K. H. Gaukler, eds. Berlin: Springer, p. 93.

Hall, T. A., Rockert, H. O. E., and Saunders, R. L. (1972). *X-Ray Microscopy in Clinical and Experimental Medicine.* Springfield, IL: Charles Thomas.

Hall, T. A., Anderson, C. H., and Appleton, T. (1973). *J. Microsc.* **99**:177–182.

Halloran, M., and Kirk, R. G. (1979). In *Microbeam Analysis in Biology.* C. P. Lechene, and R. R. Warner, eds. New York: Academic Press, pp. 571–590.

Hansma, P. K., Eilings, V. B., Marti, O., and Bracker, C. E. (1988). *Science* **242**:209–216.

Hanszen, K. J. (1971). In *Advances in Optical and Electron Microscopy.* R. Barer, and V. E. Cosslett, eds. London: Academic Press, Vol. 4, pp. 1–84.

Hanyu, Y., Ichikawa, I., and Matsumoto, G. (1992). *J. Microsc:* in press.

Harmer, J. R. (1953). *J. R. Microsc. Soc.* **73**:128–134.

Hartman, R. E., and Hartman, R. S. (1971). In *Proc. 29th Ann. Meet. EMSA.* C. J. Arceneaux, ed. Baton Rouge, LA: Claitors Publishing, pp. 74–75.

Hartmann, U., Nunner, B., Korber, Ch., and Rau, G. (1991). *Cryobiology* **28**:115–130.

Harvey, D. M. R. (1980). *Scanning Electron Microsc.* **2**:409–421.

Harvey, D. M. R. (1982). *J. Microsc.* **127**:209–221.

Harvey, D. M. R. (1986). *Scanning Electron Microsc.* **2**:953–973.

Harvey, D. M. R., Hall, J. L., and Flowers, T. J. (1976). *J. Microsc.* **107**:189.

Harvey, D. M. R., Stelzer, R., Brandtner, R., and Kramer, D. (1985). *Physiol. Plant.* **66**:328–338.

Hax, W. M. A., and Lichtenegger, S. (1982). *J. Microsc.* **126**:275–283.

Hayat, M. A. (1981). *Fixation for Electron Microscopy.* New York: Academic Press.

Hayat, M. A. (1986). *Basic Techniques for Transmission Electron Microscopy.* New York: Academic Press.

Hayes, A. R., and Stein, A. (1989). *Cryoletters* **10**:257–268.

Hayes, T. L., and Koch, G. (1975). In *Scanning Electron Microscopy.* O. Jahari, ed. Chicago, IL: IITRI Research Institute, pp. 35–42.

Hayward, S. B., and Stroud, R. M. (1981). *J. Mol. Biol.* **151**:491–517.

Hazlewood, C. F., Chang, D. C., Nichols, B. L., and Wolssner, D. E. (1974). *Biophys. J.* **14**:584–606.

Heath, B. I. (1984). *J. Microsc.* **135**:75–82.

Heide, H. G. (1965). *Lab. Invest.* **14**:396–401.

Heide, H. G. (1982). *Ultramicroscopy* **7**:299–300.

Heide, H. G. (1982). *Ultramicroscopy* **10**:125–154.

Heide, H. G. (1984). *Ultramicroscopy* **14**:271–278.

Heide, H. G., and Grund, S. (1974). *J. Ultrastruct. Res.* **18**:259–268.

Heide, H. G., and Urban, K. (1972). *J. Phys. E (Sci. Instrum.)* **5**:803–808.

Heide, H. G., and Zeitler, E. (1985). *Ultramicroscopy* **16**:151–160.

Hemley, R. J., Chen, L. C., and Mao, H. K. (1989). *Nature* **338**:638–640.

Henderson, R. (1990). *Proc. R. Soc. London B* **241**:6–8.

Henderson, R., and Glaeser, R. M. (1985). *Ultramicroscopy* **16**:139–150.

Henderson, R., Vigers, G., and Raeburn, C. (1991). *Ultramicroscopy* **35**:45–54.

Henderson, R., Baldwin, J. M., Ceska, T. A., Zemlin, T., Beckmann, E., and Downing, K. H. (1990). *J. Mol. Biol.* **213**:899–929.

Henglein, A. (1984). *Ultramicroscopy* **14**:195–200.

Herken, R., Fussek, M., and Thies, M. (1988). *Histochemistry* **89**:277–282.

Hermann, R., and Muller, M. (1991). *J. Electron. Microsc. Tech.* **18**:440–449.

Herter, P., Hentschel, H., Zierold, K., and Walter, P. (1991). *J. Microsc.* **161**:375–385.

Heuser, J. E., and Saltpeter, S. R. (1979). *J. Cell Biol.* **82**:150–73.

Heuser, J. E. (1977). In *Proc. 35th Ann. Meet. EMSA.* G. W. Bailey, ed. Baton Rouge, LA: Claitor's Publishing, pp. 676–679.

Heuser, J. E. (1981). *Trends Biochem. Sci.* **6**:64–68.

Heuser, J. E. (1983). *J. Mol. Biol.* **169**:155–195.

Heuser, J. E. (1989). *J. Electron Microsc. Tech.* **13**:244–263.

Heuser, J. E., and Kirschner, M. W. (1980). *J. Cell Biol.* **86**:212–234.

Heuser, J. E., Reese, T. S., Dennis, M. J., Jan, Y., Jan, L., and Evans, L. (1979). *J. Cell Biol.* **81**:275–300.

Heyman, R. V., and Saubermann, A. J. (1987). *J. Electron. Microsc. Tech.* **5**:315–345.

Hippe, S. (1987). *Histochemistry* **87**:309–315.

Hippe, S., During, K., and Kreuzaler, F. (1989). *Eur. J. Cell Biol.* **50**:230–234.

Hirokawa, N., and Heuser, J. E. (1981). *J. Cell Biol.* **91**:399–409.

Hirokawa, N., and Tilney, L. G. (1982). *J. Cell Biol.* **95**:249–261.

Hirsh, A. G., Williams, R. J., and Meryman, H. T. (1985). *Plant Physiol.* **79**:41–56.

Hirsh, A. G. (1987). *Cryobiology* **24**:214–228.

Hobbs, L. W. (1975). In *Surface and Defect Properties of Solids*, Vol. 4. M. W. Roberts, and J. M. Thomas, eds. London: Chem. Soc., pp. 199–200.

Hobbs, L. W. (1979). In *Introduction to Analytical Electron Microscopy.* J. J. Hren, J. I. Goldstein, and D. C. Joy, eds. New York: Plenum Press, pp. 437–480.

Hobbs, L. W. (1984). In *Quantitative Electron Microscopy.* J. N. Chapman, and A. J. Craven, eds., pp. 400–445. Scottish Universities Summer School.

Hobbs, P. V. (1974). *Ice Physics.* Oxford: Clarendon Press.

Hobot, J. (1990). *J. Struct. Biol.* **104**:169–177.

Hobot, J., and Newman, G. R. (1991). In *Scanning Electron Microsc. Supplement No. 5.* G. Roomans, ed. Chicago: SEM Inc.

Hobot, J., Bjornsti, M. A., and Kellenberger, E. (1987). *J. Bacteriol.* **169**:2055–2062.

Hodson, S., and Marshall, J. (1970). *J. Microsc.* **91**:105–113.

Hodson, S., and Marshall, J. (1972). *J. Microsc.* **95**:459–465.

Hodson, S., and Williams, L. (1976). *J. Cell Biol.* **20**:687–698.

Hoerr, N. L. (1936). *Anat. Rec.* **65**:293–317.

Holland, L. (1956). *Vacuum Deposition of Thin Films.* New York: John Wiley.

Holt, D. B., and Saba, F. H. (1985). *Scanning Electron Microsc.* **3**:1023–1045.

Homo, J. C., Booy, F., Labouesse, P., Lepault, J., and Dubochet, J. (1984). *J. Microsc.* **136**:337–340.

Honjo, G., Kitamura, N., Shimaoka, K., and Mihama, K. (1956). *J. Phys. Soc. Jpn* **11**:527–538.

Hook, G. R., Bastacky, J., Conhaim, R. L., Staub, N. C., and Hayes, T. L. (1987). *Scanning* **9**:71–79.

Hope, M. J., Wong, K. F., and Cullis, P. R. (1989). *J. Electron. Microsc. Tech.* **13**:277–287.

Hopkins, D. M., Jackson, A. D., and Oates, K. (1991). *J. Electron. Microsc. Tech.* **18**:176–182.

Horowitz, R. A., Giannasca, P. J., and Woodcock, C. L. (1990). *J. Microsc.* **157**:205–224.

Howard, R. C., and Sheffield, E. (1987). *Paper Technol. Ind.* **Mar-1987**:425–427.

Hubel, A. (1985). Electrical transients produced during the freezing of NaCl–water solutions. Master's Thesis, Dept. Mech. Eng. MIT Cambridge, MA.

Huebener, R. P. (1988). *Adv. Electron. Electron Phys.* **70**:1–78.

Huebener, R. B., and Seifert, H. (1984). *Scanning Electron Microsc.* **3**:1053–1063.

Humbel, B., and Muller, M. (1988). In *The Science of Biological Specimen Preparation.* M. Muller, R. P. Becker, A. Boyde, and J. L. Wolosewick, eds. Chicago: SEM Inc., pp. 175–183.

Humbel, B., Marti, Th., Muller, M. (1983). *Beitr. Elektronenmikrosk. Direktabbild. Oberflächen* **16**:585–594.

Humbel, B. M., Weber, W., and Merkl, R. (1991). *J. Electron. Microsc. Tech.* **17**:450–455.

Humphreys, S. H. (1989). *J. Electron. Microsc. Tech.* **13**:300–308.

Hunt, C. J. (1984). In *Science of Biological Sample Preparation.* J. P. Revel, T. Barnard, and G. Haggis, eds. Chicago: SEM Inc., pp. 123–130.

Hunt, C. J., Taylor, M. J., and Pegg, D. E. (1982). *J. Microsc.* **125**:177–186.

Hunziker, E. B., Herrmann, W., Schenk, R. K., Mueller, M., and Moor, H. (1984). *J. Cell Biol.* **98**:267–276.

Hutchinson, T. E., Bacaner, M., Broadhurst, J., and Lilley, J. (1974). In *Microprobe Analysis Applied to Cells and Tissues.* T. A. Hall, P. Echlin, and R. Kaufman, eds. New York: Academic.

Hutchinson, T. E., Johnson, D. E., and MacKenzie, A. P. (1978). *Ultramicroscopy* 3:315–324.

Ichikawa, M., Sasaki, K., and Ichikawa, A. (1989). *J. Electron. Microsc. Tech.* 12:88–94.

Ingram, M. J., and Hogben, C. A. (1967). *Anal. Biochem.* 18:54–57.

Ingram, M. J., and Ingram, F. D. (1983). *Scanning Electron Microsc.* 3:1249–1254.

Ingram, F. D., and Ingram, M. J. (1984). In *Science of Biological Specimen Preparation.* J. P. Revel, T. Barnard, and G. H. Haggis, eds. Chicago: SEM Inc., pp. 167–174.

Ingram, P., LeFurgey, A., Davilla, S., Lamvik, M., Kopf, D., Mandel, D., and Lieberman, M. (1987). In *Analytical Electron Microscopy—1987.* D. Joy, ed. San Francisco: San Francisco Press, pp. 179–183.

Ingram, P., LeFurgey, A., Davilla, S., Sommer, J. R., Mandel, L. F., Liebermann, M., and Herlong, J. R. (1989). In *Microbeam Analysis 1989.* P. E. Russell, ed. San Francisco: San Francisco Press.

Inoue, T. (1986). In *Science of Biological Specimen Preparation.* M. Muller, R. P. Becker, A. Boyde, and J. J. Wolosewick, eds. Chicago: SEM Inc., pp. 245–256.

Inoue, T., and Koike, H. (1989). *J. Microsc.* 156:137–147.

Inoue, T., and Osatake, H. (1984). *J. Electron Microsc.* 33:356–362.

International Experimental Group (1986). *J. Microsc.* 141:385–391.

Isaacson, M. S. (1977). In *Principles and Technique of Electron Microscopy: Biological Applications,* Vol. 6. M. A. Hayat, ed. New York: Van Nostrand.

Isaacson, M. S. (1989). *Ultramicroscopy* 28:320–323.

Israelachvili, J. N. (1985). *Chem. Scr.* 25:7–14.

Iwasaki, K., Toriyama, K., Nunome, M., Fukaya, M., and Muto, H. (1977). *J. Phys. Chem.* 81:1410–1417.

Iwatsuki, M. (1987). *JEOL News* 25:10–15.

Jaffe, J. S., and Glaeser, R. M. (1984). *Ultramicroscopy* 13:373–378.

Jaffe, J. S., and Glaeser, R. M. (1987). *Ultramicroscopy* 23:17–21.

Jakubowski, U. (1985). *Ultramicroscopy* 17:379–382.

Jakubowski, U., and Mende, M. (1991). *J. Microsc.* 161:241–252.

Jakubowski, U., Baumeister, W., and Glaeser, R. M. (1989). *Ultramicroscopy* 31:351–356.

Jap, B.K., Downing, K. H., and Walian, P. J. (1990). *J. Struct. Biol.* 103:57–63.

Jeffree, C. E., Read, N. D., Smith, J. A. C., and Dale, J. E. (1987). *Planta* 172:20–37.

Jeffries, C. J. (1988). *Cryoletters* 9:274–285.

Jeng, T. W., and Chiu, W. (1984). *J. Microsc.* 136:35–44.

Jeng, T. W., and Chiu, W. (1987). *Ultramicroscopy* 23:61–66.

Jeng, T. W., Talmon, Y., and Chiu, W. (1988). *J. Electron. Microsc. Tech.* 8:343–348.

Jenkins, K. A., Immediato, M. J., and Heidel, D. F. (1990). *Scanning* 12:34–40.

Jesior, J. C. (1985). *J. Ultrastruct. Res.* 90:135–144.

Jesior, J. C. (1986). *J. Ultrastruct. Res.* 95:210–217.

Johari, G. P., Hallbrucker, A., and Mayer, E. (1987). *Nature* 330:552–553.

Johnson, D., Izutsu, K., Cantino, M., and Wong, J. (1988). *Ultramicroscopy* 24:221–236.

Jones, G. J. (1984). *J. Microsc.* 136:349–360.

Jones, D., McHardy, W. J., and Alexander, C. (1987). *Scanning Microsc.* 1:1423–1429.

Joy, D. C. (ed.) (1985). *J. Microsc.* 140:282–349.

Joy, D. C. (1991). *J. Microsc.* 161:343–356.

Joy, D. C., Joy, C. S., and Armstrong, D. A. (1989). In *Electron Probe Microanalysis: Applications in Biology and Medicine.* K. Zierold, and H. Hagler, eds. Heidelberg: Springer, pp. 129–138.

Kaeser, W. (1989). *J. Microsc.* 154:273–278.

Kalab, M. (1981). *Scanning Electron Microsc.* 3:453–472.

Kalab, M., and Modler, H. W. (1985). *Food Microstruct.* 4:89–98.

Kamiya, N., and Kuroda, K. (1957). *Proc. Jpn Acad.* 33:403–413.

Kan, F. W. K., and Nancy, A. (1988). *J. Electron. Microsc. Tech.* 8:363–370.

Kandasamy, M. K., Parthasarathy, M. V., and Nasrallah, M. E. (1991). *Protoplasma* 162:187–191.

Kann, M. L., and Fouquet, J. P. (1989). *Histochemie* 91:221–226.

Karp, R. D., Silcox, J. C., and Somlyo, A. V. (1982). *J. Microsc.* **125**:157–165.

Katoh, M., and Matsumoto, G. (1980). *J. Electron Microsc.* **29**:197–198.

Kaufmann, R. (1982). In *Microbeam Analysis*. K. F. J. Heinrich, ed. San Francisco: San Francisco Press, pp. 341–358.

Kellenberger, E. (1987). In *Cryotechniques in Biological Electron Microscopy*. R. A. Steinbrecht, and K. Zierold, ed. Berlin: Springer-Verlag, pp. 35–63.

Kellenberger, E. (1991). *J. Microsc.* **161**:1183–204.

Kellenberger, E., and Chiu, W. (1982). *Ultramicroscopy* **10**:165–178.

Kellenberger, E., and Kistler, J. N. (1979). In *Advances in Structure Research, Unconventional Electron Microscopy for Molecular Structure Determination*. W. Hoppe, and R. Mason, eds. Vol. III, pp. 49–79. Braunschweig: Vieweg Verlag.

Kellenberger, E., Carlemalm, E., Villiger, W., Roth, J., and Garavito, R. M. (1980). *Low Denaturation Embedding for Electron Microscopy of Thin Sections*. Chemische Werke Lowi G.m.bH. Waldkraiburg.

Kellenberger, E., Haner, M., and Wurtz, M. (1982). *Ultramicroscopy* **9**:139–150.

Kellenberger, E., Carlemalm, E., and Villiger, W. (1986). In *The Science of Biological Specimen Preparation*. M. Muller, R. P. Becker, A. Boyde, and J. L. Wolosewick, eds. Chicago: SEM Inc., pp. 1–20.

Kirk, R. G., Knoff, L., and Lee, P. (1991). *J. Microsc.* **161**:445–453.

Kiss, J. Z., Giddings, T. H., Staehlin, L. A., and Sack, F. D. (1990). *Protoplasma* **157**:64–74.

Kistler, J., and Kellenberger, E. (1977). *J. Ultrastruct. Res.* **59**:70–82.

Klein, M. L., and Venables, J. A. (eds.) (1976). *Rare Gas Solids*, Vol.s I and II. London: Academic Press.

Kleinschmidt, A., Lang, D., and Zahn, R. K. (1960). *Naturwissenschaften* **47**:16–22.

Knapek, E., and Dubochet, J. (1980). *J. Mol. Biol.* **141**:147–161.

Knauf, M., and Mendgen, K. (1988). *Biol. Cell* **64**:363–370.

Knoll, G., Verkleij, A. J., and Plattner, H. (1987). In *Cryotechniques in Biological Electron Microscopy*. R. A. Steinbrecht, and K. Zierold, eds. Berlin: Springer-Verlag, pp. 258–271.

Knoll, G., Braun, C., and Plattner, H. (1991). *J. Cell Biol.* **113**:1295–1304.

Knox, J. M., Schwartz, G. S., and Diller, K. R. (1980). *J. Biochem. Eng.* **102**:91–97.

Kochs, M., Schwindke, P., and Korber, C. (1989). *Cryoletters* **10**:401–420.

Kochs, M., Korber, Ch., Nunner, B., and Heschel, I. (1991). *J. Heat Mass Trans*: in press.

Kohl, H., Rose, H., and Schnabl, H. (1981). *Optik* **58**:11–24.

Kopstad, G., and Elgsaeter, A. (1982). *Biophys. J.* **40**:163–170.

Korber, Ch. (1981). Das gefrieren wassriger Losungen in biologischer Substanzen. Doct. Thesis. Math-Naturwiss Fakult. RWTH Aachen, Germany.

Korber, Ch. (1988). *Q. Rev. Biophys.* **21**:229–298.

Korber, Ch., Scheiwe, M. W., and Wollhover, K. (1983). *Int. J. Heat Mass Transfer* **26**:1241–1253.

Korber, Ch., Schiewe, M. W., Woolhover, K. (1984). *Cryobiology* **21**:307–316.

Korber, Ch., Rau, G., Cosman, M. D., and Cravalho, E. G. (1985). *J. Cryst. Growth.* **63**:649–662.

Korber, Ch., Englich, S., Schwindke, P., Schiewe, M. W., Rau, G., Hubel, A., and Cravalho, E. G. (1986). *J. Microsc.* **141**:263–276.

Korber, Ch., Englich, S., and Rau, G. (1991). *J. Microsc.* **161**:313–326.

Kouchi, A. and Kuroda, T. (1990). *Nature* **334**:134–135.

Kouchi, A. (1989). *Nature* **330**:550–551.

Kourosh, S., and Diller, K. R. (1984). *J. Microsc.* **135**:39–48.

Koyro, H. W., and Stelzer, R. (1988). *J. Plant. Physiol.* **133**:441–446.

Kramers, H. A. (1923). *Phil. Mag.* **46**:836–871.

Kretzer, F. (1973). *J. Ultrastruct. Res.* **44**:146–178.

Krog, J. O., Zachariassen, K. E., Larsen, B., and Smidsrod, O. (1979). *Nature* **282**:300–301.

Kuhs, W. F., and Lehmann, M. S. (1986). In *Water Science Reviews*. F. Franks, ed. Cambridge: Cambridge University Press, Vol. 2, pp. 1–65.

Kunisch, H. (1880). Inaugural Dissertation. Univ. of Breslau.

Kusamichi, M., Monodane, T., Tokunaga, M., and Koike, H. (1990). *J. Electron Microsc.* **39**:477–486.

Lamvik, M. K. (1990). In *Proc. 12th Int. Cong. EM. Seattle*, Vol. 2, pp. 404–405. San Francisco: San Francisco Press. Vol. 2, pp. 404–405.

Lamvik, M. K. (1991). *J. Microsc.* **161**:171–182.

Lamvik, M. L., and Davilla, S. D. (1988). *J. Electron Microsc. Tech.* **8**:349–354.

Lamvik, M. K., Kopf, D. A., and Robertson, J. D. (1983). *Nature* **301**:332–334.

Lamvik, M. K., Kopf, D. A., and Davilla, S. D. (1987). *J. Microsc.* **148**:211–217.

Lamvik, M. K., Davilla, S. D., and Klatt, L. L. (1989). *Ultramicroscopy* **27**:241–250.

Lancelle, S. A., Callaham, D. A., and Hepler, P. K. (1986). *Protoplasma* **131**:153–165.

Lang, E. W., and Ludermann, H. D. (1980). *Ber. Bunsenges. Phys. Chem.* **84**:462–470.

Langanger, G., and De May, J. (1988). *J. Electron. Microsc. Tech.* **8**:391–399.

Langen, R., Poppe, C. H., Schramm, H. J., and Hoppe, W. (1975). *J. Mol. Biol.* **93**:159–171.

Langmore, J. P., and Athey, B. D. (1987). In *Proc. 45th Ann. Meet. EMSA*. G. W. Bailey, ed. San Francisco: San Francisco Press, pp. 652–653.

Lauchli, A., Spurr, A. R., and Wittkopp, R. W. (1970). *Planta* **95**:341–352.

Lawton, D. M., Gardner, D. L., Oates, K., and Middleton, J. F. S. (1989). *J. Electron. Microsc. Tech.* **11**:90–91.

Lazoff, D., and Lauchli, A. (1991a). *Planta* **184**:327–333.

Lazoff, D., and Lauchli, A. (1991b). *Planta* **184**:334–342.

Leapman, R. D., and Andrews, S. B. (1991). *J. Microsc.* **161**:3–20.

Leapman, R. D., and Andrews, S. B. (1992). *J. Microsc.* in press.

Leapman, R. D., and Ornberg, R. L. (1987). In *Proc. 45th Ann. Meet. EMSA*. G. W. Bailey, ed. San Francisco: San Francisco Press, pp. 660–661.

Leapman, R. D., and Ornberg, R. L. (1988). *Ultramicroscopy* **24**:251–268.

Leapman, R. D., and Swyt, C. R. (1989). In *Microbeam Analysis—1989*. P. E. Russell, ed. San Francisco: San Francisco Press, pp. 89–93.

Leapman, R. D., Fiori, C. E., and Swyt, C. R. (1984). *J. Microsc.* **133**:239–253.

Lechene, C. P. (1970). In *Proc. 5th Conf. Microbeam Analysis Soc.* P. Lublin, ed. San Francisco: San Francisco Press, pp. 32–34.

Lechene, C., Strunk, T., and Warner, R. (1975). *Microbeam Anal.* **1975**:49A–49E.

Lechner, G. (1974). *Proc. 8th Intl. E.M. Congress. Canberra.* Vol. 3, p. 58.

Lee, R. M. K. W. (1984). In *Science of Biological Specimen Preparation.* J. P. Revel, T. Barnard, and G. H. Haggis, eds. Chicago: SEM Inc., pp. 61–70.

Leforstier, A., and Livolant, F. (1991). *Biol. Cell.* **71**:115–122.

Lefranc, G., Knapek, E., and Dietrich, I. (1982). *Ultramicroscopy* **10**:111–124.

LeFurgey, A., Bond, M., and Ingram, P. (1988). *Ultramicroscopy* **24**:185–220.

LeFurgey, A., Davilla, S., Kopf, D. A., Sommer, J. R., and Ingram, P. (1992). *J. Microsc*: in press.

LeFurgey, A., Mandel, L. J., and Ingram, P. (1989). In *Microprobe Analysis—1989*. P. Russell, ed. San Francisco: San Francisco Press, pp. 69–72.

Lehmann, H., Kramer, A., and Schulz, D. (1990). *Ultramicroscopy* **32**:26–34.

Leibo, S. P., McGrath, J. J., and Cravalho, E. G. (1978). *Cryobiology* **15**:1241–1253.

Leigh, R. A., Chater, M., Storey, R., and Johnston, A. E. (1986). *Plant Cell Env.* **9**:595–604.

Leigh, R. A., Ahmad, N., and Wyn Jones, R. G. (1981). *Planta* **153**:34–41.

Leisegang, S. (1954). In *Proc. 3rd Int. Conf. E.M. London*, pp. 175–188.

Lepault, J. (1985). *J. Microsc.* **140**:73–80.

Lepault, J., and Dubochet, J. (1980). *J. Ultrastruct. Res.* **72**:223–236.

Lepault, J., and Pitt, T. (1984). *EMBO J.* **3**:101–105.

Lepault, J., Booy, F. P., and Dubochet, J. (1983). *J. Microsc.* **129**:89–102.

Lepault, J., Freeman, R., and Dubochet, J. (1983). *J. Microsc.* **132**:RP3–RP4.

Lepault, J., Erk, I., Nicolas, G., and Ranck, J.-L. (1991). *J. Microsc.* **161**:47–56.

Leunissen, J. L. M., and Verkleij, A. J. (1989). In *Immunogold Labeling in Cell Biology.* A. J. Verkleij, and J. L. M. Leunissen, eds. Boca Raton, FL: CRC Press, pp. 95–114.

Leunissen, J. L. M., Pluygers, H. W., and Elbers, P. F. (1979). *J. Microsc.* **117**:355–364.

Leunissen, J. L. M., Elbers, P. F., Leunissen-Bijvelt, J. J. M., and Verkleij, A. J. (1984). *Ultramicroscopy* **12**:345–352.

Lichtenegger, S., and Hax, W. M. A. (1980). In *Proc. 7th Eur. Cong. E.M.* P. Brederoo, and W. de Priester, eds. Leiden, pp. 652.

Lickfield, K. G. (1985). *J. Ultrastruct. Res.* **93**:101–115.

Liebhafsky, H. A., Pfeiffer, H. G., and Zemany, P. D. (1960). In *X-Ray Microscopy and Microanalysis*. A. Engstrom, V. E. Cosslett, and H. Pattee, eds. Amsterdam: Elsevier/North-Holland, p. 321.

Linders, P. W. J., Stols, A. L. H., Van de Vorstenbosch, R. A., and Stadhouders, A. M. (1982). *Scanning Electron Microsc.* **4**:1603–1615.

Lindroth, M., Fredriksson, B. A., and Bell, P. B. (1991). *J. Microsc.* **161**:229–240.

Ling, G. N. (1970). *Int. J. Neurosci.* **1**:129–152.

Ling, G. N. (1984). In *Search of the Physical Basis of Life*. New York: Plenum Press.

Ling, G. N., and Negendank, W. (1980). *Persp. Biol. Med.* **23**:215–239.

Ling, G. N., Ochsenfeld, M. M., Walton, C., and Bersinger, T. J. (1980). *Physiol. Chem. and Phys.* **12**:3–10.

Linner, J. G., Bennett, S. C., Harrison, D. S., and Steiner, A. L. (1986). In *Science of Biological Specimen Preparation*. M. Muller, R. P. Becker, A. Boyde, and J. J. Wolosewick, eds. Chicago: SEM Inc., pp. 165–174.

Linner, J. G., Livesey, S. A., Harrison, D. S., and Steiner, A. L. (1986). *J. Histochem. Cytochem.* **34**:1123–1135.

Lipp, G., Korber, Ch., Englich, S., Hartmann, U., and Rau, G. (1987). *Cryobiology* **24**:489–503.

Livesey, S. A., and Linner, J. G. (1987). *Nature* **327**:255–256.

Livesey, S. S., del Campe, A. A., McDowall, A. W., and Stasny, J. T. (1991). *J. Microsc.* **161**:123–456.

Loesser, K. E., and Franzini-Armstrong, C. (1990). *J. Struct. Biol.* **103**:48–56.

Lopachin, R. M., Lowery, J., Eichberg, J., Kirkpatrick, J. B., Cartwright, J., and Saubermann, A. J. (1988). *J. Neurochem.* **51**:764–775.

Lopachin, R. M., Lopachin, V. R., and Saubermann, A. J. (1990). *J. Neurochem.* **54**:320–332.

Lubbock, R., Gupta, B. F., and Hall, T. A. (1981). *Proc. Natl. Acad. Sci. USA* **78**:3624–3628.

Lucken, U. (1990). In *Proc. 12th Int. Cong. EM. Seattle*, Vol. 1, pp. 508–509.

Lucy, J. A. (1970). *Nature* **227**:814–817.

Lupu, F., and Constantinescu, E. (1989). *J. Electron. Microsc. Tech.* **11**:76–82.

Luquet, M. P., Cochet, N., Bouabdillah, D., Pulvin, S., and Clausse. (1991). *Cryoletters* **12**:191–196.

Lustyik, G., and Zs-Nagy, I. (1985). *Scanning Electron Microsc.* **1**:323–337.

Lustyik, G., and Zs-Nagy, I. (1988). *Scanning Microsc.* **2**:289–299.

Luyet, B. J. (1965). *Ann. N.Y. Acad. Sci.* **125**:502–521.

Luyet, B. H., and Gibbs, M. C. (1937). *Biodynamica* **25**:1–18.

Luyet, B. J., and Rasmussen, D. (1968). *Biodynamica* **10**:167–191.

Lyman, C. E., Newbury, D. E., Goldstein, J. I., Williams, D. B., Romig, A. D., Armstrong, J. T., Echlin, P., Fiori, C. E., Joy, D. C., Lifshin, E. (1990). *Scanning Electron Microscopy, X-Ray Microanalysis and Analytical Electron Microscopy: A Laboratory Workbook*. New York: Plenum Press.

Lyon, R., Appleton, K., Swindin, J. J., Abott, J. J., and Chesters, J. (1985). *J. Microsc.* **140**:81–91.

MacFarlane, D. R. (1987). *Cryobiology* **24**:181–195.

MacFarlane, D. R., and Forsyth, M. (1990). *Cryobiology* **27**:345–358.

MacKenzie, A. P. (1965). *Ann. N.Y. Acad. Sci.* **125**:522–530.

MacKenzie, A. P. (1972). *Scanning Electron Microsc.* **2**:273–280.

MacKenzie, A. P., and Rasmussen, D. H. (1972). In *Water Structure at the Water–Polymer Interface*. H. H. G. Jellink, ed. New York: Plenum Press, pp. 146–172.

MacKenzie, A. P., Kuster, T. A., and Luyet, B. J. (1975). *Cryobiology* **12**:427–439.

MacKenzie, A. P. (1977). *Phil. Trans. R. Soc. B* **278**:167–188.

Maisell, L. I., and Glang, R. (eds.) (1970). *Handbook of Thin Film Technology*. New York: McGraw Hill.

Makita, T., Hatsuoka, M., Watanabe, J., Sasaki, K., and Kiwaki, S. (1980). *Cryoletters* **1**:438–444.

Malone, M., Leigh, R. A., and Tomos, A. D. (1991). *J. Exp. Bot.* **42**:305–309.

Mandelkow, E. M., Rapp, R., and Mandelkow, E. (1986). *J. Microsc.* **141**:361–373.

Mandelkow, E.-V., Mandelkow, E., and Milligan, R. (1991). *J. Cell Biol.* **114**:977–991.

Margaritis, L. H., Elgsaeter, A., and Branton, D. (1977). *J. Cell Biol.* **72**:47–56.

Marshall, A. T. (1975). *Micron* **5**:272–280.

Marshall, A. T. (1977). *Micron* **8**:193–200.

Marshall, A. T. (1977). *Microsc. Acta* **79**:254–266.

Marshall, A. T. (1980). *Scanning Electron Microsc.* **II**:395–408.

Marshall, A. T. (1980). In *X-Ray Microanalysis in Biology*. M. A. Hayat, ed. Baltimore: University Park Press, pp. 207-240.

Marshall, A. T. (1980). *SEM-1980*, Vol. 2, pp. 335-348.

Marshall, A. T. (1980). In *X-Ray Microanalysis in Biology*. M. A. Hayat, ed. Baltimore: University Park Press, pp. 167-205.

Marshall, A. T. (1981). *Scanning Electron Microsc.* **2**:327:343.

Marshall, A. T. (1982). *Scanning Electron Microsc.* **1**:243-260.

Marshall, A. T. (1982). *Micron* **13**:315-316.

Marshall, A. T. (1983). *Cell Tissue Res.* **23**:215-277.

Marshall, A. T. (1984). *Scanning Electron Microsc.* **2**:493-504.

Marshall, A. T. (1987). In *Cryotechniques in Biological Electron Microscopy*. R. A. Steinbrecht, and K. Zierold, eds. Berlin: Springer, Chap. 13.

Marshall, A. T. (1988). *J. Electron Microsc. Tech.* **9**:57-64.

Marshall, A. T., and Carde, D. (1983). *J. Microsc.* **134**:113-116.

Marshall, A. T., and Condron, R. J. (1985). *J. Microsc.* **140**:109-118.

Marshall, A. T., and Condron, R. J. (1985). *J. Microsc.* **140**:99-108.

Marshall, A. T., and Condron, R. J. (1987). *Micron. Microsc. Acta* **18**:23-26.

Marshall, A. T., and Kent, M. (1991). *J. Microsc.* **162**:123-456.

Marshall, A. T., and Wright, A. (1972). *Micron* **4**:31-45.

Marshall, A. T., and Wright, O. P. (1991). *J. Microsc.* **162**:341-354.

Marshall, A. T., Carde, D., and Kent, M. J. (1982). *Micron* **13**:313-314.

Marshall, A. T., Carde, D., and Kent, M. (1985). *J. Microsc.* **139**:335-337.

Marshall, A. T., Hyatt, A. D., Phillips, J. G., and Condron, R. J. (1985). *J. Comp. Physiol. B* **156**:213-227.

Martinez, L. B., and Wick, S. (1991). *J. Electron. Microsc. Tech.* **18**:305-314.

Maruyama, K., and Okuda, M. (1982). *J. Electron Microsc.* **31**:253-256.

Maruyama, K., and Okuda, M. (1985). *J. Microsc.* **139**:265-274.

Mason, C. W., and Rochow, T. G. (1934). *Indust. Eng. Chem.* **6**:367-375.

Mason, B. J. (1957). *Physics of Clouds*, (1st ed.), p. 481. Oxford: Clarendon Press.

Mathias, S. F., Franks, F., and Trafford, K. (1984). *Cryobiology* **21**:123-132.

Matricardi, V. R., Moretz, R. C., and Parsons, D. F. (1972). *Science* **177**:268-270.

Mayer, E. (1985). *J. Appl. Phys.* **58**:663-667.

Mayer, E. (1988). *Cryoletters* **9**:66-77.

Mayer, E. (1985). *J. Microsc.* **140**:3-15.

Mayer, E., and Bruggeller, P. (1982). *Nature* **298**:715-718.

Mayer, E., and Hallbrucker, A. (1987). *Nature* **325**:601-602.

Mazur, P. (1963). *J. Gen. Physiol.* **47**:347-349.

Mazur, P. (1970). *Science* **168**:939-949.

Mazur, P., Leibo, S. P., and Chu, E. H. Y. (1972). *Exp. Cell Res.* **71**:345-355.

McCaa, C., and Diller, K. R. (1987). *Cryoletters* **8**:168-175.

McCully, M. E., and Canny, M. J. (1985). *J. Microsc.* **139**:27-33.

McDowall, A. W., Hofmann, W., Lepault, J., Adrian, M., and Dubochet, J. (1984). *J. Mol. Biol.* **178**:105-111.

McDowall, A. W., Smith, J. M., and Dubochet, J. (1986). *EMBO J.* **5**:1395-1402.

McDowall, A. W., Chang, J. J., Freeman, R., Lepault, J., Walter, C. A., and Dubochet, J. (1983). *J. Microsc.* **131**:1-9.

McGann, L. (1978). *Cryobiology* **15**:382-390.

McGrath, J. J. (1983). *Cryobiology* **21**:81-92.

McGrath, J. J. (1985). In *Heat Transfer in Medicine and Biology: Analysis and Applications*. Vol. 2. A. Shitze, and R. C. Eberhart, eds. p. 185, New York: Plenum Press.

McGrath, J. J. (1987). In *The Effects of Low Temperatures on Biological Systems*. B. W. W. Grout, and G. J. Morris, eds. London: Edward Arnold, Chap. 6.

McGrath, J. J., Cravalho, E. G., and Huggins, C. E. (1975). *Cryobiology* **12**:540-550.

McIntyre, J. A., Gilula, N. B., and Karnovsky, M. J. (1974). *J. Cell Biol.* **60**:192-203.

McLellan, M. R., Morris, G. J., Coulson, G. E., James, E. R., and Kalinina, L. V. (1984). *Cryobiology* **21**:44–59.

McPhail, G. D., Finn, T., and Isaacson, P. G. (1987). *J. Pathol.* **151**:231–238.

Mellor, J. D. (1978). *Fundamentals of Freeze Drying.* London: Academic Press.

Menco, B. Ph. M. (1984). *Cell Tissue Res.* **235**:225–241.

Menco, B. Ph. M. (1986). *J. Electron Microsc. Tech.* **4**:177–240.

Menold, R., Luttge, G., and Kaiser, W. (1976). *Adv. Colloid Interface Sci.* **5**:281–295.

Mersey, B., and McCully, M. E. (1978). *J. Microsc.* **114**:49–76.

Mersey, B., McCully, M. E., and Fatica, E. (1978). *J. Microsc.* **113**:307–310.

Meryman, H. T., and Williams, R. J. (1982). In *Crop Genetic Resources: The Conservation of Difficult Material.* L. A. Withers and J. T. Williams, eds. Series B 42, Paris: I.U.B.S.

Michel, M., Hillmann, T., and Muller, M. (1991). *J. Microsc.* **163**:3–18.

Michelmore, R., and Franks, F. (1982). *Cryobiology* **19**:163–171.

Mikula, R. J., and Munoz, V. A. (1988). *Proc. 46th Ann. Meet. EMSA*, pp. 938–939.

Miller, D. D., Bellare, J. R., Evans, D. F., Talmon, Y., and Ninham, B. W. (1987). *J. Phys. Chem.* **91**:674–685.

Miller, K. R., Prescott, C. S., Jacobs, T. L., and Lassignal, N. L. (1983). *J. Ultrastruct. Res.* **82**:123–132.

Milligan, R. A., and Flicker, P. F. (1987). *J. Cell Biol.* **105**:29–39.

Milligan, R. A., Brisson, A., and Unwin, P. N. T. (1983). *Ultramicroscopy* **13**:1–10.

Milligan, R. A., Brisson, A., and Unwin, P. N. T. (1984). *Ultramicroscopy* **13**:1–10.

Millner, R., and Cobet, U. (1972). *Phys. Med. Biol.* **17**:736–745.

Mishima, O., Calvert, L. D., and Whalley, E. (1984). *Nature* **310**:393–395.

Misra, M., Beall, H. C., Taylor, K. A., and Ting-Beall, H. P. (1990). *J. Struct. Biol.* **105**:67–74.

Molisch, H. (1897). *Das Erfrieren der Pflanzen.* Jena: Gustav Fischer, 74 pp.

Molisch, H. (Translation by D. G. Stout) (1982). *Cryoletters* **3**:331–390.

Monaghan, P., and Robertson, D. (1990). *J. Microsc.* **158**:355–363.

Monroe, R. G., Gamble, W. J., La Farge, C. G., Gamboa, R., Morgan, C. L. Rosenthal, A., and Bullivant, S. (1968). *J. Ultrastruct. Res.* **22**:22–36.

Moor, H. (1971). *Phil. Trans. R. Soc. London Ser. B* **261**:121–134.

Moor, H. (1973). In *Freeze-Etching Techniques and Applications.* E. L. Benedetti, and P. Favard, eds. Paris: Soc. Franc. de Micros Electron, Chap. 3.

Moor, H. (1987). In *Cryotechniques in Biological Electron Microscopy.* R. A. Steinbrecht, and K. Zierold, eds. Berlin: Springer-Verlag, pp. 176–191.

Moor, H. (1987). In *Cryotechniques in Biological Electron Microscopy.* R. A. Steinbrecht, and K. Zierold, eds. Chap. 8. Berlin: Springer-Verlag.

Moor, H., and Riehle, U. (1968). In *Electron Microscopy 1986*, Vol. 2. *Proc. 4th Eur. Reg. Conf. Electr. Microsc.* S. Bocciarelli, ed. Rome, pp. 33–34.

Moor, H., Muhlethaler, K., Waldner, H., and Frey-Wyssling, A. (1961). *J. Biophys. Biochem. Cytol.* **10**:1–13.

Morel, F., and Roinel, N. (1969). *J. Chim. Phys. Phys. Chim.* **66**:1084–1091.

Moreton, R. B., Echlin, P., Gupta, B. F., Hall, T. A., and Weis-Fogh, T. (1974). *Nature* **247**:113.

Morgan, A. J. (1985). *X-Ray Microanalysis in Electron Microscopy for Biologists.* Royal Microscopical Society Microscopy Handbook 05. Oxford: Oxford University Press.

Morris, G. J. (1981). In *Effects of Low Temperatures on Biological Membranes.* G. J. Morris, and A. Clark, eds. London: Academic Press, pp. 241–262.

Morris, G. J. (1987). In *The Effects of Low Temperatures on Biological Systems.* B. W. W. Grout, and G. J. Morris, eds. London: Edward Arnold, 500 pp.

Morris, C. J., and McGrath, J. J. (1981a). *Cryoletters* **2**:341–352.

Morris, C. J., and McGrath, J. J. (1981b). *Cryobiology,* **18**:390–398.

Morris, G. J., and Watson, P. F. (1984). *Cryoletters* **5**:352–372.

Moss, P. A., Howard, R. C., and Sheffield, E. (1989). *J. Microsc.* **156**:343–351.

Muller, M., Marti, T., and Kriz, S. (1980). In *Proc. 7th Eur. Cong. E.M.*, Vol. 2, P. Brederoo, and W. de Priester, eds. Leiden, pp. 720–721.

Muller, T., Guggenheim, R., Duggelin, M., and Scheidegger, C. (1991). *J. Microsc.* **161**:73–84.

Muller, W. H., Van der Krift, T. P., Knoll, G., Smaal, E. B., and Verkleij, A. J. (1991). *J. Microsc.* **164**:29–42.

Mulvaney, R., Wolff, E. W., and Oates, K. (1988). *Nature* **321**:247–249.

Murase, N., Echlin, P., and Franks, F. (1991). *Cryobiology,* **28**:364–375.

Murata, F., Susuki, S., Tsuyama, S., Imada, M., and Furihata, C. (1985). *Histochem. J.* **17**:967–980.

Murray, J. M. (1986). *J. Ultrastruct. Mol. Res.* **95**:196–209.

Murray, J. M., and Ward, B. (1987). *J. Electron. Microsc. Tech.* **5**:285–290.

Murray, J. M., and Ward, R. (1987). *J. Electron. Microsc. Tech.* **5**:279–284.

Murray, P. W. Le R., Robards, A. W., and Waites, P. R. (1989). *J. Microsc.* **156**:173–182.

Nagele, R. G., Koscuik, M. C., Wang, S. M., Spero, D. A., and Lee, H. (1985). *J. Microsc.* **139**:291–301.

Narten, A. H., Venkatesh, C. G., and Rice, S. A. (1976). *J. Chem. Phys.* **64**:1106–1121.

Negendank, W. (1986). In *Science of Biological Specimen Preparation.* M. Muller, R. P. Becker, A. Boyde, and J. L. Wolosewick, eds. Chicago: SEM Inc., pp. 21–32.

Negendank, W., and Edelmann, L. (eds.) (1989). *The State of Water in the Cell. Scanning Electron Microscopy.* Chicago: AMF O'Hare.

Nemethy, G., and Scheraga, H. A. (1962). *J. Chem. Phys.* **36**:3382–3400.

Nermut, M. V. (1977). In *Principles and Techniques of Electron Microscopy,* Vol. 7. M. A. Hayat, ed. New York: Van Nostrand Reinhold Co., pp. 79–98.

Neugeberger, D. Ch., and Zinsheim, H. P. (1979). *J. Microsc.* **117**:313–321.

Newbury, D. E., Joy, D. C., Echlin, P., Fiori, C. E., and Goldstein, J. I. (1986). *Advanced Scanning Electron Microscopy and X-Ray Microanalysis.* New York: Plenum Press.

Nicholson, W. A. P., Gray, C. C., Chapman, J. N., and Robertson, B. W. (1982). *J. Microsc.* **125**:25–40.

Niedermeyer, W. (1982). *J. Microsc.* **125**:299–308.

Niedermeyer, W., and Wilkie, M. (1982). *J. Microsc.* **126**:259–273.

Niedrig, H. (1978). *Scanning Electron Microsc.* **1**:841–848.

Noel, T. R., Ring, S. G., and Whittam, N. A. (1990). *Trends Food Sci. and Technol.* September: pp. 62–67.

Nordestgaard, B. G., and Rostgaard, J. (1985). *J. Microsc.* **137**:189–207.

O'Toole, E. T., Wray, G. P., Kremer, J. R., and McIntosh, J. R. (1990). In *Proc. 12th Int. Cong. EM. Seattle.* Vol. 1, pp. 500–501.

Oates, K., and Potts, W. T. W. (1985). *Micron Microsc. Acta* **16**:1–4.

Ockenden, I., and Lott, J. N. A. (1990). *Can. J. Bot.* **68**:646–650.

Ornberg, R. L. (1986). In *Science of Biological Specimen Preparation.* M. Muller, R. P. Becker, A. Boyde, and J. L. Wolosewick, eds. Chicago: SEM Inc., pp. 135–139.

Ornberg, R. L., and Reese, T. S. (1981). In *Microprobe Analysis of Biological Systems.* T. E. Hutchinsom, and A. P. Somlyo, eds. New York: Academic Press, pp. 213–228.

Osumi, M., Baba, M., Naito, N., Taki, A., Yamada, N., and Nagatani, T. (1988). *J. Electron Microsc.* **37**:17–30.

Ota, T. (1970). In *7th Int. Cong. E.M. Grenoble,* pp. 145–146.

Packer, K. J. (1977). *Phil. Trans. R. Soc. London Ser. B.* **278**:59–87.

Padron, R., Alamo, L., Craige, R., and Caputo, C. (1988). *J. Microsc.* **151**:81–102.

Panayi, P. N., Cheshire, D. C., and Echlin, P. (1977). *Scanning Electron Microsc.* **1**:463–470.

Parobek, L., and Brown, J. D. (1978). *X-Ray Spectrom.* **7**:26–30.

Parsons, D. F. (1974). *Science* **186**:407–414.

Parsons, D., Belloto, D. J., Schultz, W. W., Buja, M., and Hagler, H. K. (1984). *E.M.S.A. Bulletin* **14**:49–60.

Pauling, L. (1959). In *Hydrogen Bonding,* D. Hadzi, ed. London and New York: Pergamon Press, pp. 1–6.

Pawley, J. B. (1988). *Scanning* **10**:1–36.

Pawley, J. B., and Norton, J. T. (1978). *J. Microsc.* **112**:169–182.

Pawley, J. B., and Ris, H. (1987). *J. Microsc.* **145**:319–332.

Pawley, J. B., Walter, P., Shih, S.-J., and Malecki, M. (1991). *J. Microsc.* **161**:327–336.

Pawley, J. B., Hook, G., Hayes, T. L., and Lai, C. (1980). *Scanning* **3**:219-226.

Pawley, J. B., Walther, P., Shian-Juin Shih, and Malecck, M. (1991). *J. Microsc.* **161**:337-342.

Pearce, R. S. (1988). *Planta* **175**:313-324.

Pearce, R. S., and Beckett, A. (1985). *Planta* **166**:335-340.

Pearce, R. S., and Beckett, A. (1987). *Ann. Bot.* **59**:191-195.

Pearse, A. G. E. (1980). *Histochemistry: Theory and Applied.* Vol. 1, *Preparative and Optical Technology.* Edinburgh and London: Churchill Livingstone, 4th Edition.

Pease, D. C. (1966). *J. Ultrastruct. Res.* **14**:356-367.

Pease, D. C. (1967). *J. Ultrastruct. Res.* **21**:75-86.

Pease, D. C. (1968). In *Proc. 4th Europ. Congr. Electron Microsc. Rome*, Vol. II, p. 11.

Pease, D. C. (1973). In *Advanced Techniques in Biological Electron Microscopy.* J. K. Koehler, ed. Berlin: Springer-Verlag.

Penczek, P., Srivastava, S., and Frank, J. (1990). In *Proc. 12th Intl. Cong. EM Seattle*, Vol. 1, pp. 506-507.

Perlov, G., Talmon, Y., and Falls, A. H. (1983). *Ultramicroscopy* **11**:283-288.

Pesheck, P. S., Scriven, L. E., and Davis, H. T. (1981). *Scanning Electron Microsc.* **1**:515-524.

Peters, K.-R. (1984). *J. Microsc.* **133**:17-28.

Peters, K.-R. (1980). *Scanning Electron Microsc.* **1**:143-154.

Peters, K.-R. (1984). In *Science of Biological Specimen Preparation.* J. P. Revel, T. Barnard, and G. H. Haggis, eds. Chicago: SEM Inc., pp. 221-231.

Peters, K.-R. (1986). In *Advanced Techniques for Electron Microscopy*, Vol. III. J. K. Koehler, ed. Berlin: Springer-Verlag, pp. 101-166.

Petsko, G. A. (1975). *J. Mol. Biol.* **96**:381-392.

Phillips, T. E., and Boyne, A. F. (1984). *J. Electron. Microsc. Tech.* **1**:9-29.

Pihakaski, K., and Seveus, L. (1980). *Cryoletters* **1**:494.

Pinto da Silva, P. (1984). In *Immunolabeling for Electron Microscopy.* J. M. Polak, and I. M. Varnell, eds. New York: Elsevier, pp. 179-188.

Pinto da Silva, P. (1989). In *Immunogold Labeling in Cell Biology.* A. J. Verkleij, and J. L. M. Leunissen, eds. Boca Raton, FL: CRC Press, pp. 179-197.

Pinto da Silva, P., and Kan, F. W. K. (1984). *J. Cell Biol.* **99**:1156-1161.

Pinto da Silva, P., Moss, P. S., and Fudenberg, H. H. (1973). *Exp. Cell Res.* **81**:127-138.

Pinto da Silva, P., Kachar, B., Torrisi, M. R., Brown, C., and Parkinson, C. (1981). *Science* **213**:230-233.

Platt-Aloia, K. A., and Thomson, W. W. (1989). *J. Electron. Microsc. Tech.* **13**:288-299.

Plattner, H., and Bachmann, L. (1982). *Int. Rev. Cytol.* **79**:237-304.

Plattner, H., and Knoll, G. (1984). In *The Science of Biological Specimen Preparation.* J. P. Revel, T. Barnard, and G. H. Haggis, eds. Chicago: SEM Inc., pp. 139-146.

Pogorelov, A., Allachverdov, B., Burovina, I., Mazay, G., and Pogorelov, V. (1991). *J. Microsc.* **161**:123-456.

Pollard, T. D., Maupin, P., Sinard, J., and Huxley, H. E. (1990). *J. Electron. Microsc. Tech.* **16**:160-166.

Pooley, L., and Brown, T. A. (1990). *Proc. R. Soc. Lond.* B **241**:112-115.

Porter, K. R. (1984). *J. Cell. Biol.* **99**:3s-12s.

Potts, W. T. W., and Oates, K. (1983). *J. Exp. Zool.* **227**:349-359.

Prasad, V. V. B., Burns, J. W., Marietta, E., Estes, M. K., and Chiu, W. (1990). *Nature* **343**:467-479.

Preston, J. S., Whitecross, M. I., and Price, G. D. (1982). *Micron* **13**:311-312.

Preston, J. S., Whitecross, M. I., and Price, G. D. (1982). *Micron* **13**:321-322.

Probst, W., Zellmann, E., and Bauer, R. (1989). *Ultramicroscopy* **28**:312-314.

Quamme, G. A. (1988). *Scanning Microsc.* **2**:2195-2205.

Rachel, R., Jakubowski, U., and Baumeister, W. (1986). *J. Microsc.* **141**:179-191.

Rall, W. F., Reid, D. S., and Farrant, L. (1980). *Nature* **286**:511-514.

Rall, W. F., Mazur, P., and McGrath, J. J. (1983). *Biophys. J.* **41**:1-12.

Rash, J. E. (1979). In *Freeze-Fracture: Methods, Artifacts and Interpretation.* J. E. Rash, and C. S. Hudson, eds. p. 153.

Rash, E. R. (1983). *Trends Neurosci.* **6**:208-212.

Rash, J. E., and Hudson, C. S. (1979). *Freeze-Fracture: Methods, Artifacts, and Interpretations*. New York: Raven Press.

Rasmussen, D. H. (1982). *J. Microsc.* **128**:167–174.

Rasmussen, D. H. (1982). *J. Cryst. Growth* **56**:45–55.

Rasmussen, D. H., and MacKenzie, A. P. (1973). *J. Chem. Phys.* **59**:5003–5013.

Read, N. D., and Jeffree, C. E. (1991). *J. Microsc.* **161**:59–72.

Read, N. D., Porter, R., and Beckett, A. (1983). *Can. J. Bot.* **61**:2059–2078.

Read, P. L. (1965). *Vacuum* **13**:271.

Rebhun, L. I. (1961). *J. Biophys. Biochem. Cytol.* **9**:381–392.

Rebhun, L. I. (1972). In *Principles and Techniques of Electron Microscopy*, Vol. 2. M. A. Hayat, ed. New York: Van Nostrand Reinhold, p. 3.

Rebiai, R., Rest, A. J., and Scurlock, R. G. (1983). *Nature* **305**:412–413.

Reed, S. J. B. (1975). *Electron Microprobe Analysis*. Cambridge: Cambridge University Press.

Reichelt, R., and Engel, A. (1984). *Ultramicroscopy* **13**:279–294.

Reichelt, R., and Engel, A. (1985). *J. Microsc. Spectrosc. Electron* **10**:491–498.

Reid, N. (1974). In *Practical Methods in Electron Microscopy*. A. M. Glauert, ed. Amsterdam: North-Holland, p. 3.

Reid, D. S. (1978). *J. Microsc.* **114**:241–248.

Reid, N., and Beesley, J. E. (1991). *Sectioning and Cryosectioning for Electron Microscopy*, Vol. 13, *Practical Methods in Electron Microscopy*. A. M. Glauert, ed. Amsterdam: Elsevier.

Resing, H. A. (1972). *Adv. Mol. Relax. Processes* **3**:199–226.

Richardson, M., McGuffee, L. J., and Hatton, M. W. C. (1988). *J. Ultrastruct. Res.* **98**:199–211.

Richter, K., and Dubochet, J. (1990). In *Proc. 12th Intl. Cong. E.M. Seattle*, Vol. 3, pp. 488–489.

Richter, R., Gnagi, H., and Dubochet, J. (1991). *J. Microsc.* **163**:19–28.

Rick, R., Dorge, A., Arnim, E., and Thurau, K. (1978). *J. Membrane Biol.* **39**:257–268.

Rick, R., Dorge, A., Gehring, K., and Thurau, K. (1979). In *Microbeam Analysis in Biology*. C. P. Lechene, and R.R. Warner, eds. New York: Academic Press, pp. 517–534.

Rick, R., Dorge, A., and Thurau, K. (1982). *J. Microsc.* **125**:239–247.

Ridge, R. W. (1990). *J. Electron Microsc.* **39**:120–124.

Riehle, U. (1968). Über die Vitrifizierung verdunnter wassriger Losungen. Doctoral Thesis, Zurich Juris Druck Verlag.

Riehle, U. (1968). *Chemie Ing. Technol.* **40**:213–218.

Rigler, M. W., and Patton, J. S. (1984). *J. Microsc.* **134**:335–336.

Rix, E. A., Schiller, A., and Taugner, R. (1976). *Histochemistry* **50**:91–104.

Robards, A. W., and Crosby, P. (1979). *Scanning Electron Microsc.* **2**:325–343.

Robards, A. W., and Crosby, P. (1983). *Cryoletters* **4**:23–32.

Robards, A. W., and Umrath, W. (1978). *Proc. 9th Int. Cong. E.M. Toronto*, Vol. 2, pp. 138–139.

Robards, A. W., and Sleytr, U. B. (1985). *Low Temperature Methods in Biological Electron Microscopy*, Vol. 10, *Practical Methods in Electron Microscopy*. A. M. Glauert, ed. Amsterdam: Elsevier.

Robards, A. W., Wilson, A. J., and Crosby, P. (1981). *J. Microsc.* **124**:143–150.

Roberts, I. M. (1975). *J. Microsc.* **103**:113–119.

Roberts, I. M., and Duncan, G. M. (1981). *J. Microsc.* **124**:295.

Roberts, S., Grout, B. W. W., and Morris, G. J. (1987). *Cryoletters* **8**:122–129.

Robinson, R. A., and Stokes, R. H. (1968). *Electrolyte Solutions*, 2nd edition. London: Butterworth.

Robson, D. J., McHardy, W. J., and Petty, J. A. (1988). *J. Exp. Bot.* **39**:1617–1621.

Roinel, N. (1981). *J. Microsc.* **126**:253–258.

Rontgen, W. C. (1892). *Ann. Phys. Chim. (Wien)* **45**:91–54.

Roomans, G. M. (1981). *Scanning Electron Microsc.* **2**:344–356.

Roomans, G. M. (1988). *Scanning Microsc.* **2**:311–317.

Roomans, G. M. (1990). *Scanning Microsc.* **4**:1055–1063.

Roomans, G. M., Wroblewski, J., and Wroblewski, R. (1988). *Scanning Microsc.* **2**:937–946.

Roos, N., and Barnard, T. (1985). *Ultramicroscopy* **17**:335–344.

Roos, N., and Barnard, T. (1986). *Scanning Electron Microsc.* **2**:703–711.

Roos, N., and Morgan, A. J. (1990). *Cryopreparation of Thin Biological Specimens for Electron Microscopy; Methods and Applications.* Royal Microscopical Society Microscopy Handbooks No. 21. Oxford Science Publications.

Rose, H. (1974). *Optik* **39**:416–423.

Rosenbaum, R. L. (1968). *Rev. Sci. Instrum.* **39**:890–899.

Ross, G. D., Morrison, G. H., Sacher, R. F., and Staples, R. C. (1983). *J. Microsc.* **129**:221–228.

Roth, J. (1982). In *Techniques in Immunocytochemistry.* G. R. Bullock, and P. Perutz, eds. London: Academic Press, pp. 107–133.

Roth, J., Bendayan, M., Carlemalm, E., and Villiger, W. (1980). *Experientia* **36**:757–765.

Roth, J., Taajes, D. J., and Tokuyasu, K. T. (1990). *Histochemistry* **95**:123–136.

Roth, L., Bender, M., Carlemalm, E., Villiger, W., and Garavito, M. (1981). *J. Histochem. Cytochem.* **29**:666–671.

Rotmann, B., and Papermaster, B. W. (1966). *Proc. Natl. Acad. Sci. USA* **55**:134–141.

Rowe, R. C., and McMahon, J. (1989). *Microsc. Anal.* **March**:33–35.

Ru-Long, S., and Pinto da Silva, P. (1990). *Eur. J. Cell. Biol.* **53**:122–130.

Ruben, G. C. (1989). *J. Electron. Microsc. Tech.* **13**:335–354.

Ruben, G. C., and Marx, K. A. (1984). In *Proc. 42nd Ann. Meet. EMSA.* G. W. Bailey, ed. San Francisco: San Francisco Press, pp. 684–685.

Rubinsky, B., and Ikeda, M. (1985). *Cryobiology* **22**:55–68.

Rubinsky, B., Lee, C. Y., Bastacky, J., and Onik, G. (1990). *Cryobiology* **27**:85–97.

Rustgi, S. N., Peemoeller, H., Thompson, R. T., Kydon, D. W., and Pintar, M. M. (1978). *Biophys. J.* **22**:439–452.

Ryan, K. P., and Liddicoat, M. I. (1987). *J. Microsc.* **147**:337–340.

Ryan, K. P., and Purse, D. H. (1985). *J. Microsc.* **140**:47–54.

Ryan, K. P., Purse, D. H., and Wood, J. W. (1985). *Mikroskopie* **42**:225–229.

Ryan, K. P., Purse, D. H., Robinson, S. G., and Wood, J. W. (1987). *J. Microsc.* **145**:89–96.

Ryan, K. P., Bateson, J. M., Grout, B. W. W., Purse, D. H., and Wood, J. W. (1988). *Cryoletters* **9**:418–425.

Ryan, K. P., Bald, W. B., Neumann, K., Simonsberger, P., Purse, D. H., and Nicholson, D. N. (1990). *J. Microsc.* **158**:365–378.

Sachs, J. (1892). *The Formation of Crystals During Freezing.* Collected Treatises, Vol. 1. Leipzig.

Saito, K. (1981). *Scanning Electron Microsc.* **3**:553–559.

Sakai, A., and Larcher, W. (1987). *Frost Survival in Plants: Responses and Adaptation to Freezing Stress.* Berlin: Springer-Verlag.

Sakata, S., Hotsumi, S., and Watanabe, H. (1991). *J. Electron Microsc.* **40**:67–69.

Sargeant, P. T., and Roy, R. (1968). *Mat. Res. Bull.* **3**:265–280.

Sargent, J. A. (1983). *J. Microsc.* **129**:103–110.

Sargent, J. A. (1986). *Ann. Bot.* **58**:183–185.

Sargent, J. A. (1988). *Scanning Microsc.* **2**:835–849.

Sass, H. J., Massalski, A., Beckmann, E., Buldt, D., Dorset, D., Van Heel, M., Rosenbusch, J. P., Zeitler, E., and Zemlin, F. (1988). In *Proc. 46th Ann. Meet. EMSA.* G. W. Bailey, ed. San Francisco: San Francisco Press.

Saubermann, A. J. (1980). *Scanning Electron Microsc.* **2**:421–430.

Saubermann, A. J. (1986). *Bull. EMSA.* **16**:65–69.

Saubermann, A. J. (1988). *Scanning* **10**:239–244.

Saubermann, A. J. (1988). *Bull. EMSA* **16**:65–69.

Saubermann, A. J. (1988). *Scanning Microsc.* **2**:2207–2218.

Saubermann, A. J., and Echlin, P. (1975). *J. Microsc.* **105**:155–191.

Saubermann, A. J., and Heyman, R. V. (1985). *Proc. Microbeam Anal. Soc. 1985*, pp. 121–122.

Saubermann, A. J., and Heyman, R. V. (1987). *J. Microsc.* **146**:169–182.

Saubermann, A. J., and Scheid, V. L. (1985). *J. Neurochem.* **44**:825–834.

Saubermann, A. J., Riley, W. D., and Echlin, P. (1977). *Scanning Electron Microsc.* **1**:347–356.

Saubermann, A. J., Echlin, P., Peters, P. D., and Beeuwkes, R. (1981). *J. Cell Biol.* **88**:257–267.

Saubermann, A. J., Beeuwkes, R., and Peters, P. D. (1981). *J. Cell Biol.* **88**:268–278.

Saubermann, A. J., Dobyan, D. C., Scheid, V. L., and Bulger, R. E. (1986). *Kidney Int.* **29**:675–681.

Saubermann, A. J., Scheid, V. L., Dobyan, D. C., and Bulger, R. E. (1986). *Kidney Int.* **29**:682–688.

Saubermann, A. J., Riley, W. D., and Beeuwkes, R. (1977). *J. Microsc.* **111**:39–49.

Sautter, C. (1986). In *Science of Biological Specimen Preparation.* M. Muller, R. P. Becker, A. Boyde, and J. L. Wolosewick, eds. Chicago: SEM Inc., pp. 215–228.

Sawyer, L. C., and Grubb, D. T. (1987). *Polymer Microscopy.* London: Chapman Hall.

Saxton, W. O., and Baumeister, W. (1984). *Ultramicroscopy* **13**:57–70.

Sceats, M. G., and Rice, S. A. (1982). In *Water: A Comprehensive Treatise*, Vol. 7. F. Franks, ed. New York: Plenum Press, pp. 83–214.

Schatz, M., and Van Heel, M. (1990). *Ultramicroscopy* **32**:255–264.

Scheidegger, C., Gunthardt-Goerg, M., Matyssek, R., and Hatvani, P. (1991). *J.Microsc.* **161**:85–96.

Scheiwe, M. W., and Korber, Ch. (1982). *J. Microsc.* **126**:29–44.

Scheiwe, M. W., and Korber, Ch. (1982). *Cryoletters* **3**:265–274.

Scheiwe, M. W., and Korber, Ch. (1982). *Cryoletters* **3**:275–284.

Scheiwe, M. W., and Korber, Ch. (1983). *Cryobiology* **20**:257–273.

Scheiwe, M. W., and Korber, Ch. (1984). *Cryobiology* **21**:93–105.

Scheiwe, M. W., and Korber, Ch. (1987). *Cryobiology* **24**:473–485.

Scherzer, O. (1949). *J. Appl. Phys.* **20**:20–29.

Schiller, A., Rix, E., and Taugner, R. (1978). *Histochemistry* **59**:9.

Schnell, R. C., and Vail, G. (1972). *Nature* **236**:163–165.

Schroder, R. R., Hofmann, W., and Menetret, J.-F. (1990). *J. Struct. Biol.* **105**:28–34.

Schwabe, K. G., and Terracio, L. (1980). *Cryobiology* **17**:571–583.

Schwartz, G. J., and Diller, K. R. (1982). *Cryobiology* **19**:529–538.

Schwartz, G. J., and Diller, K. R. (1983). *Cryobiology* **20**:542–552.

Schweiwe, M. W. (1981). Untersuchungen zum Verfahren der Langzeitkonservierung lebender Blutzellen durch Gefrieren. Dr.-Ing.-Thesis RWTH, Aachen.

Schweiwe, M. W., Korber, Ch., Schwindke, P., Chmiel, H., and Rau, G. (1978). *Chem. Ing. Tech.* **50**:236.

Seifert, H. (1983). *Cryogenics* **22**:657–660.

Severs, N. J. (1989). *J. Electron. Microsc. Tech.* **13**:175–203.

Severs, N. J. (1991). *J. Microscopy* **161**:109–134.

Seveus, L. (1979). The subcellular distribution of electrolytes. A methodological study using cryoultramicrotomy and x-ray microanalysis. Thesis, University of Stockholm, Sweden.

Seveus, L., and Tarras-Wahlberg, C. (1988). In *The Science of Biological Specimen Preparation.* M. Muller, R. P. Becker, A. Boyde, and J. L. Wolosewick, eds. Chicago: SEM Inc., pp. 129–134.

Shabana, M. (1983). Cryomicroscopic investigation and thermodynamic modelling of the freezing of unfertilized hamster ova. M.S. Dissertation, Dept. Mech. Eng. Michigan State Univ.

Shabana, M., and McGrath, J. J. (1988). *Cryobiology* **25**:338–354.

Shah, S. J., Diller, K. R., and Aggarwal, S. J. (1987). *Cryobiology* **24**:163–168.

Sheehan, J. G., and Scriven, L. E. (1988). *Cryo-SEM of Particulate Suspensions.* Proc. 46th Ann. Meeting EMSA. G. Bailey, ed., San Francisco: San Francisco Press, pp. 218–219.

Sheehan, J. G. (1990). In *Proc. 14th Intl. Cong. EM. Seattle*, Vol. 2, pp. 418–419.

Shepard, M. L., Goldston, C. S., and Cocks, F. H. (1976). *Cryobiology* **13**:9–23.

Shiozaki, M., and Shimada, Y. (1990). *J. Electron Microsc.* **39**:18–25.

Shuman, H., Somlyo, A. V., and Somlyo, A. P. (1976). *Ultramicroscopy* **1**:317–339.

Siegel, G. (1972). *Z. Naturforsch. A* **27**:325–332.

Sigee, D. C. (1988). *Scanning Microsc.* **2**:925–935.

Silvester, N. R., Marchese-Ragona, S., and Johnston, D. N. (1982). *J. Microsc.* **128**:175–186.

Simon, G. T., Thomas, J. A., Chorneyko, K. A., and Carlemalm, E. (1987). *J. Electron. Microsc. Tech.* **6**:6317–6324.

Simpson, W. L. (1941). *Anat. Rec.* **80**:173–189.

Singh, J. (1979). *Protoplasma* **98**:329–341.

Sitte, H. (1981). *Git. Labor-medizin* **4**:317–323.

Sitte, H. (1982). *Proc. 10th Int. Cong. E.M. Hamburg*, Vol. 1, p. 9.

Sitte, H. (1984). In *The Science of Biological Specimen Preparation for Microscopy and Analysis.* J. P. Revel, T. Barnard, and G. H. Haggis, eds. Chicago: SEM Inc., pp. 97–107.

Sitte, H. (1984). *Zeiss Inf. MEM Mag. E.M.* **3**:25–31.

Sitte, H., Neumann, K., and Edelmann, L. (1986). In *Science of Biological Specimen Preparation.* M. Muller, R. P. Becker, A. Boyde, and J. L. Wolosewick, eds. Chicago: SEM Inc., pp. 103–118.

Sitte, H., Neumann, K., Hassig, H., and Edelmann, L. (1986). *Cryoprocessing of Biological Specimens. IV. An Accessory for Lowicryl Embedding with UV Light Polymerization to the Reichert-Jung Cryosubstitution Unit CS-Auto Design, Function and Results.* Product Leaflet. Cambridge Instruments, Cambridge.

Sitte, H., Edelmann, L., and Neumann, K. (1987). In *Cryotechniques in Biological Electron Microscopy.* R. A. Steinbrecht, and K. Zierold, eds. Berlin: Springer-Verlag, pp. 88–113.

Sjostrand, F. S. (1956). *Proc. Intl. Conf. E.M. Stockholm,* pp. 120.

Sjostrand, F. S. (1977). *J. Ultrastruct. Res.* **59**:292–319.

Sjostrand, F. S. (1982). *J. Microsc.* **128**:279–286.

Sjostrand, F., and Barajas, L. (1968). *J. Ultrastruct. Res.* **25**:121–155.

Sjostrand, F. S., and Halma, H. A. (1978). *J. Ultrastruct. Res.* **64**:261–269.

Sjostrand, F. S., and Kretzer, F. (1975). *J. Ultrastruct. Res.* **53**:1–17.

Sjostrom, M., Squire, J. M., Luther, P., Morris, E., and Edman, A.-C. (1991). *J. Microsc.* **163**:29–42.

Skaer, H. Le B. (1982). *J. Microsc.* **125**:137.

Skaer, H. Le B., Franks, F., Asquith, M. H., and Echlin, P. (1977). *J. Microsc.* **110**:257–270.

Skaer, H. Le B., Franks, F., and Echlin, P. (1979). *Cryoletters* **1**:61.

Skaer, H. Le B., Franks, F., and Echlin, P. (1982). *Cryobiology* **15**:589–603.

Slayter, H. S. (1980). *Scanning Electron Microsc.* **1**:171–182.

Sleytr, U. B., and Robards, A. W. (1977). *J. Microsc.* **111**:77.

Sleytr, U. B., and Robards, A. W. (1977). *J. Microsc.* **110**:1–26.

Sleytr, U. B., and Robards, A. W. (1982). *J. Microsc.* **126**:101–122.

Sleytr, U. B., and Umrath, W. (1974). *J. Microsc.* **101**:187–199.

Sleytr, U. B., and Umrath, W. (1976). *Proc. 6th Eur. Cong. E.M. Jerusalem,* Vol. 2, pp. 50–51.

Sleytr, U. B., Groesz, H., and Umrath, W. (1981). *Acta Histochem. Suppl.* **23**:29–39.

Small, J. A., Heinrich, K. F. J., Fiori, C. E., Myklebust, R. L., Newbury, D. E., and Dilmore, M. F. (1978). *Scanning Electron Microsc.* **1**:445–458.

Somlyo, A. P., and Shuman, H. (1982). *Ultramicroscopy* **8**:219–234.

Somlyo, A. P., Shuman, H., Somlyo, A. V. (1976). In *Analytical Electron Microscopy.* J. Silcox, ed. Ithaca: Cornell University, pp. 114–117.

Somlyo, A. V., Bond, M., Silcox, J. C., Somlyo, A. P. (1985). In *Proc. 43rd Ann. Meet. Elec. Micros. Soc. Am.* G. W. Bailey, ed. San Francisco: San Francisco Press, pp. 10–13.

Somlyo, A. P., Bond, M., and Somlyo, A. V. (1985). *Nature* **314**:623–625.

Spurr, A. R. (1975). *J. Microsc. Biol. Cell* **22**:287–290.

Stace, T. (1988). *Nature* **331**:116–117.

Stang, E. (1988). *J. Microsc.* **149**:77–79.

Statham, P. (1987). In *Analytical Electron Microscopy—1987.* D. Joy, ed. San Francisco: San Francisco Press, pp. 187–190.

Statham, P. J., and Pawley, J. B. (1978). *Scanning Electron Microsc.* **1**:469–481.

Statham, P. J. (1988). *Scanning* **10**:245–252.

Steere, R. L. (1969). *Cryobiology* **5**:306–323.

Steere, R. L. (1973). In *Freeze-Etching: Techniques and Applications.* E. L. Benedetti, and P. Favard, eds. Paris: Soc. Franc. Micros. Electron, Chap. 18.

Steere, R. L., and Erbe, E. F. (1979). *J. Microsc.* **117**:211.

Steere, R. L., and Moseley, J. M. (1969). *Proc. 27th Ann. Meet. E.M.S.A.,* C. J. Arceneaux, ed. Baton Rouge: Claitors Publishing, pp. 202–203.

Steinbrecht, R. A., and Muller, M. (1987). In *Cryotechniques in Biological Electron Microscopy.* R. A. Steinbrecht, and K. Zierold, eds. Berlin: Springer-Verlag.

Steinbrecht, R. A., and Zierold, K. (1984). *J. Microsc.* **136**:69–75.

Steinbrecht, R. A. (1980). *Tissue Cell* **12**:73–88.

Steinbrecht, R. A. (1982). *J. Microsc.* **125**:187–192.

Steitz, T. A., Ludwig, M. L., Quiocho, F. A., and Lipscomb, W. N. (1967). *J. Biol. Chem.* **242**:4662–4670.

Stenn, K., and Bahr, G. F. (1970). *J. Ultrastruct. Res.* **31**:536–543.

Stephenson, J. L. (1953). *Bull. Math. Biophys.* **15**:411–429.

Stephenson, J. L. (1956). *J. Biophys. Biochem. Cytol.* **2**:45–51.

Steponkus, P. L., and Dowgert, M. F. (1981). *Cryoletters* **2**:42–47.

Stewart, M., and Lepault, J. (1985). *J. Microsc.* **138**:53–60.

Stewart, M., and Vigers, G. (1986). *Nature* **319**:631–636.

Stolinski, C., and Breathnach, A. S. (1975). *Freeze-Fracture Replication of Biological Tissues. Techniques, Interpretation and Applications.* London: Academic Press.

Storey, R., and Walker, R. R. (1987). *J. Expt. Bot.* **38**:1769–1780.

Strain, J. J., Kopf, D. A., LeFurgey, A., Ingram, P., Hawkey, L. A., and Davilla, S. (1989). *Microbeam Analysis* **1989**:97–102.

Studer, D., Michel, M., and Muller, M. (1989). *Scanning Microsc. Suppl.* **3**:253–269.

Stumpf, W. E., and Ross, L. J. (1967). *J. Histochem. Cytochem.* **15**:243–256.

Sugiyama, M., Kubo, H., and Fukushima, K. (1990). In *Proc. 12th Intl. Cong. EM. Seattle*, Vol. 4, pp. 82–83.

Sumner, A. (1988). *J. Electron. Microsc. Tech.* **9**:99–112.

Sutanto, E. (1988). *Proc. 46th Ann. Meet. EMSA*, G. Bailey, ed. San Francisco: San Francisco Press, pp. 104–105.

Sybers, H. D., Myre, C. D., and Myre, M. V. (1983). *Scanning Electron Microsc.* **2**:769–776.

Symons, M. C. R. (1982). *Ultramicroscopy* **10**:41–44.

Talmon, Y. (1980). *Proc. 38th Ann. Meet. EMSA*, G. Bailey, ed. Baton Rouge: Claitors Publishing, pp. 618–619.

Talmon, Y. (1982). *J. Microsc.* **125**:227–237.

Talmon, Y. (1984). *Ultramicroscopy* **14**:305–316.

Talmon, Y. (1987). In *Cryotechniques in Biological Electron Microscopy*. R. A. Steinbrecht, and K. Zierold, eds. Berlin: Springer-Verlag, Chap. 3.

Talmon, Y., and Miller, W. G. (1978). *J. Colloid Interface Sci.* **67**:284–291.

Talmon, Y., and Thomas, E. L. (1977). *J. Microsc.* **111**:151.

Talmon, Y., and Thomas, E. L. (1977). *Scanning Electron Microsc.* **1**:265–272.

Talmon, Y., and Thomas, E. L. (1978). *J. Microsc.* **113**:69–75.

Talmon, Y., and Thomas, E. L. (1979). *J. Mat. Sci.* **14**:1647–1650.

Talmon, Y., Davies, H. T., Scriven, L. E., and Thomas, E. L. (1979). *J. Microsc.* **117**:321–332.

Talmon, Y., Siegel, D., Burns, J., and Chestnut, M. (1988). *Biophys. J.* **53**:125a.

Talmon, Y., Burns, J. I., Chestnut, M. H., and Siegel, D. P. (1990). *J. Electron. Microsc. Tech.* **14**:6–12.

Tanaka, K. (1980). *Intl. Rev. Cytol.* **68**:97–127.

Tanaka, K., and Naguro, T. (1981). *Biomed. Res.* **2**:63–70.

Tatlock, G. J. (1982). *Ultramicroscopy* **10**:87–96.

Tatlock, G. J., Hurd, T. J., Banfield, R. E. W., and Rushton, P. J. (1984). *Proc. 8th Eur. Reg. Conf. Electron Microscopy, Budapest*, Vol. 1, p. 97.

Taylor, P. G., and Burgess, A. (1977). *J. Microsc.* **111**:51–64.

Taylor, K. A. (1978). *J. Microsc.* **112**:15–25.

Taylor, K. A., and Glaeser, R. M. (1973). *Rev. Sci. Instrum.* **44**:1546–1547.

Taylor, K. A., and Glaeser, R. M. (1974). *Science* **106**:1036–1037.

Taylor, K. A., and Glaeser, R. M. (1976). *J. Ultrastruct. Res.* **55**:448–456.

Taylor, K. A., Milligan, R. A., Raeburn, C., and Unwin, P. N. T. (1984). *Ultramicroscopy* **13**:185–190.

Taylor, M. J. (1987). In *The Effects of Low Temperatures on Biological Systems*. B. W. W. Grout, and G. J. Morris, eds., pp. 3–71. London: Edward Arnold.

Tazaki, K., Fyfe, W. S., and Iwatsuki, M. (1988). *Nature* **333**:245–247.

Thom, F., and Matthes, G. (1988). *Cryoletters* **9**:300–307.

Thornberg, W., and Mengers, P. E. (1957). *J. Histochem. Cytochem.* **5**:47–52.

Tiffe, H. W., Matzke, K. H., and Thiessen, G. (1979). *J. Microsc.* **16**:385–390.

Ting-Beall, H. P., Burgess, F. M., and Robertson, J. D. (1986). *J. Microsc.* **142**:311–316.

Tobler, M., and Freiburghaus, A. U. (1990). *J. Microsc.* **160**:291–298.

Tokuyasu, K. T. (1973). *J. Cell Biol.* **57**:551–565.

Tokuyasu, K. T. (1980). *Histochem. J.* **12**:381–403.

Tokuyasu, K. T. (1986). *J. Microsc.* **143**:139–150.

Tokuyasu, K. T. (1989). *Histochem. J.* **21**:163–171.

Tokuyasu, K. T., and Okamura, S. (1959). *J. Biophys. Biochem. Cytol.* **6**:305–309.

Toner, M., Cravalho, E. G., Karel, M., and Armant, D. R. (1991). *Cryobiology* **28**:55–71.

Tosuyama, S., Suganuma, T., Murata, F., and Enokizono, N. (1984). *Med. J. Kagoshima Univ.* **36**:327–334.

Toyoshima, C., and Unwin, P. N. T. (1988). *Ultramicroscopy* **25**:279–292.

Toyoshima, C. (1989). *Ultramicroscopy* **30**:439–444.

Trinick, J., and Cooper, J. (1990). *J. Microsc.* **159**:215–222.

Trinick, J., Cooper, J., Seymour, J., and Egelman, E. H. (1986). *J. Microsc.* **141**:349–360.

Troyer, D. (1974). *J. Microsc.* **102**:215–218.

Turnbull, D. (1969). *Contemp. Phys.* **10**:473–488.

Turner, J. N., Valdre, U., and Fukami, A. (1989). *J. Electron. Microsc. Tech.* **11**:258–271.

Tvedt, K. E., Halgunset, J., Kopstad, G., and Haugen, O. A. (1988). *J. Microsc.* **151**:49–59.

Tvedt, K. E., Halgunset, J., Kopstad, G., and Haugen, O. A. (1989). *J. Electron. Microsc. Tech.* **13**:264–265.

Uhlmann, D. R. (1972). *J. Non-Crystalline Solids* **7**:337–348.

Umrath, W. (1975). *Arnzeim Forsch.* **25**:450–462.

Umrath, W. (1978). *Mikroskopie* **33**:11–19.

Umrath, W. (1983). *Mikroskopie* **40**:9–19.

Unwin, P. N. T., and Henderson, R. (1975). *J. Mol. Biol.* **94**:425–440.

Unwin, P. N. T., and Muguruma, J. (1971). *J. Appl. Phys.* **42**:3640–3641.

Unwin, P. N. T., and Zampighi, G. (1980). *Nature* **283**:545–549.

Unwin, P. N. T., Toyoshima, C., and Kubalek, E. (1988) *J. Cell Biol.* **107**:1123–1138.

Valdez, G. A., Mazni, O. B., Kanagawa, H., and Fujikawa, S. (1990). *Cryoletters* **11**:351–358.

Valdre, U., and Goringe, M. J. (1965). *J. Sci. Instrum.* **34**:582–592.

Van Bergen en Henegouwen, P. M. P. (1989). In *Colloidal Gold: Principles, Methods and Applications*, Vol. 1, M. A. Hayat, ed. pp. 191–216. New York: Academic Press.

Van Doorn, W. G., Thiel, F., and Boekestein, A. (1991). *Scanning* **14**:37–40.

Van Harreveld, A., and Crowell, J. (1964). *Anat. Rec.* **149**:381–395.

Van Harreveld, A., and Trubatch, J. (1979). *J. Microsc.* **115**:243–256.

Van Stevenick, R. F. M., Van Stevenick, M. E., Wells, A. J., and Fernando, D. R. (1990). *J. Pl. Physiol.* **137**:140–146.

Van Venetie, R., Hage, W. J., Bluemink, J. G., and Verkleij, A. I. (1981). *J. Microsc.* **123**:287–292.

Van Venrooij, G. E. P. M., Aertsen, A. M. H. J., Hax, W. M. A., Ververgaert, P. H. J. Y., Verhoeven, J. J., and Van den Vorst, H. A. (1975). *Cryobiology* **12**:46–51.

Venables, J. A. (1963). *Rev. Sci. Instrum.* **34**:582–584.

Venables, J. A., Spiller, G. D. T., and Hanbrucken, M. (1984). *Rep. Prog. Phys.* **47**:399–459.

Verkleij, A. J., and Leunissen, J. L. M. (1989). *Immunogold Labeling in Cell Biology.* Boca Raton, FL: CRC Press.

Verschoor, A., Frank, J., Radermacher, M., Wagenknecht, T., and Boublik, M. (1984). *J. Mol. Biol.* **178**:677–698.

Vinson, P. K. (1987). Fluids and cryo-transmission electron microscopy. Doctoral Thesis, University of Minnesota.

Vinson, P. K. (1987). In *Proc. 45th Ann. Meet. EMSA*, pp. 644–645. G. W. Bailey, ed. San Francisco: San Francisco Press.

Vinson, P. K. (1988). In *Proc. 46th Ann. Meet. EMSA.* G. W. Bailey, ed. San Francisco: San Francisco Press, pp. 112–113.

Vogel, R. H., Provencher, S. W., Von Bonsdorff, C. H., Adrian, M., and Dubochet, J. (1986). *Nature* **320**:533–535.

Volker, W., Frick, B., and Robenek, H. (1985). *J. Microsc.* **138**:91–93.

von Zglinicki, T. (1988). *Scanning Microsc.* **2**:1791–1804.

von Zglinicki, T., and Bimmler, M. (1987). *J. Microsc.* **146**:77–85.

von Zglinicki, T., and Uhrik, B. (1988). *J. Microsc.* **151**:43–47.

von Zglinicki, T., and Zierold, K. (1989). *J. Microsc.* **154**:227–235.

von Zglinicki, T., Bimmler, M., and Purz, H. (1986). *J. Microsc.* **141**:79–90.

von Zglinicki, T., Bimmler, M., Krause, W. (1987). *J. Microsc.* **146**:67–76.

von Zglinicki, T., (1991). *J. Microsc.* **161**:149–158.

Wade, R. H., Prollet, F., Margolis, R. L., Garel, J.-R., and Job, D. (1989). *Biol. Cell* **65**:37–44.

Walcerz, D. B., and Diller, K. R. (1991). *J. Microsc.* **161**:297–312.

Walrafen, G. E. (1964). *J. Chem. Phys.* **40**:3249–3256.

Walther, P., Hentschel, J., Herter, P., Muller, T., and Zierold, K. (1990). *Scanning* **12**:300–307.

Walzthony, D., Moor, H., and Gross, H. (1981). *Ultramicroscopy* **6**:259–266.

Ward, B., and Murray, J. M. (1987). *J. Electron. Microsc. Tech.* **5**:275–277.

Ward, R. J., Menetret, J.-F., Pattus, F., and Leonard, K. (1990). *J. Electron. Microsc. Tech.* **14**:335–341.

Warley, A. (1990). *J. Microsc.* **157**:135–147.

Warner, R. R., and Coleman, J. R. (1975). *Micron* **6**:79–84.

Warner, R. A. (1986). *J. Microsc.* **142**:363–369.

Watanabe, S., Saski, J., Wada, T.,Tanaka, K., and Otsuka, N. (1988). *J. Electron. Microsc.* **37**:89–91.

Webb, J., and Jackson, M. B. (1986). *J. Exp. Bot.* **37**:832–841.

Weibull, C. (1986). *J. Ultrastruct. Mol. Res.* **97**:207–209.

Weibull, C., and Christiansson, A. (1986). *J. Microsc.* **142**:79–86.

Weibull, C., Carlemalm, E., Villiger, W., Kellenberger, E., Fakan, J., Gautier, A., and Larsson, C. (1980). *J. Ultrastruct. Res.* **73**:233–244.

Weibull, C., Christiansson, A., and Carlemalm, E. (1983). *J. Microsc.* **129**:201–207.

Weibull, C., Villiger, W., and Carlemalm, E. (1984). *J. Microsc.* **134**:213–216.

Weibull, C., Villiger, W., and Bohrmann, B. (1990). *J. Struct. Biol.* **104**:139–143.

Weigand, K. M. (1906). *Plant World* **9**:25–34.

Welch, J. F., and Speidel, H. K. (1989). *Cryoletters* **10**:309–314.

Wells, B. (1985). *Micron. Microsc. Acta* **16**:49–53.

Welter, K., Muller, M., and Mendgen, K. (1988). *Protoplasma* **147**:91–99.

Wendt-Gallitelli, M. F., and Wolberg, H. (1984). *J. Electron. Microsc. Tech.* **1**:151–174.

Wepf, R., Amrein, M., Burkli, U., and Gross, H. (1991). *J. Microsc.*: in press.

Wergin, W. P., and Erbe, E. F. (1989). *Scanning* **11**:293–303.

Whalley, E., and McLaurin, G. E. (1984). *J. Opt. Soc. Am.* **A1**:1166–1170.

Whalley, E. (1983). *J. Phys. Chem.* **87**:4174–4179.

Wharton, D. A. (1992). *J. Microsc.*: in press.

Wharton, D. A., and Rowland, J. J. (1984). *J. Microsc.* **134**:299–306.

Wheeler, E. E., Gavin, J. B., and Seelye, R. N. (1975). *Stain Technol.* **50**:331–343.

White, D. L., Andrews, S. B., Faller, J. W., and Barrnett, R. J. (1976). *Biochim. Biophys. Acta* **436**:577–592.

Whitecross, M. I., Price, G. D., and Preston, J. S. (1982). *J. Microsc.* **128**:RP3.

Wiencke, C., Stelzer, R., and Lauchli, A. (1983). *Planta* **159**:336–341.

Wildhaber, I., Gross, H., and Moor, H. (1982). *J. Ultrastruct. Res.* **80**:367–373.

Wildhaber, I., Gross, H., and Moor, H. (1985). *Ultramicroscopy* **16**:321–330.

Willemer, H. (1975). In *Freeze Drying and Advanced Food Technology*. S. A. Goldblith, L. Rey, and W. Rothmayer, eds. London: Academic Press, p. 461.

Williams, R. C. (1954). In *Biological Applications of Freezing and Drying to Electron Microscopy*. R. J. C. Harris, ed. New York: Academic Press, pp. 303–328.

Williams, M. A. (1977). In *Practical Methods in Electron Microscopy*, Vol. 6. A. M. Glauert, ed. Amsterdam: Elsevier/North-Holland.

Williams, D. B. (1987). *Practical Analytical Electron Microscopy in Materials Science*. Mahwah, NJ: Electron Optics Publishing Group, Philips Electronic Instruments.

Williams, M. H., Vesk, M., and Mullins, M. G. (1987). *Micron Microsc. Acta* **18**:27–31.

Williamson, B., and Duncan, G. H. (1989). *New Phytol.* **111**:81–88.

Willison, J. H. M., and Rowe, A. J. (1980). In *Practical Methods in Electron Microscopy*, Vol. 8. A. M. Glauert, ed. Amsterdam: Elsevier/North-Holland.

Wilson, A. J. (1989). *Microsc. Anal.* **March**:37–38.

Wilson, A. J., and Robards, A. W. (1982). *J. Microsc.* **125**:287–297.

Winkler, H., Wildhaber, I., and Gross, H. (1985). *Ultramicroscopy* **16**:331–339.

Wiskner, B. C., Ward, K. B., Lattman, E. E., and Love, W. E. (1975). *J. Mol. Biol.* **98**:179–194.

Wolf, B. (1987). *J. Electron. Microsc. Tech.* **7**:185–189.

Wolf, B., and Schwinde, A. (1983). *Microsc. Acta.* **87**:301–306.

Wolf, B., Neher, R., and Schwinde, A. (1987). *Micron Microsc. Acta* **18**:121–129.

Wolf, B., Hauschildt, S., Dinger, V., Lammlin, C. (1991). *Microsc. Anal.* **January**:25–28.

Wolff, E. W., Mulvaney, R., and Oates, K. (1988). *Ann. Glaciol.* **11**:194–197.

Wollenberger, A., Ristau, O., Schoffa, G. (1960). *Pfluger's Arch. Physiol.* **270**:399–412.

Wolosewick, J. J., and Porter, K. R. (1979). *J. Cell Biol.* **82**:114–139.

Woods, A. M., and Beckett, A. (1987). *Can. J. Bot.* **65**:2007–2016.

Woods, P. S., Ledbetter, M. C., and Tempel, N. (1991). *J. Electron. Microsc. Tech.* **18**:183–191.

Worsham, R. E., Harris, W. W., Mann, J. E., Richardson, E. G., and Ziegler, N. F. (1974). *Proc. 32nd EMSA.* C. J. Arceneaux, ed. Baton Rouge: Claitors Publishing, pp. 412–413.

Wroblewski, J., and Roomans, G. M. (1984). *Scanning Electron Microsc.* **4**:1875–1882.

Wroblewski, R., and Wroblewski, J. (1984). *Histochemistry* **81**:469–475.

Wroblewski, R., Wroblewski, J., Anniko, M., and Edstrom, L. (1985). *Scanning Electron Microsc.* **1**:447–454.

Wroblewski, R., Wroblewski, J., and Roomans, G. M. (1987). *Scanning Microsc.* **1**:1225–1240.

Wroblewski, J., Sagstrom, S., Mulders, H., and Roomans, G. M. (1989). *Scanning Microsc.* **3**:861–864.

Wroblewski, J., Wroblewski, R., and Roomans, G. M. (1988). *J. Electron. Microsc. Tech.* **9**:83–98.

Yang, D. S. C., Sax, M., Chakrabartty, A., and Hew, C. L. (1988). *Nature* **333**:232–237.

Yase, K.noue, T., and Okada, M. (1990). *J. Electron. Microsc.* **39**:454–458.

Yeager, M., Dryden, K. A., Olson, N. H., Greenberg, H. B., and Baker, T. S. (1990). *J. Cell Biol.* **110**:2133–2144.

Zachariassen, K. E., and Hammel, H. T. (1988). *Cryobiology* **25**:143–147.

Zalokar, M. (1966). *J. Ultrastruct. Res.* **15**:469–479.

Zasadzinski, J. A. N. (1988). *J. Microsc.* **150**:137–149.

Zasadzinski, J. A. N., and Bailey, S. M. (1989). *J. Electron. Microsc. Tech.* **13**:309–334.

Zasadzinski, J. A. N., Schneir, J., Gurley, J., Elings, V., and Hansma, P. K. (1988). *Science* **329**:1013–1014.

Zierold, K. (1976). In *Proc. 6th Eur. Cong. E.M. Israel.* Vol. 2, pp. 223–225.

Zierold, K. (1980). *Microsc. Acta* **83**:25–37.

Zierold, K. (1982). *Ultramicroscopy* **10**:45–54.

Zierold, K. (1984). *Ultramicroscopy* **14**:201–210.

Zierold, K. (1984). *J. Phys. (Paris)* **45**:C2447–C2450.

Zierold, K. (1985). *J. Microsc.* **140**:65–71.

Zierold, K. (1986). *Scanning Electron Microsc.* **2**:713–724.

Zierold, K. (1986). In *The Science of Biological Specimen Preparation.* M. Muller, R. P. Becker, A. Boyde, and J. L. Wolosewick, eds. Chicago: SEM Inc., pp. 119–127.

Zierold, K. (1987). In *Cryotechniques in Biological Electron Microscopy.* R. A. Steinbrecht, and K. Zierold, eds. Berlin: Springer-Verlag, pp. 132–148.

Zierold, K. (1988). *J. Electron. Microsc. Tech.* **9**:65–82.

Zierold, K. (1989). *Microbeam Anal.* **1989**:109–111.

Zierold, K. (1991). *J. Microsc.* **161**:357–366.

Zierold, K., and Hagler, H. (1989). *Electron Probe Microanalysis: Applications in Biology and Medicine.* Heidelberg: Springer-Verlag.

Zierold, K., and Schafer, D. (1987). *Verh. Dtsch. Zool. Ges.* **80**:111–118.

Zierold, K., and Steinbrecht, R. A. (1987). In *Cryotechniques in Biological Electron Microscopy.* K. Zierold, and R. A. Steinbrecht, eds. Berlin: Springer-Verlag, Chap. 15.

Zierold, K., Tobler, M., and Muller, M. (1991). *J. Microsc.* **161**:RP1.

Zingsheim, H. P. (1984). *J. Microsc.* **133**:307–317.

Zs-Nagy, I. (1983). *Scanning Electron Microsc.* **3**:1255–1268.

Zs-Nagy, I. (1988). *Scanning Microsc.* **2**:301–309.

Zs-Nagy, I. (1989). *Scanning Microsc.* **3**:473–482.

Zs-Nagy, I., and Casoli, T. (1990). *Scanning Microsc.* **4**:419–428.

Zs-Nagy, I., Pieri, C., Guili, C., Bertoni-Freddari, C., and Zsa-Nagy, V. (1977). *J. Ultrastruct. Res.* **58**:22–33.

Zs-Nagy, I., Lustylik, G., Zs-Nagy, V., Zarandi, V., and Bertoni-Freddari, C. (1981). *J. Cell Biol.* **90**:769–777.

Index